Lecture Notes in Mechanical Engineering

Lecture Notes in Mechanical Engineering (LNME) publishes the latest developments in Mechanical Engineering—quickly, informally and with high quality. Original research reported in proceedings and post-proceedings represents the core of LNME. Volumes published in LNME embrace all aspects, subfields and new challenges of mechanical engineering. Topics in the series include:

- Engineering Design
- Machinery and Machine Elements
- Mechanical Structures and Stress Analysis
- Automotive Engineering
- Engine Technology
- Aerospace Technology and Astronautics
- Nanotechnology and Microengineering
- Control, Robotics, Mechatronics
- MEMS
- Theoretical and Applied Mechanics
- Dynamical Systems, Control
- Fluid Mechanics
- Engineering Thermodynamics, Heat and Mass Transfer
- Manufacturing
- Precision Engineering, Instrumentation, Measurement
- Materials Engineering
- Tribology and Surface Technology

To submit a proposal or request further information, please contact the Springer Editor of your location:

China: Dr. Mengchu Huang at mengchu.huang@springer.com
India: Priya Vyas at priya.vyas@springer.com
Rest of Asia, Australia, New Zealand: Swati Meherishi at swati.meherishi@springer.com
All other countries: Dr. Leontina Di Cecco at Leontina.dicecco@springer.com

To submit a proposal for a monograph, please check our Springer Tracts in Mechanical Engineering at http://www.springer.com/series/11693 or contact Leontina.dicecco@springer.com

Indexed by SCOPUS. The books of the series are submitted for indexing to Web of Science.

More information about this series at http://www.springer.com/series/11236

Károly Jármai · Katalin Voith
Editors

Vehicle and Automotive Engineering 3

Proceedings of the 3rd VAE2020, Miskolc, Hungary

Springer

Editors
Károly Jármai
Faculty of Mechanical Engineering and Informatics, Institute of Energy Engineering and Chemical Machinery
University of Miskolc
Miskolc, Hungary

Katalin Voith
Faculty of Mechanical Engineering and Informatics, Institute of Energy Engineering and Chemical Machinery
University of Miskolc
Miskolc, Hungary

ISSN 2195-4356 ISSN 2195-4364 (electronic)
Lecture Notes in Mechanical Engineering
ISBN 978-981-15-9528-8 ISBN 978-981-15-9529-5 (eBook)
https://doi.org/10.1007/978-981-15-9529-5

This Springer imprint is published by the registered company Springer Nature Singapore Pte Ltd.
The registered company address is: 152 Beach Road, #21-01/04 Gateway East, Singapore 189721, Singapore

Preface

After the 2nd International Conference on Vehicle and Automotive Engineering, there was a need to organize the next one. There were over 150 thousand downloads from the Springer Web site. Unfortunately, the COVID-19 crisis made it more difficult. Finally, we decided to organize an online conference to avoid any danger with COVID-19.

The production of the automotive and vehicle industry and its component suppliers, manufacturer of machines and equipment, and the connected mechanical engineering and process engineering industry have increased greatly in the last decades. The quick transportation of persons and goods is more and more important, as also the quick production of the vehicles. People would like to reach the destination as quickly as possible. This is the case in Hungary, where the improvement of car and vehicle industry was great in the last decades. Great car producers settled here like Mercedes Benz, Audi, Suzuki, Opel and small and medium enterprises connected to car element production have developed greatly.

The aim of the 3rd International Conference on Vehicle and Automotive Engineering at the University of Miskolc, Hungary, is to provide a good opportunity for the discussion of professional topics in this field for both academic and industrial experts. The online version of the conference makes it possible to avoid any danger with COVID-19.

The main requirements for cars and car elements are safety, manufacturability and economy. Safety against different loads such as permanent and variable actions is guaranteed by design constraints on stresses, deformations, stability, fatigue, eigenfrequency and noise, whereas manufacturability is considered by fabrication constraints. The economy is achieved by minimization of the cost.

The main topics of the conference are as follows:

- Alternative powertrains & Autonomous vehicles
  Hybrid vehicles, Electric vehicles, Fuel cell vehicles, Autonomous & connected vehicles, Artificial intelligence, Internet of Things (IoT), Applications in Smart Cities, Future Trends and Emerging Technologies

- Materials & Manufacturing
  Advanced materials and innovations in manufacturing; Metal parts forming, joining and casting technologies; Coating, wear, corrosion protection and surface engineering; Fatigue, fracture, failure and testing of materials and structural parts; Prototype building; Flexible processes.
- Sustainability & Logistics
  Standards and regulations; Design for environment; Virtual design and testing; Inspection and maintenance; Life cycle assessment; Recycling; Supply chain and logistics.
- Design & Noise
  Geometric modelling; Design and reconstruction of vehicle structures, and surfaces; Evaluation and correction of vehicle surfaces; Computer graphics and image processing in visualization and design; 3D printing and prototyping in vehicle development; Engine noise and tyre noise; Other sources of noise; Measurement techniques, simulation and analysis.
- Optimization
  Topology optimization, shape optimization, sizing; Optimization methods, cost calculation.
- Welding
  Different welding technologies; Application of ultra-high strength steels; Application of welding in vehicle industry.

It is a great pleasure to organize this conference, to give participants opportunity to show and discuss the new research results online in a friendly atmosphere.

The organizers wish all participants successful online meetings to collect new ideas and get new acquaintances.

August 2020

Károly Jármai
Katalin Voith

Acknowledgement

The editors would like to acknowledge the co-operation and help of the following organizations

- Ministry of Innovation and Technology, (ITM),
- Hungarian Investment Promotion Agency (HIPA),
- North Hungarian Automotive Cluster (NOHAC),
- Hungarian Welding Association (MAHEG),
- Hungarian Steel Structure Association (MAGÉSZ),
- Hungarian Welding Technology and Material Testing Association (MHtE),
- The EFOP-3.6.1-16-2016-00011 "Younger and Renewing University – Innovative Knowledge City – institutional development of the University of Miskolc aiming at intelligent specialisation" project implemented in the framework of the Széchenyi 2020 program. The realization of this project is supported by the European Union, co-financed by the European Social Fund.

And last but not the least

University of Miskolc, Hungary, which hosts the conference.
The editors would like to acknowledge the help of the following persons:
László Kota, assistant professor, software developer,
Gábor Lassú, research fellow,

and the reviewers:

AL ALI Mohamad, AL-FATLAWI ALAA Abdulzahra Deli, ANDREI Andras, BAGINÉ KOVÁCS Szilvia, BARNA Imre, BENCS Péter, BENOTSMANE Rabab, BODOLAI Tamás, BOGNÁR Gabriella Vadászné, BOLLÓ Betti, CZIFRA Árpád, CSERVENÁK Ákos, DÖRR Nicole, ERDŐS Antal, FÁBIÁN Enikő Réka, GÁSPÁR Marcell, GHAFIL HAZIM Nasir, HANULA Barna, HODULOVA Erika, HORÁK Péter, HORAUER Martin, HORVÁTHNÉ VARGA Ágnes, HRICZÓ Krisztián, JÁLICS Károly, JÁRMAI Károly, KOLLÁNYI Tibor, KOVÁCS György, KULCSÁR Gyula, LAKATOS István, MANNHEIM Viktória, MÓGER Róbert, ORBÁN Ferenc, PÁCZELT István, PETRIK Máté, POKORÁDI László,

SAHUL Miroslav, SKAPINYECZ Róbert, SPISÁK Bernadett, SZABADOS György, SZABÓ Balázs, SZABÓ J. Ferenc, SZARKA Angéla Váradiné, SZÁVA János, SZEIDL György, SZITA Klára Tóthné, MANKOVITS Tamás, TÍMÁR Imre, TOLLÁR Sándor, TOMPA Tamás, TOPAÇ Mehmet Murat, TÓTH László, GHICA Valeriu Gabriel, VAN TYNE Chester, VARGA Erika Baksáné, VARGA Gyula, VIŇÁŠ Ján, VIRÁG Zoltán.

August 2020

Károly Jármai
Katalin Voith

Contents

About the Editors

Dr. Károly Jármai is a professor at the Faculty of Mechanical Engineering at the University of Miskolc, where he graduated as a mechanical engineer and received his doctorate (dr.univ.) in 1979. He teaches design of steel structures, welded structures, composite structures and optimization in Hungarian and in the English languages for foreign students. His research interest includes structural optimization, mathematical programming techniques and expert systems. Dr. Jármai wrote his C.Sc. (Ph.D.) dissertation at the Hungarian Academy of Science in 1988, became a European Engineer (Eur. Ing. FEANI, Paris) in 1990 and got his habilitation (dr.habil.) at Miskolc in 1995. Having successfully defended his doctor of technical science thesis (D.Sc.) in 1995, he subsequently received awards from the Engineering for Peace Foundation in 1997 and a scholarship as Széchenyi professor between the years 1997 and 2000. He is the co-author of five books in English *Analysis and Optimum Design of Metal Structures*, *Economic Design of Metal Structures*, *Design and optimization of metal structures*, *Optimization for robot modelling with MATLAB* and three monographs in Hungarian, and has published over 698 professional papers, lecture notes, textbook chapters and conference papers. He has about 795 independent citations. He is a founding member of International Society for Structural and Multidisciplinary Optimization (ISSMO), a Hungarian delegate, vice chairman of commission XV and a subcommission chairman XV-F of IIW (International Institute of Welding). He has held several leading positions in GTE (Hungarian Scientific Society of Mechanical Engineers) and has been the president of this society at the University of Miskolc since 1991. He was a visiting researcher at Chalmers University of Technology in Sweden in 1991, visiting professor at Osaka University in 1996-97, at the National University of Singapore in 1998 and at the University of Pretoria several times between 2000 and 2005. He was the vice rector of the university between 2013 and 2017 at the field of strategy and research. He is member of the editorial board of several national and international journals. He is an honorary doctor of the Technical University of Kosice, Slovakia.

Dr. Katalin Voith is Senior Research Fellow at the University of Miskolc, Faculty of Mechanical engineering and Informatics. She graduated there in 1990 as a metallurgical engineer and received her PhD in 1993. She succeeded a postgraduate 2-year program of Human Management at the University of Miskolc, Faculty of Economics, in 2017. After graduation in 1990, she worked as a Research and Development Engineer in KÖBAL (Kőbányai Könnyűfémmű), Budapest, Hungary.

In 1996, she moved to Belgium, where she worked at first as administrative assistant to the director, for an American Association: AACC American Association of Cereal Chemists. The main tasks were to organize workshops with attendees and lecturers from all over the world and organizing commercial exhibits at international conferences and trade shows in Europe. During the next decade, she worked as a founder member of the company FALEX Tribology NV. Beside the administrative management of the company, she was also involved in international projects and played an active role in the test work of the laboratory. Her job also included the editing and preparation of the newsletters, the company Web site and other commercial brochures of the company. All workshops and open-door-days were organized by her.

After moving back to Hungary in 2010, she worked as Senior Associate for Local Economic Development for the city of Miskolc (Miskolc Holding Zrt.). The job involved managing prospective investments in the region. This included preparing and giving presentations, mostly in English.

Since March 2011, she has been working at the University of Miskolc: first at the Faculty of Materials Science and Engineering where her major task was to organize, develop and manage the foreign MSc and PhD students and their study programs. She changed to the Faculty of Mechanical Engineering and Informatics as a Senior Research Fellow in 2018 and is currently still working there. Her main research topic is LCA. Her tasks are to increase the international visibility of the faculty through organizing international conferences and finding possible partners for European research projects.

Alternative Powertrains and Autonomous Vehicles

Development of an Advanced Durable Test Target for Autonomous Emergency Brake Testing

Márton Pataki[1,2](✉) and Zsolt Szalay[1,2]

[1] Budapest University of Technology and Economics, Budapest 1111, Hungary
[2] Automotive Proving Ground Zala Ltd., ZalaZONE sqr. 17, Zalaegerszeg 8900, Hungary
marton.pataki@auto.bme.hu

Abstract. Rear-end collisions expose a significant part of all crashes with injuries. The driver assistance autonomous braking function intends to mitigate the effect of the rear-end collisions. Usually, the terminology refers to this function as AEBs (Automated Emergency Braking system). The goal of this assistance to prevent accidents in emergencies where detection-based emergency braking intervention is activated automatically without the driver's operation. For testing of these functions dedicated target test tools are used which allow tests without considerable damage on the vehicles. There are several available devices today for this purpose which have a significant purchase price and these targets normally can be used in daytime conditions. Since during these investigations of AEB performance, harsh impacts can occur the target has to be robust to endure them. With the help of a comprehensive overview, we are investigating the current test procedures and benchmark actual used test tools for car to car rear-end collision situations. Based on this specification a brand new prototype target has been manufactured. In addition, a unique feature - not included in officially approved targets - has been mounted such as tail-lights and brake lights that extends the possibilities of the test conditions. The purpose of this paper is to present a realistic target that is suitable for night tests and has a tough construction for long term usage and is at the same time a cost-efficient solution. This article demonstrates the developed test target compliance with our requirements defined by us based on inputs coming from industrial experiences, furthermore has to be proved by more different tests in both day and night conditions too. So far the industrial AEB test protocols contained only daylight tests but with the usage of this target, we had the opportunity to investigate AEB performance under limited vision condition which is a big challenge for cameras. The tests were executed with different types of passenger vehicles. The research also points to improvement opportunities based on these test results.

Keywords: Target development · Rear-end collision · Automated emergency braking system · Testing under night condition

K. Jármai and K. Voith (Eds.): VAE 2020, LNME, pp. 3–17, 2021.
https://doi.org/10.1007/978-981-15-9529-5_1

1 Introduction

Large-scale segment of the automotive design activities focuses on the development and testing of state of the-art vehicles equipped with Advanced Driving Assistance Systems (ADAS) and autonomous vehicles appearing in the future.

Currently available vehicles on the market possess automated features that correspond typically to SAE classification Level 2. This means the vehicles belong to the Level 2 support the driver in lateral and longitudinal control e.g. steering and acceleration/braking intervention. One of the first such ADAS was the Automated Emergency Braking system (hereafter referred to only as AEBs).

The majority of the traffic crashes is the rear-end collisions of a car to car interactions. According to the statistics of NHTSA the rear impact account for about 29% of the crashes with injuries or fatal accident while it is slightly more than 15% of them in Hungary. Other researches emphasized that the causes of these rear-end impacts are stationary vehicles, the moving forward vehicles with constant speed or the slowing cars. 80% of these crashes occur with stopped vehicles furthermore the study highlighted that these incidences happen with day-light and dry conditions on straight roads. Other research done by Volvo shows that rear-end collisions account for 20% of accidents with fatal injuries in West Europe. These statistics clearly show the relevancy of these scenarios thus several decisive independent organizations have begun testing these functions preventing these risky events [1–3].

In the years after 2010, vehicle developers have begun to put a lot of effort into developing such AEBs systems to avoid these accidents. Nevertheless, for a long time, there were not any procedures that would have standardized the examination of these functions to some degree. First, an independent organization, Euro NCAP, started developing test scenarios and evaluation procedures for testing these active safety systems. The organization began publishing the first testing and assessment protocols related to AEBs in 2014 after almost two years of preparation and study process involving key players dealing with these functions. In this year, the first target dummy test equipment was released and approved by Euro NCAP which can surrogate the other real participant of these rear-end crash situations [4]. According to Euro NCAP's real-world performance results show that these systems can decrease passenger vehicle accidents by up to 27% [4].

Since these assistance functions present already for a couple of years so more statements are available which can provide an approximate picture based on experiences about the performance and positive effects of these functions. The results show that FCW (Forward Collision Warning) functions is capable of itself to decrease rear-end crashes by 23%, while FCW completed with AEBs reduces them by 39%. The rate of rear-end crashes with injuries decreases by 42% with FCW and AEB combined function [6].

As there are only a few target device developer companies in the market so it is not surprising the relatively high purchase price and circumstantial support and maintenance service in this segment. In the following, this paper presents the development of a solution in detail that is significantly cheaper than industrial devices available on the market and meets the requirements of industrial test procedures. In order to get the inputs to the design process, it is essential to review most of the test scenarios defined by the industry and map out the competitor's solutions on the market. When setting requirements, extra features are also defined that increase applicability and make the dummy tool even more realistic in terms of perception issues.

2 General AEB Test Procedures

The OEM and Tier supplier companies with development and testing capacity intend to test not only according to their protocols but they also prefer procedures from other independent organizations especially test scenarios defined by Euro NCAP. There are some other AEB protocols released by other international organizations e.g. ISO (International Organization for Standardization), SAE (Society of Automobile Engineers), or the UN ECE but considered the protocols definitely the scenarios published by Euro NCAP are most precise and elaborated descriptions. Furthermore, the feedbacks from the automotive industry unequivocal show the importance of Euro NCAP test scenarios. At the beginning this organizations started to assess the safety performance of the brand-new cars on the market only from the passive safety point of view e.g. crash tests, seat belt and whiplash effect but in the last 6 years implemented examinations and rates the performance of these passenger cars from active safety aspects as well. Besides this organization, there are two other relevant associations that have begun such tests of cars.

However, the newest test protocols related to AEBs are expanded with new basic emergencies in which the VUT's turning path is crossed by other participants e.g. approaching target vehicle from the oncoming lane. This newest protocol has only this case belonging to the non-rear collision situations within the group of AEB car-to-car scenarios. Generally, the available test procedures determine such situations which require other vehicles in the same lane or path in front of VUT (Vehicle Under Test) i.e. tested vehicle. The ISO standards have some basic scenarios with two traffic vehicles but only one vehicle moves in front of our VUT, so only this traffic actor has to be replaced by a crashable dummy target vehicle. ISO published two distinct standards to investigate the AEBs and FCWs but the scenarios are very similar. ISO 15623 was named "Forward vehicle collision warning system" which does not contain only the test execution but defines requirements for the operation of ADAS function. The ISO 22839 is about the AEBs, which named as "Forward vehicle collision mitigation system" similarly to the other ISO standard that is dealing with the required operating parameters of the function [7, 8].

The American IISH is also an NCAP member owning also protocols to test the rear-end situations but its proposal relates to the American sold cars. It requires just slightly different scenarios where the passenger vehicles are observed such a rear-end situation where the other target vehicle is a static object and the test runs are executed only at speed of 20 and 40 km/h [9].

Regulations described by UN ECE is more than a simple proposal or standard compare to the description defined by SAE or ISO [10]. The newest vehicles placed on the market in certain countries in which these regulations have been applied must be tested according to this regulation. In this year will be released the UN ECE Regulation 152 related to testing of AEBs. The draft of this new protocol is already available and already clear what scenarios are expected. Previously a very similar test description was published under Reg. 131 but it applies only to heavy-duty vehicles. The UN ECE Reg. 152

requires a "Warning and activation test with static vehicle target" and a similar scenario with a moving vehicle. In the first case, the VUT has to be driven at various speeds of 20, 42 and 60 km/h and in case of moving target the VUT's required velocity is at 30 and 60 km/h while the target device has to be moved at speed of 20 km/h. NHTSA American organization has also studies and researches to observe these ADAS functions. We summarized shortly the speeds, accelerations of the actors during scenarios with static, moving with constant speed and decelerated target by the scenarios [11–13].

Table 1. Relevant rear-end collision test scenarios

Category	Euro NCAP	UN ECE	ISO	IIHS
Static	 V_{GVT} = 0 km/h a_{GVT} = 0 m/s^2 V_{VUT} = 20-80 km/h	 V_{GVT} = 0 km/h a_{GVT} = 0 m/s^2 V_{VUT} = 20,42,60 km/h	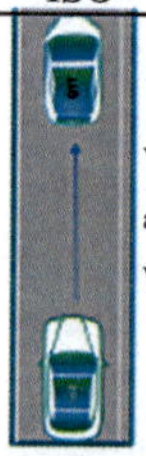 V_{GVT} = unk a_{GVT} = unk V_{VUT} =unk	 V_{GVT} = 0 km/h a_{GVT} = 0 m/s^2 Headway=150m, 200m V_{VUT} = 20,40 km/h
Moving	 V_{GVT} = 20 km/h a_{GVT} = 0 m/s^2 V_{VUT} = 20-80 km/h	 V_{GVT} = 20 km/h a_{GVT} = 0 m/s^2 V_{VUT} = 30,60 km/h	 V_{GVT} = unk a_{GVT} = unk V_{VUT} =unk	X
Braking	 V_{GVT} = 50 km/h a_{GVT} = -2 m/s^2,-6 m/s^2 Hedway=12m, 40m V_{VUT} = 20-80 km/h	X	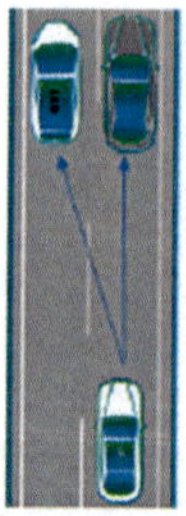 V_{GVT} = unk a_{GVT} = unk V_{VUT} =unk	X

As shown in Table 1, the Euro NCAP has the most test-cases and the biggest ranges of speed from VUT point of view. Considering the requirements and the conditions related to the weather, test track, preparation of the VUT, measurement, tolerances and the description of the test execution the Euro NCAP requires the most stringent conditions. The Euro NCAP makes a specific recommendation for the target equipment which is approved by the organization. For the official Euro NCAP testing, all the conditions shall meet the requirement.

3 Test Targets Available on the Market

The first approved target used for rear-end collision tests was the EVT (Euro NCAP Vehicle Target) developed by ADAC in 2013 [14]. The usage of the equipment started in the following two years but nowadays the recommended tool according to the latest updated protocol is the GVT (Global Vehicle Target) [15]. The EVT is a towed target concept, which represent only the relevant backside of the target while the GVT takes the shape of a full passenger car represents a Ford Fiesta [17]. The EVT's target is an inflatable balloon car with the appropriate reflective strips and trihedral corner reflector which is mounted to a base profile of the target balloon body which can roll out in case of undesirable impact along with the towing ladder frame structure. Originally the connection between the ladder frame and the target is solved magnetic pairs. According to the feedback from automotive partners, the magnetic surface quickly contaminated by sweepings or iron filings and afterward they can no longer lock which would be necessary to resist the momentum originated the braking of the towing vehicle. The Euro NCAP protocol requires two different target decelerations (2 m/s^2 and 6 m/s^2) and for purpose of the adjusting the resistance of the lock the gap should be modified with adding or removing the magnetic slabs [14]. So it can work between a well-defined discrete range hence it is slightly onerous to test with on-demand any decelerations. The GVT is a much more sophisticated solution for testing usable even for customized complex maneuvers not only rear-end impact scenes. The target part itself built from foam panels with the appropriate reflection and these are covered by a foam shell layer with printed images represent a Ford Fiesta. In order to move the desired path, a self-propelled platform is used with trajectory following capability, which can withstand driving even up to 30 tons truck. Through its low profile, it produces reduced disturbing reflection. There is already a standard for this device and Euro NCAP also defined a technical specification for this type of solution. After the appearance of this concept on the market, a study was published with the goal to investigate the characterization of reflectivity and the geometry. During this research, the researchers pointed out the effect of the wear resulting from 100 collisions. They investigated the "Soft 360 target" developed by DRI and distributed also by AB dynamics Ltd and another very similar target with the name of "4Active-C2" produced by 4Active systems [19, 20]. The results of the experiments highlighted that the 100 collisions executed at speed of 50 km/h had a greater effect on the change of the geometry than the reflectivity. In more cases, the geometry deviations exceeded the 10 mm, which is grater then the standard would require mentioned already previously [21].

In case of the "Soft 360 target" to better radar signature the radar reflective parts were hidden in the bumper, the doors and the bonnet, in addition, the license plate was printed on the reflective layer as well as in the solution developed by 4Active systems. In these self-propelled solutions, care must be taken to eliminate the reflections of their lower profile propulsion systems. In order to reduce this irregular reflection, they apply a coverage of the platform around with a skirt made from radar absorber material. In concept of "Soft 360 target", each element is fixed to the platform via Velcro while the product of the 4Active system is solved with the help of zippers and fasteners. Both manufacturers have their own self-driven platforms with similar capabilities, just like the UFO platform developed by the Austrian company Humanetics Austria Gmbh.

Overseas the NHTSA initiated to develop an own so-called SSV (Surrogate Strikable Vehicle) target which is a towed concept as well as the EVT [22, 23]. The difference between them is the target itself. The SSV made from carbon-fiber formed as a shell figured Fiesta unlike the other balloon EVT which has a printed vinyl shows a VW Touran picture [24, 25].

Fig. 1. Available targets on the market

The SSV's target form is the exact copy of a real Ford Fiesta which is mounted with a flexible foam bumper to attenuate the effect of the impact as the left bottom corner of the Fig. 1 shows. This product primarily used by NHTSA's labs. The shell shape target part of the device is mounted on a rolling cart which is covered by radar absorber foams.

There are some other developments or own solutions developed by the company for internal usage which are not officially approved but suitable for internal or pre-tests. One of this kind of development is published in Michigan, they created a foam car with sedan-shape design. In this concept, they applied mostly polystyrene which is intended to give the rigidity besides that it is also used to shield the reflective parts of the towing module. In this target were built other two softer foams into the bumper for the damping of collision and the bumper was covered with a polyurethane sheet plate. For the appropriate reflection, the colleagues of their radar team proposed to use 2 pieces trihedral corner reflector behind the tail-lights and the place of the rear wheels and an extra corner reflector was located above the registration license plate. All of the reflectors were manufactured with a 3 cm long inner-edge. In addition, they glued a further 4 cm width reflective aluminum strip behind the polyurethane cover of the bumper [22].

During another notable development was produced from an open-cell foam slab represented a Hyundai Sonata passenger vehicle. They covered the slab with a printed foil and simulated the tail-lights with retroreflected glued stickers. In order to achieve a good reflection for radar sensing two pieces 15 cm $\times$ 130 cm laminated PVC strip (Energy shield 200 oz) were mounted under the outer layer in two different height (55 cm and 88 cm). They located two radar corner reflectors with different dimensions (4 cm and 20 cm) and observed the effect for the sensor of a Toyota's AEBs. During the experiment, they also operated with changing the contrast of the contour edges for the camera sensing. For this purpose, they glued white and back strips around the contour of the slab. The best combination contained the larger corner reflector and the normal contrast settings without additional strips [26].

There are several different targets available on the market, these are typically foam panel built or inflatable balloon devices. These can be used typically limited way due to their reflectivity properties furthermore because these devices are mostly static placement objects [27, 28].

4 Advanced Test Target Requirement

The Euro NCAP the ISO and the UN ECE have an exact specification for the target to be used during the test. The ISO and the UN ECE refer to the same ISO 19206 standard which describes the required technical parameters. The specification of this standard covers the following properties: color, radar reflection, vehicle category, stability during wind.

In terms of the color, this standard required lighter color like grey or white but the point is that the target has to be well contrasted to the ground. According to the standard, the target has to posses further design features like reflective elements representing tail lights and registration plate. For windy conditions, the standard requires avoiding the flutters at a target speed of 50 km/h and a side wind up to 10 m/s. Regarding the dynamic parameters during testing the device has to be suitable up to 50 km/h towing speed and maximum 6 m/s^2 deceleration in braking scenarios. The contour and the shape have to represent a B category i.e. small car (like Ford fiesta or Opel Corsa) or C category i.e. a medium car (like Opel Astra, VW Golf or Ford Focus). In terms of radar reflections, it requires two different values for the rear bumper and one other for the rear axle. The total radar cross-section of the bumper has to be 30 m^2 while the reflector in the rear axle location has to have 20 m^2. Within this, it is also recommended by the standard that a 1,5 m wide bar with 20 m^2 and trihedral corner reflector located in the center-line of the bumper resulting the remaining 10 m^2 radar cross-section. In the standard, there is a requirement for the radius of the arc of the bumper which has to be 5m.

The Euro NCAP defines their EVT and GVT within its own protocols or technical bulletin documents. The specification of the EVT gives a proposal for an exact trihedral corner reflector with 55 mm edge-length. Besides the dimension of the reflector, it provides suggestions to locate it in vertical and horizontal directions. The same information is available for the reflective strips within this protocol. The outer shell is vinyl made from PVC with printed design elements like a back window, tail-lights, and registration plate.

Such as the most available devices on the market, they have only passive tail-lights which are in the best case mounted with reflective strips or in the simplest cases only painted design elements which are not suitable to use them under dark condition. For this reason, we lay down the need for an active rear light as a requirement, which also must possess an extra brake light function during deceleration scenarios.

In accordance with the test cases presented above, we defined the following criteria:

- The target shall be impactable at relative speed of 50 km/h
- The target shall be moveable at least 50 km/h speed
- The target shall have white or grey color
- The target shall have 3 trihedral corner reflectors, two of them shall be located in the position of the rear axles on both side the third will be mounted in the bumper. The bumper shall have a bar in horizontal direction.

- The target shall be covered by vinyl and have a back window with seats and silhouette of the headrests.
- The target shall have an additional option to activate real tail-lights with brake light feature for night scenarios.
- The target has to be able to withstand at least 30 impacts without significant damage

5 Design

Under the design process, we had to determine the basic concept with the way of the target movement and the solution which makes it enough sturdy and durable to bear the impacts. While the detailed design processing we had to calculate and simulate certain solutions. Firstly, in this chapter, the basic concept will be presented which can fulfill the defined requirements listed above in the previous chapter and after that, we summarize the design of each most important part.

5.1 Concept

Since our goal is the realization of a cost-efficient durable target for rear-end impact testing furthermore our purpose is an elaboration of concept which restricts the chance of failure to simply reparable mechanical issues so the last closing results become our final proposal. Our approach is to implement a device that is easy to use as a hatchback, combi or MPV category vehicle. In order to solve this multi-usage functionality, we conceived a concept that has a base in the structure all the time and top of that it has a swappable foam part shapes the form of the vehicles category mentioned above. To move the target a towing ladder frame part was designed mounted with hitch coupler to join the towing vehicle. The ladder frame must be enough narrow and low so that the vehicle could take the ladder frame between the two wheels in case of overrunning. Basically, the towing vehicle controls the target speed via the connection between the ladder frame and target which joint has to be able to unlock in case of impact before it would cause any damage to the target. For realistic visual and radar purposes, we had to ensure the appropriate real outlook and the well-defined radar reflective material and items within the target with the proper dimensions (Fig. 2).

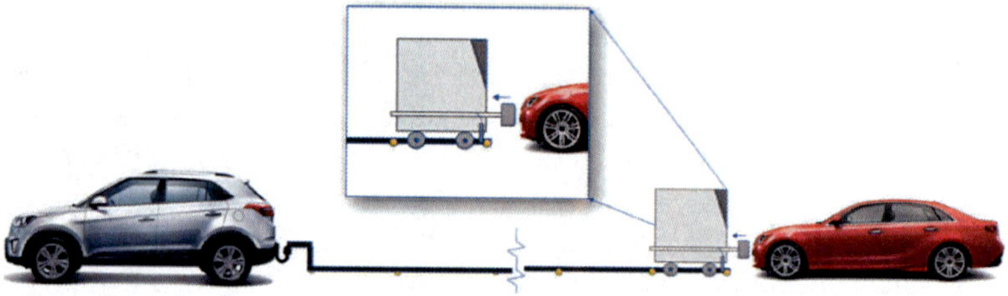

Fig. 2. Concept of the new target device

5.2 Design of Details

Towing Ladder Structure

The requirements of the towing part of the system are the length, the height, the width and the appropriate rigidity. In terms of length, we must define enough long ladder frame in order to exceed the brake distance in case of maximum VUT speed which could be even 80 km/h. To determine the brake distance, we simulated the worst case when the VUT approaches the target at speed of 80 km/h and applies a maximum braking after reaching the target. We executed the simulation with Virtual Crash 3 simulation software. For better visualization, we modelled and imported the simplified 3D model of our concept. Finally, we determined a 30 m long ladder towing part divided into 6 m long parts because it can be transported on a trailer. The illustration of the simulation is shown in Fig. 3.

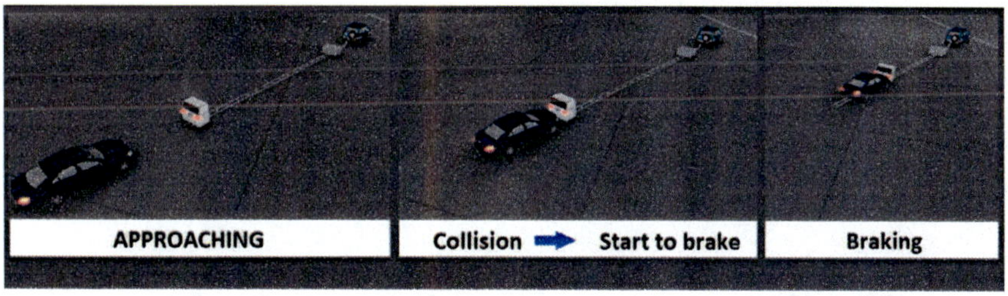

Fig. 3. Simulation to determine the necessary length of the ladder structure

Besides the dimension of the ladder must be enough narrow and low in order fitting under most serial passenger vehicles. The rigidity and the rolling wheels, which underset the frame itself are also essential parameters which ensure the ground clearance and the stability of the system. For this purpose, we had to mount the sufficient bracing and casters on the proper position determined by FEM simulation. During the design of the ladder frame, FEM simulations were run to optimize the number and length of the brace insert to frame structure reducing the detection. The final frame of the towing structure is shown in Fig. 4.

Target Device

The structure of the carrier has to be enough robust to cope with all of the force effects come from the impact of VUT. The carrier part of the target is the basic frame. We built up the body of the target on this frame. The key role of this carrier to be led by the ladder in straight ahead when the target gets shoved away by a VUT. Horizontal spring pre-tensioned casters guide the carrier in the straight direction in order to reduce the friction between the ladder and the carrier. To avoid the damages, we designed an unlock mechanism pulled by a sliding bumper thus securing that the target itself is released before the peak force would push ahead carrier. The carrier part must be sufficiently rigid to withstand the load from the body of the target and the dynamic effects come up during the collision. We used the results of simulation composed in Virtual Crash

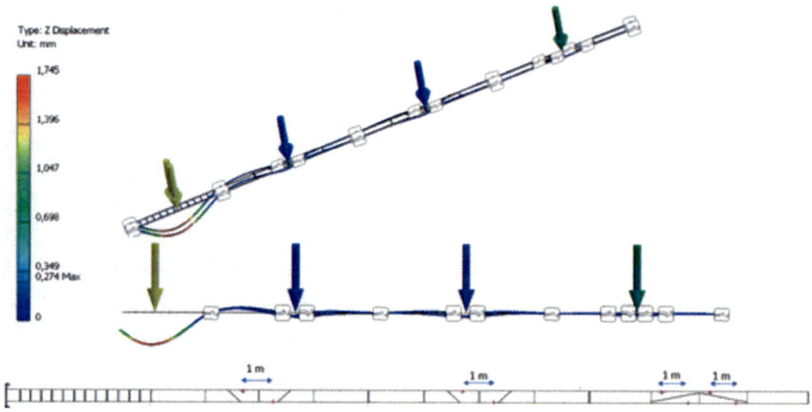

Fig. 4. FEM simulation to optimize the rigidity and the number of the casters

simulation tool. We had to design mechanisms, which ensure that the already pushed unlocked bumper not slide back and release down the latch of the lock resulting a sudden stop of sliding. We succeed to solve this problem with spring-tensioned part rolling a predefined ramp withstand to slide from the effect of impetus. The main mechanisms of the target shown in Fig. 5.

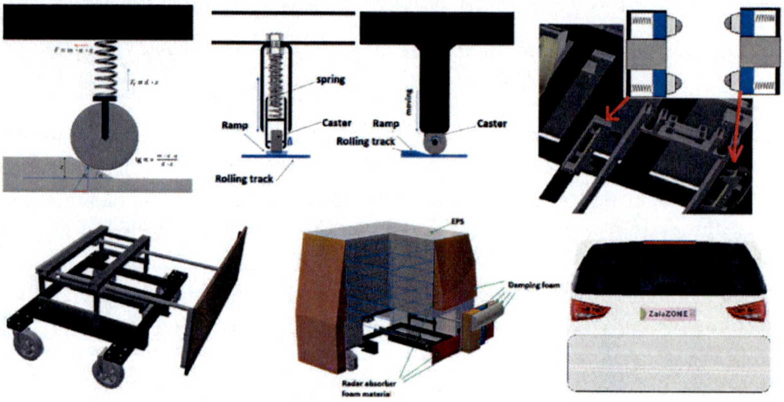

Fig. 5. The major parts of the target device

The target body with the covering is responsible for the representation of a real passenger vehicle. In addition to being similar to a real car, it must be protected from the damage of the VUT and target itself. For this purpose, we defined a sufficiently elastic material for the body part, which is low-cost EPS (Extruded Polystyrene) material available direct from the market. No false reaction could come back from the structure of the carrier so that we needed to shield all of the reflective parts e.g. steel parts. For this purpose, we had to attach an absorption material layer on the back of the target creating a pure reflection free surface. After this, the implementation of real radar reaction is the objective. According to the literature, a tetrahedron shape radar reflector can result a real passenger car radar cross-section. The other researches provided recommendations for the exact location of the reflectors so we placed these corner reflector on the back outer surface of the target 60 cm in height.

For the optical compliance, we designed and covered a vinyl shell from PVC to provide a real car design. The basic shape was formed out from the damping foam body material and the printed tilt give a proper appearance. Finally, we provided two options for design, one of them represents a sedan shape the other ensures an MVP category target shape. In order to change between the designs the only operation that we have to do is to swap the cover tilt and back foam layer attached to the main EPS foam body. Considering the improvable options to enhance the better visual perception we defined neutral look for with white base color. The design includes a real registration number plate. For the correct radar reflection, we needed to design the appropriate edge length of the corner reflector. Based on the technical description of the target in the Euro NCAP protocol recommended a trihedral corner reflector with 55 mm edge-length. According to the ISO 22839 there is an equation to determine the geometry dimensions of a reflector which provides a real car size reflection. For verification of 55 mm edge-length, we executed a check calculation for the necessary dimension. As the result came into harmony with the recommended value, we considered acceptable this edge-length. This radar reflectors have been inserted to the place suggested previously above. Besides the reflector, further two reflective strips have been bounded on the outer layer of the body plate which fosters not only the visual sensing but it can improve the radar result as well. To increase applicability we designed a dedicated circuit for the lights. The objective was that we achieve a system, which can be triggered by the deceleration of the towing vehicle. Our first solution was to implement a micro-controller with accelerometer sensors and we tuned the sensor to trigger a relay if the deceleration reaches a customizable value. We were able to tune the lower and upper limits of the deceleration. But it did not prove reliable due to the shock effects, and manual shifting of the towing vehicle, etc. So we applied a change in the concept we connect directly to the towing vehicle's electricity socket belong to the hook. The tail-lights can operate via simple switch default way on demand but the brake-lights operate in accordance with the applied braking of the towing vehicle.

Costs were calculated according to a simple method since it is the development of a prototype device. Besides the material and commercially available items we added an estimation manufacturing costs multiplied by the overhead cost and the spent time of manufacturing respectively, like cutting, grinding, milling, drilling, and welding. The total cost does not exceed the 2000 €.

6 Tests

First of all, we investigated the towing capability of up to 80 km/h. While towing the construction we measured the lateral deviation from the defined path. We checked the traipse of the towed structure with a flying drone. The maximum deviation not exceeding the required 0,1 m as shown in Fig. 6. Afterward, we accomplished brake tests to prove that the target does not roll forward even from the effect of its inertia. We applied a

6 m/s^2 speed reduction. The un-lock mechanism remained locked during the braking. The subject of the following observation was to monitor the operation of the un-lock mechanism as a response to the impact.

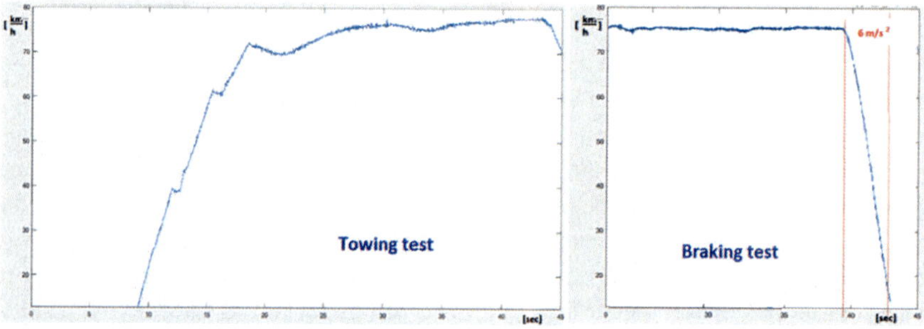

Fig. 6. Towing testing: only towing and the braking to observe the lock system

We impacted the target step by 10 km/h incremental steps up to 50 km/h relative impact speed. In every case, the lock released the ladder frame and withstand the heavy impact without any damage. We plan to carry out 30 impacts in 6 series. So within one series, we crashed it at speed of 10, 20, 30, 40 and 50 km/h. It is illustrated in Fig. 7. Finally, before we would have started the last round in the 4th round we suffered some damage in the construction of the left side tensioning spring module. Nevertheless, we were able to continue the test after a relatively minor repair.

Fig. 7. Collision tests

The next completed examination focused on the compliance of the reflection. We implemented the tests with two different cars equipped AEB systems. One of them detected and provided feedback about the target the other one in some cases did not give any response while approaching the target. The implemented test procedures depicted in Fig. 8.

To observe the tail- and brake-light functionality we organized a test during night condition. Criteria was that the luminosity shall be less than 1lx. We used the target with default tail-lights switched on. The brake-lights was triggered by the towing vehicle. The lights of the target can be switched off and removable and the users can use only a simple reflector instead of the active lights. The summary of the night test is in Fig. 9.

Fig. 8. AEB tests to investigate the perception of relevant properties

Fig. 9. Extra tail and brake light feature test under night condition

7 Conclusion

During this investigation, we explored that industrial test procedures that are focusing AEB driver assistance systems mitigated on the rear-end collision situation. These inputs were an essential aggregation for the requirement list of the development furthermore we extended this information with experiences from papers and studies of the concurrently available products on the market. The goal of this comprehensive development process was that to implement a target device, which can fulfil all the major requirement of the relevant standard, in addition, could provide an extra feature make it possible to test during night condition. Finally, the produced tool became an order of magnitude cheaper than the reviewed other products on the market. We can mention a weakness of this concept is the side tensioning spring module which should be strengthened and the other disadvantages of this concept are that it is still just a prototype and not ready to manufacture more pieces within a short time because this study was not focused on the machining tasks. For the better observation and understanding supposed to use methodology with electronic EDR placed in the VUT often used by accident reconstruction analysis. For better observation and understanding of the impact suggested applying that methodology with electronic EDR placed in the VUT often used by accident reconstruction analysis [29, 30].

Acknowledgement. The project has been supported by the European Union, co-financed by the European Social Fund EFOP-3.6.2–16-2017–00002.

We owe a debt of gratitude for the possibility and facility of ZalaZONE made it possible to execute our tests. Also, we thank Dániel Vlasics and Zoltán Makk for their persistent and valuable works.

References

1. Lee, S.E., Llaneras, E., Klauer, S., Sudweeks, J.: Analyses of rear-end crashes and near-crashes in the 100-car naturalistic driving study to support rear-signaling countermeasure development, Report, NHTSA, pp 1–125 (2007)
2. Statistical yearbook of traffic accident 2015, Hungarian Central Statistical Office. Dr. Vukovich, G., Budapest (2016)
3. Volvo Trucks: European accident research and safety report (2013)
4. Schram, R., Williams, A., Van Ratingen, M.: Euro NCAP's first step to assess autonomous emergency braking (AEB) for vulnerable road users. In: 24th International Technical Conference on the Enhanced Safety of Vehicles (ESV), Gotheburg, Sweden, (2015)
5. Cicchino, B.J.: Effectiveness of forward collision warning systems with and without autonomous emergency braking in reducing police-reported crash rates. Accid. Anal. Prevent. Part A **99**, 142–152 (2016)
6. Vehicle safety 2018, European Road Safety Observatory, European Commission (2018). https://ec.europa.eu/transport/road_safety/sites/roadsafety/files/pdf/ersosynthesis2018-vehiclesafety.pdf, Accessed 01 Feb 2020
7. International Organization for Standardization, Standards, ISO 15623 standard: Intelligent transport systems — Forward vehicle collision warning systems — Performance requirements and test procedures (2013). https://www.iso.org/standard/56655.html, Accessed 01 Feb 2020
8. International Organization for Standardization, Standards, ISO 22839 standard - Intelligent transport systems — Forward vehicle collision mitigation systems — Operation, performance, and verification requirements (2013). https://www.iso.org/standard/45339.html, Accessed 03 Feb 2020
9. IIHS, About our tests, Test protocols and technical information, Current Protocols, Autonomous Emergency Braking Test Protocol (Version I) (2013). https://www.iihs.org/media/a582abfb-7691-4805-81aa16bbdf622992/2036166062/Ratings/Protocols/current/test_protocol_aeb.pdf, Accessed 09 Feb 2020
10. SAE International, Standards, SAE J3087: Automatic Emergency Braking Performance Testing (2017). https://www.sae.org/standards/content/j3087_201710/, Accessed 16 Feb 2020
11. UNECE, Working Parties and Documents , World forum for Harmonization of vehicle regulation WP.29, United Nations: Uniform provisions concerning the approval of motor vehicles with regard to the Advanced Emergency Braking Systems (AEBS) (2013). https://www.unece.org/fileadmin/DAM/trans/main/wp29/wp29regs/2013/R131e.pdf, Accessed 26 Feb 2020
12. UNECE, Working Parties and Documents, Working Party on Automated/Autonomous and Connected vehicles (GRVA), Working Documents, United Nations: Uniform provisions concerning the approval of motor vehicles with regard to the Advanced Emergency Braking System (AEBS) for M1 and N1 vehicles (2019). https://www.unece.org/fileadmin/DAM/trans/doc/2019/wp29grva/GRVA-04-51.pdf, Accessed 20 June 2020
13. Forkenbrock, G.J., Snyder, A.: NHTSA's 2014 automatic emergency braking test track evaluation, Report, Report No: DOT HS 812 166 (2015)
14. Volker, S.: Development of a test target for AEB systems. In: 23rd International Technical Conference on the Enhanced Safety of Vehicles (ESV), Paper Number 13–0406 , Seoul, Korea, South (2013)

15. Euro NCAP, For Engineers, Protocols, Safety Assist: Test protocol-AEB car-to-car systems (2019). https://www.euroncap.com/en/for-engineers/protocols/safety-assist/, Accessed 25 Jan 2020
16. Euro NCAP, For Engineers, Protocols, Safety Assist:: Test protocol-AEB car-to-car systems (2015). https://cdn.euroncap.com/media/17719/euro-ncap-aeb-test-protocol-v11.pdf, Accessed 27 Jan 2020
17. Gover, C., Matthew, A., Jordan, S.: 3D target for future vehicle testing. In: 25th International Technical Conference on the Enhanced Safety of Vehicles (ESV), Detroit Michigan, USA (2017)
18. International Organization for Standardization: ISO 19206–1 Road vehicles - Test devices for target vehicles, vulnerable road users and other objects, for assessment of active safety functions - Requirements for passenger vehicle rear-end target
19. AB dynamics , Product, Track testing, ADAS Targets, Guided Soft Target. https://www.abdynamics.com/en/products/track-testing/adas-targets/guided-soft-target, Accessed 28 Jan 2020
20. 4Active Systems, Products, Dummies, 4ActiveC2. https://www.4activesystems.at/en/products/dummies/4activemc.html, Accessed 02 Feb 2020
21. Lindgren, M., Spetz, J., Nord, S.: Characterization of reflectivity and geometry for soft car target. In: Proceedings of the 29th CIE SESSION 29th International Commission on Illumination, Washington, USA (2019)
22. LeBlanc, D.J., Gilbert, M., Stachowski, S., Blower, D., Flannagan, C.A., Karamihas, S., Buller, T.W., Sherony, R.: Advenced surrogate target development for evaluating precollision systems. In: Paper Number 13-0204, 23rd International Technical Conference on the Enhanced Safety of Vehicles (ESV), Seoul Korea, South (2013)
23. Wolf Composite, Projects, NHTSA Strikeable Surrogate Vehicle (SSV). https://wolfcomposites.com/portofolio/nhtsa-strikeable-surrogate-vehicle/, Accessed 29 Feb 2020
24. Messring Gmbh. https://www.messring.de/en/products/active-safety/euro-ncap-vehicle-target/, Products, Acitve safety, Vehicle test system, Euro NCAP Vehicle Target, Accessed 01 Mar 2020
25. Moshon Data. https://www.moshondata.com/en/flex-moshon-towing-systems, Ltd: Products, Automotive testing, Moshon-Flex, Accesse 03 Mar 2020
26. Yang, M., Xing, P., Flynn, T., Tsuge, B., Lawrence, J., Siegmund, P.G.: The effect of target features on Toyota's Autonomous Emergency Braking System (2018). In: SAE International, SAE Technical Paper 2018-01-0533 (2018)
27. Moshon Data. https://www.moshondata.com/en/moshon-data-slab-air-tight-target-md-s, Products, Automotive testing, AEB Automotive, Slab Air-tight Target MD-S, Slab Foam Vehicle Target MD-SF, Accessed 19 May 2020
28. Inlfatable cars. https://www.imagine-inflatables.com/Giant-Inflatables/Cars/, Imagine Inflatable: Inflatable cars, Accessed 21 May 2020
29. Vida, G., Bodollo, I.: Presentation of modern accident reconstruction procedures - case study. In: VAE 2018: Vehicle and Automotive Engineering, Lecture Notes in Mechanical Engineering, vol. 2, pp 487–494 (2018)
30. Pinter, K., Szalay., Z.: Comparison of data required for accident reconstruction based on crash test. In: VAE 2018: Vehicle and Automotive Engineering, Lecture Notes in Mechanical Engineering, vol. 2, pp 476–486 (2018)

Trajectory Tracking Control of an Autonomous Car

Zakariás Erõsdi(✉), Gergely Bári, and János Papp

John von Neumann University, Izsáki út 10, Kecskemét 6000, Hungary
zacherosdi@gmail.com

Abstract. Development in the field of autonomous vehicles has been rising over the past decade. Most autonomous driving systems must be able to perform a wide variety of tasks, such as driving across an intersection, following another car, or negotiating complicated city streets.

Even though these seem to be complex problems, they can be further decomposed into three major fields of development. Detection, where the car first has to observe the different objects in its environment. Decision making, as the car has to have a high-level strategy based on the detected objects, deciding where it wants to go, and identifying obstacles to avoid. Finally, the actual control values, the needed vehicle speed and steering angle value have to be defined, so the car can actually perform according to the high-level decisions.

Current work presents possible control methods for trajectory tracking of self-driving cars. At first, a proper mechanical model, the well-known bicycle model is demonstrated. While this model represents the behavior of a vehicle within a linear range, the added effects of different tire models are also shown. The model is transformed into a state-space with distance and angular deviation from the trajectory as state variables. This form of the bicycle model can be used for controller design.

Four different control methods are presented, the Stanley controller, that won the DARPA Challenge in 2005, the first competition where driverless cars were competing against each other. A linear quadratic regulator (LQR), a model predictive controller (MPC) and a model predictive controller with input and state constraints. By testing the implemented controllers with the described model it is shown, that although the LQR method seems to be the one with the best dynamics, the constrained MPC can handle the proposed task more robust, as we can be sure that the calculated values by this method can be realized by an actuator.

Keywords: Autonomous · Car · Driverless · Trajectory · MPC · LQR

1 Introduction

Studies about autonomous vehicles have started around the nineties and their number has been growing ever since. We can see the technology to appear in our every-day's personal vehicles, enough if we just take a look at the current ADAS systems like

K. Jármai and K. Voith (Eds.): VAE 2020, LNME, pp. 18–29, 2021.
https://doi.org/10.1007/978-981-15-9529-5_2

lane departure warning system, anti-lock braking system, blind spot monitor, automatic parking, and the list just goes on [1]. The final goal of the development is clear: making a fully autonomous (level 5) car, that can navigate itself on the roads without any kind of human surveillance [2].

This task can be divided into two separate subtasks: perception, and actuation. First, the car has to get some knowledge about its surroundings, then it has to make a decision, to give command signals to the actuators. The current study focuses on the second part of the task. We consider a given trajectory [10], that the car must follow. This can be done in different ways for different circumstances. For lower speeds, when the car's behavior can be estimated with linear models the solution might be easier, but if we aim for level 5 autonomy, using nonlinear models is necessary. For example, during a sudden braking maneuver, or a sudden steering maneuver for the evasion of an accident. To urge the developers to examine the car's behavior in critical conditions we can see the theme of autonomous racing to slowly spread, enough if we just think about the DARPA Challenge in 2005, the beginning of Roborace in 2016 or about a study conducted at ETH Zürich, where they made fully autonomous RC cars, that could follow the desired trajectory with 3 m/s. If we would scale this up to the sizes of a normal car, this would mean 465 km/h [3].

The current study focuses on a model-based approach. The aim is to build up a model, that is suitable for developing trajectory follower algorithms, and to compare different control strategies with this model.

2 Mechanical Model

During the making of a model, taking a reference is a must. A model is a representation of the real life, so a comparison of the results of a model with a reference must be done, to know if the model fulfills our needs.

2.1 Carmaker Model

In this current work, a precise model of a car has been built in the carmaker software, especially designed for precise vehicle dynamics simulations. Unfortunately, this model is too complex for control purposes, as a simpler model is needed, that can be computed online, on a microcontroller. Therefore, this complex, carmaker model is considered as a reference for the simplified model.

2.2 Linear Bicycle Model

The bicycle model is a model widely used for control purposes as it only has two state variables, yet it describes car dynamics rather well. Although this model uses the following assumptions [9] (Figs. 1 and 2):

- The behavior of two wheels on one axis can be described as a single wheel
- Pitch and roll movement of the car is neglected (no weight transfer is considered)
- Velocity is a constant parameter

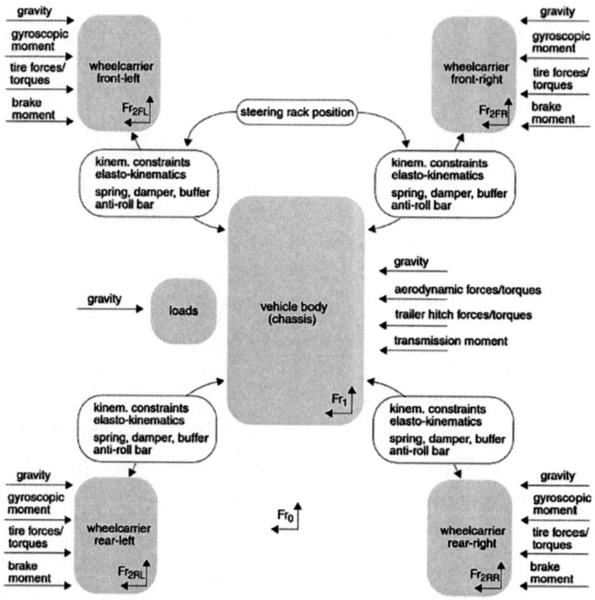

Fig. 1. Effects considered by the carmaker model [4]

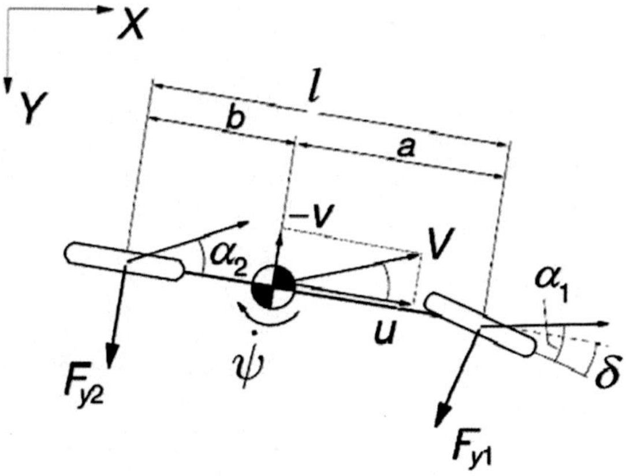

Fig. 2. Bicycle model [8]

- No aerodynamics effects
- The asphalt is considered homogeneous
- Small steering angles

Equations of motion:

$$m(\dot{v} + u \cdot \psi) = F_{y1} + F_{y1} \tag{1}$$

$$I\dot{\psi} = a \cdot F_{y1} - b \cdot F_{y2} \tag{2}$$

Tire side slip angles:

$$\alpha_1 = \delta - \frac{1}{u}(v + a \cdot r), \alpha_2 = -\frac{1}{u}(v - b \cdot r) \quad (3)$$

Linear cornering stiffnesses:

$$F_{y1} = C_1\alpha_1, F_{y2} = C_2\alpha_2 \quad (4)$$

These equations transformed into state-space representation:

$$\begin{bmatrix} \dot{v} \\ \ddot{\psi} \end{bmatrix} = \begin{bmatrix} \frac{-(C_1+C_2)}{m \cdot u} & \frac{(b \cdot C_2 - a \cdot C_1)}{m \cdot u} - u \\ \frac{(b \cdot C_2 - a \cdot C_1)}{I \cdot u} & \frac{C_2 . b^2 - C_1 a^2)}{I \cdot u} \end{bmatrix} \cdot \begin{bmatrix} v \\ \dot{\psi} \end{bmatrix} + \begin{bmatrix} \frac{C_1}{m} \\ \frac{a \cdot C_1}{I} \end{bmatrix} \cdot \delta \quad (5)$$

This model was compared with the carmaker model, using the following parameters. Trajectory tracking being the main goal, the main things considered were the vehicles X and Y coordinates during different maneuvers.

The model was built in Matlab, solved in discrete timesteps, using the classical Runge-Kutta method, with a timestep of 0.005 s (Table 1).

Table 1. Car parameters

Parameter	Value
C1	30000 N/rad
C2	30000 N/rad
m	290 kg
I	80 km m^2
a	0.77 m
B	0.77 m

The first simulation was a 5 s simulation, with the steering angle changing from 0 to 90°, with different speeds. We can see that the simplified model with the increasing of the velocity, starts to divert from the reference model (Fig. 3).

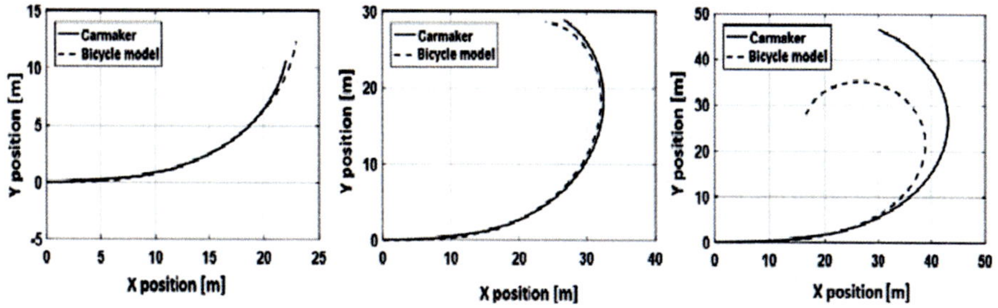

Fig. 3. Comparison of models at 20 km/h, 40 km/h and 60 km/h

The second simulation was conducted at constant 30 km/h with different steering angles at the end of the simulation (higher angle on ramp steer maneuver) We can see, that with increasing the steering angle, the models divert as well, although not as significantly as in the previous case (Fig. 4).

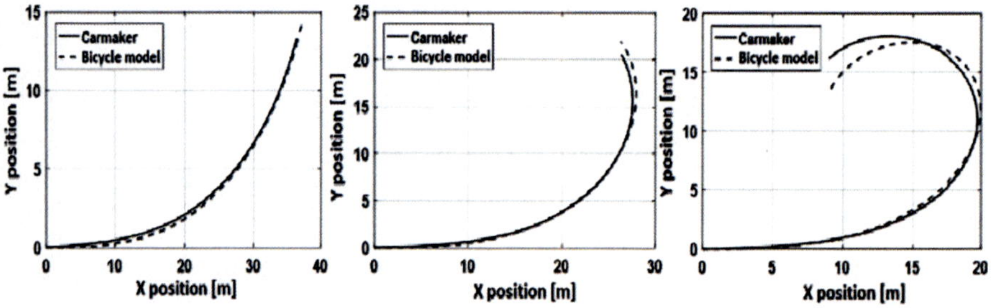

Fig. 4. Comparison of models with 45°, 90°, 180° at the end of simulation

2.3 Bicycle Model with Nonlinear Tire Model

One of the main factors that affects the behavior of the vehicle is the tire model. In the current simplified model, the tire characteristics were considered linear, with cornering stiffnesses. For further enhancing the model, adding a nonlinear tire model could significantly improve the results. For this, a Dugoff [5] tire model was created, with the following sideslip-force characteristic (Fig. 5):

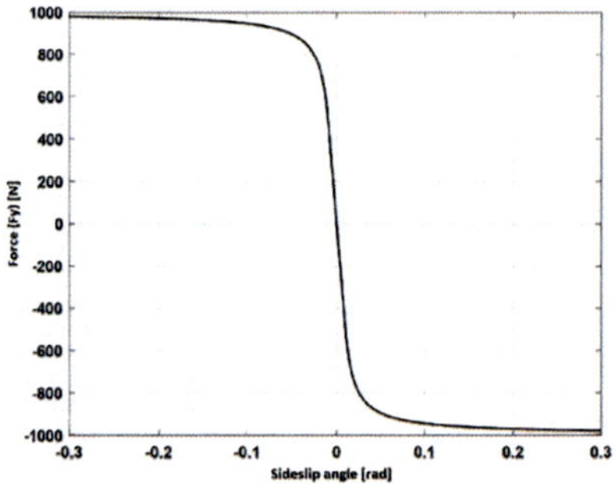

Fig. 5. Dugoff tire model

The same test cases were conducted with the bicycle model using this custom tire model as well. This way the differences between the two examined models have decreased a bit but were far from the expectation. This is mainly because of the neglected weight transfer (Fig. 6).

For an advanced trajectory tracking algorithm, a more complex model with a nonlinear tire model is necessary. For linear range though (~40 km/h) the linear bicycle model is ideal. Therefore, in the upcoming sections, different control strategies were examined on the linear model.

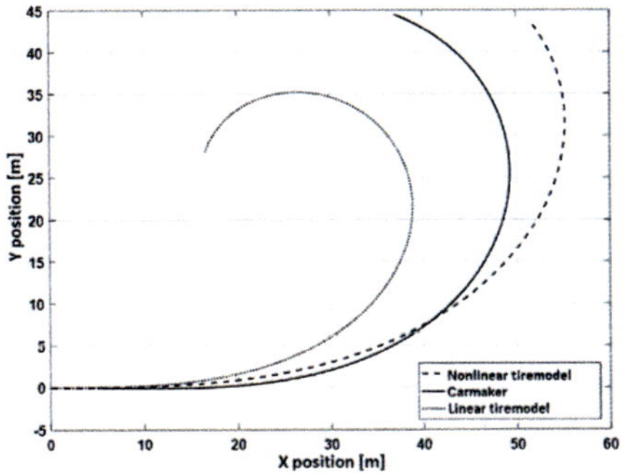

Fig. 6. Comparison of models with 60 km/h

2.4 Bicycle Model for Trajectory Control

First, the model has to be transformed into a shape, that can be used for controlling purposes. Therefore, error variables are introduced. These variables tell the deviation from the desired trajectory. The deviation distance error tells the perpendicular distance between the vehicle's actual position and the desired trajectory, while the orientation error tells the angle between the vehicle's heading and the trajectory's direction.

With these variables, the state-space can be written as follows:

$$\begin{bmatrix} \dot{e}_y \\ \ddot{e}_y \\ \dot{e}_\psi \\ \ddot{e}_\psi \end{bmatrix} = \begin{bmatrix} 0 & 1 & 0 & 0 \\ 0 & -\frac{C_1+C_2}{m \cdot u} & \frac{C_1+C_2}{m} & \frac{bC_2-aC_1}{m \cdot u} \\ 0 & 0 & 0 & 1 \\ 0 & \frac{bC_2-aC_1}{I \cdot u} & \frac{aC_1-bC_2}{I} & -\frac{a^2C_1+b^2C_2}{I \cdot u} \end{bmatrix} \cdot \begin{bmatrix} e_y \\ \dot{e}_y \\ e_\psi \\ \dot{e}_\psi \end{bmatrix} + \begin{bmatrix} \dot{e}_y \\ \ddot{e}_y \\ \cdot \\ \dot{e}_\psi \\ \ddot{e}_\psi \end{bmatrix} \cdot \delta + \; + \begin{bmatrix} 0 \\ \frac{bC_2-aC_1}{m \cdot u} \\ 0 \\ -\frac{a^2C_1+b^2C_2}{I \cdot u} \end{bmatrix} \cdot \dot{\psi}(s) \quad (6)$$

Using this model, state-space control methods can be implemented, with the errors being the state variables and steering angle being the single input. At the first iteration, we can neglect the $\dot{\psi}(s)$ parameter, as this represents the change of the desired trajectory. In the case of a trajectory with constant curvature, this is indeed zero. With feedforward techniques, this effect could be taken into consideration as well (Fig. 7).

3 Control Strategies

3.1 Stanley Control

The first control strategy examined is not yet a state-space controller. It is the steering controller that has been used on Stanley, the robot that won the famous DARPA Challenge in 2005 [6]. This method calculates the acting steering angle based on the actual velocity and the upper mentioned errors, with K as a tuning parameter that scales the importance of the errors relative to each other (Fig. 8):

$$\delta = e_\psi + arctan\frac{K \cdot e_y}{u} \quad (7)$$

This method provides an easy to compute, a solution that can drive the car back on trajectory for an initial position and orientation error.

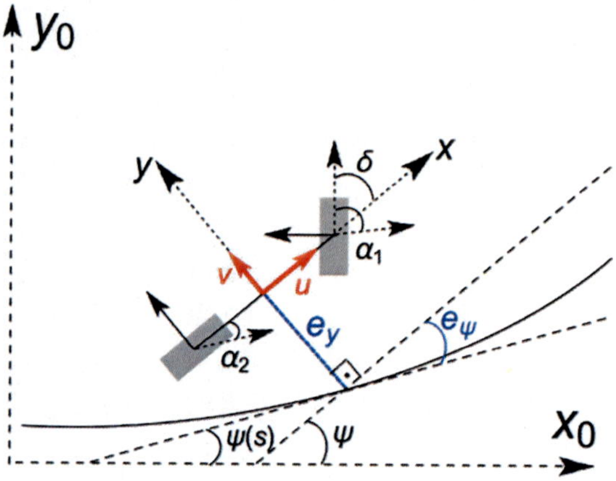

Fig. 7. Interpretation of the error variables

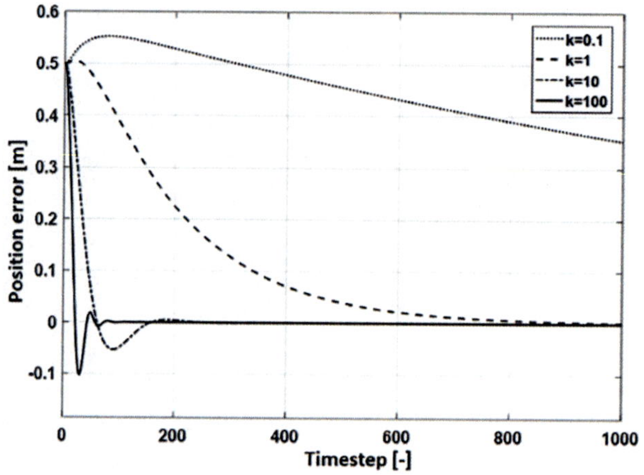

Fig. 8. Interpretation of the error variables

3.2 Linear Quadratic Regulator

The linear quadratic regulator is a state-space control method with state-feedback. The desired actuating signal (steering angle for our case) can be calculated using the following method [7]:

$$u = -Kx \tag{8}$$

$$\dot{x} = (A - B \cdot K)x \tag{9}$$

Where the input (u) is defined by minimizing the following cost function:

$$J = \int_0^\infty (x^T Qx + u^T Ru)dt \tag{10}$$

This way undesired states and inputs can be "penalized". For example, if the goal is to reduce the overshoot of the orientation angle, by increasing the weight of the orientation angle, this can be achieved. The weighting matrixes in the simulations considered the deviation distance error, the orientation error, and the const of the actuation:

$$Q = \begin{bmatrix} q_1 & 0 & 0 & 0 \\ 0 & 0 & 0 & 0 \\ 0 & 0 & q_3 & 0 \\ 0 & 0 & 0 & 0 \end{bmatrix}, \mathrm{R} = 1 \quad (11)$$

In the cost function, the relative value of each weight is considered, so the weight of the actuation was set to 1. Tuning the deviation distance error's (q_1) and the orientation error's (q_2) the following simulations were conducted (Figs. 9 and 10):

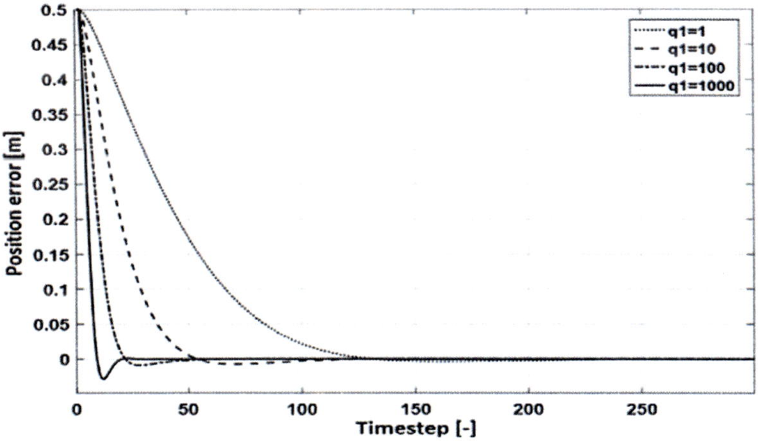

Fig. 9. LQR control with different q1, q3 being set to 1

By increasing the weight of the deviation distance error, the controller gets faster, but the overshoot increases as well. This can be compensated by increasing the orientation error's weight, but this results in slower settling time. Overall this controller is almost twice as fast as the Stanley controller, without losing stability or overshoot.

3.3 Model Predictive Controller

This method uses the same principle, but this time the model is transformed to a discrete time model:

$$x(k+1) = Ax(k) + Bu(k) \quad (12)$$

This way the cost function's form slightly differs from the const function of the LQR method, but its principle remains the same. The optimum isn't searched on an infinite horizon, only on discrete N steps (Fig. 11):

$$J(k, N) = \sum_{i=0}^{N-1} ||x(k+i|k)||_Q^2 + ||u(k+i|k)||_R^2 \quad (13)$$

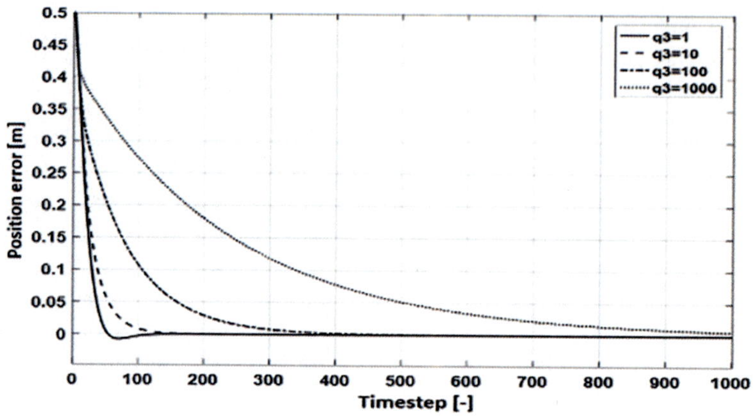

Fig. 10. LQR control with different q3, q1 being set to 100

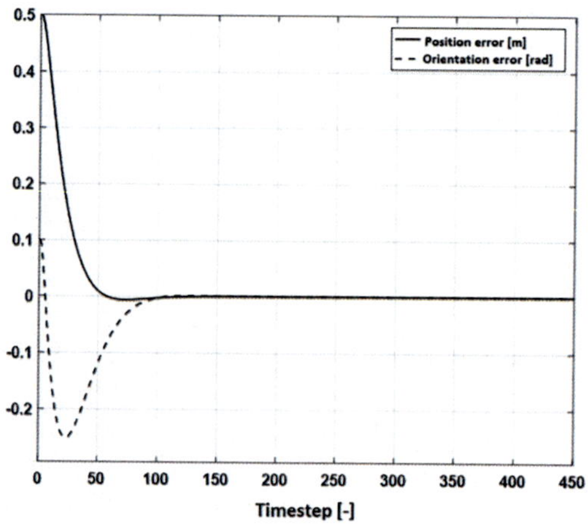

Fig. 11. Settling of position and orientation error, using MPC, N = 100

Using the same weights and a relatively large horizon (N = 100) the settling of the system with MPC controller is similar to the LQR, but large horizons take processing time, so in real, on-car applications a compromise has to be made (Fig. 12).

3.4 Constrained Model predicTive Controller

The upper mentioned strategies all were able to correct an initial error of deviation distance and angle, but the resultant control output (steering angle for our case) is usually constrained by the actual construction of the steering and suspension system. Therefore, this resultant output must be limited, which makes further calculations non-deterministic.

This can be avoided, by solving the MPC problem by taking into consideration the limitations of our system. For demonstration purposes, a limitation of a maximum 0.5 rad wheel angle was considered. The solution was calculated by using Matlab's QP solver (Figs. 13 and 14).

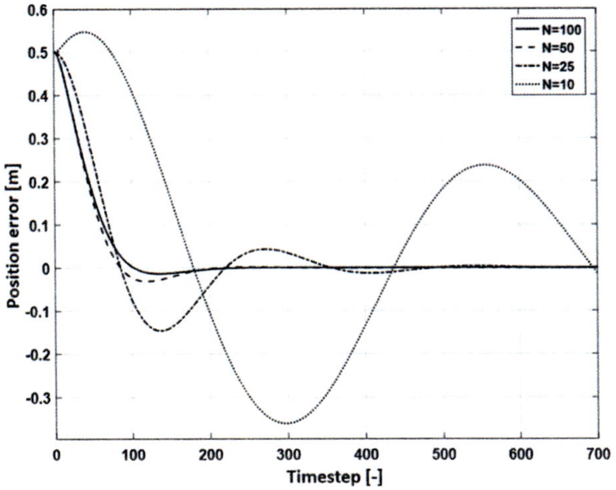

Fig. 12. Settling with different time horizon

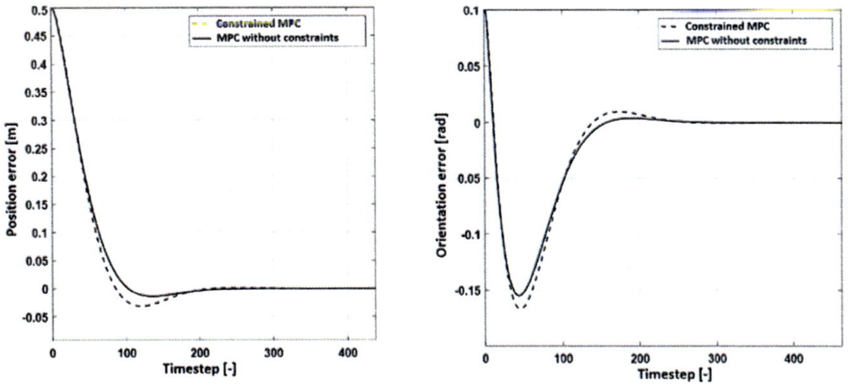

Fig. 13. Settling with different time horizon

The result quite similar to the unconstrained case, but in this case, it can be said, that the calculated actuating command can indeed be realized.

Constraints can be set for state variables as well. For example, if no overshoot is permitted, this can be set as a constraint, although this way the controller cannot guarantee that the errors will converge to zero, as the actuating command is the result of an optimization algorithm, and the optimum might not be where one would expect. This method is useful for lane keeping algorithms for example, where the geometry of the highway is precisely known, so this algorithm can ensure that the car will stay between the margins of the road.

On the picture below the solution for the hardcoded overshoot avoidance can be seen. If non of the error can get a negative value, the position error does not converge to zero (Fig. 15).

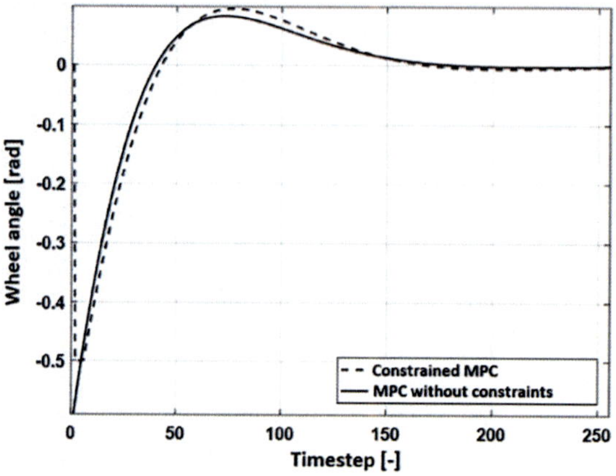

Fig. 14. Wheel angle with and without constraints

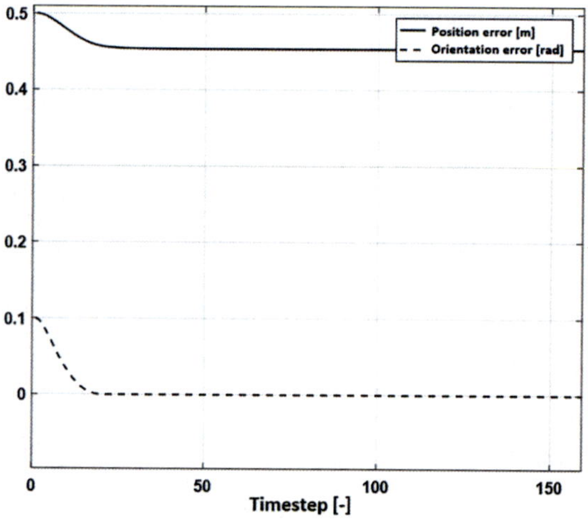

Fig. 15. State variables limited (Error must be greater than 0)

4 Conclusion

Comparing all four methods it can be said that the model-based strategies are better than the Stanley controller, for both settling time and overshoot behavior. Although the LQR seems to be the fastest, the constrained MPC is the most reliable and customizable, as we know that for each situation it will give a result that can be indeed generated by the actuators.

In this study, the low level controller, the trajectory tracker was examined. In the future, further research will be done in the field of trajectory planning, so the whole autonomous driving problem will be discussed (Fig. 16).

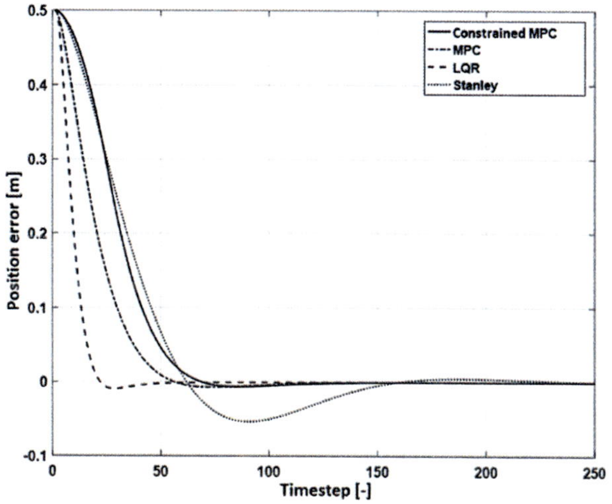

Fig. 16. Comparison of different methods

Acknowledgement. The research presented in this paper was carried out as part of the EFOP-3.6.2–16-2017–00016 project in the framework of the New Széchenyi Plan. The completion of this project is funded by the European Union and co-financed by the European Social Fund.

References

1. Van Brummelen, J., O'Brien, M., Gruyer, D., Najjaran, H.: Autonomous vehicle perception: the technology of today and tomorrow. Transp. Res. Part C Emerg. Technol. **89**, 384–406 (2018). ISSN 0968–090X
2. Federal Automated Vehicles Policy, Accelerating the Next Revolution In Roadway Safety, p. 9 (2016)
3. Alexander, L., Alexander, D., Manfred, M.: Optimization-based autonomous racing of 1:43 scale RC cars. Opt. Control Appl. Methods **36**, 628–647 (2015)
4. Carmaker User's Guide, version 6.0.
5. Kissai, M., Monsuez, B., Tapus, A., Martinez, D.: A new linear tire model with varying parameters. In: 2017 2nd IEEE International Conference on Intelligent Transportation Engineering (ICITE), pp. 108–115 (2017)
6. Thrun, S., Montemerlo, M., Dahlkamp, H., Stavens, D., Aron, A., Diebel, J., Fong, P., Gale, J., Halpenny, M., Hoffmann, G., Lau, K., Oakley, C., Palatucci, M., Pratt, V., Stang, P., Strohband, S., Dupont, C., Jendrossek, L.E., Koelen, C., Mahoney, P.: Stanley: the robot that won the DARPA Grand Challenge. J. Field Rob. **23**, 661–692 (2006)
7. stanford.edu, Linear quadratic regulator: Discrete-time finite horizon, Winter 200809 https://stanford.edu/class/ee363/lectures/dlqr.pdf
8. Vehicle Dynamics 4L150, Eindhoven University of Technology, Mechanical Engineering - Dynamics & Control, Dr. Ir. I.J.M. Besselink, lecture notes (2014)
9. Dynamical Modelling of Vehicle's Maneuvering, JK 2016, pp. 69–77. Ákos Cservenák, Tamás Szabó (2016)
10. Trajectory Planning for Automated Vehicles – A Basic Approach. VAE 2018, pp. 403–412, Ádám Nyerges, Viktor Tihanyi (2018)

Trajectory Tracking Controller Testing in Software in the Loop Environment

Zakariás Erõsdi(✉), Gergely Bári, and Gábor Sipos

John von Neumann University, Izsáki út 10, 6000 Kecskemét, Hungary
zacherosdi@gmail.com

Abstract. The future of the automotive industry is about self-driving cars, in the recent years, numerous studies has been done on the future development of autonomous vehicles. The increase can already be seen on the streets as well. Big corporations testing their latest algorithms on the highway, autonomous features like automated parking, lane keeping algorithms, ADAS (advanced driver assistance system) appearing on the newest cars and so on.

With this sudden increase in development, an advanced simulation environment is necessary, as testing each new software feature in the real world on the test-tracks or on the streets, can often be too costly or too dangerous.

The self-driving car problem is usually decomposed into three major subsystems. First the perception, with the goal of detecting different objects in the environment. Second, the high-level path planning, that defines the desired trajectory based on the target destination and detected objects in the previous layer. Finally the motion control that defines the inputs for the actuators of the car.

In an efficient simulation environment, these subsystems can be evaluated separately, while the behaviour of the whole system can still be evaluated.

This paper presents a possible way for testing advanced autonomous controller algorithms in software in the loop environment, focusing on the third subsystem, but testing the algorithm in the entire environment. First the controller itself is tuned and evaluated, by basing on a simple linear vehicle model built in Matlab. Afterwards this linear controller is tested on a nonlinear, dynamic model. Controller inputs are obtained directly from the simulated environment, while the outputs are the steering angle and vehicle velocity, so the controller works as it would on a real car. The simulation can also be driven with a predefined "driver model", so the algorithm is compared to the driver. The importance of a refined high-level strategy is shown, as in both cases, the control-problem is solved, yet the results, the actual trajectories are significantly different.

Keywords: Autonomous · Car · Driverless · Trajectory · MPC · SIL

1 Introduction

Implementation of a controller takes a lot of time and resource. While the design itself can be done on computer, testing and validating the controller must be done in real-world conditions. This whole process can be made more efficient by software (SIL)

K. Jármai and K. Voith (Eds.): VAE 2020, LNME, pp. 30–37, 2021.
https://doi.org/10.1007/978-981-15-9529-5_3

and hardware in the loop (HIL) testing. Taking out a car to the field costs money, and if something goes wrong, it can even damage the environment / the car itself. With software in the loop testing, a model of the real conditions is created, where the controller can be tuned and evaluated [1]. Software in the loop methods has various advantages compared to traditional simulation. It can be used in the design and testing phase as well [6]. Moreover, it can be expanded to hardware in the loop simulation, where not only the controller, but the actual hardware can be tested, that runs the controller [7].

Our goal is to create an end-to-end software in the loop environment for autonomous algorithm testing. Current study presents such an environment by validating a trajectory tracking controller with software in the loop methods.

2 Trajectory Tracking Controller

The controller is an MPC (Model Predictive Controller) that optimizes the steering angle over a certain time horizon. For this the following model is used [2, 3]:

$$\begin{bmatrix} \dot{e}_y \\ \ddot{e}_y \\ \dot{e}_\psi \\ \ddot{e}_\psi \end{bmatrix} = \begin{bmatrix} 0 & 1 & 0 & 0 \\ 0 & -\frac{c_1+c_2}{m \cdot u} & \frac{c_1+c_2}{m} & -\frac{bc_2+ac_1}{m \cdot u} \\ 0 & 0 & 0 & 1 \\ 0 & -\frac{bc_2-ac_1}{1 \cdot u} & \frac{ac_1-bc_2}{1} & \frac{a^2c_1+b^2c_2}{1 \cdot u} \end{bmatrix} \cdot \begin{bmatrix} e_y \\ \dot{e}_y \\ e_\psi \\ \dot{e}_\psi \end{bmatrix} + \begin{bmatrix} \dot{e}_y \\ \ddot{e}_y \\ \dot{e}_\psi \\ \ddot{e}_\psi \end{bmatrix} \cdot \delta \quad (1)$$

The state variables are the perpendicular distance error from the trajectory (e_y), the orientation error (e_ψ) and their derivatives in time (Table 1).

Table 1. Car parameters

Parameter	Value
C1	30000 N/rad
C2	30000 N/rad
m	290 kg
I	80 km m2
a	0.77 m
b	0.77 m

C1 and *C2* are being the tire stiffness, *m* the vehicle weight, *I* the vehicle inertia, *a* and *b* the distance of the center of the gravity from the front and rear track respectively.

The controller optimizes the following cost function [4]:

$$J(k,N) = \sum_{i=0}^{N-1} ||x(k+i|k)||_Q^2 + ||u(k+i|k)||_R^2 \quad (2)$$

This means that by matrix Q the deviation from the trajectory can be penalized, by matrix R the actuation, the steering can be penalized. The controller's step response was tuned in Matlab. The resulting weighting matrices were the following:

$$Q = \begin{bmatrix} 1000 & 0\,0\,0 \\ 0 & 0\,0\,0 \\ 0 & 0\,1\,0 \\ 0 & 0\,0\,0 \end{bmatrix}, \ \mathrm{R} = 1 \tag{3}$$

3 Software in the Loop Environment

Carmaker software can be integrated with Matlab / Simulink. This means that some calculations and models in the simulation can be optionally replaced by a user-made model, or the user-made model can use many state variables that are calculated by the basic Carmaker environment.

Current work presents the testing of a trajectory tracking algorithm, with constrained model predictive control discussed in our previous research in such environment. This allows the examination controller behaviour in a system with more complex physics. This way, it is possible to determine the value of individual weights - weighting matrices, since it is now possible to examine how the simplifications in the model, during controller design actually affect the performance (Fig. 1).

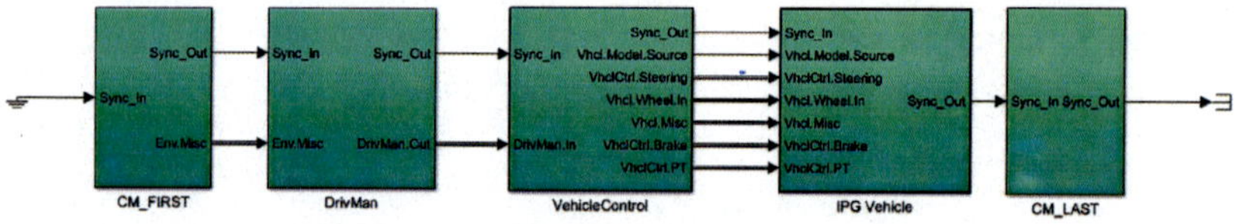

Fig. 1. Simulink model of the Carmaker environment [5]

The basic model is divided into 5 parts. The first module, CM_FIRST, describes environmental effects such as air, asphalt temperature, wind speed, air density, air denomination, track data, and many other similar parameters.

DrivMan model includes the driver model. It calculates the brake and accelerator pedal positions and the steering angle, based on environmental influences.

The VehicleControl section determines the actual inputs for the vehicle model. This determines, for example, how much braking torque a given brake pedal position produces on the wheels, how much acceleration a given accelerator pedal position produces, the driver assistance software can typically be inserted here.

The IPG Vehicle incorporates the actual mathematical model, which describes the dynamics of the car as presented in our previous work.

Finally, CM_LAST connects the Simulink model to the other modules of the Carmaker environment.

The VehicleContol part of the base model is extended in such way, that the steering angle is calculated by our own control algorithm. The speed is freely selectable. It can be the speed profile calculated by the basic carmaker model, it can be a constant or a self-made speed profile (Fig. 2).

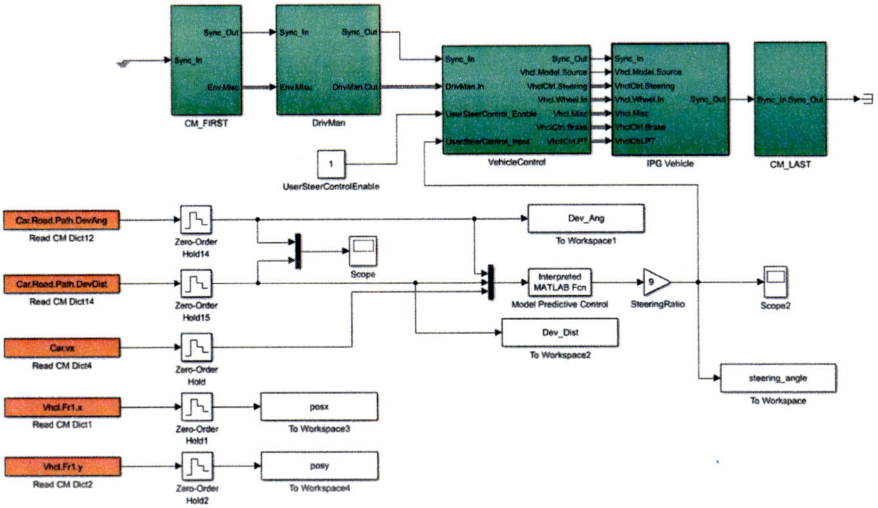

Fig. 2. Extended Simulink model

Through the UserSteerControlEnable variable, the steering angle that enters to the Vehicel Control block can be selected. It can be the one calculated by the inbuilt carmaker driver model, or it can be the one calculated by our own algorithm.

The steering angle defined by our controller is computed by the Model Predictive Control Interpreted Matlab Function, which has its inputs, the deviation from the trajectory, from the Carmaker environment.

The speed of the errors is not needed as during the simulation the weight of the errors is set to zero. The longitudinal speed of the vehicle is an input as well. The optimization algorithm calculates the ideal steering angle profile for each sampling time (Ts = 5ms) for the next N = 25 (prediction horizon) time.

4 Results

Two types of simulations were conducted, in which the steering angle required by the carmaker driver model was compared to the steering angle defined by the MPC algorithm. First a *bus-stop* maneuver, secondly a *skid pad* maneuver.

The following weighting matrices were used:

$$Q = \begin{bmatrix} 1000 & 0 & 0 & 0 \\ 0 & 0 & 0 & 0 \\ 0 & 0 & 1 & 0 \\ 0 & 0 & 0 & 0 \end{bmatrix}, \mathrm{R} = 1 \tag{4}$$

The bus stop maneuver has 4 consecutive 90° turns. With this maneuver you can test the reactivity of the car, in this case the regulator. A slow regulator would not be able to complete the maneuver. In the first round, we compared the trajectory of the car in the two cases. During the simulations, the longitudinal speed was constant at 10 km/h (Fig. 3).

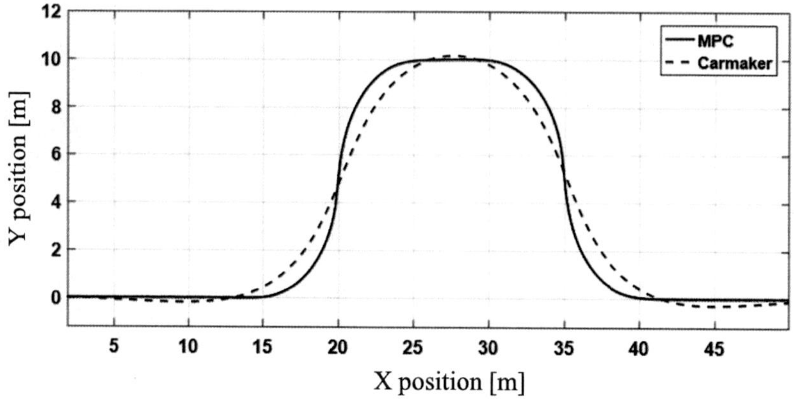

Fig. 3. The trajectory of a car during a bus stop maneuver

It can be seen that with the steering angle defined by the MPC, the car is almost in the center of the track. The 4 consecutive turns are very clear and do not slip off the trajectory. In contrast, the original Carmaker driver's trajectory differs from the center line of the track, as the main point here is not to move the car in the center, but to have the shortest possible travelling distance (Fig. 4).

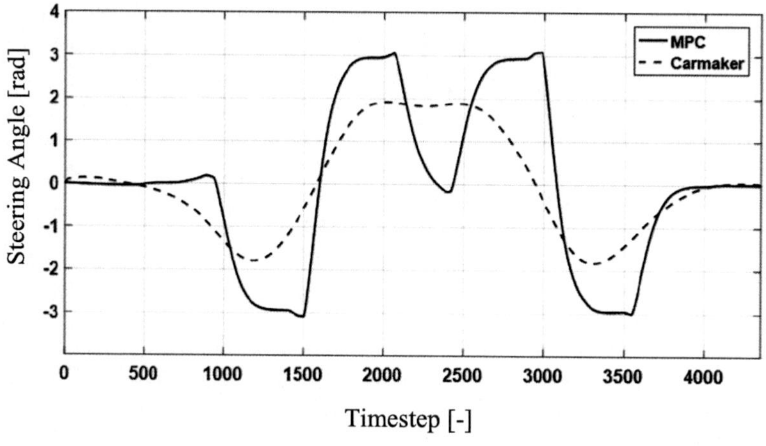

Fig. 4. Steering angle change during bus stop maneuver

Comparing the course of the steering angles over time, we can still state what has been described above. It is easy to see the MPC taking all four corners and then entering "steady-state" steering, by having a constant steering angle. On the other hand, the Carmaker basic model is more continuous with longer, but slower steering angle changes.

The *skid pad* maneuver is an 8-shaped track where the car turns two laps to the right, then turns two laps to the left, then leaves the course. Several parameters of the car can be tested by this method. Firstly, the maneuver involves the steady state behaviour during corners, the transient behaviour when changing direction, and finally the acceleration on corner exit.

During the simulation the longitudinal control of the vehicle was up to the Carmaker's driver model. The MPC steering angle calculation was compared to the Carmaker driver's steering angle calculation (Fig. 5).

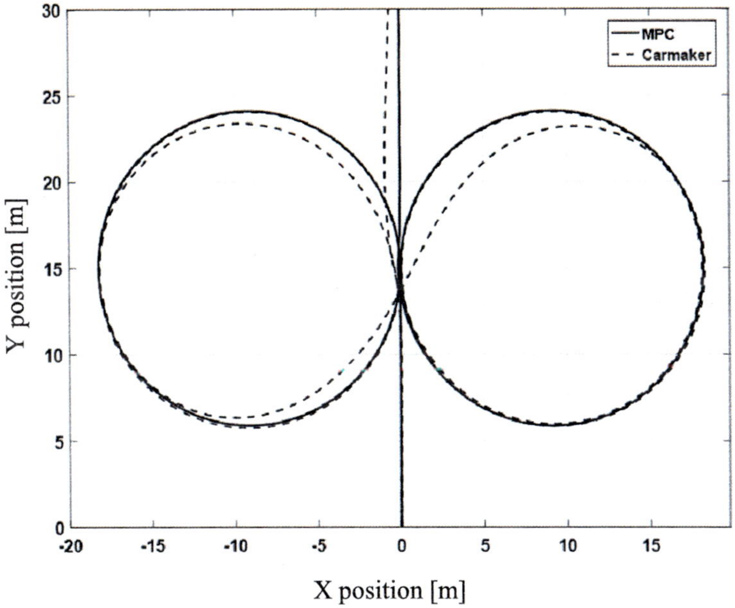

Fig. 5. Car trajectory during skid pad maneuver

In this case, you can see the original carmaker model "cutting off" the curves, and the MPC is moving in the middle. In steady state, at the apex of the curves, however, the positions are the same in both cases.

We can also observe the fundamental differences between the two strategies when looking at the steering angle over time. What is even more striking is that during the turn during acceleration, the actuator signal oscillates considerably. This is mainly because the optimization algorithm assumes a constant speed between 20 time steps, but with significant acceleration, this assumption can greatly ruin the solution. Further tuning of the weighting matrices can reduce the tendency to oscillate.

The following weights are granting no oscillation while cornering and accelerating in longitudinal direction in the same time (Figs. 6 and 7):

$$Q = \begin{bmatrix} 1\,0 & 0\,0 \\ 0\,0 & 0\,0 \\ 0\,0 & 1\,0 \\ 0\,0 & 0\,0 \end{bmatrix}, \mathrm{R} = 1 \tag{5}$$

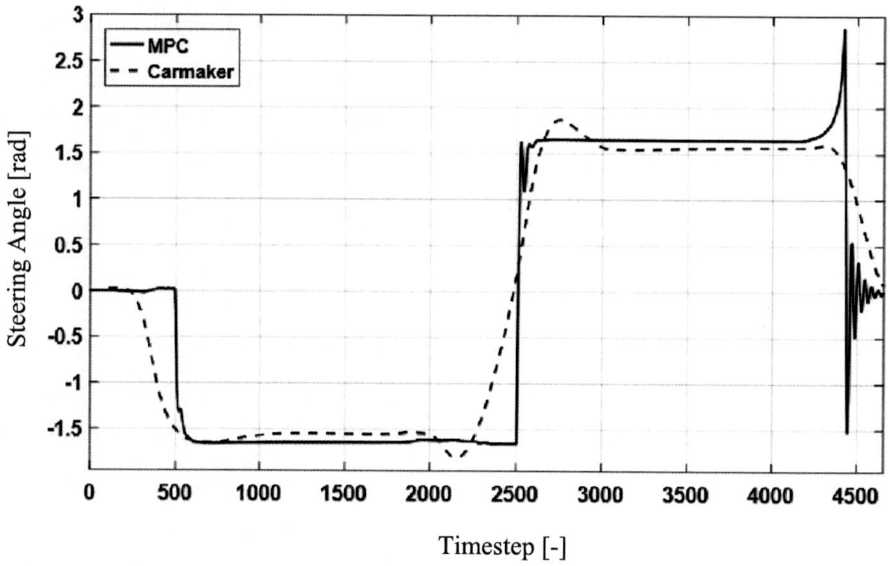

Fig. 6. Steering angle change during skid pad maneuver, original weights

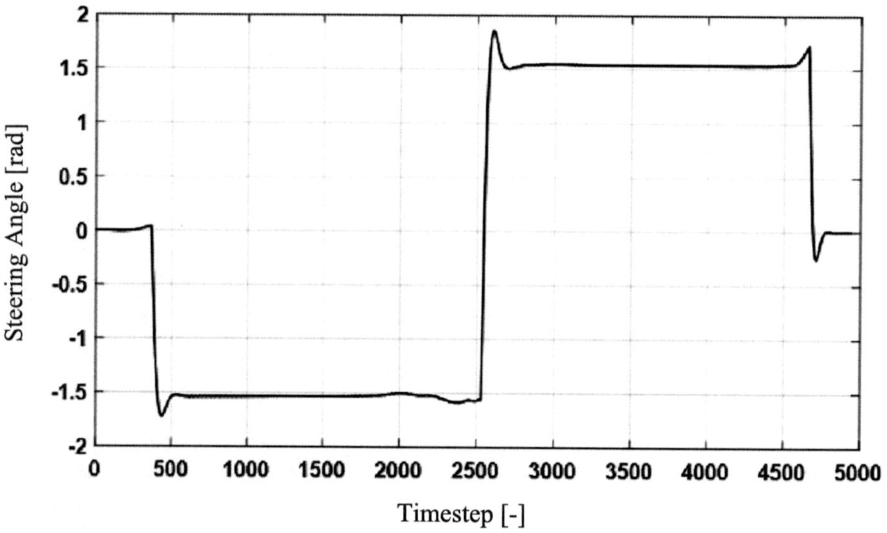

Fig. 7. Steering angle change during skidpad maneuver with new weights

5 Conclusion

Simulating the autonomous system, controller design and tuning was possible without taking the car out to real environment. This workflow can be used for rapid algorithm test. The next steps for this project is to further improve the simulation, so other phases of the autonomous system can be tested in this environment as well, like object detection and trajectory planning based on the detected objects.

Acknowledgement. The research presented in this paper was carried out as part of the EFOP-3.6.2–16-2017–00016 project in the framework of the New Széchenyi Plan. The completion of this project is funded by the European Union and co-financed by the European Social Fund.

References

1. Mohamed, A.: Tewolde Girma. Kwon Jaerock, Software-in-the-Loop Modeling and Simulation Framework for Autonomous Vehicles (2018)
2. Jarrod, M.: Snider, Automatic Steering Methods for Autonomous Automobile Path Tracking (2009)
3. Vehicle Dynamics 4L150, Eindhoven University of Technology, Mechanical Engineering - Dynamics & Control, Dr. Ir. I.J.M. Besselink lecture notes (2014)
4. Rodriguez, J., Cortes, P.: Predictive Control of Power Converters and Electrical Drives. Wiley-IEEE Press, Chichester (2012)
5. Carmaker User's Guide, version 6.0.
6. Demers, S., Gopalakrishnan, P., Kant, L.: A generic solution to software-in-the-loop. In: MILCOM 2007 - IEEE Military Communications Conference, Orlando, FL, USA, pp. 1–6 (2007). https://doi.org/10.1109/MILCOM.2007.445526810.1109/MILCOM.2007.4455268
7. Yan, Q.-Z., et al.: Chassis control system development using simulation: software in the loop, rapid prototyping, and hardware in the loop. SAE Trans. **111**, 1734–1744 (2002). *JSTOR*, www.jstor.org/stable/44719352. Accessed 19 Aug 2020

Range-Reducing Effect of Contaminants in Case of Solar Vehicles

Dávid Matusz-Kalász(✉), István Bodnár, and Rafael Ruben Boros

University of Miskolc, Miskolc, Hungary
{elkmkd,vegybod,elkruben}@uni-miskolc.hu

Abstract. Over the past decade, the numbers of solar power plants and the capacity installed have increased rapidly. In Hungary, the total installed capacity of solar power plants in 2010 barely exceeded 2 MW, in 2020 this value will already exceed 1000 MW. Hungary wants to reach 6000 MW in the capacity of solar power plants in this decade. Solar panels are not only spread in solar power plants, smaller ones have also used in laptop and mobile phone charges, as well as in electric or hybrid vehicles. Solar-powered ships appeared years earlier, and it didn't take long to create solar-powered airplane. In the case of land vehicles, light vehicles for 1–2 people appeared in the first round, but we can also find examples of family cars and vans.

Keywords: Contaminant · Solar · DC/DC converter

1 Introduction

As is well known, several factors can affect the operation of solar panels. We have no influence on certain factors in case of fixed solar panels, such as the ambient temperature, the intensity of the solar radiation and its angle of incidence. In addition to these factors, we must also consider the effect of natural pollutants as shadow formers. These factors are difficult to identify, and their impact cannot be calculated in an exact way. Surface contaminants have two main effects on solar cell operation. One effect is the shadow effect, through which the power generation capacity of the solar cell decreases. The surface contaminant reduces the active surface of the solar cell. The other effect is the thermal insulation, as the contaminants deposited on the surface of the solar cell act as a thermal insulating layer, thereby reducing the heat dissipation capacity of the solar cell and warming over. As a result of the heating, the terminal voltage of the solar cell decreases. As a result of both effects, the solar cell continues to work at a lower operating point, which essentially means an efficiency loss [1].

Since solar cells are semiconductor electronic circuits containing a p-n junction, persistently high temperatures can lead to degradation of the p-n junction. In essence, the recrystallization of cells causes a permanent decrease in efficiency and shortening of lifetime. Under extreme conditions, it can even lead to rapid burnout of the cells, i.e. immediate destruction.

K. Jármai and K. Voith (Eds.): VAE 2020, LNME, pp. 38–48, 2021.
https://doi.org/10.1007/978-981-15-9529-5_4

Sources of contamination can be natural or artificial. The most common natural sources of pollution are animal feces, leaves and pollen from vegetation, agricultural fly dust, or sand in the desert environment, and volcanic ash near volcanoes. Artificial sources of pollution include dust and soot from industry, power plants, household fires and transport, or minerals in the wash water. Surface contamination reduces efficiency. In the case of a solar cell placed on a roof, the amount of contamination cannot be determined, but under laboratory conditions (Fig. 1) we can place an exact amount on it and infer from the results obtained.

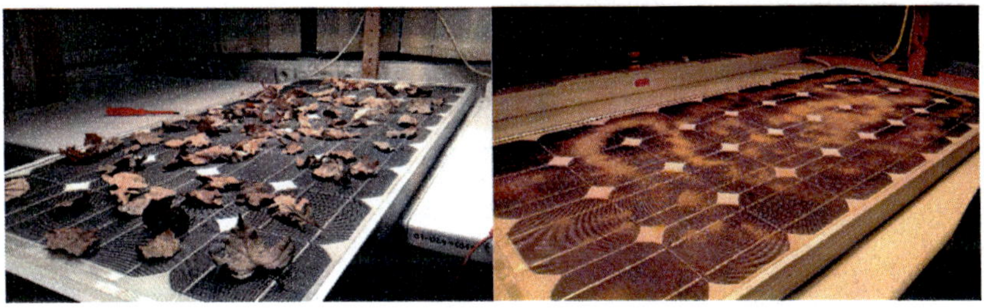

Fig. 1. Leaves as large contaminants on the surface of the solar cell.

Larger contaminants such as leaves or high particle size (e.g. sand) roll off the surface of the solar cell. Some of the fly dust comes into contact with water to form a muddy layer on the surface of the solar cells, so they partially run off, but their complete removal is a more difficult task. The most common airborne dust, such as PMM 2.5 and PMM 10, adhere electrostatically to the surface of the solar cell, making them the most difficult to remove. However, the biggest problem is animal feces, primarily bird droppings. Because they are concentrated on a small surface, they can form a thick layer of thermal insulation on the surface of the solar cell. Such contaminants, mostly affecting one cell, less often two cells, create hot spots on the solar cell. These hot spots are more prone to permanent damage, faster destruction. However, it is more difficult for bird droppings to leave the solar cell in a natural way, so their artificial removal may be justified. When removing contaminants, care must be taken to ensure that the water is hard in some geographical areas. After removing the contaminants, make sure that the water is hard in some geographical areas. As a result, the hard water evaporating from the solar cell causes surface contamination in the form of limescale or mineral deposits [2].

Hot spots are detected primarily by thermal imaging. In a highly polluting environment, it is advisable to perform the thermal imaging examination annually, so that it is easy to filter out which cells have been damaged or defective. Faulty cells can affect the operation not only of the modules but also of the entire solar power plant. The more failed cells, the greater loss in efficiency of the system. Some manufacturers give a 10–20-year warranty on their solar panels, therefore feedback to a distributor / manufacturer within the warranty period for defects may constitute a free replacement of the module.

Contaminants reduce the operating efficiency of the solar cell, i.e. the instantaneous efficiency, which also reduces the amount of generated electricity. The extent of this reduction also affects the operation of solar-powered equipment. Depending on the environmental conditions, there may be a decrease of up to several percents in the daily energy production of solar cells. Measurements and simulations are required to quantify the effects, as there is currently no exact mathematical model to pinpoint them. In our research, we performed measurements to estimate the effects.

2 Solar Powered Cars

Since the solar panels generate direct current, it is an obvious solution to place them on top of vehicles and connect them to a 12 VDC system. Although solar panels are not enough for running an electric car, however, they can provide power to the control circuits or provide a slight increase in range. Over the past decade, many vehicle manufacturers tried to build solar cars, with greater or lesser success. Solar powered prototype is currently being developed by the Lightyear Company (Netherlands) and the Sono Motors (Germany) [5]. The parameters of solar cars can be seen is Table 1.

Table 1. Comparison of solar powered prototype vehicles.

Name of prototype	Lightyear One (Car 1)	Sono Sion (Car 2)
Manufacturer	Lightyear	Sono Motors
Total area of solar panels	5 m^2	7.5 m^2
Solar range	50–60 km	34 km
Ratio of solar usage	57%	19%
Nr. of seats	5	5

The solar panels are always used to charge the vehicle battery pack, however, electric cars run on tens of kW electric motors. On the other hand, only a few hundred watts of solar power capacity can be placed on them. It follows that solar panels alone are not enough to charge and operate the vehicle, therefore, the importance of solar cars lies primarily in the daily commuting of the urban population. It should not be overlooked that most cars are usually used by one person at a time. For this reason, it is unnecessary to produce large and powerful solar cars. It would be practical to design these vehicles for daily, low-power commuting, rather than for a family car. It is worth researching in this direction and developing cheaper solar cars [3, 4, 5].

Figure 2 contains the monthly distribution of the amount of solar energy measured in Budapest with a 45% south-facing Pyranometer. It can be observed that about 71.2% of the annual solar energy falls in the period from April to September (6 months). As the tested solar cell has an efficiency of 18.75%, the electricity generated by the solar cell would be 256.61 kWh/m^2 per year. Because the solar panels in the electric car are located almost horizontally, they are able to produce only 85% of this value. We can count on an annual energy production of 218.12 kWh/m^2.

The Car 1 has 5 m^2 and the Car 2 has 7.5 m^2 of solar panels. Accordingly, the solar panels of Car 1 produce 1090.6 kWh of electricity per year and the solar panels of Car 2 generate 1635.9 kWh of electricity. If both vehicles have a consumption of 200 Wh/km, which is the average consumption of the electric cars currently in use, the Car 1 would be able to cover (5453 km) per year and the Car 2 (8179.5 km). The energy conversion efficiency of the vehicle, which is the product of the charging, storage and DC/AC conversion efficiencies, should not be overlooked. The energy conversion efficiency varies between 80–90% depending on the charge and the type of battery used in the car [3, 4, 5].

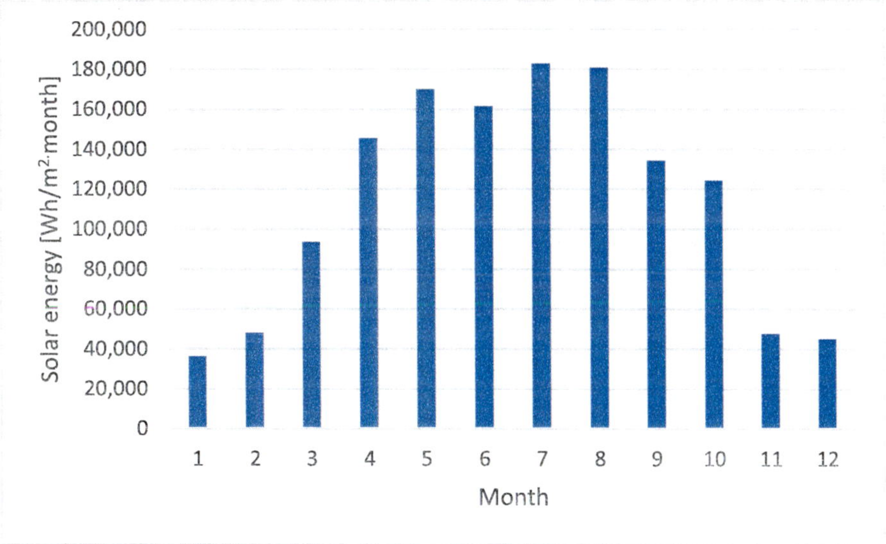

Fig. 2. The monthly distribution of the amount of solar energy in Budapest.

Figure 3 shows the monthly range with solar panels and Table 2 shows the daily range. It can be observed that in the period between April and September, Car 1 can cover an average of (647 km) per month and Car 2 (970 km). During this period, on average, Car 1 can cover (21.19 km) and Car 2 (31.78 km) from the daily energy production of solar panels [5].

Consider a conventional electric car with an average roof area of 4 m^2. Ideally, 750 Wp solar capacity can be placed on a surface of this size. The annual energy production of these solar cells is 872.479 kWh. With so many solar panels, neglecting the losses, it would be able to cover 4362 km per year. This means an average of 11.95 km per day. It should not be overlooked that solar cells are contaminated by contaminants in the environment around them. Contaminants reduce their efficiency, thus the range of solar-powered cars [5].

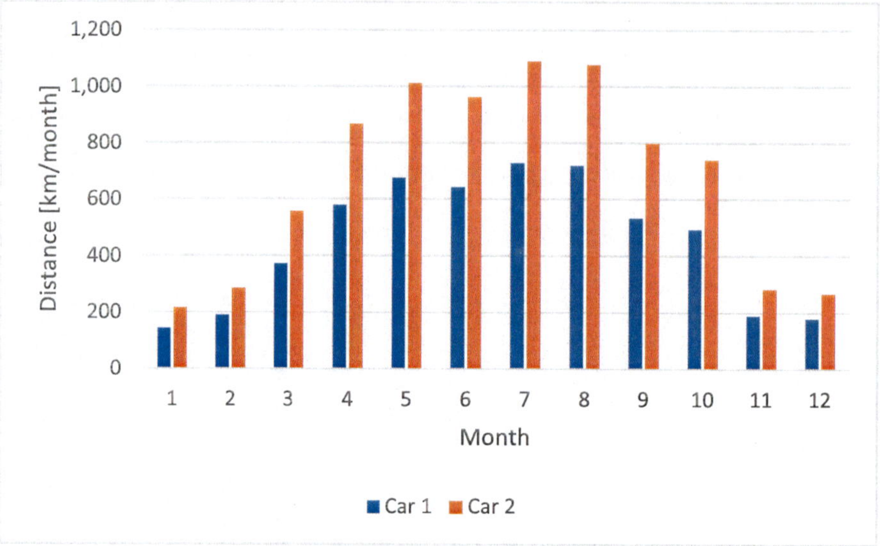

Fig. 3. Comparison of monthly solar range of solar powered vehicles.

Table 2. Quantifying daily solar range of solar powered vehicles in every month of a year.

Month	Car 1 [km/day]	Car 2 [km/day]
1	4.65	6.97
2	6.81	10.22
3	12.02	18.03
4	19.33	28.99
5	21.83	32.75
6	21.43	32.15
7	23.49	35.24
8	23.23	34.84
9	17.81	26.72
10	15.95	23.92
11	6.32	9.48
12	5.78	8.67

3 Power Electronics in Electric Cars

The transmission system for electric cars is simpler than for cars with combustion engines, as there is no need for a clutch, a large gearbox. The transmission system is driven by an electric motor. It is advisable to install electric motors in vehicles that have a high starting torque and are as constant as possible. This assumption is well satisfied by the permanent magnet synchronous motor (PMSM). The PMSM motor is usually three-phase, the rotor is made of permanent magnet, so the need for maintenance is minimal. The stator requires three-phase AC voltage for winding. The speed is proportional to the frequency of the voltage.

The power electronics circuit diagram that can be integrated with the solar system is shown in Fig. 4. The AC voltage is generated by a three-phase inverter (DC/AC converter) from the DC voltage of the battery. The switching elements of a modern inverter consist of SiC MOSFETs, so the inverter has good efficiency. The inverter is controlled by a microcomputer that implements a field-oriented drive, thus providing the highest possible torque at any speed. The inverter is a voltage source inverter, so its input requires a constant DC voltage that is connected to the intermediate DC circuit. Here, a buffer capacitor performs voltage smoothing [6, 7].

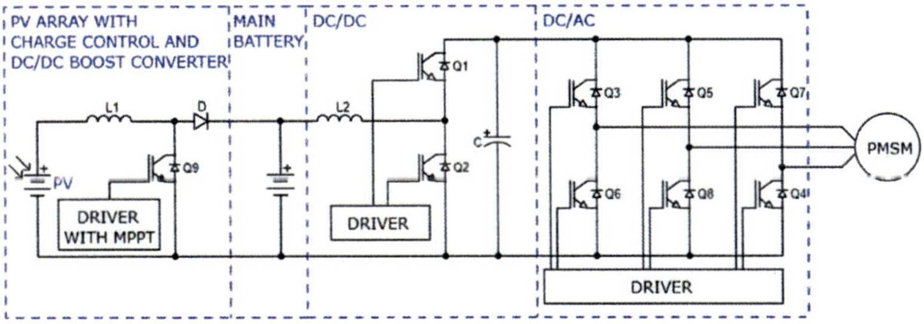

Fig. 4. Main power electronics with solar power

The battery voltage varies over a wide range between charged and discharged. In most cases, the nominal voltage of the battery is lower than the rated voltage of the motor, so a DC/DC converter must be used. In order to increase the range of the car, it is advisable to use kinetic energy in the form of electricity when braking. In regenerative mode, the PMSM machine is able to feed electricity back into the battery during braking. The amount of voltage generated must be controlled, as it is much higher than the battery voltage, so a voltage-reducing DC/DC converter is required. The two DC/DC converters can be combined to create a bidirectional DC/DC converter (buck-boost) [8].

The range of the car can be further extended by charging that battery with a solar cell. The solar cell generates a constant direct current at a constant light intensity. The solar cell cannot be connected directly to the battery, as the solar cell generates a lower voltage than the battery terminal voltage. The solar cell is a current generator, so a step-up DC/DC converter is required, which uses the Maximum Power Point Tracking algorithm (MPPT) of the solar cell to charge the battery [8].

4 Effect of Contaminants

In the first round, we performed outdoor measurements, the aim of which was to see to what extent the pollutants deposited on the solar cells reduce the performance and production of the solar cells. For the measurements, a 1.5 kW peak power solar system placed on the roof of a garage with a 20° slope was used in an urban area (Fig. 5).

Fig. 5. Measured solar panels on the roof of a garage.

On the first day, the solar panels were cleaned. Subsequently, we continuously measured the performance and energy production of the solar cell. The solar panels used as controls were washed daily. In the first days, we hardly experienced any differences in production. By day 9, the peak performance of the solar cell had decreased by 2.53%, with cumulative production losses approaching 10%. On day 9, a thunderstorm cleared the solar panels (Fig. 6). As the amount of deposited powder could not be measured, laboratory tests were started. The aim of the laboratory test was to determine how much dirt deposits cause power loss or loss of production [9].

Fig. 6. The ninth day was dirty solar panel (left) and post-rain solar panel (right).

In our laboratory tests, a KS-85 type solar panel was placed on a table of the same size during measurements. The ambient temperature was 20 °C. The area of the solar panel is 0,5 m^2. The type of solar panel is monocrystalline. The surface of the solar panel was illuminated by eight reflectors. The voltage and current of the solar panel were measured by a METRIX MX-59HD digital multimeter. We polluted the solar cell with powder having four different properties. These were sand, fly ash, soil and urban dust. The dust concentration was between 0–75 g/panel (0–150 g/m^2) at 5 g/panel scale. The weight of powder doses was measured by a Voltcraft PS-200B jeweler scale. The method of scattering was the sieving. The maximal grain size was 0,5 mm. Due to the sieving; pollutants can spread almost perfectly over the entire surface. In this way, we can approximate the real deposition much better by eliminating large differences in grain size (selection of larger grains). When the solar cell is exposed to light along with heat radiation, causing a certain amount of voltage drop. The pollutant accumulates the heat, thereby enhancing its negative impact. The greatest reduction in voltage was observed for ash, the smallest nonlinearity was measured with sand [10, 11].

The results of the voltage drop can be seen in Fig. 7.

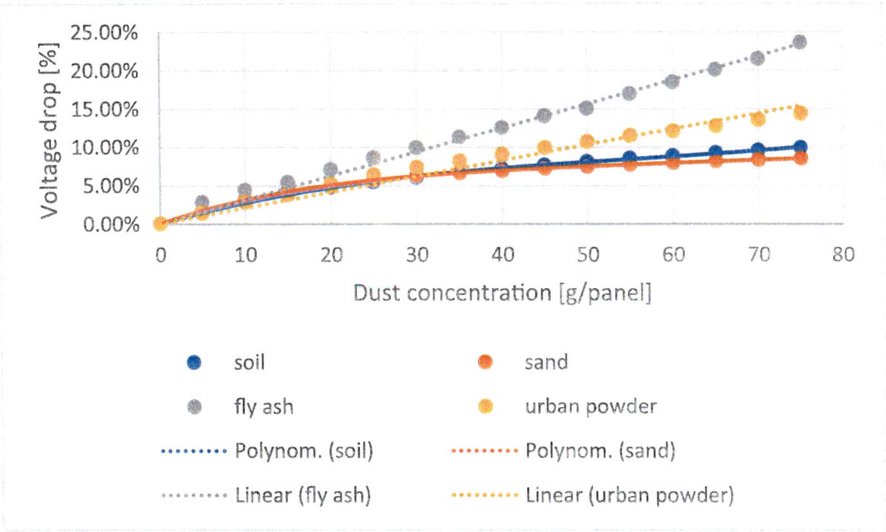

Fig. 7. The measured voltage drop caused by contaminants.

Similar trends can be observed in the case of voltage and current decreases. The decrease of illumination can cause a decrease of current, which is further exacerbated by surface contamination. As pollutants spread on the surface in a thinner but more extensive layer, their shading effect deepened the amperage reduction. The greatest reduction in amperage was observed for ash as in case of voltage (Fig. 8) [10, 11].

The decrease in power provided by the solar panel is a significant problem, while electric motors have a constant significant energy demand. Since the product of these two quantities is the electrical power, similar trends (Fig. 9) can be observed in the case of voltage and current decreases [12].

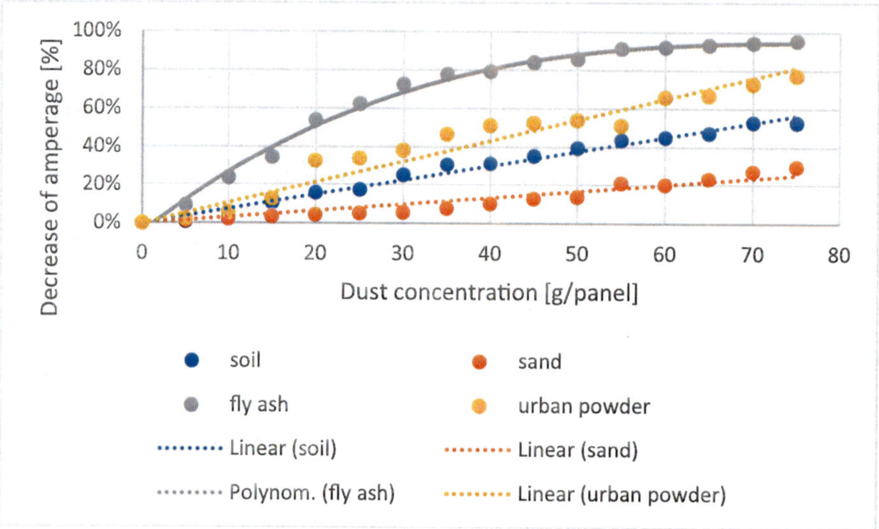

Fig. 8. The measured amperage reduction caused by contaminants.

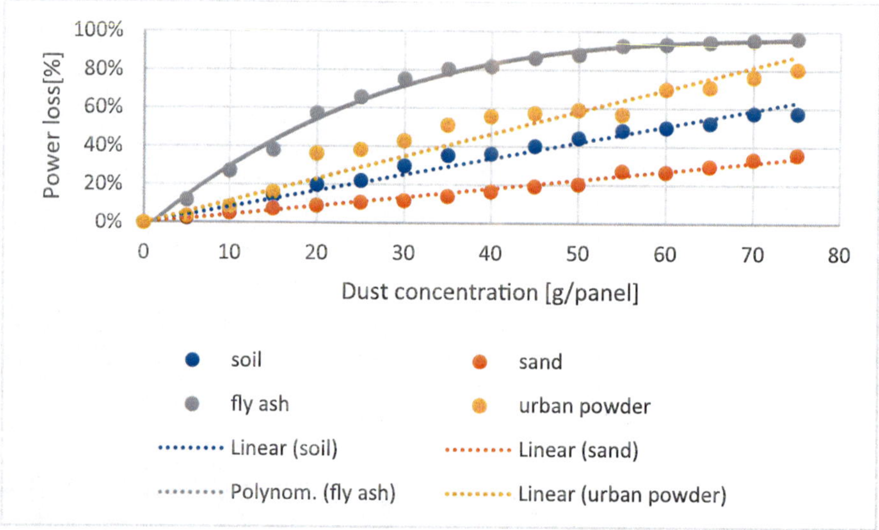

Fig. 9. Power-reducing effect of contaminants.

5 Conclusions

The highest specific surface area has a significant impact on the operation and lifetime of the solar cell. Lower density pollutants as ash form a thicker insulating layer on the solar panel, resulting in more heating. The ash data set shows a polynomial change as a function of concentration due to this thick cover layer. If the solar cell operates at a higher temperature, the service life is reduced due to the nature of the electronics.

The other three contaminants show a close, linear change. At low concentrations (below 15 g/panel) no significant difference was observed between the different materials. It is expected that later measurements with low concentrations are worthwhile, as in this case, we can examine a case that is closer to reality. Under real conditions, the cleaning effect of the air flow can remove a significant amount of larger grain size contaminants. Only contaminants that remain electrostatically adhere to the surface. Furthermore, wet contaminants are difficult to remove, resulting in permanent performance degradation.

At a concentration level of 10 g/panel (20 g/m^2) the power loss was about 8%. Consider an average roof area of 4 m^2, which means ideally 750 Wp solar capacity, 8% power loss causes numerous reductions in total solar range. The annual power loss of this solar car is 69,798 kWh, which is equal to 448.96 km solar range loss per year. This means an average of 0.956 km per day.

References

1. Abderrezek, M., Fathi, M.: Experimental study of the dust effect on photovoltaic panels' energy yield. Sol. Energy **142**, 308–320 (2017)
2. Gürtürk, M., Benli, H., Ertürk, N.K.: Effects of different parameters on energy – exergy and power conversion efficiency of PV modules. Renew. Sustain. Energy Rev. **92**(9), 426–439 (2018)
3. Betancur, E., Mejía-Gutiérrez, R., Osorio-Gómez, G., Arbelaez, A.: Design of structural parts for a racing solar car. In: Eynard, B., Nigrelli, V., Oliveri, S., Peris-Fajarnes, G., Rizzuti, S. (eds.) Advances on Mechanics, pp. 25–32. Lecture Notes in Mechanical Engineeringinger, Design Engineering and Manufacturing (2017)
4. Minak, G., Fragassa, C., de Camargo, F.V.: A brief review on determinant aspects in energy efficient solar car design and manufacturing. In: Campana G., Howlett R., Setchi R., Cimatti B. (eds.) Sustainable Design and Manufacturing 2017, SDM 2017, Smart Innovation, Systems and Technologies, vol. 68, pp 847–856 (2017)
5. Bodnár, I., Boros, R.R., Matusz-Kalász, D.: Solar powered electric car with VVVF drive control. GÉP **71**(3–4), 55–60 (2020)
6. Trovão, J. P., Pereirinha, P. G., Ferreira, F. J. T. E., Jorge, H. M.: Study of inductor effects in a bidirectional DC-DC converter for electrical vehicle. In: The XIX International Conference on Electrical Machines - ICEM 2010, Rome, Italy, pp. 1–6 (2010).
7. Gautam, D.S., Musavi, F., Eberle, W., Dunford, W.G.: A zero-voltage switching full-bridge DC-DC converter with capacitive output filter for plug-in hybrid electric vehicle battery charging. IEEE Trans. Power Electron. **28**(12), 5728–5735 (2013)
8. Perdigão, M., Trovao, J.P., Alonso, J.M., Saraiva, E.: Large-signal characterization of power inductors in EV bidirectional DC-DC converters focused on core size optimization. IEEE Trans. Ind. Electron. **62**(5), 3042–3051 (2015)
9. Malik, A.Q., Damit, S.J.B.H.: Outdoor testing of single crystal silicon solar cells. Renew. Energy **28**, 1433–1445 (2003)
10. Bhattacharya, T., Chakraborty, A.K., Pal, K.: Influence of environmental dust on the operating characteristics of the solar pv module in Tripura, India. Int. J. Eng. Res. **4**(3), 141–1440 (2015)

11. Rao, A., Pillai, R., Mani, M., Ramamurthy, P.: An experimental investigation into the interplay of wind, dust and temperature on photovoltaic performance in tropical conditions. In: Proceedings of the 12th International Conference on Sustainable Energy Technologies, pp. 2303–2310. The Hong Kong Polytechnic University, Hong Kong (2013)
12. Adinoyi, M.J., Said, S.A.M.: Effect of dust accumulation on the power outputs of solar photovoltaic modules. Renew. Energy **60**, 633–636 (2013)

Alternative Propulsion Buses in the Metropolitan Public Transport

István Lakatos(✉), Ferenc Szauter, Dániel Pup, and Andor Nagy

Széchenyi István University, Győr, Hungary
lakatos@sze.hu

Abstract. The lecture is analysing the possibility of an optimal energy mix through the example of a Hungarian metropolis. Using the city bus routes, we analyse and compare the traditional Diesel, CNG, and electric propulsion. An optimal energy mix is provided by using SWOT analysis.

Keywords: Energy mix · Transport budapest · Public bus

1 Introduction

The article uncovers mainly the CNG based road traffic possibilities. Meanwhile, we offered place for the other means of city bus propulsion, with special attention to purely electric, and hydrogen as the possible future energy source.

The different propelled vehicles need a diverse infrastructure, so along our path, we mentioned the layouts of fueling stations and loading stations.

The main scope is to analyse the public transport, but during the in-depth analysis we took smaller, personal and smaller truck related reports and evaluations into consideration because in certain cases the fueling stations are suitable for other means of road transport.

The article follows the logic based upon:

- technical and technological fundamentals,
- related economic viewpoints
  - CNG,
  - electric,
  - Diesel propulsion as the basis of comprehension.

Diesel technology appears as the base of comprehension for comparative and SWOT analysis, because we've been searching alternatives for this, widespread propulsion.

Along with the technical data and technological parameters, we display the propulsion methods in connection with operational, fuel supply, and legal means too.

The analysis will evaluate the application possibilities of hydrogen usage – as a propellant -, and the present application possibilities of the fuel cell propulsion.

K. Jármai and K. Voith (Eds.): VAE 2020, LNME, pp. 49–66, 2021.
https://doi.org/10.1007/978-981-15-9529-5_5

The second chapter will analyse the full CNG operation possibilities of the city of Győr's public transportation fleet. First, we evaluated the present fleet, transport network, and the relevant city structure.

2 Comparison of the CNG, Diesel Combustion, and Electric Propulsion in Terms of Technical and Financial Conditions in the Public Transport

This chapter compares the CNG and the electric propulsion based upon technical and financial assets, using SWOT analysis. Figure1 shows the main propulsion possibilities.

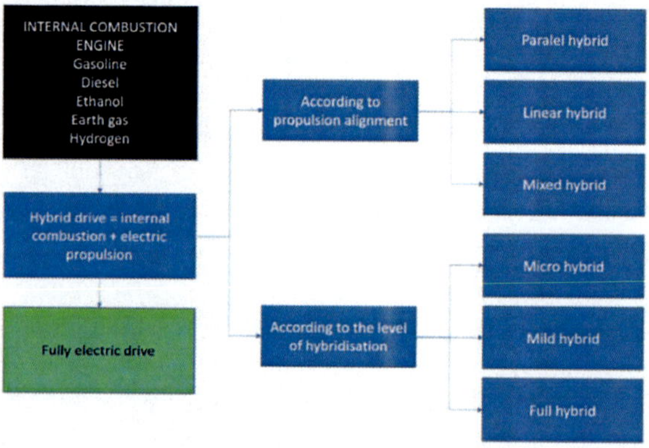

Fig. 1. Overview of propulsion systems

It is important to mark, that it is not possible to overlook the environmental load while taking care of the financial viewpoints, and the pollution. If we take out the CO_2 emission of the environmental load, then outside the product life cycle, the presiding raw material production, manufacturing, recycling, and neutralizing environmental loads must be taken into consideration too (Fig. 2).

2.1 Analysis of CNG as an Alternative Propellant

Internal combustion engines are to be operated efficiently and environment friendly by earth gas or bio gas. The main two propellants are CNG (compressed natural gas) and LNG (Liquified Natural Gas). According to its characteristics, earth gas is an "external ignition" gasoline-like propellant.

The autogas is a "dry type" gas, that does not lubricate the surface between the piston rings and the cylinder that much like a gasoline-air mixture. The wear is more extensive. Meanwhile, the quality of the lubricating oil is not decreasing that fast as by gasoline, and there is no mixture between gasoline and oil.

The main part of the vehicle's cost is from the variable costs, where the main part is the price of the propellant. Here are a few numbers compared to Diesel fuel:

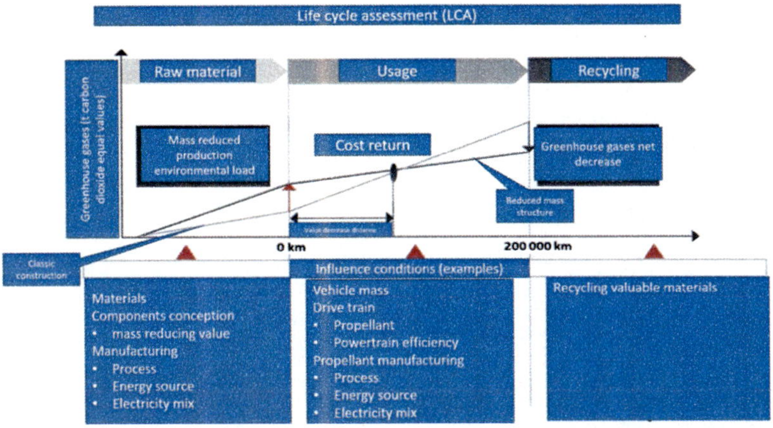

Fig. 2. Effect of different vehicle conceptions to the full CO2 emission during the product life cycle (Forrás: Audi AG)

1. The CNG vehicle prices are 6% higher than diesel fueled one.
2. The CNG vehicle's maintenance and repair costs are 2,4% higher than a diesel fueled one.
3. The CNG vehicle's propellant fuel is approximately 20–25% lower than a diesel fueled vehicle.
4. CNG vehicle has no AdBlue consumption
5. So as a conclusion, the total cost of a CNG vehicle is lower than the diesel vehicle, looking at the same running performance.

Further advantages of the application of CNG vehicles:

1. Proven reliability of the technology,
2. Economic and comfortable alternative to substitute the diesel vehicles,
3. Compared to other means of propellants, the most efficiently usable
4. It is able to fulfil the low emission requirements without any special interventions.
5. The CNG vehicle is 100% compatible with biogas operation
6. The noise pollution is 50–75% lower compared to the diesel
7. The natural gas reserves are bigger in volume on the Earth, than the fossil based propellants (biogas – huge reserves)

Technical Characteristics of the CNG Operation

CNG is usable in Otto-engines (gasoline fuel), and Diesel engines too. The engines that are using poor mixture diesel, can achieve higher performance, compared to stoichiometric Otto engines, with higher NO_x and carbon-hydrogene emission costs.

The properly setup with compressed earth gas engine can produce higher effective performance compared to the gasoline engine because the octane number of the earth gas is higher than the gasoline.

CNG tanks can be refilled using low pressure (slow refill) or high pressure (rapid refill). Differences are the price per unit on the refueling station and the refilling time.

Types of CNG-cylinders, and their weight/volume ratio:

In the EU's 27 member states, the operated CNG vehicle number started to increase from 2007, and it has reached the volume growth of 150% during 4 years. From 1 September 2014 only Euro-6 certified vehicles can be put into operation. Nowadays, the places that are capable of CNG refueling can be seen in Fig. 3.

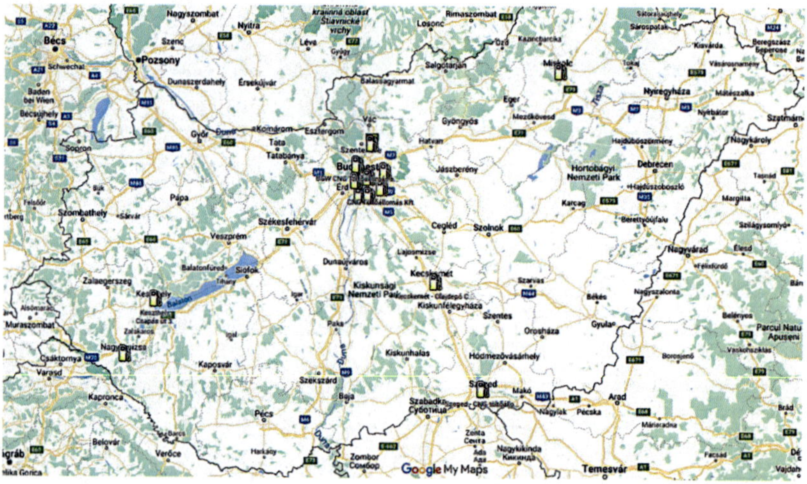

Fig. 3. CNG refilling stations in Hungary

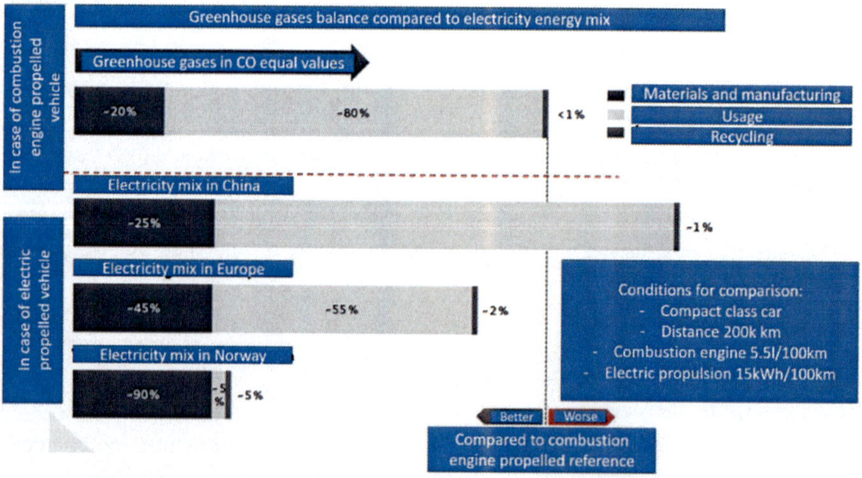

Fig. 4. The electromobility CO2 equal environmental load compared to combustion means, from the point of view of the energy mix

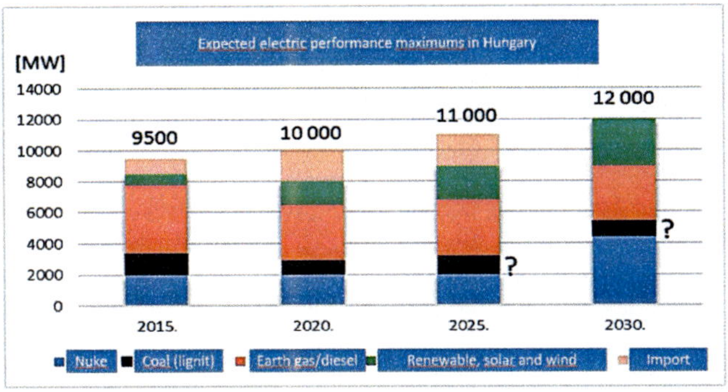

Fig. 5. The energy mix in the future in Hungary (Source: Dr. Strobl Alajos presentations 2014)

Operation and Maintenance Regulations of the Vehicles

Operating conditions are laid down on the following regulations (in terms of CNG propellant vehicles). Many of them do not have an English version as they were made before the EU membership of Hungary.

1. 6/1990 (IV.12) KöHÉM act.: "a közúti járművek forgalomba helyezésének és forgalomban tartásának feltételeiről"
2. 1/1990 (IX.29) KHVM act: "a gépjárműfenntartó tevékenység személyi és dologi feltételeiről"
3. 17/1993 (VII.1) KHVM act: "egyes veszélyes tevékenységek biztonsági követelményeiről szóló szabályzatok kiadásáról"
4. 30/2006 (VI.1) GKM act: "autógáz tartályok időszakos ellenőrzéséről, 2. Annex, for the 30/2006. (VI. 1.) GKM act: "Autógáztartályok időszakos ellenőrzése"
5. National Fire Prevention Regulation (OTSZ) (28/2011. (IX.6.) Act of Ministry of Internal Affairs
6. 51/2007. (V. 17.) GKM act: "a gázüzemű munkagép gáz-üzemanyag ellátó berendezései beszerelésének, karbantartásának, javításának, a belső égésű motorral üzemelő munkagép gázüzeműre történő utólagos átalakításának műszaki-biztonsági követelményeiről"
7. Regulation No 110 of the Economic Commission for Europe of the United Nations (UNECE) — Uniform provisions concerning the approval of specific components of motor vehicles using compressed natural gas (CNG) and/or liquefied natural gas (LNG) in their propulsion system

SWOT Analysis of CNG Operation

We present the SWOT analysis of CNG operation in an arranged Table 1 form.

In case of local, suburban, and interurban buses, the low grid of refilling stations is not a problem, as there is a possibility to deploy a CNG station because of the constant demand in the area.

One of the most defining factors is the support of the state (not only for buses and limited areas) when we talk about CNG and BIO-CNG expansion, in its present phase (natural propelled vehicles, regulatory background, taxes, infrastructure, direct support, temporary payment easing conditions upon procurement of these high priced vehicles).

To be able to uphold the preferences, it would be essential to:

- Necessity to expand the refilling infrastructure that is independent of the volume (home, industrial, public) meanwhile increasing CNG production and refilling performance.
- Reducing regulatory obstacles, supporting research
- Propellants evaluation according to "well-to-wheel" potentials
- Closer cooperation between the vehicle-, refilling technology manufacturers, and the gas companies would be necessary (Mutual research projects)

Table 1. CNG SWOT analysis

Strength of CNG	**Weaknesses of CNG**
• Cheaper then any other propellant • There are big volumes as reserve available (in Hungary too), • Its price is more stable as the diesel fuel, • Numerous advantages in terms of environment protection: lower CO_2, NO_X, and practically zero soot. • Much lower noise pollution • Renewable form: biogas (zero CO_2 emission), • Tax free in the EU until 2018 • The CNG-fuelled vehicles are in the most favorable emission class (weight tax, environment protection zones, etc.), • A serious competitor for public vehicles iin the sectors of freight or passenger transportation. • There is more and more operator experience available	• The selection variety of CNG vehicles are narrow compared to the others • Nowadays, the CNG developed engines are not suitable for diesel propellants • The CNG vehicles are more expensive than the diesel or gasoline propelled ones • Their operating range is limited due to the limitation of the cylinder size, much shorter than the similar diesel engines • The fuel market's leader companies will acknowledge it as a competitor, a very serious lobby is against it • The building cost of a CNG refilling station is high • There are a number of regulations to be kept during maintenance and operation: • Certification: (driver, fuel station clerk, maintenance staff), • Special, explosion safe tools • Leak detectors, ventilation
Possibilities of CG operation	**Hazards**
• Substituting the present – height sustainable – vehicles with much cheaper • Continuously increasing gasoline and diesel prices • Increasing ecological awareness among the patrons and the enterpreneurs • Creating new jobs • The European energy and climate politics aims are favoring the methane-based products, such as earth gas, biogas • Strengthening environmental regulations, i.eg. Stopping diesel vehicles in case of smog • Projects running that are trying to popularize CNG and biogas • New funding possibilities from the government, and the EU. • Decreasing energy dependency, especially by biogas	• Raising of the earth gas prices, by the government • Changing tax policy, and political directions • Because of the bad regulation, the DIY refilling is spreading uncontrollably: in case of several accidents, the public may divert from choosing CNG

Purely Electric Propulsion

The usage of electric motors instead of internal combustion ones offers an excellent possibility to reach the maximum performance. A traditional, internal combustion engine's efficiency is somewhere between 25–30%, which means that the energy is going to waste as heat. Instead, with the usage of electric motors, the efficiency can be pushed up to 70–80%.

Unlike the internal combustion engine, the electric motor can provide its maximum torque by very small rpm-s, while the performance of the motor is increasing with the rpm. This is valid until the motor reaches the nominal performance, where it becomes constant.

Meanwhile, the accumulators are influencing the efficiency of energy transformation within the vehicle. Most of the electric cars are equipped with an energy storage system too, which can be made out of several types of chemicals.

Considering the Used Electric Energy, and Electric Energy-Mix in Calculating the Environmental Load

Just as the figure shows below, the pure electric vehicles are not necessarily better than the combustion engines, because the electric energy which is used for reloading the accumulators, is also coming from the country network.

During the full life cycle of the vehicles, - especially the ones operated by electric charging – there is a lot laying on that what kind of electric energy mix does a country have, that shows that what kind of energy carriers are included in the electric energy in a country, on percentages (nuclear, renewable, earth gas, coal, etc.) Behind the electricity generated from the other several energy carriers, the environmental load's most part will appear by the power stations. (we can also count with the raw material mining when we talk about ecological footprint).

In Hungary, just as in Europe, from the point of view of the greenhouse effect causing gases, the electricity production's environmental load is improving. According to the plans, more and more places are taken over by renewable, and nuclear energy. Using these two, the environmental load of the electric vehicles can be significantly reduced compared to the life cycle. Also, a determining viewpoint that which part of the day do we load our electric vehicles, and what is the present electric energy mix.

Characteristics of accumulators: The accumulator is the primary energy source of an electric vehicle. It stores the electric energy in chemical form. Nowadays the only obstacle in front of the electric cars to become widespread is the technical limitations of battery technologies, especially, that all the types have different energy density, significantly lower than liquid propellants. As a result, for the same range, we need significantly heavier accumulator packs to be built in, than we would need from liquid propellant, so the overall weight of the vehicle significantly increases, which has a negative effect on the running dynamics and space inside.

SWOT Analysis of Purely Electric Propulsion

We complete the SWOT analysis of the purely electric vehicle, then two sample projects that have already been carried out in two big cities, then we continue with public transport focused SWOT for European levels.

Strengths

With the development of the electric technology, it has become cheaper to manufacture the energy storage, and transportation systems, so they became cheaper, their storage capacity is bigger and bigger, their loading speed has increased too.

Therefore, they can be used for a bigger range. Side by side the accumulator development, the reload cycle number within an accumulator group has significantly increased too, without the necessity to change. Further advantage too, that there is a bigger space being left in the vehicle, than by a combustion engine propelled car. As the technology evolves, other means of charging will come forward, such as inductive charging, or complete changing of the battery packs along the road. These are the improvements, with that, they try to extend range, life cycle, and purity the electric propulsion technology.

Weaknesses

The technology is evolving with great steps, but it is not harmless. Depending on the method, it can be high voltage or high amperage. In such cases, the vehicles must be designed safe because it is mandatory to equip the system with shock protection and circuit breakers. These protection elements are also mandatory for recharging stations too. The mean of recharging generally determines the waiting time, which can lead to profit loss in case it is being long. Also a disadvantage of the electric vehicles that comes to the surface only upon servicing, because any kind of maintenance can be done on these vehicles by certified workshops and staff. There are very few about these in Hungary, and their training takes a long time. The accumulator cells require a bigger place than by any other alternative drive. It also increases the weight of the vehicle, that these accumulators must be placed into a strengthened case or part, to prevent injury. The future developments are aiding the spreading of electric cars, but it is expensive, even with the supports today. The accumulator change, as a recharging method, brings up numerous problems. Can a specific vehicle get a new accumulator or just new vehicles can get a new accumulator, or they can use each other's weaker, or newer accu packs? As sound or noise, the emission f the electric vehicle is low, but that also has a disadvantage too, because the car is practically silent, and can easily cause injury in a pedestrian who did not hear the car coming. The cost of research and development is pretty high compared to a single unit, so people must count with a higher retail price in case of an electric vehicle.

Options

The electric vehicles are widely supported, worldwide, meaning in Hungary too. The users enjoy significant discounts, regarding tax, and other material discounts. There is no registration tax that affects every other vehicle. it is not necessary to pay tax after performance Also not necessary to pay tax after the usage of a company car if it is electric. The green registration plate can be given to purely electric vehicles according to *326/2011. (XII. 28.) Korm. Rendelet* 60.§.

Dangers

The electric car recharging technology is constantly improving, but these aren't sufficient to extend the range, to a safe 2–300 km. Besides, the quick spreading of the electric vehicles will not be easily followed by the building of recharging stations. The electric vehicles are to be used mainly for short range, local traffic, where it can be recharged in the depot overnight, and then after less than 100–150 km, repeat the overnight charge.

3 Preparing for Future Technologies, Experimental Application of Hydrogen-Propelled Buses

3.1 Hydrogene as a Propellant

Some say hydrogene is the energy carrier of the future because upon its burning, only water is produced. It is colorless, odorless gas (room temperature), it is the most common element of the universe. Its melting point is around 14K (−259 °C), the boiling point is 20K (−253 °C). It is flammable, mixed with oxygen, forms a detonating gas mix.

The energy load of hydrogen can be described in many ways, depending on its consistency, and storage ability, compared to the traditional buses' diesel fuel:

- 1 normal cubic meter (0 °C, 1,013 bar) volume hydrogen gas energy equals with 0,3 L of diesel fuel (Table 2).
- 1 L of liquid hydrogen equals 0,24 L of diesel fuel's energy, and at last
- 1 kg of hydrogen is equal to 2,79kgs (3,33 L) of diesel fuel.

Table 2. Comparison of hydrogen and traditional propellants

	Combustion heat [MJ/kg]	Heating value [MJ/kg]	Burn product
Hydrogen	141,974	119,617	H_2O
Gasoline	45,217	42,035	H_2O, CO_2 (CO)
Diesel	44,715	41,843	H_2O, CO_2 (CO)

Source: Óbuda University, 2009

There are many scientific ways to create hydrogen. It may be produced by thermochemical reaction, from water using electrolysis, or solar energy, or from microbes. The result is only partially hydrogen in any case. One of the possible solutions is what uses earth gas as a base material, from what hydrogen can be produced as a result of thermochemical reaction.

The other is the electrolysis: Base material is water (H_2O). IT uses electricity to separate Hydrogen (H_2). From Oxygen (O_2).

Similar to electrolysis is the process, where the base material is also water, but the separation is done by solar energy instead of electricity. The result is the same as well, hydrogen and oxygen are produced. This process is in early development status, a lot of solutions are need to be found, in order to evolve the procedure appropriately, but offers a long term solution for environmental protection.

Microbes, just as bacteria and microalgae also capable of producing hydrogen, using solar light, or organic materials. These researches are in very early stage but can be a long term solution for low carbon-hydrogen emission.

Hydrogen must be stored, and this is a hard job, because of the small mass of the molecule, and its combustible characteristics, so several methods are used:

In case of the hydrogen operated vehicles, two solutions are common:

- One is the compressed gas cylinders (CH_2),
- The other is the liquid hydrogen (LH_2 – liquid hydrogen).

There are many experiments with several metals, that are capable to react with hydrogen as a hybrid.

The hydrogen can be used for two methods in the vehicles:

- fuel cells, or
- combustion engines.

Fuel cells (FC – Fuel Cell) producing electricity using hydrogen and oxygen, so electric motors can be propelled with this. In ICE (internal combustion engines) hydrogen is burnt, so the propulsion is the same as other internal combustion vehicles.

Pros and Cons of Fuel Cells

Pros:

- Fully clean exhaust gas
- Electric, can be fitted to linear hybrid propulsion
- Quiet
- Good efficiency

Cons:

- Complicated design, expensive
- Hard to control
- Sensitive for temperature changes
- Requires pure hydrogen (5–9, 99,999%)
- Performance drops with age

Interest in Using Hydrogen in Europe

The European Parliament voted for the decreasing in the newly built heavy vehicles' carbon dioxide emission by 30% compared to 2019, until 2030.

Worldwide 298, in Europe 111 (2018) hydrogen refueling stations are operating. (Germany, Denmark, Norway, UK, Austria, Sweden, Netherlands, Belge, France, Italy and Switzerland.) Until 2020, experts expect to have 520 hydrogen refueling stations (150 in Germany, 65 in UK, 15 Denmark) in which, there is a major role in the renewable energy using the continent's advantages. In Hungary, 2 until 2020, then 5 until 2025, and 14 until 2030.

The cost of one hydrogen refueling station is between EUR 300 000 and 2 billion. Depending on its size, and the terrain, but this cost is decreasing continuously. Obviously, it depends on whether the hydrogen is produced centralized and later distributed, or manufactured on the spot.

4 Pilot Analysis for a Hungarian City's Public Transport Bus Fleet

The analysis was made based on punctual, directly exported data from the operators' operation control systems, using full-year operations and maintenance data.

4.1 Characteristics of Local Bus Transport

The city under the scope has a flat terrain, because of what dense cyclist traffic is present too. The city was called "the city of cyclists" until the 70's, but the increasing motorization, the number of cyclists had slightly decreased temporarily. The environment-related questions that came up, and the efforts in the last decade to build more bicycle roads seemingly increased the cycling motivation. But there are alternative answers to the problems of public transport.

The local municipality –ordered transport services are provided by a Transportation Center.

The improvement of public transport can decrease the negative effects of motorization, i.e.g.:

- the jam caused by personal means of transport (cars),
- the narrow channel of the inner city,
- increasing environmental damages.

Besides the bus fleet modernization, and supporting the public transport, the narrow inner city street's jam can only be solved by regulatory acts, such as restrictions, paid parking spots, and pedestrian zones.

The public transport is served by 102 pieces of diesel-powered bus. (51 solo, and 51 two sectioned).

They ran 126 km as a daily average, in the examined year of 2018 it took 46000 km.

The fleet contains buses mainly manufactured between 1983 and 2010, which means an average vehicle age of 20, which means that it is not a young fleet.

The daily minimum run is 125 km, the maximum is 259 km, what – balanced by the trip times – annually 22.9K, km annual minimum, and 80K km annual maximum. The total distance run by the buses is minimum 467K, and 2,19B.

Examining the 2018's average daily bus usage means 8,14 h trip, and 6,3 h standby time. This is very important, because upon planning, we calculated with electric buses as well, and here the standby time can be spent with recharging sing chargers on the frequent places. We also checked the exact routes, lines, and transport of the city outside the depots.

The gap between two buses is 10–15 min in the inner city's root network, in peak hours, which can be reduced to 5–10 min in school time. Outside peak hours, we can talk about 20–30 min gaps between buses that can be up to 1 h in case of a weekend. The situation is similar in the agglomeration. The motivation to use public transport is increased by the modern passenger information system deployed in the stops, and on the vehicle itself, and the vehicles' cleanliness.

The procurement of the new bus rolling stock, -following the external abroad examples – will mean to improve the willingness to use public transport.

The long term strategy of the city contains various alternatives, from which the modernization of the bus fleet can mean a significant cost preference, meanwhile it can increase the public's willingness to use public transport.

Aerial and route network scopes of view, it can be stated, that with the development of the industrial park, the traffic generated by industrial participants, the outer sectors have emerged to the inner city's traffic demands. We can offer distinguished attention to the inner city routes. The toll-free circle route operates from October 2011, and with 7.2kms route length, and 29 min trip time. The CITY circle route starts in every 15 min and stops by 19 bus stops. At the moment, the types that operate the route are CREDO BC 11 normal floor level, and BN12 lowered entry level buses.

4.2 Evaluation of the Present Operations Conditions (km Run, Maintenance, Economical Characteristics)

After the summary of the city's public transport and bus fleet, we started to calculate the detailed costs, where the starting data were the following:

- Total number of buses: 102 pieces
- Operating period: 10 yr
- Procurement costs of buses
- In case of non-diesel buses, we have calculated with workshop refurbishment and training of qualified personnel
- In case of used buses, we have calculated with operational costs taken from the SAP system, what practically must cover "everything". We took maintenance, repair, material costs, labor costs, which gave us the operational costs for the full life cycle.
- Upon calculating the operational costs, we took 3 categories into account. The result of these three will give the total life cycle cost from the side of the vehicle (tires, authority inspections, are not in the calculation):
  - Maintenance (i.eg. Oil change, filter changes)
  - Preventive repairs (DPF filter, brake disc, brake drum, crank belt)
  - Repairs (depending on the expected probability of failures, i.eg. Engine, transmission, AdBlue, ECU)
- Total distance ran
- Fuel, electricity, and CNG contains the costs of building of the refilling stations too
- When using the data, we took 2018 into consideration ONLY. (closed fiscal year) (Table 3)

Analysis of CNG Operation

Table 3. Diesel-CNG acquisition cost comparison

Acqusition of the vehicle	New diesel bus	New CNG bus	dimension
Procurement costs	80 000 000	88 000 000	HUF/pc
Numberof vehicles	102	102	pc
Total procurement costs	8 160 000 000	8 976 000 000	HUF
Difference if procurement costs compared to diesel, divided by new and present values		816 000 000	HUF
Workshop rebuild and training of staff	0	30 000 000	HUF

Comparing the diesel to the CNG or electric buses – we talk about 102 pieces of vehicles – the acquisition costs do matter. According to our resources, the acquisition cost of a CNG bus exceeds the diesel bus with 8B Forints, so the total cost of the fleet would be 816B forints, and in addition, workshop rebuild and training of staff for 30B (Table 4).

Table 4. Comparison of Diesel-CNG operating costs

Calculation of operating costs	New diesel bus	New CNG bus	dimension
Full life cycle operating costs	**0,21**	**0,22**	EUR/km
Full life cycle operating costs	68,04	71,28	HUF/km
Running performance (average 2018)	46 000	46 000	km/annum
Average daily running performance (average 2018)	126	126	km/day
Number of buses	102	102	pc
Operating period	10	10	year
Full life cycle cost	3 192 436 800	3 344 457 600	HUF
Maintenance cost plus, compared to diesel vehicles (new and present values separated)		**152 020 800**	**Ft/10 év**

When we talk about operating, the CNG is more costly as diesel. The maintenance/repair costs material and personnel like parts are 71.28 HUF/km, meanwhile, the new diesel buses go as low as 68 HUF/km. The next table shows 152Bn HUF extra cost over 10 year (Table 5).

The earth gas that the CNG propulsion uses, according to our sources, cost 210 HUF/kg, which is exceeded by the diesel, with an average of 60 HUF/liter. The consumption rates are more or less the same. The CNG pros will appear on the fuel costs, because the CNG that is necessary to run 100 km, costs 3000 HUF cheaper than with

Table 5. Diesel-CNG comparison of fuel costs

Fuel cost calculation	New diesel bus	New CNG bus	Dimension
Consumption (SORT1)	44,10	43,60	kg/100 km;
Price per unit	279,18	210,00	HUF/l; HUF/kg;
Other ops cost	12 312	9 156	HUF/100 km
AdBlue	100	0	HUF/100 km
Total	12 412	9 156	HUF/100 km
Average km cost	124,12	91,56	HUF/km
Savings on fuel costs		3 256	HUF/100 km
Number of buses	102	102	PC
Running performance	46 000	46 000	km/bus/year
Total savings compared to a diesel (new and used separated)		**1 527 639 190**	**HUF/10 év**

diesel, so it is 25,6%-cheaper (average). So considering the full life cycle of 10 years, and 102 pieces of buses, will result as 1,53 kBn HUF savings (Table 6).

Table 6. Diesel-CNG 10 yr comparison (TCO)

Full possession costs (10 yr of operation)	New diesel bus	New CNG bus	Dimension
Number of buses	102	102	pc
Total acquisition costs	8 160 000 000	8 976 000 000	HUF
Rebuild workshop and staff training	0	30 000 000	HUF
Maintenance, repairs	3 192 436 800	3 344 457 600	HUF
Fuel costs	5 823 634 390	4 295 995 200	HUF
Full commissioning cost	17 176 071 190	16 646 452 800	HUF
Full commissioning cost/bus	168 392 855	163 200 518	Ft
Savings during the full life cycle compared to the diesel (new and old values separated)		**529 618 390**	**HUF/10yr**

Upon calculating the total costs (TCO) the following cost elements were summarized. As long as the 10 years operations period a diesel bus costs 168.4Bn HUF, the CNG buses produce 5.2Bn savings. If we talk about the whole fleet, we have saved 529.6Bn HUF, so it can be stated that the full diesel fleet's change to CNG means considerable values (Table 7).

Evaluation of Fully Electric Operation

Table 7. Comparison of the acquisition costs of Diesel and electric buses

Acquisition cost calculation	New diesel bus	New electric bus	Dimension
Procurement costs	80 000 000	115 200 000	HUF/Pc
Number of units	102	102	PC
Total procurement costs	8 160 000 000	11 750 400 000	HUF
Total increase of procurement costs compared to diese, new and used values separated		**3 590 400 000**	**HUF**
Rebuild workshop and training	0	5 000 000	Ft

The new electric bus procurement costs are way higher than the diesel buses, but the procurement cost of 144 Bn/unit, will decrease to 115.2bn using government support of 20%. So we can count of 44% increase of costs, which earns 35.2Bn Forints per unit. In case of the 102 units, this exceeds 3,59 kBn Forints, plus an extra 5bn added for rebuild the workshop (Table 8).

Table 8. Comparison of diesel and electric buses in terms of operational costs

Operating cost calculation	New diesel bus	New electric bus	Dimension
Total life cycle operations costs	0,21	0,11	EUR/km
Total life cycle operations costs	68,04	35,64	HUF/km
Runing performance (2018 average)	46 000	46 000	km/yr
Average daily running performance (2018-as átlag)	126	126	km/day
Number of buses	102	102	pc
Operating period	10	10	yr
Full life cycle cost	3 192 436 800	1 672 228 800	HUF
Maintenance cost increase compared to diesel (new and present values separated)		**−1 520 208 000**	**HUF/10yrv**

The high acquisition costs are balanced by the other cost elements, that will appear in the planned maintenance costs. For a fulfilled km performance in case of a diesel costs 68HUF/km, when it is electric, it is only 35,64 HUF/km, which is calculated by 102 units, will reach the amount of 1.52kBn HUF savings (Table 9).

The biggest advantage of electric buses is at the cost of the electricity that they use as fuel. The table is divided into two parts, the primarily calculated electricity costs were

Table 9. Comparison of fuel costs by the diesel and electric propulsion, calculated with two unit prices

Fuel cost calculation – designated electricity costs	New diesel bus	New electric bus	Dimension
Fuel consumption (SORT1)	44,10	128,00	kWh/100 km
Price per unit	279,18	63,00	HUF/kWh
Operational cost per unit	12 312	8 064	HUF/100 km
AdBlue	100	0	HUF/100 km
TTL	12 412	8 064	HUF/100 km
Average km cost	124,12	80,64	HUF/km
Savings on fuel		4 348	HUF/100 km
Number of buses	102	102	pc
Running performance	46 000	46 000	km/bus/yr
Total cost savings compared to diesel (new and present separated)		**2 040 005 590**	**HUF/10 év**

determined as 63HUF/kWh, which contains the subsidiary cost of building a charger. So we used this data for further calculations.

The electricity that is necessary to go 100km, is outstanding with a kilometer cost of 63 HUF, which will help the company save 2,04kBn HUF just from fuel (Table 10).

Table 10. TCO comparison between diesel, and electric operation for 10 years

TCO by 10 yrs life cycle	New diesel bus	New electric bus	Dimension
Number of buses	102	102	pc
total acquisition cost	8 160 000 000	11 750 400 000	HUF
Rebuild workshop and training	0	5 000 000	HUF
Repair and maintenance	3 192 436 800	1 672 228 800	HUF
Fuel cost	5 823 634 390	3 783 628 800	HUF
TCO	17 176 071 190	17 211 257 600	HUF
TCO/unit	168 392 855	168 737 820	HUF
Savings compared to diesel (separated new and present)		**−35 186 410**	**HUF/10yr**

The table above contains the total TCO of the electric bus fleet. The repair and maintenance costs, added with the fuel costs, will produce more than 3.56kBn HUF savings, what will match with the 3.59kBn increase as procurement costs, due to the higher prices of the vehicles, but the TCO will come to a 35.2Bn maximum during 10 years of operation.

5 Conclusions

Later on, on thelong term, if the costs of the electric chargers have come back, the operation of the buses will result in a much cheaper operating system compared to new diesel or CNG buses.

Despite this, the applicability of these buses in Győr is limited, due to their range, of 120–150 km, so they would only be servicing shorter routes.

Hydrogen propelled buses will mean a better alternative on mid-term, next to the CNG or electric buses. Now, the high procurement costs and the lack of refueling infrastructure, this cannot be applicable to the complete refurbishment of the fleet.

Despite this, looking at the political and technical development directions, it is necessary to prepare for operating, storing and refilling hydrogen propelled buses. This is not only important because of the emission regulations, but an environmentally friendly technology that has zero emission locally.

Taking all of the above into consideration, we propose the following solution for selecting the vehicles for the new fleet:

- 90–95 pieces CNG buses (solo, and double section),
- 2 pc E-bus (solo): Mainly city circle (CITY) routes,
- 2 pc H_2 bus (solo): Mid-long term thinking about operational experiences, and maintenance lessons to be learned.

References

1. https://www.kormany.hu/download/a/0c/e0000/A%C3%9CINK_fin.pdf
2. https://net.jogtar.hu/
3. https://cngeurope.com/
4. https://www.cngport.hu/
5. https://www.cngport.hu/tudastar/tudnivalok-jarmuvek-cng-s-atalakitasarol.html
6. https://www.cngport.hu/media/documents/gazautoszerelo_muhely_eloirasok_khvm.pdf
7. https://www.vezess.hu/hirek/2016/05/03/a-kormany-emelne-a-benzinarat-ha-lemegy-az-olaj-ara
8. https://kitt.uni-obuda.hu/mmaws/ (CNG-vel kapcsolatos konferencia előadások)
9. Fehér Könyv – Útiterv az egységes európai közlekedési térség megvalósításához – Úton egy versenyképes és erőforrás-hatékony közlekedési rendszer felé (2011, COM).
10. Országos Tűzvédelmi Szabályzat (28/2011. (IX.6.) BM rendelet
11. Motor Jikov Group Zrt. szakmai anyagok
12. Üzemeltetői adatbázisok
13. https://europa.eu/investeu/projects/clean-buses-gelderland%E2%80%99s-roads_hu
14. https://www.fch.europa.eu/project/joint-initiative-hydrogen-vehicles-across-europe-2
15. https://www.origo.hu/auto/20180601-hidrogen-konferencia-budapesten.html
16. https://www.autoszektor.hu/hu/content/milyen-utemben-terjedhet-el-hidrogen-uzemanyagcellas-autok-hasznalata

17. https://omnibusz.blog.hu/2010/08/11/hidrogennel_hajtott_buszok
18. https://index.hu/belfold/budapest/2015/03/19/bkv_johet_a_hidrogencella/
19. https://omnibusz.blog.hu/2018/01/28/jive_2_300_uj_hidrogen_busz
20. https://e-cars.hu/2018/03/29/elkezdodhet-a-toyota-hidrogen-uzemanyagcellas-buszanak-sorozatgyartasa/
21. https://www.vezess.hu/haszongepjarmu/2019/06/10/kozep-europai-a-legujabb-hidrogencellas-autobusz/
22. https://formula.hu/auto/2019/06/13/europaban-is-elterjedhetnek-a-hidrogencellas-buszok-es-teherautok
23. www.eea.europa.eu
24. Audi AG LCA szakmai anyagok
25. Szauter Ferenc – Alternatív járműhajtások (jegyzet)
26. https://www.tankonyvtar.hu/en/tartalom/tamop412A/2011-0042_jarmufedelzeti_elektronika/ch09s03.html
27. Bloomberg New Energy Finance
28. https://www.vda.de/en/services/Publications/automation.html
29. https://EVvolumes.com
30. https://e-mobi.hu/hu/map
31. Energy Agency: World Energy Outlook 2017
32. Dr. Strobl Alajos előadásanyag 2014. – magyarországi energiamix
33. JATO Dynamics startisztikai adatbázis
34. https://docplayer.hu/3321396-Energetika-dr-toth-peter-dr-bulla-miklos-dr-nagy-geza.html
35. https://www.tmhnc.com/blog/gas-versus-electric-forklifts-which-is-better
36. https://www.euro1training.com/news/forklift-fuel-gas-electric/
37. https://www.hcoinnovations.com/analysis-series-compressed-natural-gas-cng-lift-trucks-comparison/

Structural Analysis of a Multi-axle Steering Linkage for an 8 × 8 Special Purpose Vehicle

Mehmet Murat Topaç[1](✉), Onur Çolak[2], Levent Bilal[3], Arda Tanrıverdi[1], Merve Karaca[3], and Mustafa Maviş[2]

[1] Department of Mechanical Engineering, Dokuz Eylül University, Izmir, Turkey
murat.topac@deu.edu.tr
[2] Volkan Fire Fighting Vehicles, R&D Division, Izmir, Turkey
[3] The Graduate School of Natural and Applied Sciences, Dokuz Eylül University, Izmir, Turkey

Abstract. Considered as a safety sub-system, steering linkages should work under variable loads that arise in any extreme conditions during operation of the vehicle. Therefore, they should resist failure during service conditions, which require sufficient mechanical strength against different stress values. During the mechanical design stages of a steering linkage, taking the critical loading into account for a singular steering wheel position may not represent the most challenging case. Therefore, it should be applied for the full turning ranges of the steer axle wheels in order to obtain the critical steering wheel angle that forces the elements and joints the most. In this study, the variation of joint forces in the steering mechanism of an 8x8 ARFF vehicle was investigated by using FE (finite element) analysis. Stress distributions and bearing loads on the critical structural elements were established within the full turning range of the system. Firstly, the conformity of the results obtained from FE model was validated by means of a kinetic analysis that was carried out in MATLAB® environment by using a sub-linkage of the steering mechanism. Subsequently, a detailed FE model of the multi-axle steering linkage was created in order to determine the maximal joint forces and stress variation on the connection elements in full turning ranges of the steer axle wheels. Finally, the effect of steering booster was revealed on joint forces, as well as the stress behavior of critical structural components of the mechanism.

Keywords: 8x8 vehicle · Multi-axle steering · Steering linkage · Structural analysis · Joint forces · Finite element analysis

1 Introduction

As a result of the legal restrictions on axle load in heavy vehicles, multi-axle applications are considered as a feasible option [1]. At this point, the growing needs of special-purpose vehicles lead to some developments in steering strategies of multi-axle vehicles. It is possible to have a more proper steering operation by steering more than one axle. A multi-axle steering mechanism contributes to the considerable handling ability and maneuverability of a multi-axle vehicle [2]. Thus, it provides more reliable steering

K. Jármai and K. Voith (Eds.): VAE 2020, LNME, pp. 67–78, 2021.
https://doi.org/10.1007/978-981-15-9529-5_6

besides enhancing lateral stability in some multi-axle vehicles [3–4]. Therefore, they must satisfy a reasonable and reliable steering motion. However, steering systems should also satisfy mechanical strength requirements against challenging service loads, which occur in extreme cases, since they are seen as safety sub-systems [5]. Furthermore, loads that act on different steering linkage components vary with different steering orientations, as a result of the linkage geometry. As a result, during the evaluation of steering design load, the complete steering range should be taken into account in order to observe the highest loads that apply to different components. The mechanical safety condition should be sought in that manner. An exemplary critical turning maneuver of the ARFF vehicle is shown in Fig. 1. In this work, the mechanical analysis of a multi-axle steering system of an 8x8 Aircraft Rescue and Fire Fighting (ARFF) vehicle was summarized. The vehicle prototype is shown in Fig. 2.

Fig. 1. Turning maneuver test of an 8 × 8 ARFF vehicle.

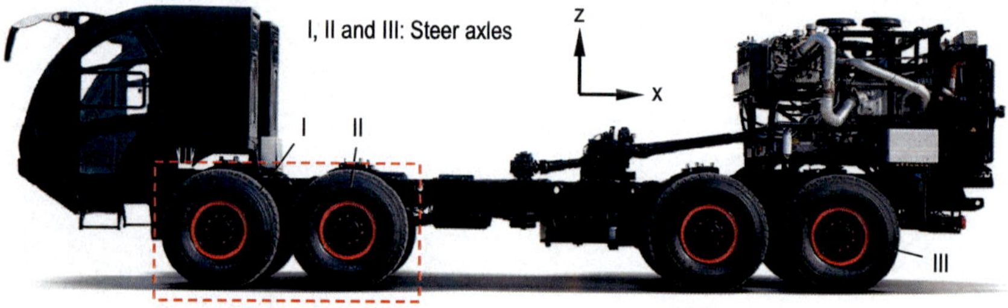

Fig. 2. The 8 × 8 ARFF vehicle prototype [6].

During the R&D studies of this vehicle, two double front axle steering linkage variants were evaluated. The mechanical assessments of the first steering linkage version are reported in [6–7]. The general view of the second version, which will be reviewed in this study is also shown in Fig. 3. In the literature, there are studies that focus on the kinematic analysis of the multi-axle steering [5, 8–12]. However, so less of those interest in the mechanical design of multi-axle steering mechanisms, 8 × 8 vehicles in particular. Therefore, this study aims to provide a contribution to the literature in this field. In the scope of this study, FE (finite element) analyses of the mechanism were carried out to establish bearing loads and stress distributions on critical parts for various turning range of pitman arm through the ANSYS® Workbench™ 19.2. Then, the values obtained from FE analyses were validated by kinematic and kinetic and analyses in MATLAB® environment. Kinetic equations on each joint were implemented in terms of pitman arm turning angle by taking the 4-bar mechanism into account. Therefore, the force characteristics in joints depending on the steering angle were determined. Finally, FE analyses of the complete system were carried out. The effects of steering booster on the characteristics of joints forces and stresses were obtained.

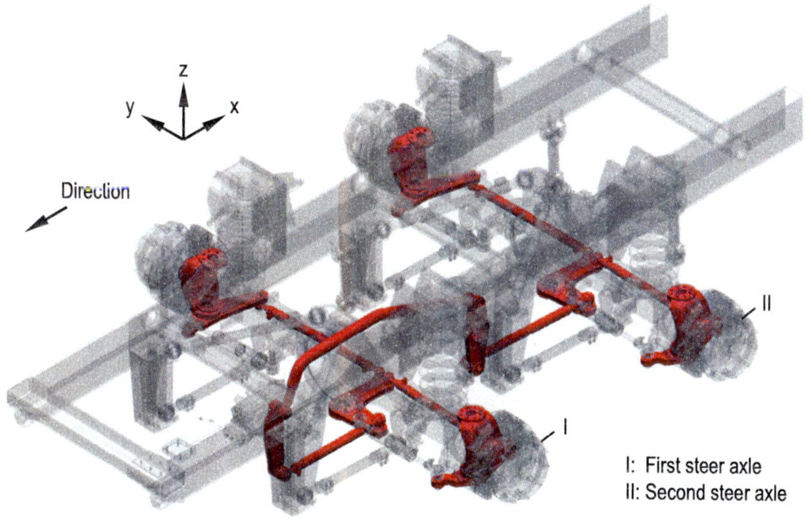

Fig. 3. The double front axle steering system.

2 System Description

The general view of the steering linkage at a fully-right turn maneuver can be seen in Fig. 4. In this system, the steering box moment input (M_S) is given to the system by the pitman arm (BAE), which transmits the moment from the steering wheel to steering system mechanically. The first steer axle receives the input through (EF) rod and L-arm (FGH), while the motion is transmitted to the second through the connecting rods (BC) and (JK), sector (or swing) arm (CDJ) and the L-arm (KLM). Here, the sector arm provides the simultaneous steering of the steer axles (I and II). The Ackermann synchronization between the first and second axles is directly dependent on the ratio between a and b. In this application, the track-rod consists of two parts (IR) and (HQ) as similar to [13]. The optimal kinematic design of this system is introduced in more detail

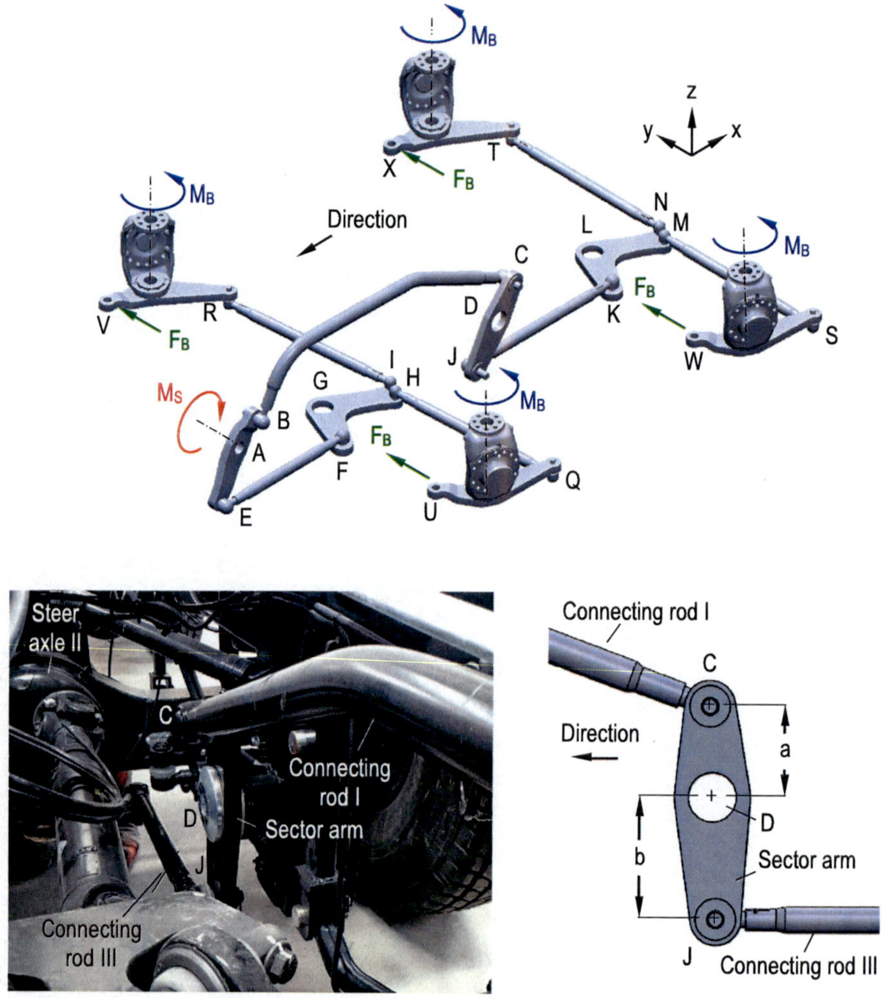

Fig. 4. Positions of the linkage elements at full right turn maneuver (above) and dimensions of the sector arm (below).

in [14]. One of the main design targets for the steering mechanism is the acquirement of reasonable Ackermann error values for given steering ranges as well as an adequate strength against failure. Thus, failure on any of the structural elements causes critical results for the vehicle. Within the turning range and under different steering box moment and booster force values, structural elements of the linkage are subjected to different bending and torsion loads depending on the steer angle. Moreover, the positions and dimensions of the structural elements themselves have effects on these loads. Hence, the determination of the stress distribution for various positions of structural elements is of great importance in terms of system safety. Within the full range of linkage, bearing loads are expected to be different dependent on the turning angle. Therefore, designing structural elements for the only neutral position of the vehicle might not be reliable.

3 Structural Analysis

3.1 Load Model

The vehicle is assumed as fully loaded and the wheels are firmly on the ground. There are several approaches that are employed during the mechanical strength evaluation of steering mechanisms, which are available in the literature and take different critical load conditions into account. In this study, bore torque approximation that represents zero-speed steering, was used as a load model since it is one of the most critical service loads, to which the system might encounter during operation. The details of the load model are discussed more detailed in [15–16]. All the same, in many cases, no clue related to the steering orientation is given for the load model. Related to that, bearing loads on system components are expected to vary for different steer angles under the same bore torque value. Therefore, investigating the system safety only for zero-steer position might not be sufficient.

3.2 FE Modelling and Validation

During the mechanical examination of the system, the bore torque approach-based structural analyses were carried out through the FE method. However, in order to check the sensitivity of the FE model, a validation study was made that includes the kinetic analysis of a sub-fraction of the mechanism. In that manner, a simplified FE model for that section of the mechanism that covers the connecting rod I, sector arm and pitman arm was created. Using this model, the force variation characteristics in the upper joint of the sector arm were evaluated depending on different values of steer angle. In the following, a kinetic model was created, which consists of three bars including the sector and pitman arms in a way that it can be considered as a four-bar mechanism for the simplicity of calculation. Hence, the deviation of the force in comparison with the FE model was evaluated. Fig. 5 shows the elements of the FE model that is generated for the validation study.

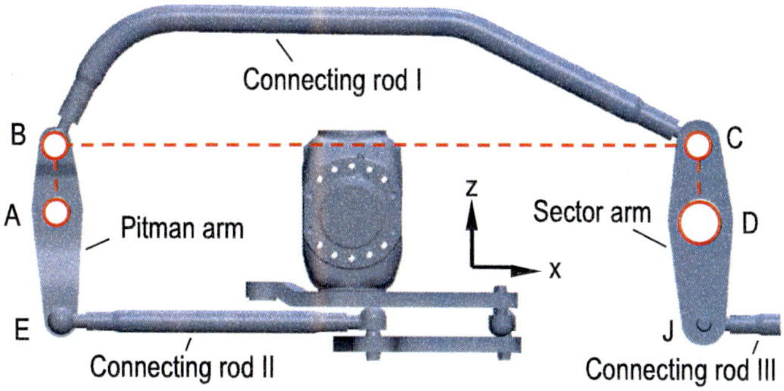

Fig. 5. Elements of the FE model.

The kinetic model of the four-bar mechanism-idealized section of the system is illustrated in Fig. 6a. In this model l_2, l_3 and l_4 are the kinematic lengths of the components. θ_2, θ_3, and θ_4 are the angular positions of bars with respect to the x-axis. The analysis of the simplified model for the variation range of a system parameter requires the definition of all geometric variables in terms of that parameter since it is a one degree-of-freedom mechanism. Therefore, θ_3 and θ_4 should be expressed in terms of θ_2 that is considered as the generalized coordinate. In order to write these expressions, position vectors are employed, which define the locations of joints with respect to a fixed reference point for the variation range of θ_2. The position vectors that are employed in the four-bar mechanism are given in Eqs. (1–5). Since they compose a non-linear set of equations, Newton-Raphson approach is used to solve the equations.

$$\mathbf{r}_{AB} + \mathbf{r}_{BC} = \mathbf{r}_{AD} + \mathbf{r}_{DC} \tag{1}$$

$$\mathbf{r}_{AB} = (l_2 \cos\theta_2)\mathbf{i} + (l_2 \sin\theta_2)\mathbf{k} \tag{2}$$

$$\mathbf{r}_{BC} = (l_3 \cos\theta_3)\mathbf{i} + (l_3 \sin\theta_3)\mathbf{k} \tag{3}$$

$$\mathbf{r}_{DC} = (l_4 \cos\theta_4)\mathbf{i} + (l_4 \sin\theta_4)\mathbf{k} \tag{4}$$

$$\mathbf{r}_{AD} = (AD_x)\mathbf{i} + (AD_z)\mathbf{k} \tag{5}$$

Once the simplified model is fully defined in terms of the system parameter θ_2, the kinetic equations are written based on the free-body diagrams of bars. In order to determine joint forces, the linear set of kinetic equations is solved through the matrix approximation. In this method, the coefficient, system input and unknown matrices **A**, **b** and **F** are written as below, respectively:

$$\mathbf{A} = \begin{bmatrix} 1 & 0 & -1 & 0 & 0 & 0 & 0 & 0 & 0 \\ 0 & 1 & 0 & -1 & 0 & 0 & 0 & 0 & 0 \\ 0 & 0 & l_2\sin\theta_2 & -l_2\cos\theta_2 & 0 & 0 & 0 & 0 & 0 \\ 0 & 0 & 1 & 0 & -1 & 0 & -\cos\theta_3 & 0 & 0 \\ 0 & 0 & 0 & 1 & 0 & -1 & -\sin\theta_3 & 0 & 0 \\ 0 & 0 & 0 & 0 & l_3\sin\theta_3 & -l_3\cos\theta_3 & 0 & 0 & 0 \\ 0 & 0 & 0 & 0 & 1 & 0 & 0 & -1 & 0 \\ 0 & 0 & 0 & 0 & 0 & 1 & 0 & 0 & -1 \\ 0 & 0 & 0 & 0 & -l_4\sin\theta_4 & l_4\cos\theta_4 & 0 & 0 & 0 \end{bmatrix} \tag{6}$$

$$\mathbf{b} = \left[0\;0\;-M\;0\;0\;0\;0\;0\;0\right]^T \tag{7}$$

$$\mathbf{F} = \left[A_x\;A_z\;B_x\;B_z\;C_x\;C_z\;P\;D_x\;D_z\right]^T \tag{8}$$

$$\mathbf{F} = \mathbf{A}^{-1} \times \mathbf{b} \tag{9}$$

The matrix solution is obtained in MATLAB® environment and the solution set is repeated through the variation range of θ_2; thus, the variation profile of joint forces is calculated. The FE and RKM (Rigid kinematic model) approaches for the evaluation of axial reaction force P under the driving torque M are compared also in Fig. 6b for the joint C. Here, the steer angle β_L is measured from the right wheel of the first steer axle that is termed as the reference wheel. Positive sign of β_L indicates left turn steering. The axial force characteristics were given as axial force ratios P/P_{ref}, where the forces at $\beta_L = 0°$ as reference values (P_{ref}). From the diagram, the result is drawn that the joint force variation that is obtained through the FE model is deviated with the simplified kinetic model of the mechanism with roughly 3%, which can be said to be acceptable. After this stage, FE analysis method is employed for the full steering linkage.

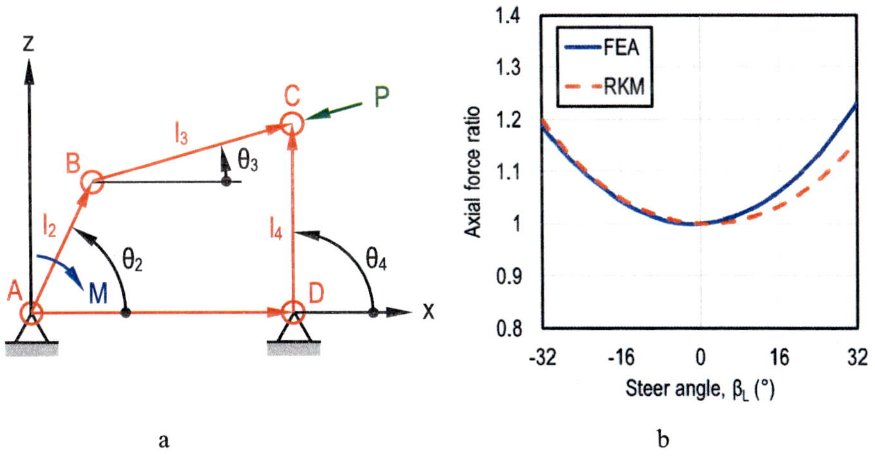

Fig. 6. a. Simplified to four-bar mechanism kinetic model of the steering linkage **b.** Comparison of the force characteristics that are obtained through FE analyses and RKM as functions of steer angle of the reference wheel.

3.3 FE Modelling of the Full Steering Linkage

The FE model of the full steering linkage is shown in Fig. 7. The model is fixed from the middle joint (A in Fig. 4) of the pitman arm, where the steering wheel input is given to the system. In the junction points, where tie-rods and connection rods are connected to the arms, spherical joints (S) were used. In the connection points between the arms and the vehicle chassis, only rotation (R) is allowed. The joint types used in the FE model of the full steering mechanism are shown in Fig. 7. The wheel torque values that are obtained according to the bore torque approximation were applied from the wheel-hub parts. In addition, the steering booster forces (F_B) were applied to the model from the track rods (Points U, V, W and X in Fig. 4). The scenario, in which the actual steering booster force is used, was termed as Case 0. Nonetheless, in order to obtain the effects of steering boosters on the overall mechanical strength of the system, the wheel torque loads were also applied with different values of the steering booster force as additional load cases, 10% (Case 1) and 20% (Case 2) lower, precisely. Selected load cases were employed for the complete steering range of the wheels in order to take the most critical positions that were taken into account during the mechanical design.

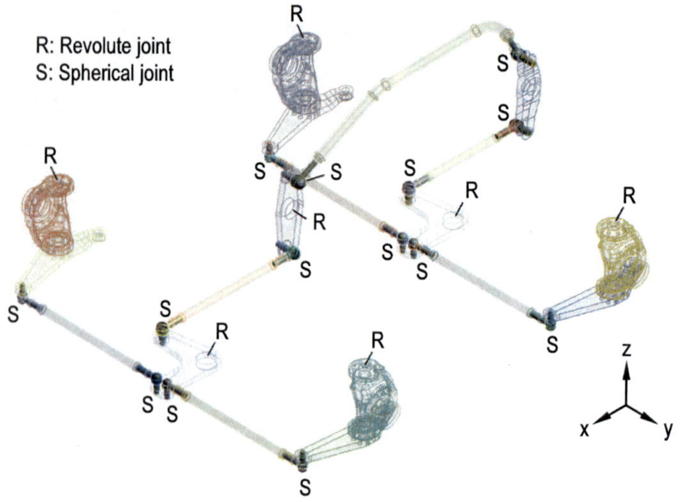

Fig. 7. Joint types used in the FE model of the steering linkage

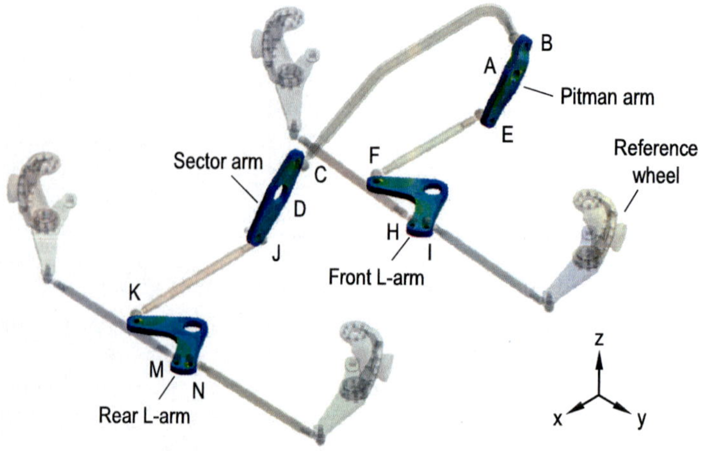

Fig. 8. Fully-left steer maneuver.

4 Results and Discussion

During the evaluation of the results, four critical components of the linkage are taken into account as exemplary; front and rear L-arms, pitman arm and the sector arm. The fully-left steer maneuver of the linkage indicating these structural components is shown in Fig. 8.

The joint force variation characteristics for the selected components were obtained throughout three cases as the function of the steer angle of the reference wheel. The forces were given as force ratios F/F_{ref} for each joint, where the forces at the non-steered position of the linkage as reference values (F_{ref}). The joint force variation characteristics are given in Fig. 9.

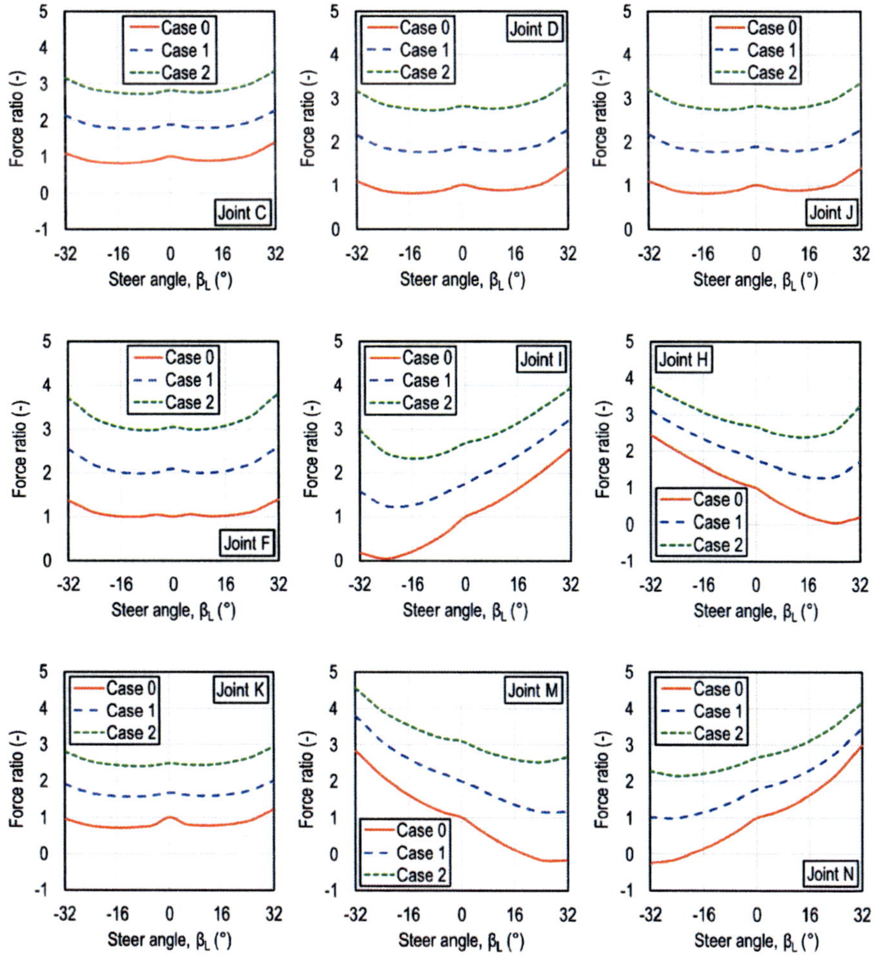

Fig. 9. Joint force variation characteristics.

From the diagrams in Fig. 9, a ratio between the maximum force F_{max} and F_{ref} is obtained for the joints. As a result, crucial increases were observed in joint forces with fairly identical variation characteristics of joint forces. From the results it is drawn that the joint forces can be as about 3 times higher compared to relevant reference values. Depending on the steering booster force decrease by 20%, it was observed that the forces at joint J as an example can also be up to 3 times higher. A remarkable point during the mechanical strength evaluation of sector arm is the negligibly small force values in a specific position of the front L-arm at joint I. In this geometric arrangement of the linkage, steering booster almost completely bears the bore torque and the joint forces are released from the sector arm, which can be seen in the diagram in Fig. 9. According to the joint force analyses, mechanical strength evaluations of components were also carried out. In a similar fashion, the von Mises stress variation characteristics of critical regions of selected components are given in Fig. 10 as the ratios of corresponding reference values. The critical regions in the neighborhood of the joints were taken into account during the evaluations.

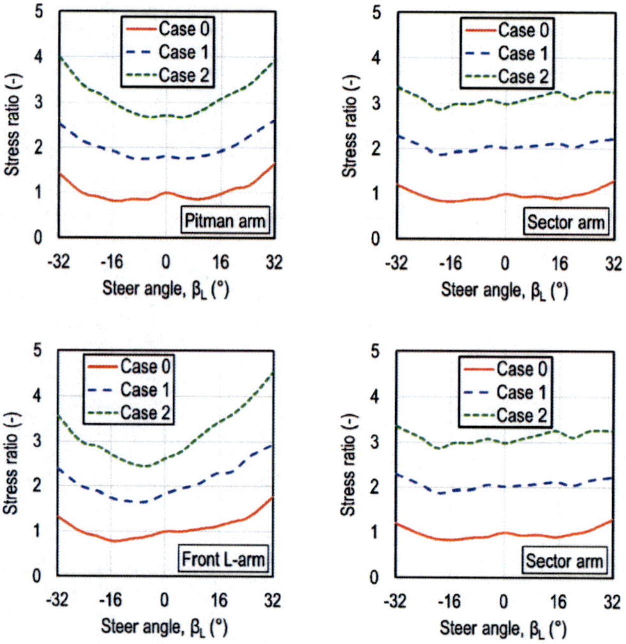

Fig. 10. Stress variation characteristics of selected components.

The maximum increase of stress ratio depending on the steer angle is also obtained as about 2. The FE analysis depending on the wheel positions revealed that the mechanical strength conditions are satisfied for the yield stress criterion the system in the most critical layout of the mechanism.

5 Conclusions

In this work, force and stress analyses were carried out for a double front axle steering mechanism of an 8x8 special purpose vehicle. In that manner, at the first stage, an FE model was created by taking a sub-fraction of the mechanism into account for the determination of system sensitivity to FE modelling. Thus, the joint force variation depending on different values of steer angle was observed for the critical connection parts. The obtained results were validated by means of an equivalent RKM for four-bar mechanism idealized part of the steering linkage in the MATLAB® environment. Once the results were correlated with two different approaches, the FE model for the complete system was generated and stress variation behavior of the mechanism was obtained as a function of steering angle with various steering booster forces. It was seen that the system satisfies the mechanical strength requirements in the most challenging simulation conditions. Some results obtained from this work are summarized below:

- The joint forces obtained from the sub-FE model and RKM analysis deviate with a maximum level of 3%.
- It was revealed that the joint forces of the linkage components under steering angle may increase by up to 3 times compared to the reference value, in which the wheels are not steered.
- It is evaluated that, throughout the steering mechanism design, taking the non-steered position of the wheels into account solely may result in misleading results and insufficient mechanical strength for the mechanism. Moreover, due to the variating nature of stress behavior, the fatigue life assessments should also be considered during the design of similar systems.
- The steering booster force has a major role in the stress values over the mechanism. In this exemplary work, it was seen that the 20% reduction of steering booster force leads to stress increase by up to 3.13 times. Therefore, it can be concluded that the selection of the proper steering booster is of great importance for heavy duty vehicle steering systems.

Acknowledgements. This study was partially supported by The Scientific and Technological Research Council of Turkey (TÜBİTAK) as a part of an R&D project (project no: 3170127). It was also supported by the TÜBİTAK within the framework of 2209-B program (Registration no: 1139B411901688). The authors are grateful for the technical support of Volkan Fire Fighting Vehicles, Izmir, Turkey.

References

1. Wu, J., Zhang, S., Yang, Q.: Deformation effect simulation and optimization for double front axle steering mechanism. In: 4th International Conference on Computer Modeling and Simulation, Singapore (2012)
2. Ahmed, W., Azim, R.A, Fatima, S.: Steering strategy for a multi-axle wheeled vehicles. In: International Mechanical Engineering Congress and Exposition, Pittsburgh (2018)

3. Bayar, K., Ünlüsoy, Y.S.: Steering strategies for multi-axle vehicles. Int J. Heavy Vehicle Syst. **15**(2/3/4), 208–236 (2008)
4. Wu, D.H., Lin, J.H.: Analysis of dynamic lateral response for a multi-axle-steering tractor and trailer. Heavy Vehicle Syst. **10**(4), 281–294 (2003)
5. Yucheng, L., Wei, Z., Guifan, Z., Cong, W.: Kinematical Models and Emulation of Multi-Axle Steering of Off-Highway Vehicles with Multi-Axle. SAE Technical Paper 952121 (1995)
6. Topaç, M.M., Karaca, M., Başdemir, A., Kuleli, B.: Buckling safety assessment for the multi-axle steering linkage of an 8x8 special purpose vehicle. Celal Bayar Univ. J. Sci. **15**(4), 321–327 (2019)
7. Topaç, M.M., Karaca, M., Kuleli, B.: A design optimization study for the multi-axle steering system of an 8x8 ARFF vehicle. In: Ntalianis, K., Vachtsevanos, G., Borne, P., Croitoru, A. (eds.) Applied Physics, pp. 342–347. System Science and Computers III, Springer, Berlin Heidelberg New York (2019)
8. Yusheng, H., Tingchao, X.: Design and Analysis of Multi-Axle Steering System of Heavy Duty Vehicle. SAE Technical Paper 931919 (1993)
9. Conghua, L., Bin, Z., Xiuhua, G., Yaping, Y.: Optimization design of multi-axle steering mechanism of multi-axle steering vehicles. Trans. Chin. Soc. Agric. Eng. **9**, 81–85 (2007)
10. Qin, G., Zhang, Y., Chen, L.: Synthesis and Analysis of the Double-Axle Steering Mechanism Considering Dynamic Loads. SAE Technical Paper 2008–01–1105 (2008)
11. Hou, Y., Hu, Y., Hu, D., Li, C.: Synthesis of Multi-Axle Steering System of Heavy Duty Vehicle Based on Probability of Steering Angle. SAE Technical Paper 2000–01–3434 (2000)
12. Chaudhuri, S., Saini, V., Singh, M.: Kinematic Analysis of Multi-Axle Steering System for Articulated Vehicle. SAE Technical Paper 2009–26–0067 (2009)
13. Matschinsky, W.: Radführungen der Straßenfahrzeuge. Springer, Heidelberg (2007)
14. Topaç, M.M., Kaplan, A., Kuleli, B., Deryal, U.: Design of a Multi-Axle Steering Mechanism for a Special Purpose Vehicle: Kinematic Design and Optimization. In: 8th International Advanced Technologies Symposium, Elazığ, pp. 1–8 (2017)
15. Sharp, R.S., Granger, R.: On car steering torques at parking speeds. Proc. Inst. Mech. Eng. Part D: J. Automob. Eng. **217**(2), 87–96 (2003)
16. Rill, G.: Road Vehicle Dynamics: Fundamentals and Modeling. CRC Press, Boca Raton (2012)

Materials and Manufacturing

Three Generations of Advanced High Strength Steels in the Automotive Industry

Miklós Tisza(✉)

University of Miskolc, Miskolc 3515, Hungary
tisza.miklos@uni-miskolc.hu

Abstract. Sheet metal forming is one in all the foremost important production processes in car manufacturing; therefore its developments are significantly determined by the demands of the automotive industry. Recent trends in car production are also characterized by applying lightweight principles. Its main priority is to fulfil both the customers' demands and also the increased legal requirements. Applying high strength steels could also be thought to be one in all the potential possibilities. Applying high strength steels have a positive response for several of the requirements: increasing the strength may result in the appliance of thinner sheets leading to significant mass reduction. Mass reduction ends up in lower consumption and increased environmental protection. Increasing strength often leads to a decrease in formability. In this paper, an outline of recent material developments within the automotive industry concerning the employment of recent generation advanced high strength steels are going to be given.

Keywords: Advanced High Strength Steels - AHSS · Automotive industry · Lightweight manufacturing

1 Introduction

Increasing global competition in car-making requires low-cost production, which is strongly connected with lightweight manufacturing. The need for lightweight manufacturing within the vehicle industry may be explained by several reasons: the continuously increasing environmental restrictions, the requirement for the reduction of harmful emissions, and the higher safety requirements should be mentioned. Fulfilling these requirements, weight reduction has a decisive role. Within the total weight of an automobile, the car body incorporates a determinant role. Sheet metal forming thought to be one in all the foremost important manufacturing processes in the production of car body elements. This is why the elaboration of new, low-cost manufacturing processes is one of the main objectives in sheet metal forming: in this respect, the lightweight production principles are of utmost significance. The two main possibilities for producing lightweight automotive parts are the applying of high strength steels or lightweight materials – especially various high strength aluminum alloys [1]. In this paper, I mainly concentrate on Advanced High Strength Steel materials. First, the main requirements

K. Jármai and K. Voith (Eds.): VAE 2020, LNME, pp. 81–94, 2021.
https://doi.org/10.1007/978-981-15-9529-5_7

and the driving forces for car manufacturing will be overviewed. It will be followed by the classification of Advanced High Strength Steels (AHSS) steels introducing some important representatives of 1st, 2nd and 3rd generation AHSS materials.

2 Main Requirements in the Automotive Industry as the Driving Forces of Car Manufacturing and Sheet Metal Development

Considering the main requirements for the automotive industry in recent decades, the main driving forces of material developments can be clearly defined, too.

The global competition in car manufacturing is very strong and furthermore, the requirements are often contradictory: for example, from the customers' side more economical, more safe and higher comfort together with better performance are the most important issues. These are further increased by legal requirements as the ever-increasing environment restrictions including the reduction of harmful emissions and also higher safety requirements. Some of these legal requirements are in accordance with the customers' demands, however, some impose further requirements. Due to the worldwide competition in car manufacturing, the automotive industry has to find the best answers to these challenges. To meet these requirements is hardly possible with conventional materials and manufacturing methods. This is often one of the most reasons that the development trends in the automotive industry are the most driving forces in material development and sheet metal forming, too.

In the fulfillment of these manifold requirements, the weight reduction has an important role: reducing the overall weight of vehicles results in lower consumption, and thus less harmful emissions together with more economical vehicles and increased environmental protection. If we analyze the potential weight reduction in various parts of a regular automobile, it can be stated that about 45% of the total weight is covered by the body parts, chassis and suspension elements [2], thus, the main focus should be placed on these components. These parts are mostly produced by sheet metal forming, therefore sheet metal forming as a key technology has a critical role in the weight reduction of automobiles.

3 Material Development Tendencies in Sheet Metal Forming Concerning the Lightweight Production Principles in Car Manufacturing

Lightweight production principles led to the intensive development of recent, new materials. Concerning steel materials, these developments resulted in the widespread application of assorted grades of high strength steels. The origin of those developments may be traced back to the mid-seventieth when the first examples of micro-alloyed steels arrived in the economic, industrial application. Since then, thanks to the continual pressure on material development several new high strength steel grades appeared and reached already the everyday industrial application. Systematic analysis of those developments may be found in several papers from various authors in the literature [3–7]. In the following sections, a systematic scientific classification of those developments will be summarized.

3.1 Classification of Steel Developments

Steel developments are classified in several alternative ways. One usual way of classification is completed per the metallurgical designation. In step with this, steels can be grouped into the subsequent types: low strength steels (including mild steels, interstitial free IF-steels), conventional high strength steels like Carbon-Manganese (C-Mn) steels, Bake-Hardenable (BH) steels, High Strength Low Alloyed (HSLA) steels, and also the newer varieties of Advanced High Strength Steels (AHSS), e.g. Dual Phase (DP) steels, Transformation Induced Plasticity (TRIP) steels), Twinning Induced Plasticity (TWIP) steels, Complex Phase (CP) steels, Martensitic (MS) steels. In recent years, several new AHSS grades have been developed, e.g. TRIP-aided Bainitic Ferrite (TBF), Quenching & Partitioning (Q&P), or different kinds of NanoSteels: of these with the primary aim supplying even higher strength parameters with significantly increased formability.

Recently, widely applied classification relies on mechanical properties – mainly strength and formability parameters as the Ultimate Tensile Strength (UTS) and Total Elongation (TE) as shown in Fig. 1. This kind of classification is often used together with the designation of steel generations' development, as well.

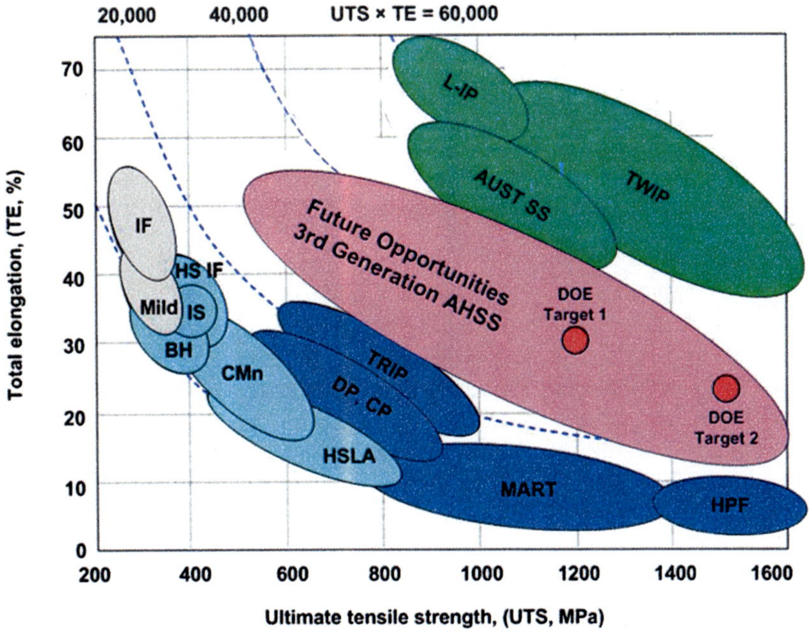

Fig. 1. Ultimate Tensile Strength (UTS) vs Total Elongation (TE) for various generations of high strength steels [4]

In Fig. 1, the relationship between strength and ductility parameters can be seen applying the classification method mentioned above. This classification added with a graphical representation, too. The product of the ultimate tensile strength and the total elongation (UTS $\times$ TE) follows a hyperbolic function, as it can be well seen from Fig. 1.

The constant (C = UTS × TE) provides a good basis for further classification of newly developed Advanced High Strength Steels.

In Fig. 1, the conventional mild steels (IF and Mild steels) formerly widely applied in Body in White (BiW) production in the automotive industry, the group of conventional high strength steels including Bake Hardening (BH), Isotropic (IS), High Strength Interstitial Free (HS IF), Carbon-Manganese (CMn) and High Strength Low Alloyed (HSLA) steels. Following the wide-spread application of conventional high strength steels, intensive development was initiated in the steel industry in close cooperation with the automotive industry to develop alternative types of Advanced High Strength Steels (AHSS) that may better fulfill the needs of lightweight principals building automotive structures.

In the next sections, some results achieved in the development of three generations of Advanced High Strength Steels will be overviewed.

4 Main Groups of Advanced High Strength Steels (AHSS)

AHSS are complex, sophisticated materials, with carefully selected chemical compositions and multiphase microstructures, achieved by precisely controlled heating and cooling processes. Various strengthening mechanisms are applied to get significantly increased strength, better formability, improved toughness, and fatigue properties to meet the various requirements that are defined for automotive body structures. We will discuss the main groups of Advanced High Strength Steels according to their development stages.

4.1 First Generation Advanced High Strength Steels (1G-AHSS)

There are several new materials grades among the 1st generation AHSS, among them the Dual-Phase-, Complex Phase-, TRIP- and MS or often termed as PHS-steels (PHS stands for Press Hardening Steels). From this 1st generation AHSS, DP- and TRIP-steels are the most characteristic members that are most widely applied in the automotive industry. Obviously, we have to mention the PHS-steels, too, but since the application of these steels requires special dedicated Hot Press Forming (HPF) processes, in this respect, we refer to an overview of this special field summarized in [8]. In this section, we will only analyze in detail the DP- and TRIP-steels.

Dual-Phase (DP) Steels

Development of Dual-Phase (DP) steels started at the beginning of the new age of steel development. Current commercially available AHSS steels have evolved from the early work on Dual-Phase steels within the late 1970s and early 1980s. Dual-Phase steels are one in all the foremost widely applied Advanced High Strength Steels in today's car making industry. This is often mainly thanks to their better strength and formability parameter combination compared to the conventional high strength steels like HSLA steels. For DP steels, high specific strength and good initial work hardening rate are characteristic besides the continuous yielding behavior, and superior ductility compared to standard steel grades. These properties make them particularly suitable for producing various body structures, closures, etc. in vehicles [9].

Dual-Phase (DP) steels generally have a ferrite matrix containing mainly hard martensite or in some cases bainite second phases as islands as shown in Fig. 2. It is very characteristic that the ferrite phase is usually continuously providing excellent ductility. During forming, the strain is concentrated within the lower strength ferrite phase surrounding the martensite islands providing a unique work hardening rate.

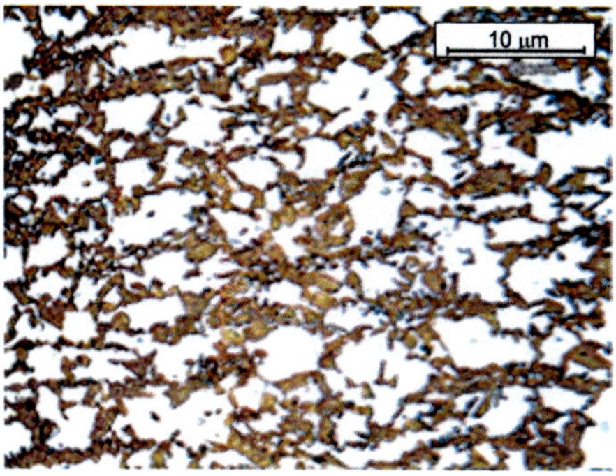

Fig. 2. Micrograph of a DP 690 steel containing martensite islands in ferrite matrix

Processing of DP Steels
There are various processing routes for producing DP steels. The time-temperature diagrams of the three most widely applied manufacturing processes can be seen in Fig. 3.

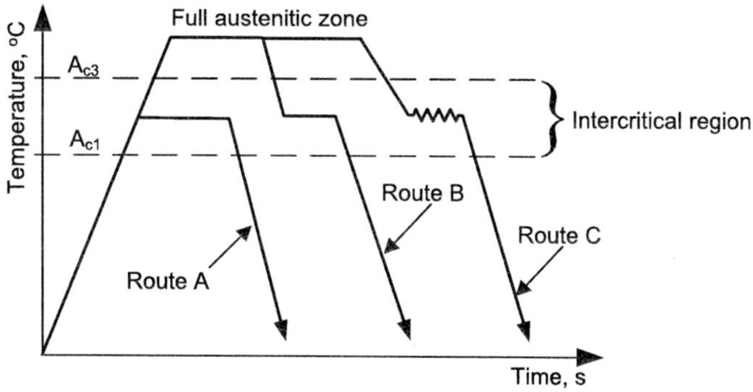

Fig. 3. Three different processing routes for producing DP steels

The first method – shown as Route A in Fig. 3 – means a rapid cooling from the intercritical temperature directly to room temperature. The resulting microstructure comprises ferrite and martensite. Higher intercritical temperatures, for the same holding period, result in larger amounts of martensite with increased tensile strength and decreased percentage elongation [10].

The other method for processing of DP steels (Route B in Fig. 3) first involves a slow cooling from the austenitic region to the desired ferrite transformation temperature, followed by quenching to room temperature for transforming the remaining austenite to martensite [11]. This processing route results in lower tensile strength and higher ductility than those gained by Route A.

The third method for producing DP steels (Route C in Fig. 3) involves hot rolling of steel in the intercritical region with a slow cooling rate, then followed by second cooling at a very fast rate and finally slow cooling (i.e. coil cooling) to room temperature. This method of cooling is known as ultra-fast cooling (UFC) and the processing route is referred to as new generation thermo-mechanical controlled processing [12]. Better properties obtained by Route C compared to those obtained by either Route A or Route B due to the higher grain refinement achieved during rolling.

Transformation Induced Plasticity (TRIP) Steels

Advanced high-strength transformation-induced plasticity (TRIP) steels are highly compatible for light-weighting car body construction with an additional advantage to reduce the safety problems. One amongst the main features of TRIP steels that the strain- or stress-induced transformation of retained austenite present within the microstructure in a sufficient amount can substantially harden the steel during deformation looking on the processing route, and so ends up in a better ductility [13].

The microstructure of TRIP steels contains retained austenite embedded in an exceedingly primary matrix of ferrite. Usually, about five volume percent of retained austenite, hard phases like martensite and bainite are present in varying amounts. Figure 4 shows the schematic microstructure of TRIP steel (TRIP 700).

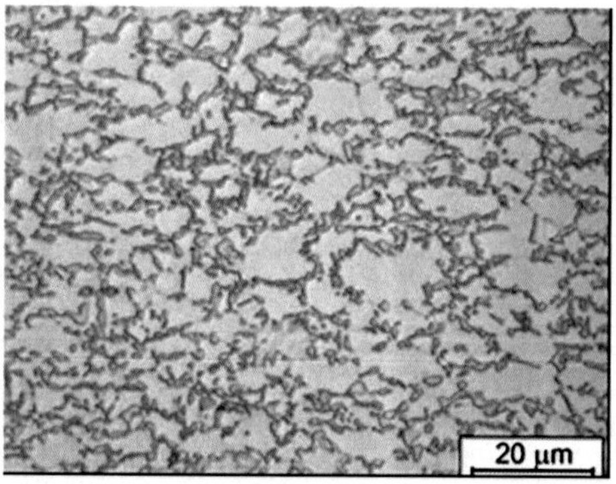

Fig. 4. Micrograph of a typical TRIP steel (TRIP 700)

TRIP steels may have been characterized by a comparatively low content of alloying elements. As an example, in TRIP 790 steel (UTS =790 MPa), the overall content of alloying elements is about 3.5 wt. Percent. Thus, the appropriate selection of suitable alloying elements and also the amount required to get the planned properties is critical during the alloy design stage. The carbon content in TRIP steels is more than in DP steels. Carbon is mostly kept within the range of 0.20–0.25% due to weldability reasons.

The upper carbon content is necessary for stabilizing the retained austenite phase to below ambient temperature. In TRIP steels, austenite stabilizers are present, mainly C, Mn, and/or Ni. These elements assist in maintaining the mandatory carbon content within the retained austenite. TRIP steels mainly contain multi-phase microstructures composed of about 50–55% ferrite, 30–35% bainite, 7–15% retained austenite, and 1–5% martensite.

The outstanding combination of ductility and strength in TRIP steels may be a result of deformation supported the transformation of retained austenite to martensite. This transformation (on deformation) of phases is termed the TRIP effect that has excellent strength and elongation combination along with high impact resistance. These characteristics predestinate TRIP steels as a decent candidate for the third generation AHSS, too. Dispersed hard second phases in soft ferrite provide high work hardening rate, as experienced in DP steels, too. Furthermore, in TRIP steels the retained austenite progressively transforms to martensite with increasing strain, thereby increasing the work hardening rate at higher strain levels [14].

Processing Methods of TRIP Steels
The basic processing route of TRIP steels consists of heating the steel to the full austenitic zone, and after the necessary soaking time cooling down to the intercritical region followed with deformation here, and quick transfer to the bainitic zone with subsequent holding there, and finally quenching to room temperature (shown in Fig. 5).

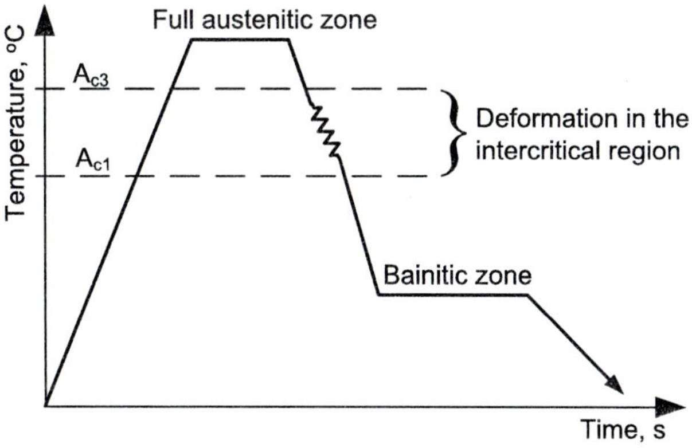

Fig. 5. Conventional processing route of TRIP steels

The deformation within the inter-critical region increases the speed of the transformation of austenite (γ) to ferrite (α). The remaining austenite is enriched with carbon content, which stabilizes the γ phase. Furthermore, this deformation increases the nucleation rate of bainite but decreases the rate of growth that leads to small plates of bainite. This also helps the enrichment of the γ phase in carbon and further increases its stability. The stability of retained austenite is enhanced by the high carbon content and the more carbon in γ phase results in more stability of γ during the TRIP effect, too, since more stable austenite needs longer time to be transformed into martensite; simultaneously, these processes contribute to the increase of the ductility, too. With this process, an improved strength–ductility combination is achieved [15]. Obviously, the described processing method for TRIP steels is more time-consuming. This is due to the need for special arrangements to deform the material at high temperature, to hold the specimen in the bainite region, and so on. This limits the use of TRIP steels in industrial applications. Some authors [16] using this route reported that rolling in the intercritical region improves TRIP steel properties by enhancing the carbon content and dislocation density, decreasing the grain size, and leading to a granular type morphology.

4.2 Second Generation of Advanced High Strength Steels (2G-AHSS)

The 2nd generation of Advanced High Strength Steels was the next step in steel development. These steels can be found in the range of $R_m \times A_{80} = 40.000–65.000$ (MPa%). Twinning Induced Plasticity, i.e. TWIP-steels are the most characteristic representatives of this group, however, there are some other material grades like high Manganese austenitic stainless steels (AUST SS) and the so-called Lightweight Induced Plasticity (L-IP) steels, too. In this section, we will only introduce the most characteristic type, i.e. the TWIP-steels.

Twinning-Induced Plasticity (TWIP) Steels

TWIP steels have a superior balance of tensile strength and elongation using the TWIP effect. The name of TWIP steel is originated from its characteristic deformation mode, i.e. the twinning induced plasticity. The twinning causes high value of the instantaneous hardening rate (n-value) as the microstructure becomes finer and finer. The resultant twin boundaries serve as grain boundaries and strengthen the steel.

TWIP steels have high manganese content (Mn = 17–24%) that results in fully austenitic microstructure even at room temperatures. TWIP steels are normally composed of Fe, Mn, or Ni (15–35%), Si (1– %), and Al (1–3%) [17]. These steels exhibit outstanding tensile strength-ductility combinations (e.g. a TWIP steel with tensile strengths above 1000 MPa may possess 50–60% ductility) [17]. The n-value may increase to a value of 0.4 that may result in 50–60% uniform elongation. In many grades in this group, the tensile strength may be even higher than 1500 MPa [18].

In TWIP steels, the strain hardening is strongly dependent on the stacking fault energy (SFE). This parameter controls the deformation behavior of the steel. Alloying elements generally decrease SFE leading to enhanced twinning behavior during deformation and hence lead to improved ductility. It is also known that SFE < 20 mJ/m^2 causes austenite to martensite conversion, and by this results in the TRIP effect. For pure twinning, SFE is desired to be greater than 20 mJ/m^2. Aluminum is added to steel to raise SFE, to retard the TRIP effect, and to result in pure twinning. Typical microstructure of a TWIP-steel is shown in Fig. 6.

TWIP steels show excellent mechanical performance with extremely high strength and ductility properties. Unfortunately, this category is not really viable for industrial applications due to its limitations: poor productivity and high production costs. The usual processing of TWIP steels includes homogenizing above the upper critical temperature and quenching to room temperature [19]. TWIP steels are often produced by homogenizing followed by deformation at temperatures above the upper critical one, with subsequent quenching to room temperature. Deformation at higher temperature

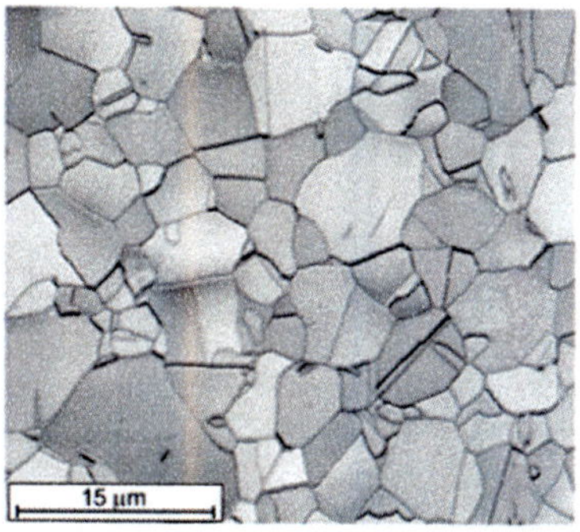

Fig. 6. Micrograph of a TWIP steel in annealed condition

provides fine grain size and a high volume fraction of twins. The finer the grain structure the more twinning occurs that improves ductility and strength. Two kinds of twins are observed in TWIP steels: (a) annealing twins caused by heat treatment, and (b) deformation twins caused by plastic deformation.

4.3 Third Generation of Advanced High Strength Steels (3G-AHSS)

The main target in developing the 3rd generation AHSS was to achieve the properties in the range between the 1st and 2nd generation AHSS with less alloying elements, hence, with less expensive processing that is suitable for early commercialization. The group of 3rd Generation AHSS (3G-AHSS) development may be clearly identified on the diagram of Tensile Strength vs Total Elongation between the 1st and 2nd generation AHSS regions (see Fig. 1).

Medium Manganese Steels combining the TRIP and TWIP effects, Quenched and Partitioned (Q&P) Steels, TRIP assisted Bainitic-Ferritic Steels (TBF) and NanoSteels are usually considered to belong into this group of AHSS.

In this section, we will mainly deal with the two most promising examples of this group, i.e. the Quenched and Partitioned (Q&P) Steels and TRIP assisted Bainitic-Ferritic Steels (TBF).

Quenched and Partitioned (Q&P) Steels

Quenched and Partitioned (Q&P) steels are one of the main results of the recent developments of 3rd generation AHSS steels. The theory of Q&P steels is partly based on the knowledge of duplex stainless steels and the quenching and partitioning process [20]. The Q&P steels usually contain carbon, manganese, silicon, nickel, and molybdenum alloying elements. The quantity of alloying elements is around 4 percent, which is far below that of within the 2nd generation AHSS. During heat-treatment of Q&P steel, quenching is interrupted and is reheated for partitioning. With this reheating process, a unique microstructure is formed containing 5 to 12 percent stable retained austenite, 20 to 40 percent ferrite, and 50 to 80 percent martensite.

The concept of Q&P process for automotive materials was first published by Speer in 2003 [21]. In Q&P process, the steel is quenched down below the M_s temperature, where austenite is not fully transformed. Thanks to the alloying concept of Q&P steels, this temperature usually is within the range of 200–350 °C. It implies that the microstructure may be a mixture of martensite and austenite. Steel is then reheated and aging is completed between 300–500 °C; this is termed the "partitioning step". During this treatment, carbon diffuses from the supersaturated martensite, providing the carbon enrichment of austenite, which increases its stability at room temperature; furthermore, it supports further TRIP effect during deformation.

However, the complex evolution of the microstructure during partitioning and the detailed mechanisms are not fully revealed, many of the Q&P evolutions are still a matter of debate. For example, the formation of bainite during partitioning cannot be completely excluded; it could explain the measured carbon enrichment in the retained austenite because the partitioning temperatures are per those for bainite formation.

Though the detailed mechanisms are not fully explained, the benefits of Q&P treatment are clearly shown by the improved mechanical properties. The range of strength that may be achieved with this new concept is between 1,000 and 1,500 MPa, with a total elongation of 20%. Moreover, because the matrix may be quite tempered martensite, damage resistance is improved compared to DP or TRIP steels with identical strength levels.

Recently, Q&P steels with 2,100 MPa tensile strength together with 9 percent uniform elongation and about 13 percent total elongation were developed. The elongation level of this steel is reminiscent of DP 980 that is a cold-formable grade. Q&P steels, initially with 980 MPa and later 1,180 MPa strength were first developed by Baosteel [22].

The development of Q&P steel grades required a very important modification of the annealing lines. Quenching and reheating step was not possible until recent years. The strong demand from the automotive market towards 3rd generation advanced high strength steels has led steel making companies investing in the upgrading of their annealing lines to make sure the processing of Q&P steel products.

Processing of Q&P Steels

Q&P steels are a series of C-Si-Mn, C-Si-Mn-Al or other similar compositions that are processed by the quenching and partitioning (Q&P) heat-treatment process. Q&P steels possess an excellent combination of strength and ductility with a final microstructure of ferrite (in the case of partial austenitization), martensite and retained austenite. This microstructure makes them suitable to use in the automotive industry as a new generation AHSS. They are suitable for cold stamping of various structures and safety parts having complicated shapes to improve fuel economy and promoting passenger safety.

It is possible to change the amount of retained austenite at room temperature and its stability with alloying elements such as carbon, manganese, nickel, etc. However, it affects the cost and may be detrimental concerning the welding properties. The third generation of AHSS grades was developed to overcome these disadvantages; one of the good examples are those 3rd generation AHSS that is based partly on the quenching-and-partitioning process (Q&P steels) and on the properties of the medium-manganese steels. In this case, the composition of steel is not adequate for keeping the retained austenite at room temperature, but annealing, cooling, and thermal processes are optimized to change the austenite's composition and decrease its Ms temperature. For medium-Mn steels, where a relatively larger manganese amount (typically 5 to 8 wt. %) is characteristic slightly simplifies the thermal treatment. The intercritical annealing provides a chance to form austenite and to increase its carbon and manganese content; then the steel is cooled down to room temperature. The complex multiphase fine-grained microstructure together with the TRIP effect arising from the progressive transformation of the retained austenite during deformation provides excellent mechanical behavior. By these processes, the UTS above 1,200 MPa and uniform elongation larger than 12% can be achieved.

There are two main versions of Quenching&Partitioning process. The basic version includes a fast quenching from the austenitization temperature down to the temperature slightly above the M_f temperature and followed by the partitioning below the M_s temperature as shown in Fig. 7.

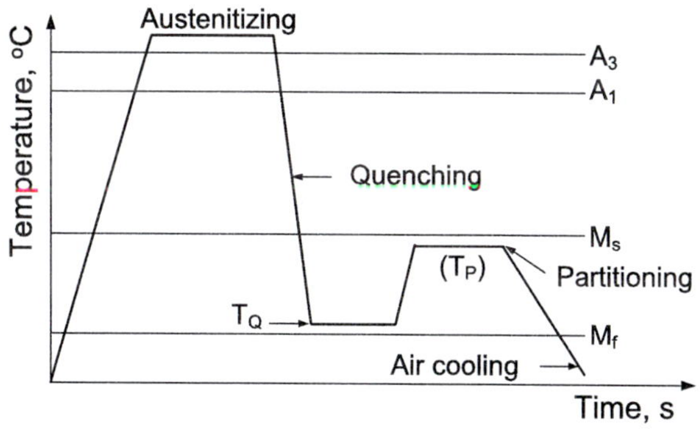

Fig. 7. Basic process of Quenching&Partitioning

There is a newer version of Quenching&Partitioning applying Double-Stabilization Thermal Cycle – DSTC as shown in Fig. 8.

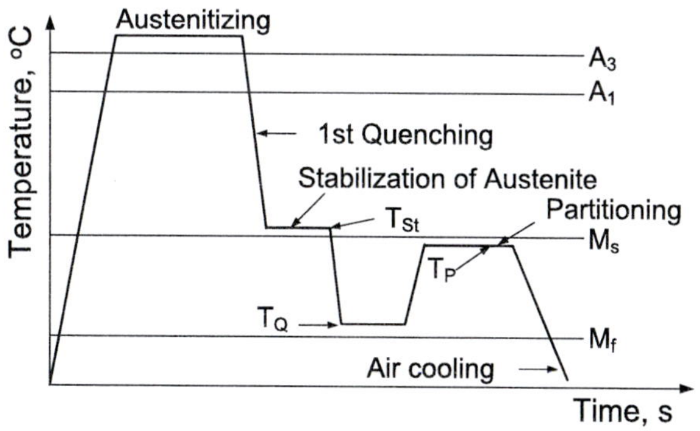

Fig. 8. Quenching&Partitioning applying Double-Stabilization Thermal Cycle (DSTC)

The main processing way of this newer Quenching&Partitioning steel grade with DSTC thermal cycle may be summarized by the following steps:

1. *Full Austenitization* is the first step in Quenching&Partitioning applying Double-Stabilization Thermal Cycle, too.
2. *1st Quenching:* the full austenitization is followed by a 1st or Initial Quenching down to the temperature slightly above the M_s temperature to avoid bainitic transformation of austenite. Applying this step, the austenite is further stabilized.
3. *Finish Quenching:* the initial quenching is followed by a finish quenching above the M_f temperature setting up the retained austenite/martensite ratio.
4. *Carbon Partitioning:* Carbon Partitioning is done slightly below the M_s temperature. During this carbon partitioning, the carbon will diffuse from the martensite to austenite thus providing even higher stability of austenite to resist its transformation to martensite.
5. *Air Cooling:* from the temperature of carbon partitioning air cooling may be applied to room temperature to get the required austenite-ferrite-martensite ratio.

TRIP Assisted Bainitic-Ferritic (TBF) Steels
TRIP assisted bainitic-ferritic (TBF) steels may be considered as a further significant development step among the 3rd generation AHSS. Their microstructure contains bainitic-ferritic matrix with retained austenite particles. Typical chemical compositions of TBF steels contain C, Si and Mn as major alloying elements. Alloy modifications include variations of the Al, Nb and Cr content [23]. The cementite formation during bainitic transformation is suppressed by the Si constituent. The added Si enhances the C content in retained austenite and it stabilizes the austenite. High Si contents of 1.5 wt % are used in these types of steels.

Processing of TBF Steels
TBF steels are produced by isothermal holding in the bainitic regions after fast cooling from the full austenitic zone. A significant benefit of these steels compared to Q&P steels processing that these can be produced by conventional heat-treatment devices, while the processing of Q&P steels requires significant modifications in the annealing lines.

5 Conclusion

In this paper, the recent developments in Advanced High Strength Steel production and application were overviewed. Considering both the customers' demand and the legal requirements, it was shown that some of these requirements are coinciding while others are contradictory. To fulfil these contradictory requirements, the application of high strength steels may be regarded as one of the most potential developments. Among these developments, the application of new Advanced High Strength Steels (AHSS) may be regarded as the most important one. In the last decades, different grades of AHSS were developed. They are classified as first, second and third generation AHSS. Some of these AHSS grades are already widely applied in the world automotive industry; some still are in the development phase. The main properties, the metallurgical background and the main processing routes of AHSS were discussed.

Acknowledgments. This work summarizes the results achieved within the project Material and process developments for the Hungarian Automotive Industry jointly financed by the European Union and the Hungarian Government (Grant No. AutoTech-4.2.2/A-11/1-KONV-2012–0029). Both financial supports are gratefully acknowledged.

References

1. Tisza, M.: Metal Forming in the Automotive Industry, 1st edn., 294. p. University Press, Miskolc (2015). ISBN 978–963–358–082–0
2. Lotus Engineering: An assessment of mass reduction opportunities for 2017–2020 model year vehicle programs. International Council on Clean Transportation, 308. p., March 2010
3. Wagener, H.W.: New developments in sheet metal forming: sheet materials, tools and machinery. J. Mat. Proc. Techn. **72**, 342–357 (1997)
4. Tisza, M.: Development of lightweight steels for automotive applications. In: Engineering Steels and High Entropy-Alloys, IntechOpen (2020). ISBN 978–1–78985–948–5
5. Matlock, D.K., Speer, J.G.: Processing opportunities for new advanced high-strength sheet steels. Mater Manuf. Process. **25**, 7–13 (2010)
6. Lesch, C., Kwiaton, N., Frank, B.: Advanced High Strength Steels (AHSS) for automotive applications: tailored properties by smart microstructural adjustments. Steel Res. Int. **88** (2017). https://doi.org/10.1002/srin.201700210
7. Nanda, T., Singh, V., Singh, V., Chakraborty, A., Sharma, S.: Third generation of advanced high-strength steels: Processing routes and properties, J. Mater. Des. Appl. (2016), August 12. doi.org/https://doi.org/10.1177/1464420716664198
8. Tisza, M.: Hot forming of boron alloyed Manganese steels. Mat. Sci Forum. **885**, 25–30 (2015). https://doi.org/10.4028/www.scientific.net/MSF.885.25
9. Li, C., Li, Z., Cen, Y., et al.: Microstructure and mechanical properties of dual phase strip steel in the overaging process of continuous annealing. Mater. Sci. Eng. A. **627**, 281–289 (2015)
10. Adamczyk, J., Grajcar, A.: Effect of heat treatment conditions on the structure and mechanical properties of DP-type steel. J. Achieve Mater. Manuf. Eng. **17**, 305–308 (2006)
11. Meng, Q., Li, J., Wang, J., et al.: Effect of water quenching process on microstructure and tensile properties of alloy cold rolled dual-phase steel. Mater Des. **30**, 2379–2385 (2009)
12. Rana, R., Liu, C., Ray, R.K.: Evolution of microstructure and mechanical properties during thermo-mechanical processing of a low-density multiphase steel for automotive application. Acta Mater. **75**, 227–245 (2014)
13. Kuziak, R., Kawalla, R., Waengler, S.: Advanced high strength steels for automotive industry: a review. Arch. Civil. Mech. Eng. **8**, 103–117 (2008)
14. Skalova, L., Divisova, R., Jandova, D.: Thermo-mechanical processing of low-alloy TRIP-steel. J. Mater Process. Technol. **175**, 387–392 (2006)
15. Shi, W., Li, L.: Thermal stability of retained austenite in TRIP steel after different treatments. J. Iron Steel Res. Int. **15**, 61–64 (2008)
16. Basuki, A., Aernoudt, E.: Influence of rolling of TRIP steel in the intercritical region on the stability of retained austenite. J. Mater. Process. Technol. **89–90**, 37–43 (1999)
17. Chung, K., Ahn, K., Yoo, D.H., et al.: Formability of TWIP (twinning induced plasticity) automotive sheets. Int. J. Plast. **27**, 52–81 (2011)
18. Allain, S., Chateau, J.P., Bouaziz, O., et al.: A physical model of the twinning-induced plasticity effect in a high manganese austenitic steel. Mater. Sci. Eng. A. **384**, 143–147 (2004)
19. Grajcar, A., Borek, W.: Thermo-mechanical processing of high-manganese austenitic TWIP-type steels. Arch. Civil Mech. Eng. **8**, 31–37 (2008)

20. Savic, V., Hector, L., Singh, H., Paramasuwom, M., et al.: Development of a lightweight third-generation advanced high-strength steel (3GAHSS) vehicle body structure. SAE Int. J. Mater. Manuf. **11**(4), 303–313 (2018). https://doi.org/10.4271/2018-01-1026
21. Speer, J.G., Edmonds, D.V., Rizzo, F.C., Matlock, D.K.: Partitioning of carbon from supersaturated plates of ferrite, with application to steel processing and fundamentals of the bainite transformation. Curr. Opin. Solid State Mater. Sci. **8**, 219–237 (2004)
22. BaoSteel: Automotive Advanced High Strength Steels. Product Manual (2013)
23. Bachmaier, A., Hausmann, K., Krizan, D., Pichler, A.: Development of TBF steels with 980 MPa tensile strength for automotive applications. In: Proceedings of International Conference on New Developments in Advanced High Strength Steels, Colorado, June 2013. https://doi.org/10.13140/RG.2.2.24907.80169

Development of Cutting Edge Radius Size of Solid Carbide Mills When Drag Finishing

Boris Pätoprstý[1](✉), Marek Vozár[1], Peter Pokorný[1], Tomáš Vopát[1], Ivan Buranský[1], Miroslav Zetek[2], Šárka Cajthamlová[2], and Vít Laudát[2]

[1] Faculty of Materials Science and Technology, Slovak University of Technology, Bratislava, Slovakia
boris.patoprsty@stuba.sk

[2] Faculty of Mechanical Engineering, University of West Bohemia, Pilsen, Czech Republic

Abstract. The article deals with the influence of time of the drag finishing process on the size of the cutting edge radius for cutting tools. In the experiments, four-tooth cemented carbide mills with diameter of 10 mm were used that were ground on Reinecker WZS60 tool grinding machine from Ceratizit CTS20D. Tools were drag finished on OTEC DF3Tools with the use of SIX 70/16 granulate and grinding oil. The time of the drag finishing process was varied from 1 to 10 min. After drag finishing, the tools were measured on the Zeiss Surfcom 5000 shape and countour measuring machine. The aim of the paper was to determine the influence of time as a parameter on the achieved cutting edge radius size r_n. Cutting edge radius size on both the face teeth and helical teeth was measured. The dependence of cutting edge radius from process time when drag finishing in the mentioned granulate was determined and described in the article.

Keywords: Drag finishing · Cutting edge preparation · Milling tools

1 Introduction

Cutting tools are one of the most important factors entering into the process of manufacturing parts by metal cutting. Virtually all tool manufacturers are optimizing the macrogeometry of their cutting tools. It is necessary in order to increase tool life when machining various materials. For the past decade, the importance of modifying the microgeometrical parameters of the cutting tools in addition to the macrogeometry has been stressed by some authors and research institutions. [1, 2, 3] While there are novel methods of edge preparation currently being developed [4, 5], drag finishing remains a viable method of modifying the microgeometry of cutting tools with some of its properties haven't been thoroughly investigated. [6, 7] Properly adjusting the parameters of the drag finishing process can lead to better results, increasing the tool life. [8, 9] Modifying the microgeometry of the cutting edge of cemented carbide tools is especially beneficial, as it results in the extension of tool life as well as better adhesion of the coating layer. [10, 11] In this paper, the importance of the time parameter of the drag finishing process is investigated and its influence on the tool's microgeometry is discussed.

K. Jármai and K. Voith (Eds.): VAE 2020, LNME, pp. 95–100, 2021.
https://doi.org/10.1007/978-981-15-9529-5_8

2 Materials and Methods

Tools were manufactured by grinding from cemented carbide rods of grade CTS20D by Ceratizit Company, which corresponds to grade K20-K40 by ISO. This cemented carbide material with the content of Co of 10% and 90% of WC achieves a hardness of 1600 by Vickers. Prior to manufacturing the mills on a CNC grinding machine, rods were cut up using Wire Electrical Discharge Machining to make workpieces with dimension Ø10h6 × 110 mm.

2.1 Manufacture of the Tools for the Experiment

Macrogeometry of the tools used for the experiment was designed for machining difficult to cut materials, specifically for machining austenitic stainless steel. Grinding machine Reinecker WZS 60 was used for manufacturing solid cemented carbide mills. Four grinding wheels from synthetic diamond by the producer URDIAMANT Slovakia s.r.o. were used to manufacture the tools. The grain size of grinding wheels was D64 by ISO 6106, which means that the average grain size was from 50 – 62 μm. Different grinding wheels were used to manufacture the flute of the tool and also the face of the tools. NUMROTOplus software was used to create NC code for the grinding machine. The result of the tool manufacturing simulation is shown in Fig. 1.

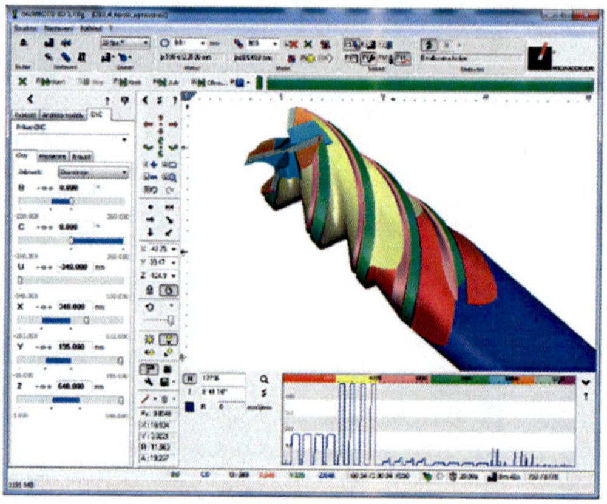

Fig. 1. Simulation of grinding the cutting tools

Geometry designed for the solid cemented carbide mills was following: Diameter of the tool – 9.947 mm, core diameter – 4.893 mm, helix angle 28.190°, Rake angle (on the face) – 7.78°, rake angle (on the face) 2 – 15.78°, flank angle – 7.56°, rake angle (on the helix) – 9.07°. In Fig. 2, the grinding process can be seen.

2.2 Modifying the Cutting Edge of the Tools by Drag Finishing

As a method for edge preparation of the manufactured tools, drag finishing was used. The machine used for the drag finishing was OTEC model DF-3 Tools. Grinding granulate

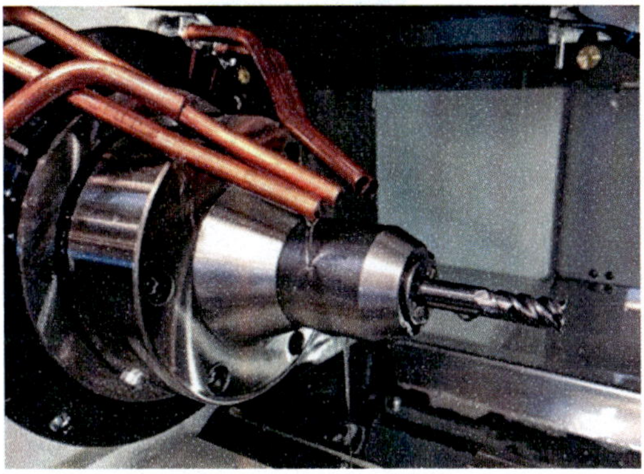

Fig. 2. Tool in the grinding process

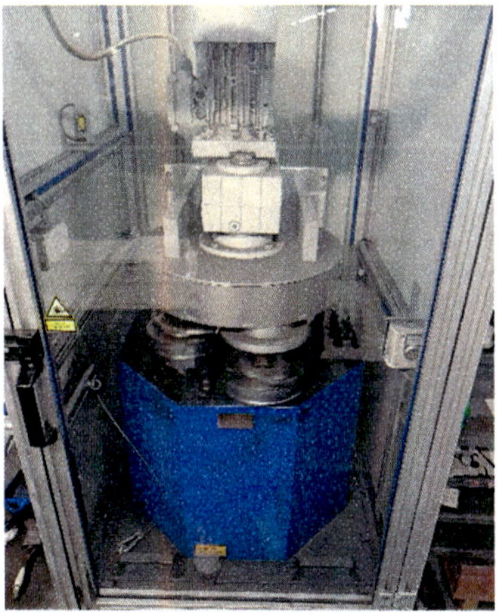

Fig. 3. Process of drag finishing on OTEC DF-3 Tools

OTEC SIX 70/16 with OTEC HL 06 grinding oil was used for the drag finishing process. Ten tools were drag finished with the time of 10 min, while every minute one of the tools was taken out of the machine, so the first tool was drag finished for one minute and the tenth tool was drag finished for 10 min. Every solid cemented carbide mill was drag finished half of the time clockwise and the other half of the time counter-clockwise. After cutting edge preparation the cutting edge radius was measured and evaluated and after that, the next batch of ten solid cemented carbide mills was also drag finished in order to verify the size increase of the cutting edge radius.

2.3 Measurement of the Tool's Microgeometry

Tools were measured after drag finishing, in order to determine the exact shape and size of the cutting edge microgeometry. The measuring machine used in this research was

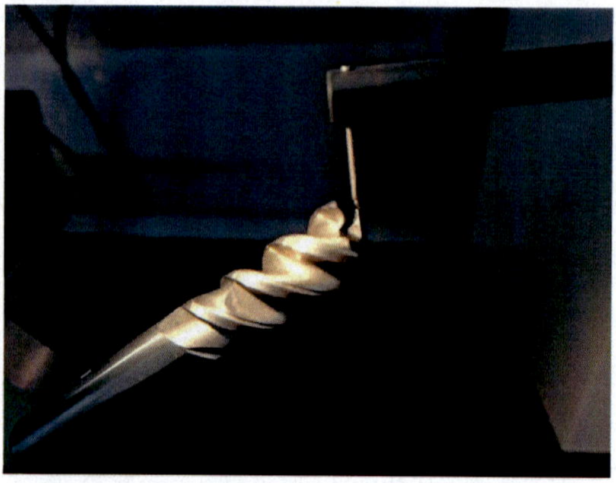

Fig. 4. Touch measurement of the tool's cutting edge on ZEISS Surfcom 5000

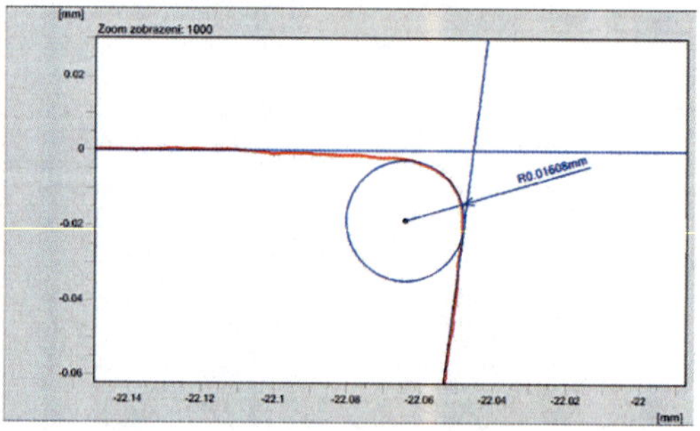

Fig. 5. Example of measurement protocol of cutting edge on ZEISS Surfcom 5000

contour and surface measuring machine ZEISS Surfcom 5000. Cutting edge radius was measured one millimetre from the tip of the tool at the end of the tool.

3 Results and Discussion

Solid cemented carbide mills were ground on Reinecker WZS 60 grinding machine, then drag finished on OTEC DF-3 Tools and the size of cutting edge radius was measured on ZEISS Surfcom 5000. The aim was to evaluate the dependence of drag finishing time on cutting edge radius, the solid cemented carbide mills were measured in one-minute intervals during drag finishing.

There were measured the cutting edge radius on the helix of the solid cemented carbide end mill and on the face of the tool. The total time of the drag finishing process was 10 min. After measurement, the results were evaluated and they were plotted into a graph of dependence of drag finishing time on the size of cutting edge radius.

In Fig. 6, it can be seen that the maximal cutting edge radius was reached after 10 min of process and the average size of cutting edge radius was 15.017 μm. As the curve begins it is steep and after 3 min the size of the cutting edge radius is increasing slowly.

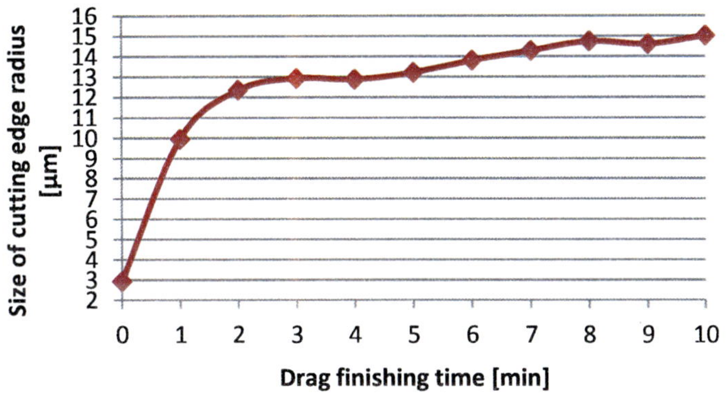

Fig. 6. Dependence of size of cutting edge radius on time on the face teeth

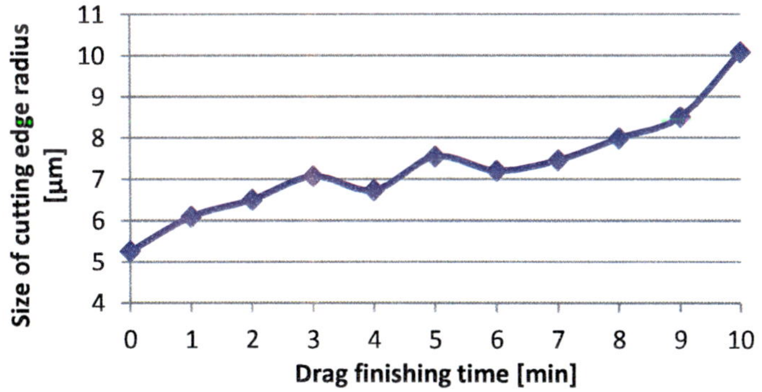

Fig. 7. Dependence of size of cutting edge radius on time on the helix teeth

The character of the second curve of dependence of the size of cutting edge radius on time on the helix teeth is not as steep as the previous curve. The average size of cutting edge radius was 10.07 μm after 10 min.

4 Conclusion

The experiment described in the article was carried out in order to investigate the influence of the drag finishing process time on the site of the cutting edge rounding on both face teeth and helical teeth. Tools with macrogeometry designed for machining austenitic stainless steel were manufactured by grinding and drag finished to achieve modified microgeometry of the cutting edges. After periodically measuring the cutting edge radii, the average values were plotted into graphs, describing the development of their size on the drag finishing time. The curve describing the size of the radii of the cutting edges on the helical teeth does not initially rise as steeply as the curve of the cutting edge radii of the face teeth. The cause of this can be that the size of the cutting edge radii on the helical teeth is almost two times higher compared to the face teeth after grinding. That is why such a rapid development of the cutting edge radii on the helical teeth took place. Adding to this, the resulting size of the cutting edge rounding on the face teeth was 15μm compared to the 10μm on the helical teeth. This difference is probably caused by the process of drag finishing itself, where the flow of the abrasive grains around the tool is not uniform, and the cutting edges on the face teeth were affected more by the abrasive than the helical teeth.

References

1. Denkena, B., Biermann, D.: Cutting edge geometries. CIRP Annals - Manufacturing Technology CIRP **63**(2), 631–653 (2014)
2. Rodriguez, C.J.C.: Cutting Edge Preparation of Precision Cutting Tools by Applying Micro-abrasive Jet Machining and Brushing, Ph.D. thesis) University of Kassel, Kassel University Press, Kassel (2009)
3. Uhlmann, E., Oberschmidt, D., Kuche, Y., Löwenstein, A.: Cutting edge preparation of micro milling tools. Procedia CIRP **14**, 349–354 (2014)
4. Vopát, T., Podhorský, Š., Sahul, M., Haršáni, M.: Cutting edge preparation of cutting tools using plasma discharges in electrolyte. J. Manuf. Process. **46**, 234–240 (2019)
5. Chan, J., Koshy, P.: Tool edge honing using shear jamming abrasive media, CIRP Annals – Manufacturing Technology (2020)
6. Hronek, O., Zetek, M., Baksa, T.: The immersion depth influences on cutting edge radius during drag finishing. In: Katalinic, B. (ed.) Proceedings of the 28th DAAAM International Symposium, pp. 1030–1036. Published by DAAAM International, Vienna, Austria (2017). ISBN 978-3-902734-11-2, ISSN 1726-9679
7. Zetek, M., Česáková, I., Švarc, V.: Increasing cutting tool life when machining inconel 718. Procedia Engineering **69**, 1115–1124 (2014)
8. Celaya, A., et al.: Influence of cutting edge radius on tool life in milling Inconel 718. In: AIP Conference Proceedings (2019)
9. Wang, W., et al. Effect of edge preparation technologies on cutting edge properties and tool performance. Int. J. Adv. Manuf. Technol. Int. J. Adv. Manuf. Technol. **106**(5–6), 1823–1838 (2020)
10. Baksa, T., Schornik, V., Adamek, P., Zetek, M.: Machining of inconel 718 using uncoated cutting tools with different cutting edge quality. In: Katalinic, B. (ed.) Proceedings of the 27th DAAAM International Symposium, pp. 0441–0446 .Published by DAAAM International, Vienna, Austria (2016). ISBN 978-3-902734-08-2, ISSN 1726-9679
11. Zlamal, T., Mrkvica, I., Szotkowski, T., Malotova, S.: The influence of surface treatment of PVD coating on its quality and wear resistant. Coatings **9**(7), 439 (2019)

Micromechanical Analysis of Glass Fiber/Epoxy Lamina

Saad Alsarayefi[1,2](✉) and Károly Jálics[1]

[1] University of Miskolc, Miskolc 3515, Hungary
machsa@uni-miskolc.hu
[2] University of Thi-Qar, Thi-Qar, Iraq

Abstract. Fiber reinforced polymer composite materials have been a major class of engineering materials. Their applications diversify between automobiles, aircraft, space vehicles, and others because of their notable offer of mechanical properties, ease of fabrication, and super design flexibility. Moreover, fiber composites are distinguished due to their light weight, excellent fatigue strength, and good corrosion and impact resistance. The current paper investigates the mechanical behavior of a unidirectional fiber reinforced polymer composite lamina consisting of fibers embedded in epoxy resin as a matrix. Micromechanical analysis is done on a square-patterned unit cell of the above composite to predict the longitudinal modulus (E_1), Transverse modulus (E_2), In-plane shear modulus (G_{12}) and Major Poisson's ratio (V_{12}). These engineering constant are evaluated to three types of fiber (E-Glass, R-Glass, and S-Glass) with various fiber volume fractions based on the theory of elasticity approach. Computer Aided Design Environment for Composites (CADEC) software is used to do the numerical analysis. This theoretical investigation helps to realize the bearing ability of unidirectional fiber reinforced composite subjected to longitudinal load by analyzing the engineering design constants.

Keywords: Micromechanics · Composite lamina · Unidirectional fiber

1 Introduction

A composite material is a combination of two or more distinct materials that formed to have enhanced mechanical properties and performance that are superior to those of the constituents if they act individually. One of the combined materials is called the reinforcement which is stiffer and stronger while the other one is called the matrix. The insured advantageous properties of composite materials have steadily extended their use in various fields of engineering [1].

The offered properties of fiber reinforced composites, especially unidirectional ones, such as high strength, light weight, non-corrosive and good impact absorbent properties, are the essential reasons for them to be widely adopted for many applications like aerospace and automobile structures [2]. When designing a composite material, the properties can be controlled by many parameters such as fiber or matrix content (volume

K. Jármai and K. Voith (Eds.): VAE 2020, LNME, pp. 101–111, 2021.
https://doi.org/10.1007/978-981-15-9529-5_9

fraction), fiber size and spacing, layer sequence and fiber orientation. The achievement of required properties is based on selecting among the above parameters. For example, the strength in the fiber direction of unidirectional composite materials is larger than this in other direction and there is a linear increment between the fiber volume fraction and the longitudinal young modulus [3]. Researchers have done a lot of work on determining the properties of composite materials by developing many analytical and numerical methods and models. Rule of mixture or inverse rule of mixture and Halphin-Tsai are approaches of the analytical methods. Numerical approaches include the unit cell methods and periodic microstructure methods [4–9].

Current work deals with the evaluation of the engineering constants such as longitudinal young modulus E_1, transverse young modulus E_2, in plane shear modulus G_{12}, and Poisson ratio V_{12} of unidirectional fiber reinforced composite. These properties are calculated by theory of elasticity based on a representative volume element (RVE) or a unit cell. The methods used are the rule of mixture (ROM), Halphin-Tsai, cylindrical assemblage model (CAM), and periodic microstructure model (PMM). A square unit cell was adopted to carry out the micromechanics calculation in this work.

1.1 Unit Cell

The possible periodic distribution of the fiber in composite materials could be of a square unit cell or hexagonal unit cell (see Fig. 1) [10].

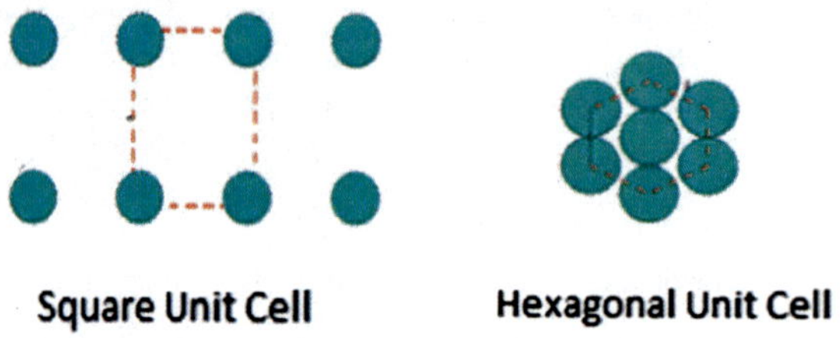

Fig. 1. Schematic representation of unit cells.

1.2 Micromechanical Analysis

The purpose of micromechanics is to study composite materials, considering the constituent materials interaction in details. It lets the analyzer to compute the characteristics of a heterogeneous composite layer or a lamina by representing it as a homogenous-anisotropic material [11]. Furthermore, it determines the stiffness and strength of a composite material by studying the stresses and strains at a micro-structural level for the fiber, matrix, and the interface of the fiber and matrix. The results of micromechanics help to understand the load sharing between the constituents, the fiber arrangement of the composite and its influence on the material, as well as helping the designer to predict the average properties of the lamina and design the material by predicting the constituent content (volume fraction), their distribution and orientation [12]. To carry out the micromechanics analysis, there are assumptions that must be taken into account such as that the fibers are distributed in a periodic fashion within the matrix, fibers are infinitely long, each layer has orthotropic properties, and the bonds between the fibers and matrix are perfect.

2 Methodology and Materials

The focus of the current work is to evaluate the engineering properties of a fiber reinforced composite material of a unidirectional fiber based on square array of RVE. The analysis is done using analytical methods such as the rule of mixture, Halphin-Tsai, cylindrical assemblage model CAM, and periodic microstructure model PMM. These methods were used individually to calculate modulus in the direction of the fibers E_1, modulus in the transverse-to-fiber direction E_2, In-plane shear modulus G_{12}, and In-plane Poisson's ratio V_{12}. All calculations are done based on the assumption that the material is anisotropic and homogenous and all formulas are applied to a unit cell [12].

2.1 Rule of Mixture

For a unit cell, the longitudinal modulus:

$$E_1 = E_f V_f + E_m V_m \tag{1}$$

Poisson's ratio:

$$V_{12} = V_f V_f + V_m V_m \tag{2}$$

Where:

E_f: Fiber elastic modulus
V_f: Fiber volume fraction
E_m: Matrix elastic modulus
V_m: Matrix volume fractio
V_f: fiber Poisson's ratio
V_m: matrix Poisson's ratio
V_{12}: Poisson's ratio for plane 1–2

2.2 Inverse Rule of Mixture

The modulus that is perpendicular to fiber direction (Transverse Modulus) E_2:

$$\frac{1}{E_2} = \frac{V_m}{E_m} + \frac{V_f}{E_f} \tag{3}$$

2.3 Halpin-Tsai Equation

The results of this equation are more accurate and beneficial for analytical evaluation and design purposes

$$E_2 = E_m\left[\frac{1+\zeta\eta V_f}{1-\eta E_f}\right] \tag{4}$$

Where:

$$\eta = \frac{\frac{E_f}{E_m}-1}{\frac{E_f}{E_m}+\zeta} \tag{5}$$

$\zeta = 2$ for circular and square fiber, $\zeta = 2\frac{a}{b}$ for rectangular fibers.

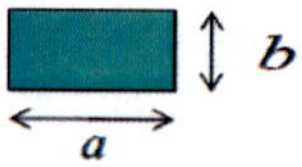

2.4 Cylindrical Assemblage Model (CAM) Formula

Shear Modulus in the 1–2 Planes G_{12}:

$$G_{12} = G_m\left[\frac{(1+V_f)+(1-V_f)\frac{G_m}{G_f}}{(1-V_f)+(1+V_f)\frac{G_m}{G_f}}\right] \tag{6}$$

Where: G_m: Matrix Shear Modulus
G_f: Fiber shear modulus

2.5 Periodic Micro-structure Model (PMM) Formula

$$G_{12} = G_m\left[1+\frac{V_f\left(1-\frac{G_m}{G_f}\right)}{\frac{G_m}{G_f}+S_3\left(1-\frac{G_m}{G_f}\right)}\right] \tag{7}$$

Where: $S_3 = 0.49 - 0.47V_f - 0.027V_f^2$.

2.6 Materials

The study is carried out on three types of materials which are E-galss/Epoxy, S-glass/Epoxy, and R-glass/Epoxy with different fiber volume fraction V_f, and the results are compared and analyzed.

Glass fiber in general is formed from bulk glass. Glass is an amorphous substance manufactured from a blend of sand, limestone, and other oxidic compounds. Thus, the main chemical constituent of glass fibers (45–75%) is silica ($SiO2$). There are various types of glass fibers have been found by controlling the chemical composition and the manufacturing processes [12]. All types can have typical glass properties such as hardness, corrosion resistance, light weight, and low cost, leading to making glass fibers one of the most used fibers in industrial composite materials. The glass fiber types are E, D, C, S, and R glass which all have similar stiffness but different strength and different resistance atmosphere degradation resistance. The types used in this study are E-glass (E for electric) which used with applications that need high strength and high chemical resistance. S- Glass (S for strength), has high strength but it is very costly compared with other types. R- Glass is the European version of American high performance S-Glass. It is distinguished by the high strength and modulus, high temperature resistance, and good fatigue stability [13].

Table 1 contains the required properties of the constituent materials.

Table 1. The properties of the used constituents of the composite materials.

Material	Modulus (GPa)	Poisson's Ratio
E-Glass	72	0.22
S-Glass	85	0.22
R-Glass	86	0.25
Epoxy	4.667	0.35

3 Results and Discussion

Current analysis tries to distinguish among three types of laminas consisting of glass fibers embedded in epoxy matrix, by evaluating the engineering constants. The three types of glass fibers are E-glass, S-glass, and R-glass. Different Micromechanics formulas based on different approached are used to predicts some of the constants and then study them by comparing the results.

The longitudinal modulus is evaluated for all types of materials by the ROM methods and a comparison among the results is made as shown in Fig. 2:

- For all suggested volume fractions V_f, the longitudinal young's modulus E_l of the E-glass/Epoxy is lower than those of R-glass/Epoxy and S-glass/Epoxy which they seemed to have very close young's modulus all over the variation of V_f (see Fig. 2).
- The longitudinal young's modulus E_1 increases linearly with increase in fiber volume fraction for all the three types of composites materials proving the fact that the strength of the unidirectional fiber composite is as high as the fiber volume fraction high since the fiber is stiffer and stronger.

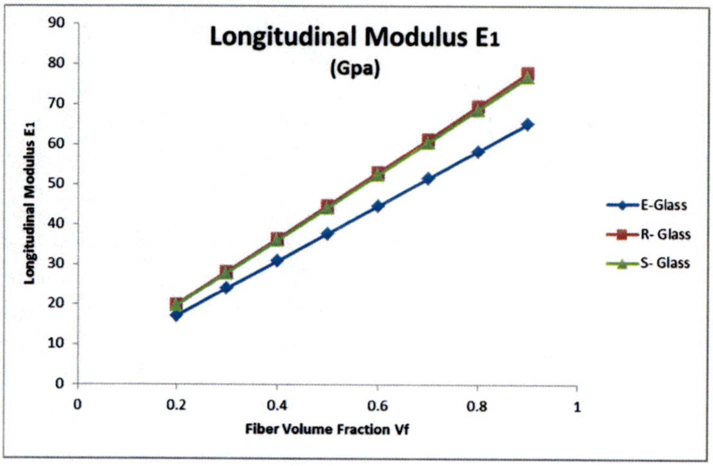

Fig. 2. Variation of E_1 with fiber volume fraction for all fibers.

As shown in Figs. 3, 4, 5, 6, 7, 8, 9, 10 and 11, that for each material the transverse modulus E_2 is calculated in two ways, rule of mixture ROM and Halphin-Tsai. The in-plane shear modulus G_{12} is evaluated by the cylindrical assemblage model CAM and the periodic microstructure model PMM. The in-plane passion's ratio for each type of material is calculated based on the ROM and PMM. A comparison is made between every two different results of each constant. Furthermore, the variation of longitudinal young modulus with the volume fraction for all types of composite materials is done. From the above graphs, some findings can be stated:

E- Glass/Epoxy Lamina

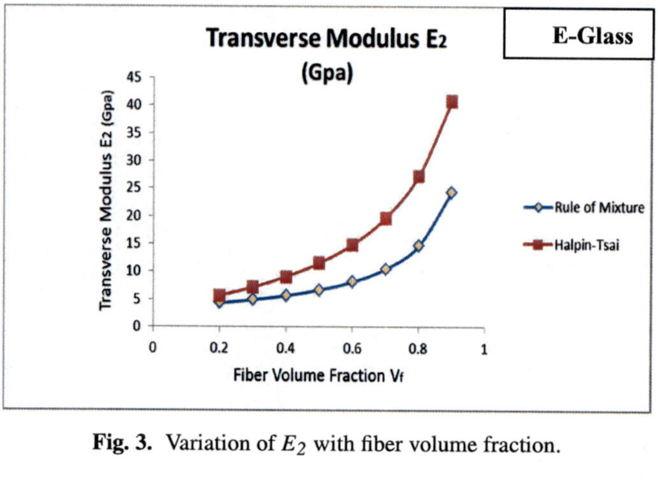

Fig. 3. Variation of E_2 with fiber volume fraction.

- The transverse young's modulus E_2 increases linearly with increase in fiber volume fraction for all the three types of composites materials up to 80% of volume fraction and rapid increase happens after that.
- For all three types of materials, the magnitude of $E2$ computed by ROM is much lower than this computed by Halphin-Tsai methods for all fiber volume fractions, pulling to the mind the fact that the ROM equation under-estimates the actual value for $E2$ while the Halphin-Tsai methods give more accurate results of it (see Figs. 3, 6, 9).

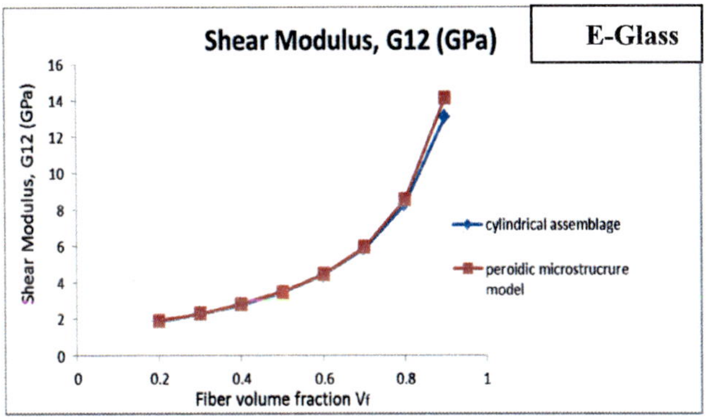

Fig. 4. Variation of G_{12} with fiber volume fraction

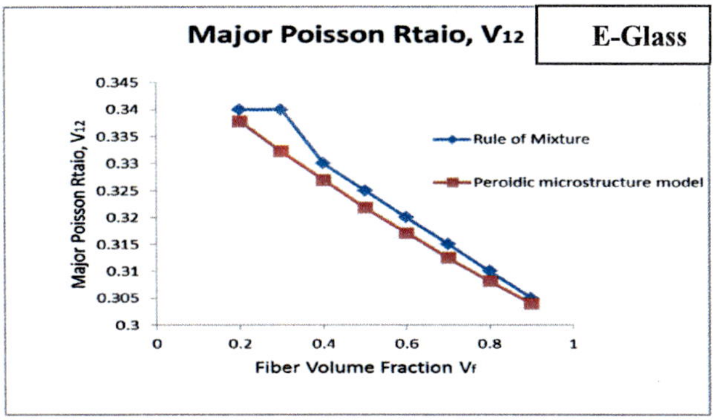

Fig. 5. Variation of V_{12} with fiber volume fraction.

R- Glass/Epoxy Lamina

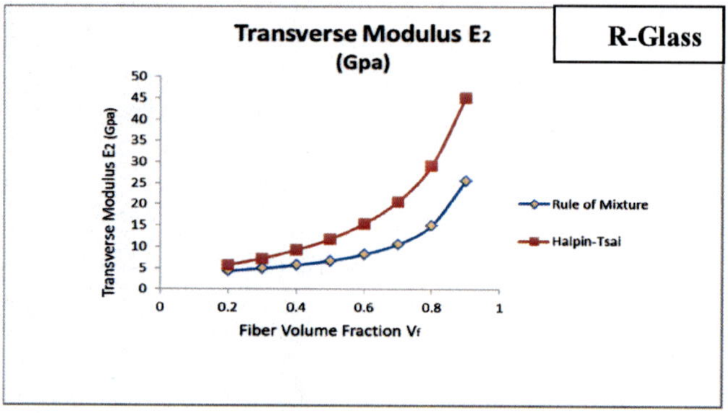

Fig. 6. Variation of E_2 with fiber volume fraction.

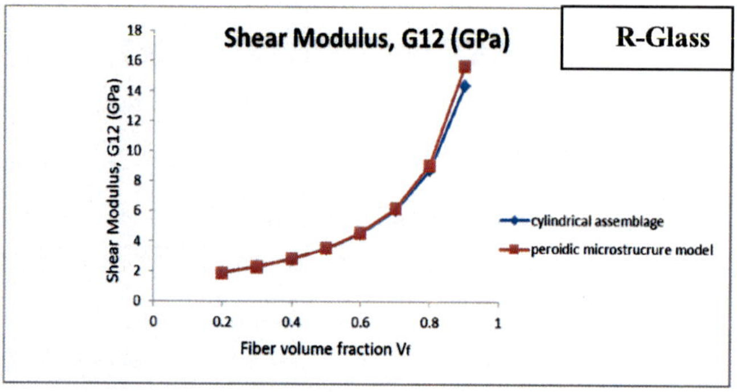

Fig. 7. Variation of G_{12} with fiber volume fraction.

- In plane shear modulus G_{12} increases linearly with an increase in fiber volume fraction for all the three types of composites materials, but, after 80% of the V_f, there is a noticeable difference of *G12* computed by CAM and PMM methods (see Figs. 4, 7, 10).

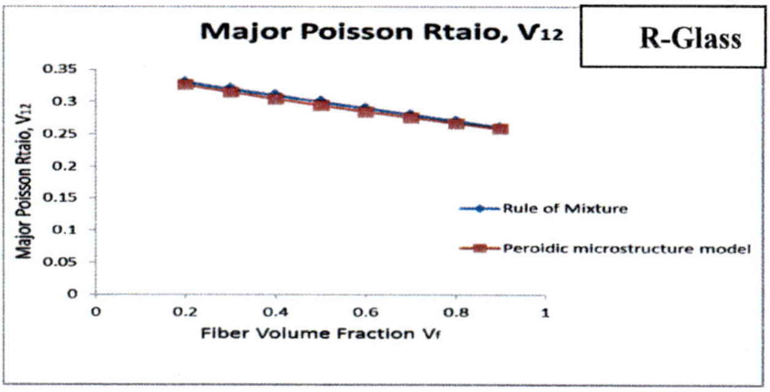

Fig. 8. Variation of V_{12} with fiber volume fraction.

S- Glass/Epoxy Lamina

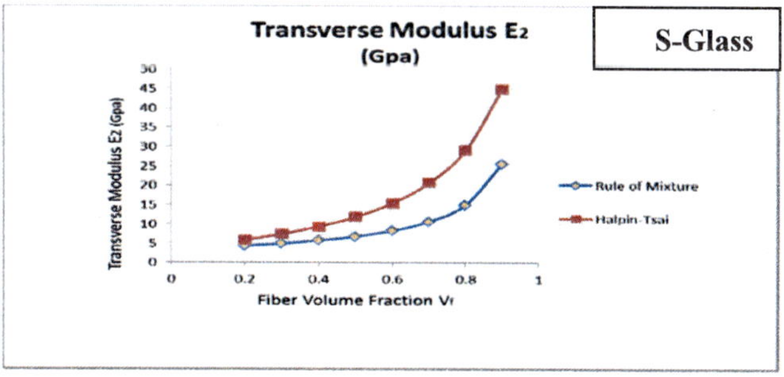

Fig. 9. Variation of E_2 with fiber volume fraction.

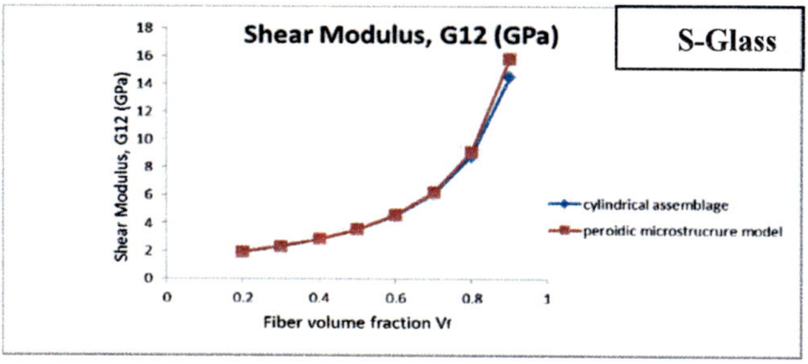

Fig. 10. Variation of G_{12} with fiber volume fraction.

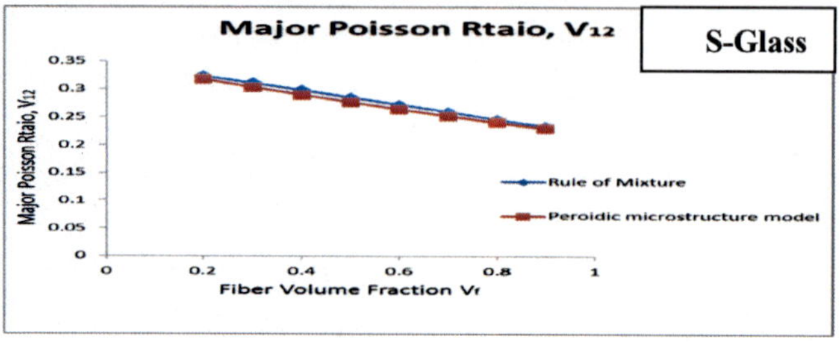

Fig. 11. Variation of V_{12} with fiber volume fraction.

- For all composites, Poisson's ratio V_{12} is decreasing with increasing the fiber volume fraction. However, for E-Glass/Epoxy, the magnitude of V_{12} computed by ROM tries to stay at the same level at the beginning of the curve, but a rapid decrease happens after that making a noticeable difference with the magnitude of this computed by PMM (see Figs. 5, 8, 11).

Generally, it can be seen that the longitudinal modulus E1 is increasing with the increase of fiber volume fraction. Also, there is a linear increment of the transverse modulus E2 with the increment of fiber volume fraction. In addition, in plane Shear modulus G12 increases linearly with an increase in fiber volume fraction for all the three types of glass fibers. However, for all composites, Poisson's ratio V12 is decreasing with increasing the fiber volume fraction.

4 Conclusion

This work is carried out to evaluate several elastic constants for three types of unidirectional fiber reinforced composite materials (E-glass/Epoxy, S-Glass/Epoxy, R-Glass/Epoxy), using some approaches of composite materials micromechanics. The focus has been given to distinguish among the three types of glass fibers. Square unit cell is used with all necessary assumptions and CADEC software is used to compute the engineering constants. Elastic moduli E_1 results are used to make a comparison between the glass fiber types and are calculated using the ROM equation. The results of E_2 are found using the ROM methods and Halphin-Tsai methods and are analyzed and compared between the three types. Moreover, G_{12} is evaluated using the two methods CAM and PMM by highlighting the results of each method. Poisson's ratio V_{12} is also calculated with ROM and PMM for all three types of composite materials. Current theoretical investigation findings assist the designers and analyzers to expand their engineering knowledge about unidirectional fiber reinforced composite and its design process parameters. Future work will include a finite element results and comparison will be done. A macro-mechanics analysis of a laminate will also be of a future interest study.

References

1. Rajak, D.K., Pagar, D.D., Menezes, P.L., Linul, E.: Fiber-reinforced polymer composites: manufacturing, properties, and applications. Polymers **11**(10) (2019). https://doi.org/10.3390/polym11101667
2. Alberto, M.: Introduction of fibre-reinforced polymers – polymers and composites: concepts, properties and processes. In: Fiber Reinforced Polymers - The Technology Applied for Concrete Repair (2013). https://doi.org/10.5772/54629.
3. Syam, P.A., Syed, A.H., Pandurangadu, V.: Micromechanical analysis of FRP composites. Int. J. Mech. Eng. Technol. **4**(2), 272–285 (2019)
4. Evans, J.T.: Analysis and Performance of Fiber Composites (Second Edition) (1992). https://doi.org/10.1016/0921-5093(92)90189-8
5. Yang, Q.S., Qin, Q.H.: Micro-mechanical analysis of composite materials by BEM. Eng. Anal. Bound. Elem. **28**(8), 919–926 (2004). https://doi.org/10.1016/S0955-7997(03)00118-8
6. Wang, A.S.D.: Micromechanics analysis of composite materials. In: Mota, S., Carlos, A., Cristóvão, M., Manuel, J.M. (eds.) Mechanics of Composite Materials and Structures, pp. 81–130. Springer, Dordrecht (1999)
7. Melro, A.R., Camanho, P.P., Andrade Pires, F.M., Pinho, S.T.: Micromechanical analysis of polymer composites reinforced by unidirectional fibres: part II-micromechanical analyses. Int. J. Solids Struct. **50**, 1906–1915 (2013). https://doi.org/10.1016/j.ijsolstr.2013.02.007
8. Zeman, J.: Micromechanical Analysis of Random Composites (2000)
9. Hussain, S., Reddy, B., Reddy, V.: Prediction of elastic properties of FRP composite lamina. ARPN J. Eng. Appl. Sci. **3**(6), 70–75 (2008)
10. Alsarayefi, S., Gafil, H., Jálics, K.: Optimisation of the fibre size for a fiber glass epoxy composite. Des. Mach. Struct. **9**(1), 5–12 (2019). https://doi.org/10.32972.dms.2019.001
11. Babu, K.S., Rao, K.M., Raju, V.R.C., Murthy, V.B.K., Kumar, M.S.R.N.: Micromechanical analysis of FRP hybrid composite lamina for in-plane transverse loading. Indian J. Eng. Mater. Sci. **15**, 382–390 (2008)
12. Barbero, E.J.: Introduction to Composite Materials Design, 2nd edn. CRC, London (2011)
13. Sathishkumar, T.P., Satheeshkumar, S., Naveen, J.: Glass fiber-reinforced polymer composites. J. Reinf. Plast. Compos. **33**(13) (2014). https://doi.org/10.1177/0731684414530790

Non-destructive Test for Control of the Surface Quality of Semi Product at the Automotive Industry

Péter Szobota(✉), Máté Sepsi, and Valéria Mertinger

University of Miskolc, Miskolc-Egyetemvaros 3515, Hungary
szobotapeter209@gmail.com

Abstract. The aim of the research is to develop a measurement method that can determine the degree of decarburization on the surface of heat-treated or even on finished parts, without any damage, with high reliability. Several times the decarburization is revealed only in the semi-finished or finished state when the surface of the component does not have the required parameters. In this case, a cut from the sample has to be used to verify the process, while the component becomes waste. Centerless X-ray diffractometers have been developed primarily for non-destructive, residual stress measurement. Internationally unique, two such diffractometers, at the Institute of Physical Metallurgy, Metal Forming and Nanotechnology and in the 3D Laboratory are available. The basic idea of the present research is to determine and to use the characteristics of the interference function detected by the diffractometer, which goes far beyond the residual stress test [1]. One such feature is the broadening of the peaks on the interference function, which correlates with the chemical composition of the diffracting phase and the micro stresses [2–5]. Changes in chemical composition occur, for example, as a negative consequence of the heat treatment in component during decarburization.

Keywords: Residual stress · Decarburizing · Case depth · X-ray stress measurement

1 Introduction

The atmosphere in the heat treatment furnace contains various gases such as carbon dioxide, carbon monoxide, hydrogen, nitrogen, oxygen, and water steam [6]. These gases reduce the carbon content dissolved in iron, as they react with it to a certain depth on its surface, changing its composition and consequently its strength [1, 7–10]. There are many factors have influence on decarburization and its depth, such as the chemical composition of the atmosphere in the furnace, and the thermal cycle during all manufacturing steps [11, 12]. Decarburization can occur not only in the solid-state but also in the molten state, for example during selective laser melting (SLM) [13]. To avoid these cases, minimizing the decarburization process is important, depending on the application, the degree of decarburization is regulated properly for many components. Another negative

K. Jármai and K. Voith (Eds.): VAE 2020, LNME, pp. 112–122, 2021.
https://doi.org/10.1007/978-981-15-9529-5_10

effect of decarburization is that additional cost-increasing technologies need to be used to compensate for the softening for example surface spinning strengthening (3S) treatment [14] or water jet peening [15]. This means that measuring the degree of decarburization is important. According to the practical approach of the definition, it considers decarburized the layer thickness that loses its functional properties. In some exceptional cases the ductile, but lower carbon content surface can benefit for specific applications for example shock absorption [16], or texture control [17].

The most frequently used methods for measuring decarburization are metallographic examination and hardness measurement, which have the serious disadvantage that they might need to be carried out by destructive (sample cutting) techniques. There are only a few exceptional cases where non-destructive technique can be used to determine decarburized layer thickness for example magnetic hysteresis measurements [18–20]. The main goal of this research is to obtain information about the progress of decarburization by X-ray diffraction without sample cutting. During the research, decarburized layers were produced to different degrees, and then the broadening of the peaks of the interference function was determined during residual stress measurement. Multi-step electrochemical polishing was used for depth measurement. Metallographic inspection and hardness tests were also performed to determine the correlation between decarburization and profile broadening. The three methods complement and highlight the benefits of X-ray diffraction studies. The purpose of the depth studies in the present case was to show how sensitively respond the full width at half maximum (FWHM) of the diffraction profile to the changing carbon content gradient. With the measurement method validated in this way, the surface decarburization can be determined without destruction.

2 Experimental

C45 quality hot-rolled bar steel as the raw material was chosen for the experimental specimens, because it is widespread in the automotive industry in terms of its practical use. The geometric design of the specimens, as shown in Fig. 1. Was chosen so that the placement and insertion of the specimens in the furnace, as well as the heat treatment and subsequent measurements, could be safely reproduced and easily performed.

Decarburized layers of different thicknesses were produced by heat treating at 900 °C in an air atmosphere, in a resistance heating furnace with different heating (holding) times: sample 1: 0.5 h; sample 3: 1 h; sample 9: 2 h. A contact K-type (Ni-Cr) thermocouple was used to determine the warm-up time of the specimens, which was typically 0,25 h. The heating time refers as an isothermal holding time and does not include the warm-up time. The subsequent water quenching was used to cool down the samples. The oxide layer formed during the heat treatments was removed by chemical etching with a mixture of hydrochloric acid and water (50–50 volume %).

A Stresstech Xstress Robot X-ray diffractometer was used to determine the residual stress and FWHM value. The first measuring point was made on the surface and then the electrochemical etching was carried out in many steps to remove the material (See Fig. 2). The electrochemical etching was performed with a Struers MoviPol-5 machine, and the removed layer thickness was determined by a digimatic indicator with 1 μm resolution and 3 μm accuracy, the material removing step size was typically 50–100 μm. The location of the tests is indicated in Fig. 1 XRD test was done after each etching step. The etching steps were continued until the value of the FWHM no longer changed.

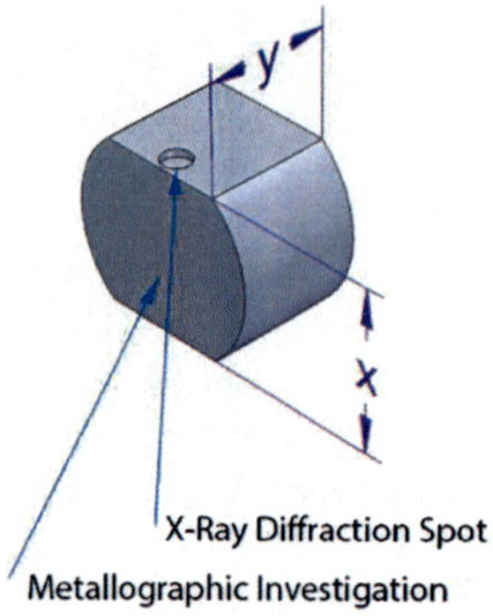

Fig. 1. Geometry of the experimental samples. The places of the X-Ray diffraction and metallographic investigation are mentioning, x = 21 mm, y = 17 mm.

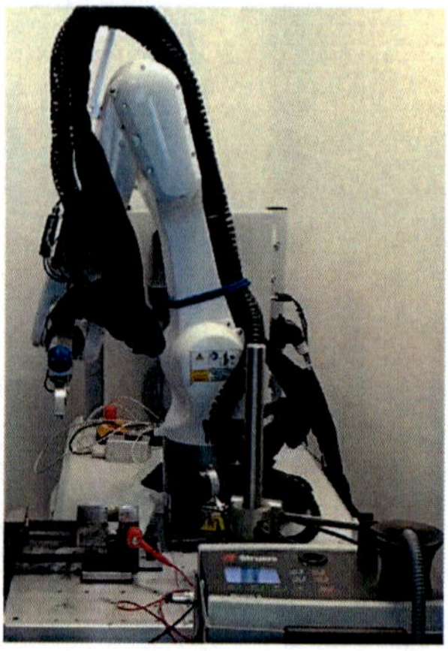

Fig. 2. The Stresstech Xstress Robot and the measuring setup for accurate determination of the removed material.

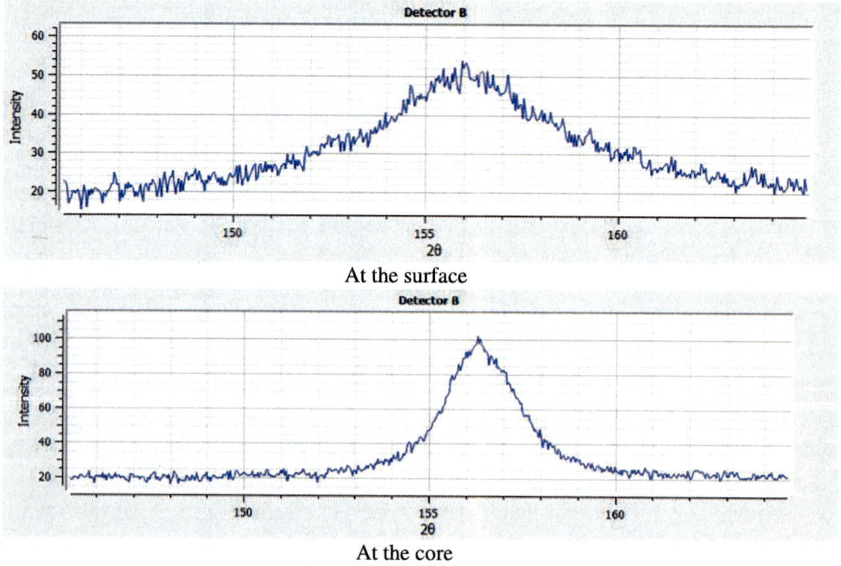

Fig. 3. Broadening of the ferrite {211} reflection

For the study, the residual stress measurement routine of the equipment was used with Cr X-ray tube, 30 kV accelerating voltage with 8 mA beam current, and modified Ψ measurement mode, 11/11 tilting (relating to the sample surface) were applied at each measuring step. The calculated stress component was parallel to the "y" (See

Fig. 1) at each point. Residual stress and FWHM were determined of the ferrite {211} diffraction using Gaussian fitting. Figure 3 shows example to the interference function at the decarburized surface and at the core of the sample. Xtronic software was used to calculate the stress and FWHM data at each measuring step. Stress was calculated based on the 11/11 tilting positions data using the $\sin^2\psi$ calculation, the FWHM data was calculated as an average of the data of the 22 tilting positions.

Evaluation of the XRD data revealed that the routine used in the residual stress test to assign the measured interference function to a Gaussian function did not give adequate results. Therefore, we performed further studies in the case where two Gaussian functions were used for the fitting process (double curve fitting process). An example of this is shown in Fig. 4.

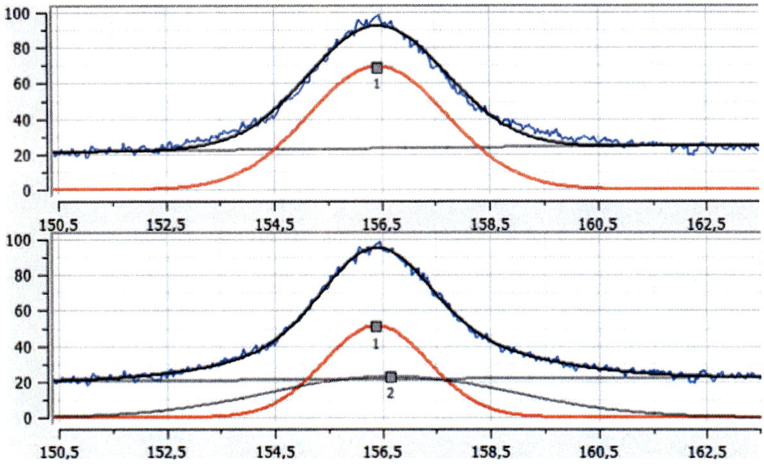

Fig. 4. Fitting of ferrite {211} reflection (intensity in function of 2θ) with one or with two Gaussian curves (double curve fitting process- at intensity maximum numbered 1 and 2). In the latter case P2 peak represents the bigger broadening

After the X-ray diffraction measurements, a cross-sectional grinding was performed using the standard sample preparation method (See Fig. 1). Four stages of grinding, polishing and etching in 2% Nital were performed. Because our study focuses also on the surface of the sample, we also mounted the samples because of the rounding on its edge. Optical microscopic images of the prepared samples were taken with a Zeiss Axio Imager M1m optical microscope.

Hardness measurements of the prepared samples were performed by Instron Tukon 2100B (Wilson Instruments) Vickers microhardness tester with a load of 50 g. The first measured point was taken at a distance of 20 μm from the surface and 100 μm step size was used for additional points.

3 Results

The typical microstructure features are shown in Fig. 5 for each sample. The effect of 0.5-h and 1-h heat treatment is not visible on the optical microscopic images, fully martensitic microstructure can be observed. In contrast, after 2 h of heat treatment, due to decarburization process ferrite, perlite and bainite also appear in the area close to the surface. Moving closer to the core the volume fraction of martensite increases.

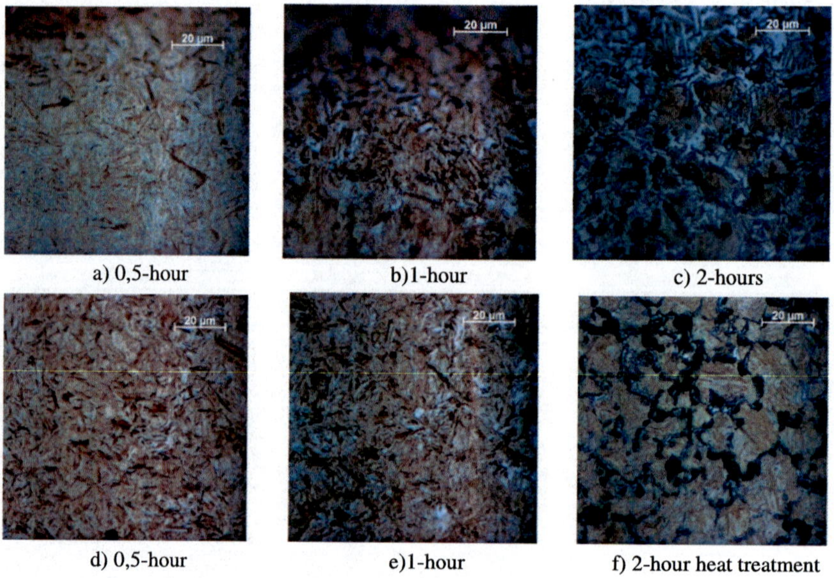

Fig. 5. Optical microscopic features with different heat treatments. a), b), c) near the surface d), e), f) near the core of the samples

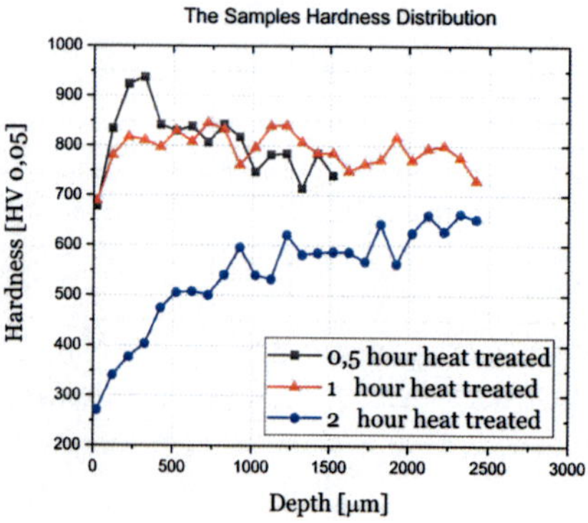

Fig. 6. The hardness distribution of the samples

The hardness results for the three samples are shown in Fig. 6. The samples that have been heat treated for 0.5 and 1 h, have similar curves, although with a large standard deviation. In contrast, the value on the surface of the 2-h sample is very small (280 HV), which increases continuously in function of the distance, not even reaching the value of the other two samples at a depth of 2500 µm. The 2-h heat treated sample clearly shows the softening effect of decarburization [12].

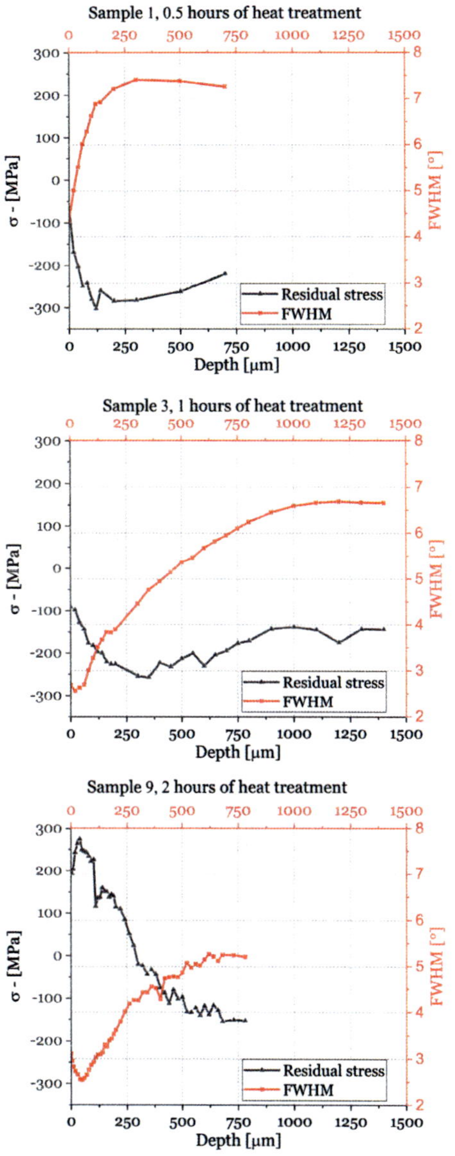

Fig. 7. The distribution of the residual stress and FWHM

The residual stress and the FWHM data in function of distance from the surface of the three samples are shown in Fig. 7. The 0,5 and the 1 h-long heat-treated sample exhibit compressive residual stress at the surface, which is increases in function of distance and shows a compressive stress maximum around 200 and 300 μm respectively. The 2 h long heat treatment process resulted tensile stress at the surface which decreases and turned into compressive stress in function of depth.

The FWHM value increases as a function of distance in all three samples until it reaches a saturation value, although the rate of increase shows significant differences. At a depth of 750 μm, the sample treated for 0.5 h had already saturated to a value of about 7.3°, the broadening of the sample treated for 1 h was 6°, while the samples treated for 2 h showed a broadening of only 5°. For the 2-h sample, near surface data, both stress and FWHM, show some non-monotony. This effect is attributed to the imperfect removal of the oxide layer.

FWHM data for the 1,5- and 2-h heat treated samples were also determined when not only one but two Gaussian curves were used for fitting (double curve fitting process). Fig. 8, 9 and 10 show the FWHM data, the intensity maximum and the volume fraction of P2 data respectively after the double curve fitting process, where the P2 curve performs the larger broadening (See Fig. 4).

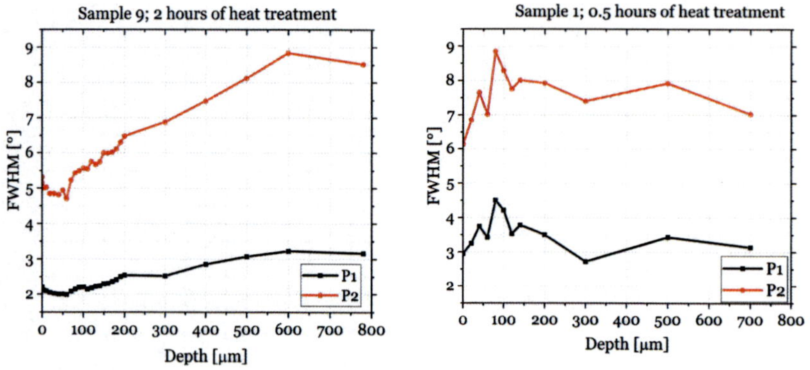

Fig. 8. The FWHM data in case of double curve fitting process for ferrite {211}. P2 represents the larger broadening curve.

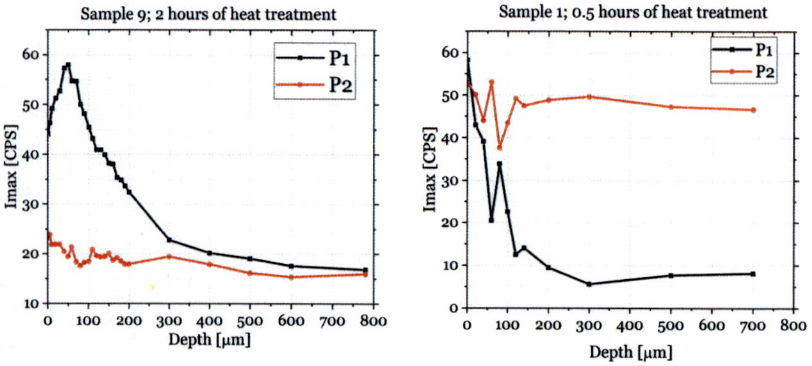

Fig. 9. The intensity maximum data in case of double curve fitting process for ferrite {211}. P2 represents the larger broadening curve.

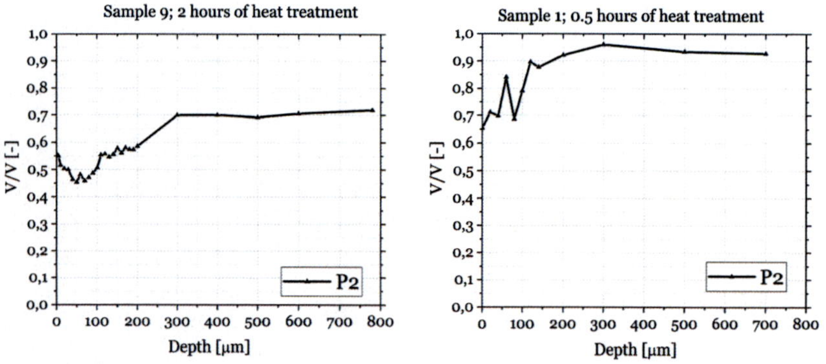

Fig. 10. The volume fraction of P2 curve in case of two duple curve fitting process for ferrite {211}. P2 represents the larger broadening curve.

4 Conclusions

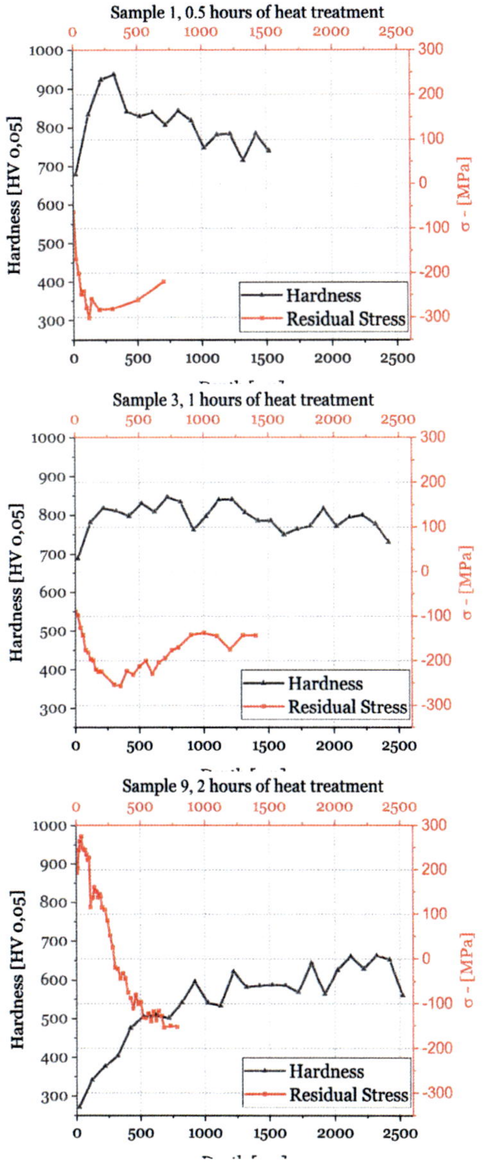

Fig. 11. The distribution of the hardness and residual stress

The results obtained by the different measurement methods are compared by plotting in a common diagram. Fig. 11 shows the change in residual stress and hardness values. A clear observation for all three samples is that the increase in hardness is accompanied by a change in the residual stress in the direction of the compressive stress. Although optical microscopy could not detect decarburization in the microstructure of after 0.5 and 1-h treatment, changes in residual-stress and hardness also suggest that the surface of these samples was also decarburized, but not so to the extent that it causes change in the microstructure features. Sample after 2 h treatment clearly shows softer microstructure elements such as ferrite and perlite due to decarburization cause not only a lower value of hardness but also the appearance of very dangerous tensile stresses. The presence of tensile stress plays a decisive role in the formation of cracks and especially in their propagation.

Figure 12 shows the change in hardness and FWHM together. The two parameters clearly move together, but the FWHM reacts more sensitively in the case where the resulting microstructure is martensitic but with lower carbon content or even ferritic due to decarburization. That is, according to our assumptions, the FWHM is the parameter that shows the presence of decarburization with greater certainty than the hardness or even the microstructure.

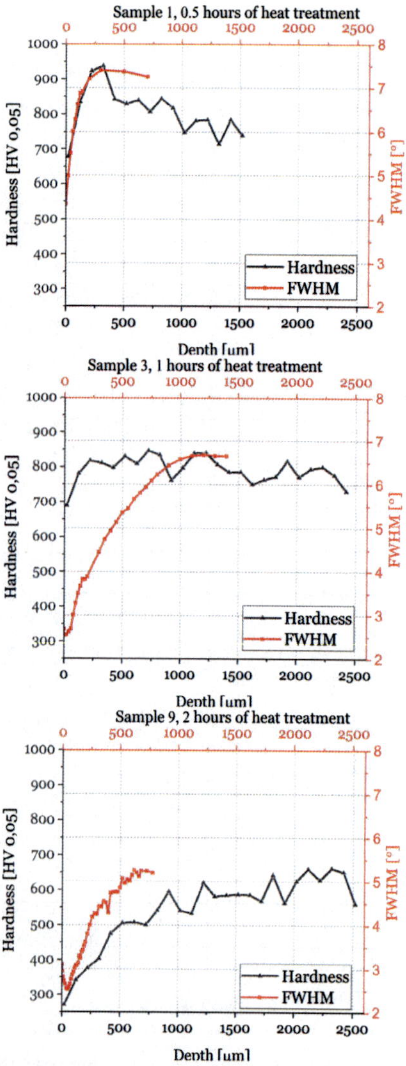

Fig. 12. The distribution of the hardness and FWHM

If we examine the change in FWHM and residual-stress together (Fig. 7), we can see that the change in the two parameters is inversely proportional in the sense that the increase in FWHM is accompanied by a change in residual-stress in the negative, i.e. compressive, direction.

Because the three samples were cooled in the same way, the different stress and FWHM results are due to the microstructure differences due to changes in composition, and the stresses are not thermal but transformational. What also strengthens our theory is that the change in FWHM, in our case its increase is due to the higher martensite content and the higher carbon content, which is a clear evidence of decarburization in the near-surface layer. That is, the change in FWHM is a reliable, well-measurable parameter to detect the presence of decarburization.

Considering the results that the double Gauss fitting provided, it can be assumed that although the instrumental broadening is significant, it can be observed that microstructural phases with different carbon contents and thus different third-order stresses can be separated. The tendency of Fig. 10 to increase the volume fraction of the phase with higher radial broadening coincides with the tendency of the highly decarburized sample in Fig. 5 to move from the ferrite-rich part to the perlite, bainite-martensitic, martensitic region, i.e. from phases loaded with a smaller defect structure to larger and from the larger crystallite size to the smaller one.

5 Summary

In the course of our research work, a new, non-destructive test method for prediction of the decarburization process in steels was presented. Centerless X-ray diffractometers developed to measure residual stresses and able to record the interferential function and its change in the large (above 120°) 2θ angular range with high resolution, the information content of this signal goes beyond the determination of first-order stresses. The third-order stresses can be traced back to the presence of zero and one-dimensional defects in the crystal structure, causing the diffraction signal to broaden which can also be measured with high resolution with these devices. Since the carbon content in the near-surface layers changes during the decarburization process, so does the density of the crystallographic defects, which causes the detected signal to broaden. In many cases, the residual stress measurements are usually determined not only on the surface but also by electrochemical polishing in the subsurface layers, so the instrumental conditions for the determination of the decarburized layer thickness on X-ray diffraction bases are given. Decarburized specimens were prepared with different holding times at 900 °C, on which the residual stress measurements were performed with a Xstress Robot centerless diffractometer. In addition to the residual-stress measurement, other characteristics of the interference function, such as the change in FWHM, were also examined. In addition to X-ray diffraction analysis, hardness distribution and microstructure analysis were also performed. It was found that the X-ray diffraction test method indicates the fact of decarburization even if it is only to the extent that it does not cause change in the microstructure features. Our preliminary assumption has also been proved that the FWHM reliably indicates the presence of decarburization. One of the most important messages of our results is that the microstructure change resulting from decarburization, i.e. the appearance of lower hardness ferrite and perlite, not only reduces the hardness but also causes the appearance of very dangerous tensile stresses. The appearance of surface tensile stress after hardening due to decarburization is extremely detrimental to the formation and propagation of cracks. This is a much bigger problem than softening caused by a decrease in carbon content. Additional fitting analysis was also used to evaluate the interferential function more in details. Manual double Gauss fitting was applied to solve inaccuracies in the software. With this fitting technique the phases in different microstructure feature with different carbon content and thus different third-order stresses can be separated. Our future plans are to extend our investigations to the extremes of decarburization, and to check the calibration curve with carburizing processes.

Acknowledgements. The described article was carried out as part of the EFOP-3.6.1-16-00011 "Younger and Renewing University - Innovative Knowledge City - institutional development of the University of Miskolc aiming at intelligent specialisation" project implemented in the framework of the Szechenyi 2020 program. The realization of this project is supported by the European Union, co-financed by the European Social Fund and was also supported through the National Research, Development and Innovation Office - NKFIH K119566 project.

References

1. Sepsi, M., Mertinger, V., Benke, M.: Sample cutting-free pole figure measuring method for centreless diffractometers in modified X mode. Mater. Charact. **151**, 351–357 (2019)
2. Totten, G.E.: Steel Heat Treatment Metallurgy and Technologies. CRC Press Taylor & Francis Group, Cleveland (2007)
3. Krawitz, A.D.: Introduction to Diffraction in Materials Science and Engineering. Wiley, New York (2001)
4. Yi, J., Gharghouri, M., Bocher, P., Medraj, M.: Distortion and residual stress measurements of induction hardened AISI 4340 discs. Mater. Chem. Phys. **142**, 248–258 (2013)

5. Rocha, A.S., Hirsch, T.: Fast in situ X-ray diffraction phase and stress analysis during complete heat treatment cycles of steel. Mater. Sci. Eng. A **395**, 195–207 (2005)
6. Vander Voort, G.F.: Understanding and measuring decarburization. Adv. Mater. Process. **173**, 22–27 (2015)
7. Haimbaugh, R.E.: Practical Induction Heat Treating. ASM International, Ohio (2001)
8. Zhao, X.J., Guo, J., Wang, H.Y., Wen, Z.F., Liu, Q.Y., Zhao, G.T., Wang, W.J.: Effects of decarburization on the wear resistance and damage mechanism of rail steels subject to contact fatigue. Wear **364–365**, 130–143 (2016)
9. García Navas, V., Gonzalo, O., Quintanaa, I., Pirling, T.: Residual stresses and structural changes generated at different steps of the manufacturing of gears: effect of banded structures. Mater. Sci. Eng. A **528**, 5146–5157 (2011)
10. De la Rosa, C.E.F., Trejo, M.H., Roman, M.C., Lopez, E.A.: Effect of decarburization on the residual stresses produced by shot peening in automotive leaf springs. ASM Int. **25**(7), 2596–2603 (2016)
11. Zhang, C., Zhou, L., Liu, Y.: Surface decarburization characteristics and relation between decarburized types and heating temperature of spring steel 60Si2MnA . Int. J. Miner. Metall. Mater. **20**(8), 720–724 (2013)
12. Zhang, C., Xie, L., Liu, G., Chen, L., Liu, Y., Li, J.: Surface decarburization behavior and its adverse effects of air-cooled forging steel C70S6 for fracture splitting connecting rod. Metals Mater. Int. **22**(5), 836–841 (2016)
13. Zhao, X., Song, B., Zhang, Y., Zhu, X., Wei, Q., Shi, Y.: Decarburization of stainless steel during selective laser melting and its influence on Young's modulus, hardness and tensile strength. Mater. Sci. Eng. A **647**, 58–61 (2015)
14. Ren, C.X., Wang, D.Q.Q., Wang, Q., Guo, Y.S., Zhang, Z.J., Shao, C.W., Yang, H.J., Zhang, Z.F.: Enhanced bending fatigue resistance of a 50CrMnMoVNb spring steel with decarburized layer by surface spinning strengthening. Int. J. Fatigue **124**, 277–287 (2019)
15. Ijiri, M., Yoshimura, T.: Sustainability of compressive residual stress on the processing time of water jet peening using ultrasonic power. Heliyon **4**, e00747 (2018)
16. Lefevre-Schlick, F., Bouaziz, O., Brechet, Y., Embury, J.D.: Compositionally graded steels: the effect of partial decarburization on the mechanical properties of spherodite and pearlite. Mater. Sci. Eng. A **491**, 80–87 (2008)
17. Oldani, C.R.: Decarburization and grain growth kinetics during the annealing of electrical steels. Scripta Mater. **35**(11), 1253–1257 (1996)
18. Perevertov, O., Stupakov, O., Tomas, I., Skrbek, B.: Detection of spring steel surface decarburization by magnetic hysteresis measurements. NDT&E Int. **44**, 490–494 (2011)
19. Kahrobaee, S., Kashefi, M., Alam, A.S.: Magnetic NDT technology for characterization of decarburizing depth. Surf. Coat. Technol. **205**(16), 4083–4088 (2011)
20. Stupakov, O., Perevertov, O., Tomas, I., Skrbek, B.: Evaluation of surface decarburization depth by magnetic Barkhausen noise technique. J. Magn. Magn. Mater. **323**, 1692–1697 (2011)

Investigation into Applicability of Automotive Quality Regulations in the Steel Industry

Béla Kondás(✉) and Csaba Deák

University of Miskolc, Miskolc 3515, Hungary
{kondas.bela,deak.csaba}@uni-miskolc.hu

Abstract. The supply chain of the automotive industry mainly consists of processing and assembling companies. Raw material producers, such as manufacturers of plastic-, rubber-, aluminium- and steel industry, are exceptions to some extent. The continuous process industry has its own peculiarities. This study is searching for the limits of the applicability of automotive quality requirements concerning the rolled product manufacturing segment of the steel industry, where the quality of the end product determines the ability of production, the usability and the durability of the parts assembled in automobiles. In the present study firstly the peculiarities of the steel- and automotive industry is presented, based on technical literature. Then the quality rules of the automotive industry are outlined in detail with special focus on their aims. The study is based on IATF 16949 standards and the relevant customer specific requirements. Finally, the limits and difficulties of the application and compliance of these rules and regulations are discussed.

Keywords: Automotive · Steel industry · Quality assurance

1 Introduction

The supply chain of the automotive industry mainly consists of processing and assembling companies.

Raw material producers, such as manufacturers of plastic-, rubber-, aluminium- and steel industry, are exceptions to some extent. The continuous process industry has its own peculiarities. This study is searching for the limits of the applicability of automotive quality requirements concerning the rolled product manufacturing segment of the steel industry, where the quality of the end product determines the ability of production, the usability and the durability of the parts assembled in automobiles.

The significance of the theme is undoubted since the proper application of quality tools (i.e. Core Tools) in automotive industry still poses a serious challenge to quality professionals today. The usage of these tools can only lead to the desired results if they are applied by experts with the necessary knowledge who would assess the features and specifications of the target area. Third-party auditors have observed in some cases quality tools are based on false practices or beliefs. In order to discover and then reduce such cases, International Automotive Task Force issued a standard in 2016 (IATF 16949) in

K. Jármai and K. Voith (Eds.): VAE 2020, LNME, pp. 123–136, 2021.
https://doi.org/10.1007/978-981-15-9529-5_11

which it defined that both internal and external auditors must prove their knowledge of the so called "Core Tools" [1].

In the present study firstly the peculiarities of the steel- and automotive industry is presented, based on technical literature. Then the quality rules of the automotive industry are outlined in detail with special focus on their aims. The study is based on IATF 16949 standards and the relevant customer specific requirements. Finally, the limits and difficulties of the application and compliance of these rules and regulations are discussed and our proposal for the solution is presented.

2 Background

2.1 The Relationship Between Steel Industry and Automotive Industry

The use of aluminium parts is spreading today, however, steel parts are still ineluctable in the automotive industry. In his study, SINGH [2] collected arguments for the use of steel parts, such as optimal quality/price ratio, decreasing specific weight, strength, formability and furthermore its energy absorbing capability, which has special importance in parts for chassis. Parts of axles, steering and chassis are manufactured from rolled steel products in many cases.

The current study focuses on rolled steel product that is the final product of the steel industry that can be the followings:

- hot and cold rolled coils, strips and sheets
- hot rolled bars, rods and sections
- bright steels

The supply of such products can be direct or via several suppliers for the automobile manufacturer (OEM) as Fig. 1 shows.

2.2 Characteristics of Rolled Steel Products

Product requirements (mostly standards) of rolled steel products define the following main attributes of the product apart from the dimensional traits:

2.2.1 Chemical Composition

Product requirements define the value (limits) of the main chemical elements that a product contains. European standards apply two specifications. One refers to the result of ladle analysis, the other to the result of product analysis. The results may deviate from one another due to segregation during the solidification process at casting. The test certificate belonging to the delivered rolled product shows the result of ladle analysis that is conducted in the steel mill. In case of customer claims only product analysis can be done as root cause analysis during which sample is taken from the delivered rolled product [3].

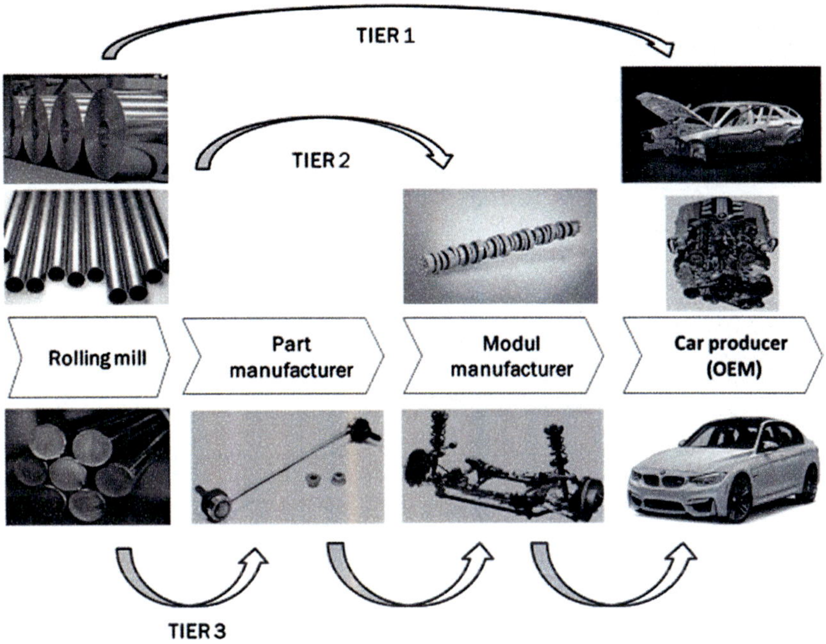

Fig. 1. Stages of steel product producers in the supply chain

2.2.2 Mechanical Properties

The mechanical properties of rolled products can generally be described by tensile strength, yield strength, elongation and impact energy. In some types of material, hardness can also be distinctive. As in the case of chemical composition, the relevant product standards also define the values and limits of tolerance in terms of mechanical properties. Due to the anisotropic nature of steel (as mentioned above), the mechanical properties may vary significantly, even in the same product. Therefore, testing of materials permits repeated sampling in case the value is outside of the specification. At this point, it is worth mentioning the mathematical model of distribution of mechanical properties. SAKAI [4] studied the distribution diagram of several materials in his research. Studying standard Japanese structural steels, he concluded that distribution of tensile strength is normal, in some cases lognormal, yield strength is rather lognormal whereas the distribution of hardness is steadily normal. When examining values of impact energy, Studying British structural steel was scrutinised. According to his results, within the temperature range of plus 20 °C and minus 20 °C the distribution of impact energy is lognormal. At around minus 60 °C the result shows 2 parametric Weibull distribution. GÜNDEL [5] was examining steel sections between 1984 and 2007 in Germany, Luxemburg and Italy and he found that the type of manufacturing process has an effect on the distribution model of the yield strength of the product. STRAUSS [6] examined 18–18 thousand samples of sheets made of structural steel in Austria and the Czech Republic. In his study, he found that the distribution of yield strength depends on the thickness of the material.

2.3 Peculiarities of Steel Industry

Products of the steel industry are manufactured by complex and expensive equipment in a closely linked process steps (see Fig. 2).

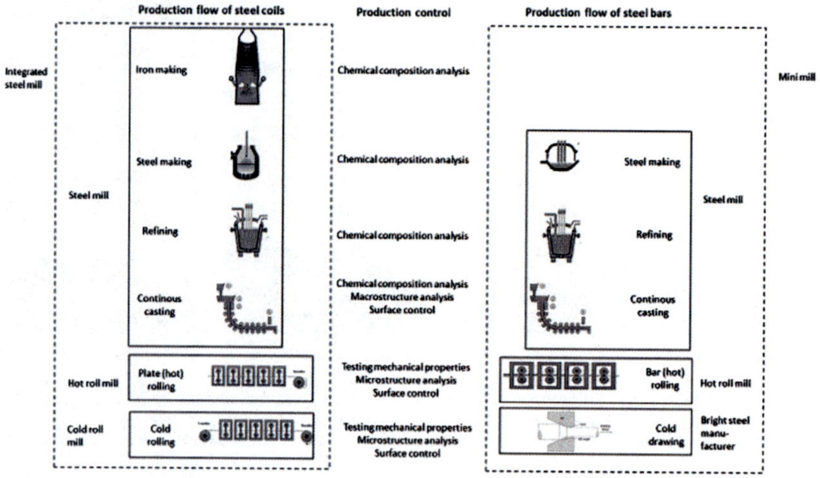

Fig. 2. Process flow and production control of steel products

The characteristic feature of manufacturing is that raw materials (e.g. iron ore or steel scrap) are rather heterogeneous and the final product manufacturing itself has a high energy demand. Therefore, the production planning of the final products (rolled products) and semi-final products (billet, bloom, slab) is based on energy saving, customer requirements and the shortest possible manufacturing- and storage time. The sequence planning of the ordered materials in the rolling mill is based on sizes and steel grades.

These are combined into rolling time schedule, on which the planning of billet or slab is based [7]. Steel works endeavour to produce the largest amount of liquid steel (so called: batch) possible. The production of smaller amount means higher cost for the steel manufacturers and cause idle times in the processing [8]. The typical batch capacity of European steel manufacturers is around 90–100 tons in the case of electric steel and around 300 tons in the case of converter steel [9]. This capacity of steel mills must be absorbed by orders. The different, extremely complex product specifications, which are so typical of the automotive industry, make the production planning so difficult because when producing liquid steel, only those products can be arranged into a batch that has very similar properties or the requirements of material standards are overlapping. Due to these circumstances in many cases not all customer requirements can be fulfilled in 100% especially in the case of small tolerance or small quantity of special steel grades.

If during the batch calculation it turns out that customer requirements cannot be fulfilled in 100%, the customer is strongly asked by the steel producer to accept the deviation [10, 11]. The environmental conditions of steel industry are a lot rougher compared to the automotive industry. The production and test equipment is exposed to high temperature, extreme load and dust. ESSIG [12] studied the lifetime of working rolls and back-up rolls in a rolling mill. He found that back-up rolls are overhauled biweekly whereas working rolls are overhauled weekly. Certainly, this period very much depends on the product portfolio and produced quantity. According to RAY [13] the working rolls in the cold roll mill have to be scraped after 15 thousand to 35 thousand tons of produced quantity.

2.4 Peculiarities of Automotive Industry

The development in the European industry has been enormous since the 1990s. The life cycle of models has been reduced from 9 years to 6 years [14]. It demands intense innovation that forced car manufacturers to divide the resources. Therefore, car manufacturers have built up immense supplier chains in the last decades so that the assembled parts could be designed and produced professionally. Due to this several of the first stage suppliers have nearly grown as big as the car manufacturers themselves. Their activities have been globalised and their strategic decision-making has been centralised. The total number of passenger-cars produced worldwide continuously increased until 2017, apart from the effect of the 2008 economic crisis [15]. To achieve this result a perfectly well-built logistic background was indispensable. Every member of the supply chain was expected to cooperate smoothly so that the car manufacturers could fulfil the customers' requirements on time in proper quality. It is only possible if every member of the chain shares the same work-principles.

The fulfilment of the Quality requirements is ensured by strict contract forced by car manufacturers. Since each member of the supply chain works with minimal buffer stock, any shortage of parts due to quality or any other problems can cause downtime and further expenses. The quality requirements regarding operation and services but not the product requirements are analysed in this study [14–18].

2.5 Quality Requirements in the Automotive Industry

2.5.1 IATF 16949 Quality Management System Standard

The International Automotive Task Force published its quality management system standard in 2016 with the objective to promote the building and maintenance of such quality management systems which improves customer satisfaction, continuously analyses and recognises the (potential) risks to operation meanwhile the standard requires to comply with the legislation. An outstanding aim of the application of the standard is to focus on avoiding faults instead of correcting them. The authors of the standard impose shared responsibility on the participants of the supply chain, driven by their intention. In practice, this means that each supplier must pro-actively cooperate with his customer, from the preparation of the offer all the way to the after-care following the delivery of the final product. In terms of its basics, the standard is relying upon ISO 9001, however, supplementary requirements are added and, in some places, the existing provisions are made more stringent. One of the most significant new elements in the standard is that it includes a compulsory integration of the customer-specific requirements in the scope of application of the specific quality management system [1]. The number of steel companies having IATF 16949 certificate is shown in Fig. 3.

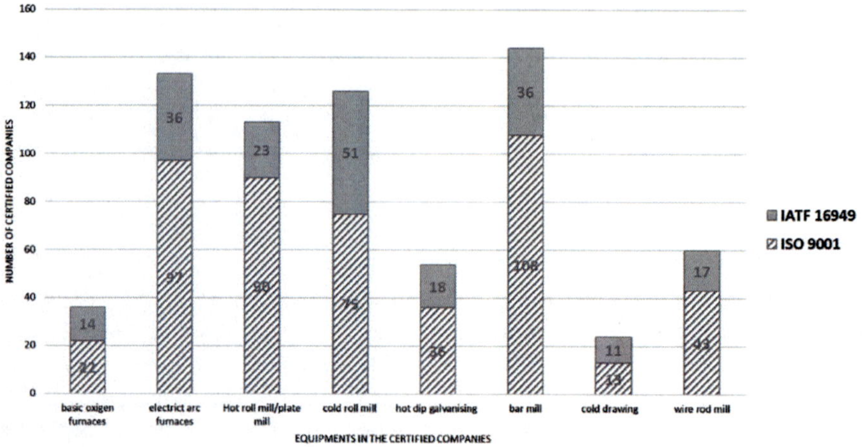

Fig. 3. Number of certified steel companies in EU 27. Source: Own results of the authors according to data from [9]

2.5.2 Customer Specific Requirements

According to the statement of IATF [19], each of the customer requirements being supplementary or interpretation requirements related to any specific chapter of the IATF 16949 standard shall be classified as customer-specific requirements. Almost all participants of the supply chain prepare a catalogue of customer-specific requirements. The most frequent names are 'Supplier Manual' or 'Quality Assurance Agreement'.

2.5.3 Automotive Industry Standard

The collective name of quality techniques most frequently used in the automotive industry is "Core Tools". The interpretation and rules on the application of such techniques are presented by the industry standards, available in two identified groups, first is published by the Automotive Industry Action Group (AIAG) and the second is by the German Association of the Automotive Industry (VDA) (see Fig. 4).

While the earlier ones have been used in the automotive industry with unchanged contents, VDA continuously collects the experience and updates its publications matching the variable demands. More recently efforts have been made to unify the two families of requirements by subjects, which was already successful in case of FMEA in 2019. Due to the extremely broad range of the subject, this article addresses now only the essence of the AIAG specifications and their applicability in the steel industry.

Advanced Product Quality Planning (APQP)

APQP is a project management tool for the definition and implementation of all activities which ensure that the product to be manufactured is going to meet the customer requirements. By means of its application, it is possible to achieve the efficient utilisation of resources, thereby the pre-calculated cost level and the project deadlines can be observed. The AIAG APQP reference manual defines five project steps. 1. Define of objectives, 2. Product design and development, 3. Process design and development,

AIAG Requiremets	Topic	VDA Requirements
APQP Handbook	Project management	RGA Handbook
FMEA Handbook	Risk analysis	FMEA Handbook
MSA Handbook	Measurement analysis	VDA 5 Handbook
SPC Handbook	Process control	VDA 4 Handbook
PPAP Handbook	Customer approval	VDA 2 Handbook

Fig. 4. Group of automotive industry standards

4. Product and process validation, 5. Evaluation of feedbacks, definition of corrective measures [20]. Quality management planning is also required by IATF 16949 standard in its section 8.3.2.1. However, the application of the AIAG APQP (or VDA RGA) rules is noted as a recommendation, also allowing the use of other alternative tools [1].

Failure Mode and Effect Analysis (FMEA)
FMEA is a multi-disciplinary analytical method with the aim to define and assess and eliminate, as far as possible the potential risks of the product design and of the manufacturing processes. By means of its application, the quality of the products improves, the failure costs decline and the customer satisfaction improves consequently [21]. The fault possibilities significant with respect to product function (DFMEA) and production process (PFMEA) are taken into account in the course of the analysis, the probability of their occurrence, the detectability of the fault and the severity of the impact of the fault are assessed. The use of FMEA as a compulsory element is specified by the standard IATF 16949 itself, in its sections 8.3.5.1 and 8.3.5.2 [1]. In his paper MIGALSKA [22] describes the successful rolling mill application of the PFMEA method. It is highlighted that the number of faulty products manufactured was decreased by the use of the tool. However, the criteria are not stated by which the severity of the various faults was assessed.

Measurement System Analysis (MSA)
MSA is a method to determine whether the measuring system applied is suitable for the decision-making. The objective is to determine the proportion of the deviation of the measuring system to the deviation of the product as well as to determine the source of deviations. By applying this method, the risk is reduced that a wrong decision is made in the course of quality control, that is, a faulty product is classified as pass or a good product is classified as fail, based upon the outcome of the assessment. Several methods are listed in the reference manual MSA for the execution of the analysis, and among these, the most widely used is the Gauge Repeatability & Reproducibility (GRR)

test [29]. Measurement System Analysis is also required by IATF 16949 standard in section 7.1.5.1.1 [1]. However, the application of the AIAG MSA rules is explicitly not requested [1]. In his paper, KLAPUT [23] presents his findings concerning the use of GRR in a steel plant. It is noted in the paper that multiple box and whisker box analyses should preferably complement the method in order to obtain a more precise result.

Statistical Process Control (SPC)

The two main tools of statistical process control include the capability study and the application of the control charts. The capability study determines the rate of deviation by which manufacturing can produce specific product property, as well as the location of the deviation in the field of tolerance. The control charts investigate whether the condition declared as capable exists also in the long-term? Its purpose by random sampling inspection is to test whether the manufactured population is within the set specifications based upon the result of the sample? [24]. The compulsory implementation of the capability study is specified by the IATF 16949 standard in its section 9.1.1.1., with the concession that in case of manufacturing methods, where the compliance of the product cannot be demonstrated through capability studies, the customer and the supplier may agree on alternative methods [1]. Papers were contributed by several authors about the application of the SPC studies in the steel industry. A comprehensive treatise is provided by MAMZIC [25] about the specific features of introducing SPC in plants of continuous production process. In his paper, it is concluded that in plants of such type, whenever possible, the continuous- measuring should be chosen instead of the random inspection (SPC). In his paper, HOMER [26] reports on his experience gained in a cold roll mill in Germany. In this plant, the thickness of the rolled sheet was measured by computer-assisted continuous data collection and the results were assessed on X-bar- range control chart. In accordance with his conclusion, cost-efficient operation could also be enhanced in addition to improving the quality of manufactured products. It is interesting to note that in general, if a control card is used to supervise production, then it is not customary to perform 100% inspection, while in this case, both were applied according to the paper. MOHANARAO et al. [27] present the successful application of a control chart applied to supervise the process parameters of a blast furnace. It is clear from the findings above that SPC can be widely applied, it may focus both on process and product parameters.

Production Part Approval Process (PPAP)

In section 8.3.4.4. the IATF 16949 standard specifies that the manufacturers are required "The organization shall establish, implement, and maintain a product and manufacturing approval process conforming to requirements defined by the customer(s)" [1]. Product approval, or 'initial sampling' as it is known in the professional dialogue, is nothing but the manufacture of a certain quantity of products by the manufacturer before the start of serial delivery, of which samples are taken and tested with respect to all the specifications of the customer. Once the results are satisfactory, the sample is submitted to the buyer together with the PPAP report for approval. The format and contents of the PPAP report are specified by the reference manual AIAG PPAP. Compulsory annexes to the minutes, among others, are the documents certifying the application of the key core tools, such as:

- FMEA
- CP
- MSA
- Results of capability studies

The annexes required by the reference manual shall always be made by the manufacturer with full contents, however, the customer may define a more restricted scope for their submission. The stages shall be defined in the so-called PPAP levels. The customer shall review the report submitted as well as the samples delivered after approval, and should the product specifications (technical drawing, standard, etc.) of the product or the manufacture of the product (equipment, tools, process flow, location of equipment) may change in the course of serial delivery by the manufacturer, or eventually the raw materials used for the manufacture of the products, then a new first sample needs to be approved. The reference manual provides that no PPAP process needs to be started for bulk products (such as sheets, coils, bars) unless otherwise instructed by the customer [28].

3 Experimental Results and Analysis

In the section of the paper up to this point, the main characteristics of the steel industry and automotive industry were summarised, which revealed the remarkable contradiction between the two industries. While the automotive industry is characterised by large volume manufacture repeated under almost identical conditions in short intervals, the steel industry features a single unique batch (charge) production of steadily growing volume. While all participants of the automotive industry make efforts at the exploration of customer requirements and exceeding them, steel industry, in contrast, if not a full batch is to be produced, is forced to make compromises due to economic considerations in some cases, with respect to the fulfilment of customer demands. The values of the process parameters in the manufacture of automotive parts can be considered steady, whereas in the steel industry the process parameters of the batch compositions in accordance with the requirements changing from case to case vary among the broad boundaries of process parameters.

According to the experience, we have gained so far, steel industry companies can satisfy the requirements of IATF 16949 in view of the points stated in the Sanctioned Interpretations and FAQ publications. The challenge is represented by the customer specific requirements due to the points described earlier, as these publications are drafted on the basis of the manufacturing characteristics of part-production companies in the first place, as they provide the backbone of the supply chain. Therefore, the possibilities of meeting the customer specific requirements are investigated below.

The customer specific requirement catalogue of 30 large companies using rolled steel products in Europe was investigated in the course of our research. These companies are active in the various areas of the industry, from the manufacture of electronic parts, through elements used in the interior design. Regardless of the industry, it could be stated that the most frequent specification in the catalogues referred to the obligatory use of some Core Tools (see Fig. 5).

It was striking that the application of Core Tools has been demanded by most of the automotive industry customers to this day mainly on the basis of AIAG requirements, therefore, the possibilities of fulfilling these requirements will be investigated in the following.

3.1 Assessing the Application of APQP Requirements

An important element of Quality Planning is the feasibility study, through which a given steel company evaluates whether they can produce the ordered steel grade or product size. Results must clearly be communicated to the customer, especially if not all customer requirements can be fulfilled. APQP is applicable in the steel industry without any problems.

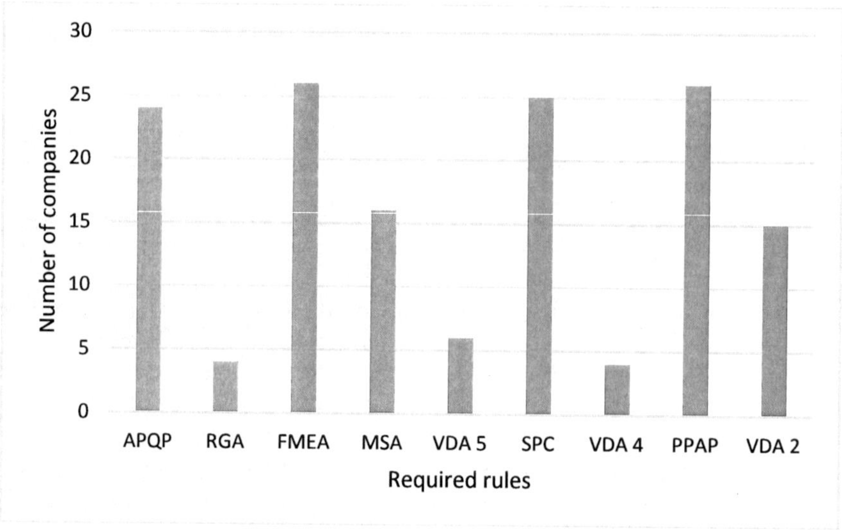

Fig. 5. Required quality tools acc. Customer specifications

3.2 Assessing the Application of FMEA Requirements

In order to assess risks, it is required to know the manufactured product's function. Without such information, it is impossible to correctly assess the possible failures as well as the severity of the risks. We should highlight the fact that rolled steel products are input materials of the part- manufacturing industry, therefore the input materials' function is not yet known at the time of the manufacture. Their function is being developed in the part production phases. As a result, it is essential that the customer informs in advance the steel producer or rolling mill about the function of the part that is intended to be manufactured, and its method of processing. If such information exchange does not happen, the steel producer or rolling mill can do one thing: it proceeds with a 'default application' that would normally apply to that given product (e.g. welding, turning, bending, etc.) such proceeds can only result a so called general PFMEA. If the given rolled product or steel grade is designed or developed by the steel company, design FMEA and process FMEA could be performed, however, such a case rarely occurs.

3.3 Assessing the Application of MSA Requirements

Earlier, we discussed the difficulties of testing steel products. For the reasons outlined in Sect. 2.3, the tool has an essential role, especially for carrying out visual and dimensional checks, including tests to determine material continuity defects. The possibility of using additional tools should be investigated.

3.4 Assessing the Application of SPC Requirements

It is one of the most important tools of automotive quality assurance, as many people expect that its application ensures that the output of mass production, which is repeated at short intervals (less than one year), always represents the same quality. Therefore, almost all customers outline the mandatory use of this tool in their customer-specific requirements. However, it is worth noting, that IATF 16949 allows SPC's replacement by other alternative tools. This is due to the characteristic of continuous process industry like the steel industry. Requirements of rolled steel products in the automotive industry are getting more complex. Standard steel products are less used by Part Designers. As steel manufacturing is a very energy intensive process, at the batch calculation several customer requests have to be combined in order to save costs. Manufacturing process parameters are selected to meet unified customer needs. As the pairing of customers for charge calculation may change from time to time, the parameters previously used at the production may vary at a given steel grade. Since the batch production process parameters are always chosen for a given alloy, it is possible that process parameters that vary within wide limits are selected for the production of the same steel grade. Today, most steel companies have a very high level of continuous data collection process, which will further improve upon digitalisation advances. Such a process applies to both the process parameters and product characteristics. From the collected data sets, capability studies and control charts are generated. Such studies and charts are used for internal data analysis in order to well customise the material mixes to satisfy different customer needs. SPCs can maximise its full potential if the same steel grade is produced in a full batch and this is repeated at short intervals over the long term. In this case, there is a sufficient amount of comparable data, although it rarely occurs. If it does occur, careful evaluation is required due to the reasons outlined in Sect. 2.3. In conclusion, several steel producers and rolling mills are putting in place continuous product monitoring systems, which is an excellent substitute for the use of SPC tools.

3.5 Assessing the Application of PPAP Requirements

In addition to SPC, PPAP is another important tool for automotive quality assurance. Its main role is to keep the main factors constant that affect product quality during manufacturing. These factors are shown in Fig. 6.

Factors	Continous Casting	Hot Rolling	Cold Rolling	Bar Drawing
Steel Grade	✓	✓	✓	✓
Input Material Source	✓	✓	✓	✓
Production Site	✓	✓	✓	✓
Production Equipments	*	*	*	*
Working Tools	X	X	X	*
✓	feasible			
X	not feasible			
*	feasible with conditions			

Fig. 6. Investigation stability of key factors during serial production

The table shows that ensuring the stability of production is the most difficult task. According to the PPAP requirements, classical initial sampling has two main boundaries in the steel industry:

- due to the fact that Tools (e.g. working rolls in the rolling mill) require regular replacement outlined in Sect. 2.3.
- the constancy of production equipment cannot always be guaranteed especially in steel factories where, for the same production phase, there is multiple equipment available (see Fig. 7).

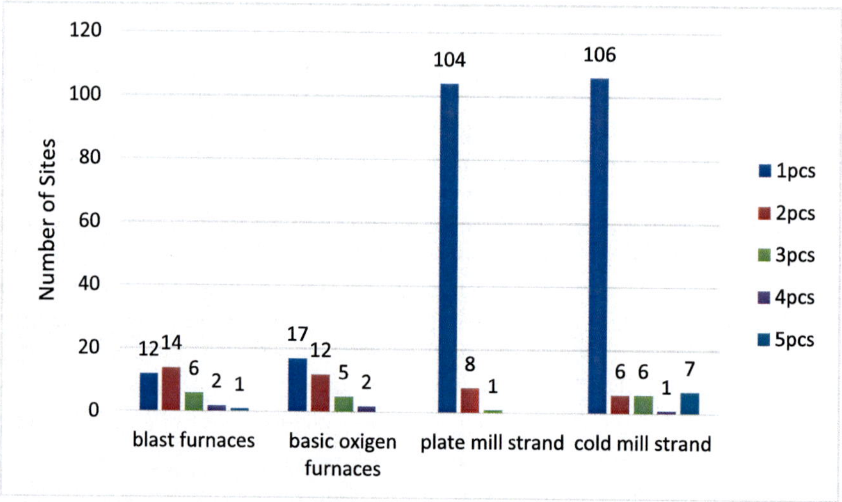

Fig. 7. Number of technical equipment by plate production sites. Source: Own results of the authors according to data from [9]

If initial sampling is required, it is essential to agree on a fix production site of the steel product supplier as well as the source of the raw material which was used as an input. On the other hand, in order to maintain a high-quality standard, it is not only enough to assess the first-tier suppliers of the material but to analyse the entire supply chain route of the product. For example, in the case of a company producing bright bars, it is not only enough to record the source of the hot-rolled bar but also the source of the CC billet. This is due to the fact that steel companies do not operate a full vertical integration, everywhere. Steel mills, hot and cold rolling mills and bright steel producers can work separately from each other, therefore their raw material purchases can be supplied from different sources, which can result a significant difference in quality.

4 Conclusions and Perspectives

In this article, we have reviewed the characteristics of the automotive and the steel industry with their main guiding principles as well as we highlighted their main contradictions. We, also, outlined the automotive industry's main quality regulations and examined their application on the steel industry. The outcome identified that 'Core Tools' not always achieve stable product quality, therefore, it is necessary to sometimes use additional tools. We have to highlight the limitations on the use of SPC and PPAP in the steel industry. As we have seen before, the use of these tools is stated in each customer's customer-specific requirements. Therefore, it is essential for supplier quality assurance professionals to adjust the application of these quality tools to local conditions and, where appropriate, work with steel professionals to develop alternative solutions that ensure stable product quality. These solutions have to be registered in a separate agreement between the two parties as an annex to the customer-specific requirements. Quality assurance professionals need to define new rules on the first sampling of steel products that are more specific than the already stated rules described in Annex F of the PPAP Manual.

References

1. International Automotive Task Force: Quality Management System Requirements for Automotive Production and Relevant Service Parts Organisations, 1st edn. IATF (2016)
2. Singh, M.K.: Application of steel in automotive industry. Int. J. Emerg. Technol. Adv. Eng. **6**(7), 246–253 (2016)
3. Szabadíts, Ö.: Acélok, Öntöttvasak, 1st edn. MSZT Szabványkiadó, Budapest (2005)
4. Sakai, T., Nakajuma, M., Tokaji, K., Hasegawa, N.: Statistical distribution patterns in mechanical and fatigue properties of metallic materials. Mater. Sci. Res. Int. **3**(2), 63–74 (1997)
5. Gündel, M.: Herleitung des Überfestigkeitsbeiwerts auf Basis statistischer Kennwerte europischer Baustähle. Stahlbau **89**(H1), 28–37 (2020)
6. Strauss, A., Kala, Z., Bergmeister, K., Hoffmann, S., Novak, D.: Technologische Eigenschaften von Stählen im europischen Vergleich. Stahlbau **75**(H1), 55–60 (2006)
7. Óvári, A.: Vaskohászati Kézikönyv, 1st edn. Műszaki Könyvkiadó, Budapest (1985)
8. Ito, K., Shioya, M., Ago, M., Kurokawa, T., Kobayashi, H., Mori, J.: Production planning and scheduling technology for steel manufacturing process. Nippon Steel & Sumitomo Metal Technical report no.121, March 2019

9. Owen, M.: Iron & Steelworks of the World, 26th edn. Fastmarkets, London (2019)
10. Isoherranen, V., Kess, P.: Production planning optimization and challenges in steel production: SSAB process review. Int. J. Model. Oper. Manag. **6**(1/2), 19–32 (2016)
11. Lu, S., Shu, H., Johansson, C., Xie, L.: A production planning model for a steel plate fabrication plant with flexible customization and manufacturing. In: Proceedings of the 19th World Congress, The International Federation of Automatic Control, Cape Town, South Africa, 24–29 August 2014 (2014)
12. Essig, R.M.: Maintenance of steel mill rolls for flat rolled products by roll turning lathe, pp. 39–42. McMaster University Hamilton, Ontario (1980)
13. Ray, A.K., Mishra, K.K., Das, G., Chaudhary, P.N.: Life of rolls in a cold rolling mill in a steel plant. Eng. Failure Anal. **7**, 55–67 (2000)
14. Svéhlik, Cs.: Kihívások és Trendek a Világ Autógyártási Struktúrájában, pp. 149–176. Nyugtmagyarországi Egyetem, Sopron (1980)
15. Kondás, B.: Autóipari Hatások a Hengerelt Acéltermékek Európai Kínálatára. Bányászati és, Kohászati Lapok **153**(1) (2020)
16. Magyar, M., Hlédik, E.: Investigation the business network of the Hungarian automotive suppliers. In: A Hatékony Marketing- EMOK 2018 Tudományos Konferencia, pp. 926–937 (2018)
17. Morauszki, K.S.: Autóipari Beszállítói Értékelési és Kiválasztási Kritériumrendszer Vizsgálata és Elemzése Minőségügyi Aspektusból. Szent Istbán Egyetem, Gödöllő (2019)
18. Hasan, M.S., Gao, J., Wasif, M., Iqbal, S.A.: An integrated decision making framework for automotive product development with the supply chain. In: 8th International Conference on Digital Enterprise Technology (2014). Procedia CIRP 25, pp. 10–18
19. International Automotive Task Force: Häufig gestellte Fragen. https://www.iatfglobaloversight.org/wp/wp-content/uploads/2019/11/IATF-16949-FAQs_Oct2019.pdf
20. Chrysler Corporation: Ford Motor Company and General Motor Corporation: Advanced Product Quality Planning and Control Plan Reference Manual, 2nd edn. AIAG, Michigan (2008)
21. Automotive Industry Action Group: Failure Methods and Effect Analysis, 1st edn. AIAG, Michigan (2019)
22. Migalska, M.: The FMEA method used in process approach to the quality management of sheet metal process. In: Metal 2003, May 2003, pp. 20–22 (2003)
23. Klaput, P., Vykidal, D., Tosenovsky, F., Halfarova, P., Plura, J.: Problems of application of measurement system analysis in metallurgical production. Matabk **55**(3), 535–537 (2016)
24. DaimlerChrysler Corporation: Ford Motor Company and General Motor Corporation: Statistical Process Control Reference Manual, 2nd edn. AIAG, Michigan (2005)
25. Mamzic, C.: Guidelines for the application of statistical process control in the continuous process industries. Meas.+Control **28**, 69–73 (1995)
26. Homer, G.R.: Factory-wide SPC in a German cold rolling mill. J. Microcomput. Appl. **14**, 313–326 (1991)
27. MohanaRao, O.R., VenkataSubbaiah, K., NarayanaRaou, K., SrinivasaRaou, T.: Application of univariate statistical process control charts for improvement of hot metal quality. Int. J. Eng. Res. Appl. **3**(3), 635–641 (2013)
28. DaimlerChrysler Corporation: Ford Motor Company and General Motor Corporation: Production Part Approval Process Reference Manual, 4th edn. AIAG, Michigan (2006)
29. Chrysler Group: Ford Motor Company and General Motor Corporation: Measurement System Analysis Reference Manual, 4th edn. AIAG, Michigan (2010)

Sustainability and Logistics

Artificial Aging of Ultra-low Viscosity Lubricant Samples on a Programmable Oil Aging Rig

András Lajos Nagy(✉) and Ibolya Zsoldos

Széchenyi István University, Egyetem tér 1, Győr 9026, Hungary
nagy.andras1@sze.hu

Abstract. An artificial lubricant aging rig was developed in order to simulate aging processes of automotive lubricants. This article presents the development of the aging apparatus and its control system as well as results of artificial aging of SAE 0W-20 grade automotive lubricant with a modified thermal cycling procedure. Friction and wear measurements on a high frequency reciprocating rig were conducted to describe the lubricating properties of the artificially aged samples. Select oil samples were analyzed through FTIR spectroscopy.

Keywords: Artificial lubricant aging · Methodology · Friction and wear

1 Introduction

Lubrication plays a vital role in complex engineering systems through ensuring long-term operation of sliding pairs. The lubricant of an internal combustion engine has to withstand extremely harsh conditions in addition to severe mechanical and thermal loads. Oxidation and nitration through air and exhaust gases [1, 2], as well as dilution and chemical degradation through fuel [3–6] and precipitated water vapor all contribute to the aging of an engine oil.

The acidity of combustion by-products from fuels with bio-derived ethanol content could possess a threat to metal surfaces in contact with the lubricant. Although field test results showed no significant increase with the introduction of acetic acid to the engine oil, laboratory corrosion tests concluded that iron and copper machine parts can be prone to corrosion when the acidity of the lubricant increases. Therefore, it can be hypothesized, that an additional increase in bio-ethanol content of gasoline fuels could lead to elevated levels of corrosion under real-life use cases [7].

The degradation processes of a lubricant under long-term load can lead to ambiguous results when regarding frictional losses and wear resistance with an aged lubricant. This can be attributed to changes in the chemical composition during aging, which can lead to alterations in the formation of boundary layers on the lubricated surfaces [8].

Another aspect of friction and wear due to oil degradation is the presence of water and humidity in the environment surrounding the lubricated contact. Elevated water content and humidity can lead to meagre wear resistance in boundary lubricated contacts with lubricants containing zinc dialkyl dithiophosphate antiwear additives. This phenomenon

K. Jármai and K. Voith (Eds.): VAE 2020, LNME, pp. 139–147, 2021.
https://doi.org/10.1007/978-981-15-9529-5_12

could be a result of depolymerisation of longer phosphate chains and the formation of a thinner reaction layer on the surface [9]. This observation is confirmed by [10] and [11].

To investigate the effect of selected alternative fuel formulations on engine oil aging a lab-scale artificial lubricant aging rig was developed [12]. Aging experiments with a fully formulated synthetic engine oil were conducted under continuous oxidation and cyclic thermal load based on CEC L-48 [13]. The produced aged samples were subjected to friction and wear testing on an SRV5 high frequency reciprocating rig from Optimol Instruments Prüftechnik with a ball-on-disc model system. Measurements were carried out according to the ISO 19291:2016 standard for the determination of tribological quantities for oils and greases [14]. Results showed a significant scattering of the coefficient of friction and average wear scar diameter values between identically aged oil samples. The aging rig was revised in order to produce more stable results through the introduction of a precise temperature control system and airflow regulation in the sample container during aging.

2 Lab-Scale Artificial Oil Aging Rig

2.1 Initial Setup of the Oil Aging Rig

The oil aging rig is based on a FALC BE-6 6-station laboratory heater equipment originally designed for Soxhlet extraction. The heating mantles accept round bottom flasks up to 250 ml and use individual power regulator circuits for each flask with a maximum power throughput of 180 W per mantle. The original design does not incorporate any means of temperature measurement of the mantle surfaces or the liquid inside the flasks. Six discrete power cut-off switches allow the selection of the maximal power throughput of each mantle in five stages independently. A target temperature range is given for each stage by the manufacturer of the apparatus.

The first setup involved six 250 ml round bottom flasks with six Drechsel bottle heads – a glass bottle head with two glass tubes running through, one connecting the head directly to the environment, the other extending into the flask – to allow for pressurized air to be lead through the contents of the flask. The inlet of each bottle head was connected with silicone tubing to a laboratory pressurized air supply system which was restricted from 7 bars to around 1,1 bar pressure. The outlets were led through a two-stage gas washing bottle setup composed of two Drechsel bottles.

In order to determine the temperature of the mantles, a K-type thermocouple was wedged between each flask and its corresponding mantle for the first test run as a rudimentary approach. The thermocouples were connected through six MAX31850 thermocouple amplifiers to an Arduino MEGA board. The controller board is equipped with a datalogger shield, which provides a real time clock and an SD-Card interface, which can handle memory cards with FAT16 and FAT32 file systems, enabling temperature log files of up to 4 GB in size. A 2,8″ TFT-LCD screen with capacitive touch sensing provides a basic GUI for user input and feedback of current temperature and state information about the running experiment. The initial setup of the aging rig implemented no closed-loop temperature control.

2.2 Early Experimental Results on a Ball-On-Disc Model System

To determine the lubricity of artificially aged oil samples, ball and disc specimens made from 100Cr6 steel were subjected to friction and wear experiments according to the ISO 19291:2016 standard. A load of 300 N is applied to the ball, which oscillates on a 1 mm long linear path with 50 Hz frequency. The test is conducted with a lubricant volume of 0.3 ml. Specimens are heated to 100 °C, which is the only deviation from the standard procedure. The duration of the test is 120 min after a 30-s running-in period. Each oil sample was tested in three consecutive test runs. Details of the experiments as well as a discussion of the results were published in [12].

A brief look at the coefficient of friction (CoF) curves (Fig. 1) suggests that the system stabilizes after 3600 s based on experimental data from measurements with the fresh oil (Ref. Oil) and the aged sample from the 4th heater (Sample 4). With this consideration in mind, the comparison of CoF values averaged between 3600 and 7230 s (Fig. 2) confirms that there is a significant difference between the aged samples. The large deviation of CoF values for Sample 2 and the difference between test runs with Sample 3 also hint to the unstable nature of the test method under inferior lubricating conditions.

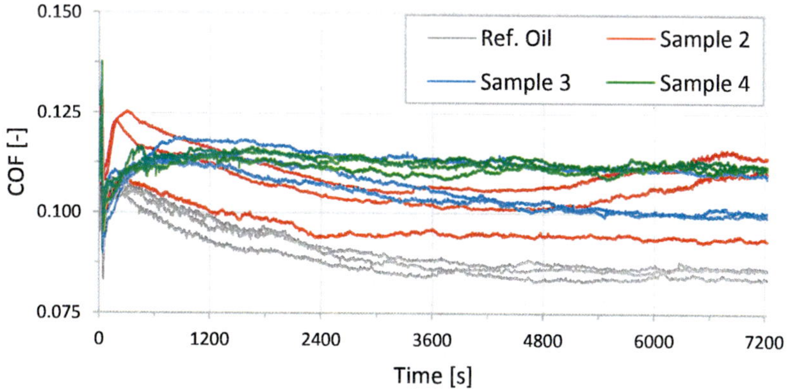

Fig. 1. Coefficient of friction curves of selected aged samples against the fresh oil (Ref. Oil)

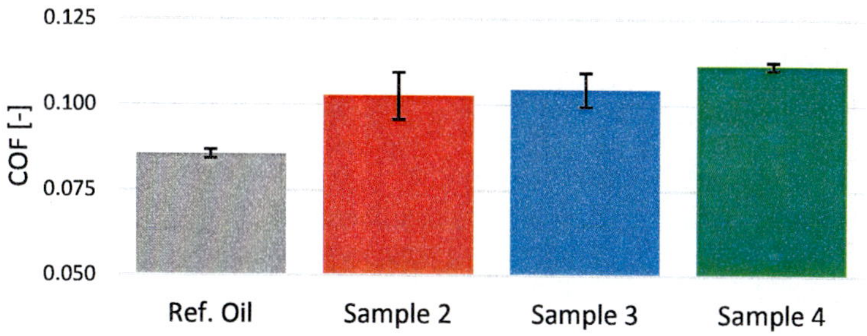

Fig. 2. Averaged coefficient of friction for each tested sample with ± 1σ deviation

The experienced variation of friction results as well as corresponding wear values between aged samples was deemed dissatisfactory for the assessment of thermo-oxidative oil degradation phenomena. Hence, a precise temperature control system was developed for this application.

2.3 Physical Setup and Control System Development

Physical System Development. For precise temperature control, the measurement of the sample temperature directly inside the fluid must be realized. The initial setup utilized a one-neck flask design, which was accommodated with a Drechsel bottle head. The external thermocouples were substituted with fine-wire thermocouples (Ø0,25 mm), which were guided through the silicone tubing on the inlet side of the Drechsel bottle heads. This allowed the measurement of temperatures inside the flasks but introduced noticeable noise into the temperature measurement due to the incompatibility of the thermocouple amplifiers with the higher resistance of the fine-wire thermocouples.

Since it was not possible to feed common size thermal sensors through the bottle heads, the one-neck round bottom flasks were substituted with three-neck flasks with identical ground glass joints. The left joints of the flasks were occupied by silicone rubber caps. The rubber caps were fitted with regular Ø1,5 mm K-type thermocouples. The middle joints were occupied by caps as well, which can be easily removed for sample taking and contaminant dosing. The Drechsel bottle heads were connected to the right joint of each flask, allowing laboratory air to circulate through the sample lubricant. These modifications enable the introduction of a precise closed-loop control system for sample temperature supervision.

In addition to the implementation of a closed-loop temperature control system, the admixing of the oxidizing agent (laboratory air) was also reevaluated. In order to control the amount of air passing through the lubricant samples, individual flow meters were introduced into the tubing preceding the inlet of each bottle head. The flow meters utilize a single-tube rotameter principle to determine the volumetric flow of the gas passing through them. The flow meters are calibrated to instrument air – compressed air for pneumatic control systems – and are able to measure flows ranging from 1 L/min to 10 L/min.

Control System Development. Temperature control is realized through six power width modulated bidirectional triode thyristor (AC dimmer) circuits. These circuits communicate via SPI (Serial Peripheral Interface) on a two-wire bus with the Arduino control board. The control software of the aging rig is implemented in the Arduino language, a programming language similar to C. The main software loop consists of three components:

- read and store values from each K-Type thermocouple amplifier,
- run PID algorithm and set output value for each AC dimmer,
- refresh temperature values on screen.

Temperature values are read out in series through SPI with a sampling rate of 1 Hz. In order to avoid measurement errors, the control system uses five consecutive measurements to calculate an average temperature value. This value is given to the PID algorithm, which calculates the control terms according to the setpoint temperature and the latest average temperature value for the corresponding thermocouple. The calculated terms are then used to determine the duty cycle, which is passed through the PWM (Power Width Modulation) output to the AC dimmer circuits. The most recent average temperature values are displayed on the LCD screen to inform the user about the current condition of the samples. The temperature values of each sample are stored on the SD-Card every 60 s. Since PID control algorithms function efficiently around their setpoint, a logical control is implemented at the start of each heating cycle. The maximum allowable PWM output is set for the active heating mantles, while the corresponding sample reaches a temperature of 20 °C below the setpoint.

A lower and upper heating boundary is also implemented due to fire safety considerations. If any of the samples are unable to reach the setpoint during a given time period, the corresponding mantle is shut off permanently. To avoid overheating, the PWM output of the corresponding mantle is set to zero in the case of a sample registering temperatures over 180 °C.

As opposed to static setpoint laboratory hotplates, the control system of the aging rig can be programmed to follow a given function for the desired sample temperature. This enables the implementation of a variable temperature aging cycle, which can mimic temperature conditions present in an internal combustion engine during mixed use conditions. A two-stage thermal cycle was utilized in order to test the reliability of the control system. A cycle time of 24 h was selected with a total aging time of 96 h (4 cycles). During the first 12 h of a cycle a temperature setpoint of 160 °C was defined (tempering phase). The heater mantles were shut off during the second stage of each cycle (polymerization phase). A second aging experiment was conducted with the modified setup to determine the performance of the improved aging rig.

3 Experimental Results with the Improved Aging Rig

3.1 Temperature Measurement and Control Accuracy

Figure 3 indicates temperature measurement values of the initial and modified setup of the aging rig.

Colored solid lines represent temperatures of the 2nd, 3rd and 4th heating mantle recorded during the first test run of the initial setup. Gray solid lines represent temperatures of all six mantles recorded during the test run of the modified setup with the closed-loop temperature control algorithm. Both measurements were set up to reach a sample temperature of 160 °C. The initial setup (v0) utilized an analogous timer mains switch in order to realize the 12-h long heating phase and to repeat it three times without manual intervention.

As opposed to the erratic temperatures of the 4th mantle and the overshoot of mantles 3 and 4 during the test run with the initial setup, the modified aging rig can reach the desired setpoint temperature in a shorter time and can precisely hold it for the desired time period. Reaching the setpoint faster under the same conditions is possible with the modified setup since the original power regulator circuit did not allow for an automatic tuning of the output power in relation to the sample temperature. The modified setup uses

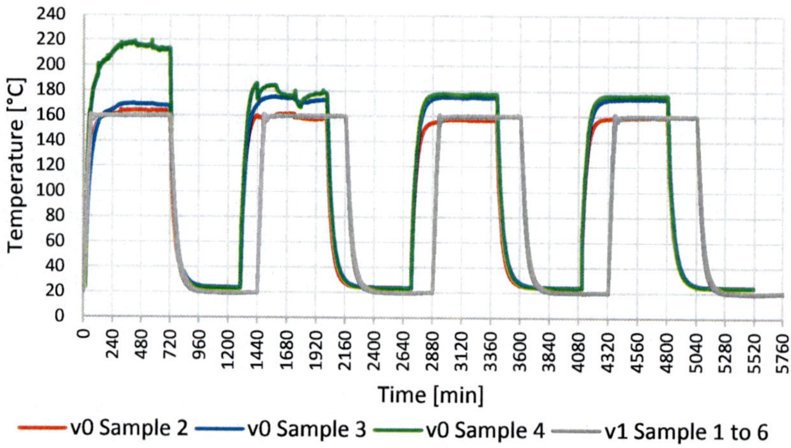

Fig. 3. Temperature over aging time of selected samples during testing of the initial (v0 – without closed-loop control) and modified (v1) setup of the artificial oil aging rig.

higher power for heating during the run-up phase and moderate power for maintaining sample temperature during the tempering phase.

3.2 Friction and Wear Experiments and Chemical Analysis

As concluded in a previous article, the standard procedure according to [14] cannot be universally applied to used or artificially aged lubricant samples. The load and lubrication conditions enacted on the sliding pair by this procedure are designed to assess the anti-wear and extreme pressure additives in a novel lubricant formulation using a neat lubricant. Milder changes in the composition of the lubricant due to oxidation and thermal degradation and the resulting changes in friction and wear in a sliding pair cannot be distinguished under the conditions of this experimental method. Hence, an experimental setup with milder conditions was introduced, under the following considerations:

- Ball and disc samples according to [14],
- Continuous oil supply instead of nearly starved lubrication condition,
- Test load reduced to 200 N,
- Test time reduced to 90 min,
- Running-in period at test start extended to 5 min,
- Transition period between running-in and test conditions altered to 5 min,
- Lubricant and disc sample temperature at 100 °C.

The alteration of the friction and wear test parameters makes a direct comparison between oil samples aged with the initial and modified setup impractical. Instead of explicit CoF and wear values, the deviation of these values is compared.

Comparing only the mean values shows the difference in the experimental setup of the friction and wear test. A shorter test with lighter load results in a reduced amount of wear. The continuous lubrication introduces a larger oil volume on the surface, which can form a thicker, more stable hydrodynamic film. This could lead to elevated frictional losses. Contrary to the expectations, the comparison of the deviations reveals that the altered experimental setup for wear testing does not have a significant effect on the stability of the friction and wear test results. Both AWSD (averaged wear scar diameter) and CoF values show a comparable amount of scattering to results with the previous setup, a significant difference between the results cannot be distinguished (Fig. 4).

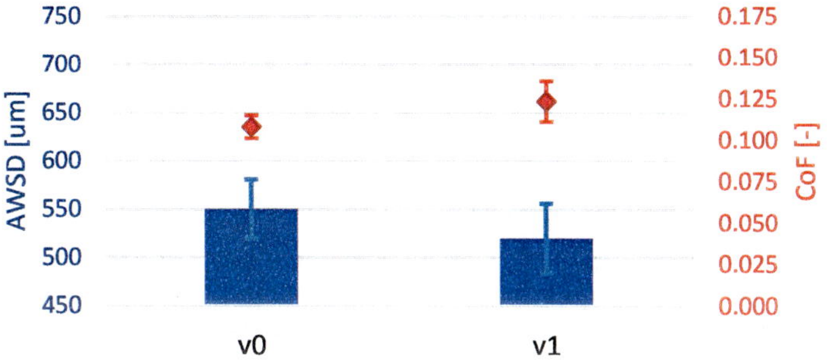

Fig. 4. Mean of Averaged Wear Scar Diameter (AWSD – bar graph) and Coefficient of Friction (CoF – rhomboid marker) values of friction and wear experiments comparing oil samples aged with the initial (v0) and modified (v1) aging rig with ± 1σ deviation.

In order to confirm the suitability of the developed aging rig for the simulation of certain aspects of engine oil aging, the samples aged with the modified aging rig were sent to chemical analysis through Fourier-transform infrared spectroscopy (FTIR). Oxidation and nitration levels were determined according to DIN 51453 [15], zinc dialkyl dithiophosphate (ZDDP) anti-wear additive, as well as phenolic and aminic antioxidant content were determined according to [16] and [17]. Averaged values of all six samples for each chemical property are shown in Table 1 with their corresponding deviations.

Table 1. Averaged values and standard deviation of chemical properties of artificially aged engine oil (v1 aging rig setup, 96-h cycle)

	Oxidation [A/cm]	Nitration [A/cm]	ZDDP [%]	Phenol. AO [%]	Amin. AO [%]
Ref. Oil	0	0	100	100	100
Avg	4,1	0,6	10,8	82,2	83,0
Dev (abs)	0,1	< 0,1	0,2	1,1	0,6
Dev (%)	2,7	4,8	1,6	1,4	0,8

The chemical analysis shows a mild oil degradation except for the ZDDP anti-wear additive content that shows a level of depletion comparable to engine oils with 7000 km mileage or higher [16]. This phenomenon could also be a result of the interaction between the metal sleeve of the thermocouples and the zinc-phosphate additive, which should be investigated and taken into account before further experimentation. The significantly decreased ZDDP content compared to the base levels could also be the cause of high deviation in AWSD values experienced with samples produced through the modified aging experiment.

In general, the aging experiment with the revised aging rig configuration produces highly comparable sample degradation in terms of oxidation, nitration and additive depletion. The deviation of values is under 2,5% for additive traces and under 5% for oxidation and nitration values. The presented trends are in agreement with the values published in [17].

4 Conclusion

An artificial oil aging rig was developed based on a conventional Soxhlet extractor. The presented design is capable of simulating dynamic conditions during the aging process through an Arduino-based closed-loop temperature controller and logger device. Samples aged with the modified device were compared to samples degraded with the original setup through friction and wear experiments. The comparison revealed the unsuitability of the ball-on-disc model system for the determination of tribological quantities of artificially aged lubricants and to reveal subtle changes in oil composition. In order to compare samples from the modified aging experiment with each other, FTIR spectroscopy measurements were carried out. The following conclusions can be drawn based on the presented data:

- the modified aging rig is capable of a precise temperature control around the desired setpoint,
- the modified aging rig is capable of producing samples with comparable chemical properties during the same aging procedure,
- the used aging procedure results in a mild-to-moderate degradation of the lubricant, mainly affecting the ZDDP anti-wear additive content of the investigated lubricant.

Acknowledgements. The authors would like to thank Dr. Nicole Dörr and Mr. Ádám Agócs of the Austrian Excellence Centre for Tribology (AC2T research GmbH) for their support with the chemical analysis of the artificially aged lubricant samples.

References

1. Johnson, M.D.: Laboratory assessment of the oxidation and wear performance capabilities of low phosphorus engine oils. In: Spring Fuels & Lubricants Meeting & Exhibition, SAE International (2001)
2. Lacey, P.I.: Effect of oil drain interval on crankcase lubricant quality. In: SAE Powertrain & Fluid Systems Conference & Exhibition, SAE International (2003)
3. Costa, H.L., Spikes, H.: Effects of ethanol contamination on friction and elastohydrodynamic film thickness of engine oils. Tribol. Trans. **58**(1), 158–168 (2015)

4. Khuong, L.S.: Effect of gasoline–bioethanol blends on the properties and lubrication characteristics of commercial engine oil. RSC Adv. **7**(25), 15005–15019 (2017)
5. Shayler, P.J.: Fuel transport to the crankcase, oil dilution and HC return with breather flow during the cold operation of a SI engine. In: SAE 2000 World Congress, SAE International (2000)
6. Artmann, C.: A new measurement technique for online oil dilution measurement. SAE Int. J. Fuels Lubr. **6**(3), 594–604 (2013)
7. Besser, C.: Impact of engine oil degradation on wear and corrosion caused by acetic acid evaluated by chassis dynamometer bench tests. Wear **317**(1–2), 64–76 (2014)
8. Uy, D.: Characterization of anti-wear films formed from fresh and aged engine oils. Wear **263**(7–12), 1165–1174 (2007)
9. Cen, H.: Effect of water on ZDDP anti-wear performance and related tribochemistry in lubricated steel/steel pure sliding contacts. Tribol. Int. **56**, 47–57 (2012)
10. Cen, H.: Effect of lubricant ageing on lubricants' physical and chemical properties and tribological performance; part i: effect of lubricant chemistry. Ind. Lubr. Tribol. **70**(2), 385–392 (2018)
11. Cen, H.: Effect of ageing on lubricants' physical and chemical properties and tribological performance: part II: effect of water contamination on lubricant. Ind. Lubr. Tribol. **71**(1), 48–53 (2019)
12. Nagy, A.L.: Development of an artificial aging process for automotive lubricants. In: Bihari, E. (eds.) SPRING WIND 2019, vol. 3, pp. 98–104. DOSZ (2019)
13. CEC L-48: Oxidation Stability of Lubricating Oils used in Automotive Transmissions by Artificial Ageing. CEC (1995)
14. ISO 19291:2016: Lubricants - Determination of tribological quantities for oils and greases - Tribological test in the translatory oscillation apparatus. ISO (2016)
15. DIN 51453: Testing of lubricants - Determination of oxidation and nitration of used motor oils - Infrared spectrometric method, Deutsches Institut fur Normung E.V. (2004)
16. Dörr, N.: Engine oils in the field: a comprehensive chemical assessment of engine oil degradation in a passenger car. Tribol. Lett. **67**, 68–89 (2019)
17. Besser, C.: Generation of engine oils with defined degree of degradation by means of a large scale artificial alteration method. Tribol. Int. **132**, 39–49 (2019)

Smart Contracts in the Automotive Industry

Olivér Hornyák[1](✉) and George Farid Alkhoury[2]

[1] Department of Information Engineering, University of Miskolc, Miskolc, Hungary
oliver.hornyak@uni-miskolc.hu

[2] Computer Science Engineering Master, University of Miskolc, Miskolc, Hungary

Abstract. The automotive industry has undoubtedly significantly changed our societies and daily life. It is among the most advanced, complicated, and innovative industries. The automotive industry is an essential driving factor of many other advanced industries, it requires the contribution of many other technologies like advanced manufacturing systems, cyber-physical systems, and robotics. Blockchain technology can be highly beneficial for the automotive industry, to enhance its data security, integrity and reliability, tracking and location management, enhanced connectivity, mobility-as-a-service, tamper prevention, and fraud detection. One of the emerging blockchain capabilities is smart contract enforcement and autonomy. In this article, a basic overview of smart contracts is given, early history, definition and concepts, relation with the blockchain, listing its benefits, data sources, design, and describing its importance, finally we highlighted some serious, outstanding and worth to mention steps toward activation and enabling the smart contracts and blockchain roles in the automotive industry specifically and the industry in general.

Keywords: Blockchain · Smart contracts · Automotive industry

1 Introduction

The modern automotive industry along with its cutting edge technologies is one of the most technologically advanced industries in today's world. With innovations ranging from self-driving smart cars that depend on the Internet of Things (IoT) and Artificial Intelligence (AI), hybrid engines exploiting both clean electric motor side by side with the conventional fuel-powered engine, vehicular ad hoc networks and so on. These have attracted much attention recently due to the increasing demand for traffic safety and comfortable vehicle driving experience. Similar to all other modern industries, the automotive industry is vulnerable to low efficiency at the operational level, security gaps, and problems that lead to critical cyber-attacks. Such malicious attacks could cause major losses and critical incidents which have negative effects on the cars and car parts prices and costs of maintenance and other services. It also affects automotive industry stockholders and participants in the vehicle lifecycle like car owners, staff of manufacturing facilities and enterprises, service agents, and so forth [1].

The Blockchain by definition is a decentralized database that enables distributed public ledgers that hold immutable data in a secure and encrypted way. It ensures the

K. Jármai and K. Voith (Eds.): VAE 2020, LNME, pp. 148–157, 2021.
https://doi.org/10.1007/978-981-15-9529-5_13

trustworthiness of transactions or smart contracts in such a way that can never be altered, it is also called the distributed ledger technology (DLT) [33]. Blockchain technology supports cryptocurrencies such as bitcoin but it might have a great contribution to the automotive industry that can revolutionize products, services, and processes. It provides an innovative approach and smooth distributed platform where different information and data about patents, insurance, ownership, maintenance services, and others could be securely stored and transactions could be executed accurately and reliably [6].

Blockchain technology has many other potential use cases and applications across finance, business government, and other industries. For example, in banking and finance Blockchain provides a way to securely and efficiently executing the international payments, improve the capital market, streamline trade finance, supports secure peer-to-peer transactions, and combating money laundering. At the business level, Blockchain will be an essential part of supply chain management because its immutable ledger suits tasks such as real-time tracking of goods. Blockchain will support also Healthcare system and real-estate, media, and energy domains. Blockchain could simplify and secure Government individuals' recordkeeping such as birth and death dates, marital status, or property transfers, it will have a basic role in identity management, voting and taxes also [34]. Finally, at the level of other industries Blockchain can provide multiple security benefits like Tamper-Proof Data, No single points of failures, Privacy, Identity, and Access Management, Information Security, and Smart contracts enforcement and autonomy, see Fig. 1 [1]. In this article, we are going to overview the concept of Smart Contracts and some of their promising applications in the automotive industry.

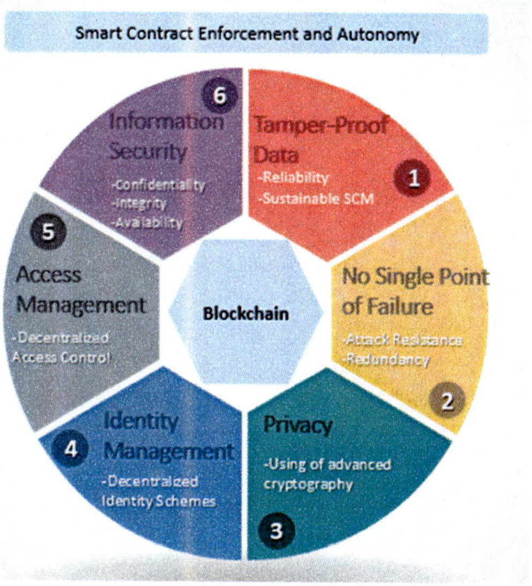

Fig. 1. Blockchain key capabilities for cybersecurity [1]

2 Smart Contracts

Contract law's history is back to the ancient Greek and Roman times. Plato [11] described agreement categories as they exist today. *Contractus litteris* (literal contracts) were part of the Roman law. A smart contract – also known as a digital contract - inherits the terms and conditions of the conventional contract. The smart contract codes the business agreement, which is then verified and signed by the parties and then uploaded to the network. A smart contract is a piece of distributed code that is stored on the blockchain and that runs independently when specific conditions and rules are satisfied. At the time of writing this paper, there are 102 million unique Ethereum addresses [15] holding more than 50 million smart contracts [16].

We can divide smart contracts in terms of legality into two different types: strong and weak [17]. Weak smart contracts can be changed without any modification cost. Strong smart contracts can not be modified, or the modification cost is so high that modification is not feasible. Traditional law enforcement can not influence the execution of smart contracts (either by involved third parties or by a judge) in the case of strong smart contracts.

A smart contract provides the capability to update, so they need mechanisms to add updates and modifications that may be required legally. For example, public blockchain data or an API (Application Programming Interface) can be used to access the latest legal terms of the smart contract. The contracting parties may be asked to update the source code of smart contracts by themselves, without involving any third-parties. Smart contract terms and conditions may be defined as unmodifiable to prevent changing the contract by one or both parties.

Smart contract as a concept is not intrinsically related to blockchain; it is often used interchangeably with a wide range of expressions, from "Digital Contract" to "Smart Legal Contract" or "Smart Contract Code". But in general, all definitions of smart contracts usually include some type of automated, self-executing transaction [2].

3 Benefits of Smart Contracts

A smart contract could be defined as a computer program that respects certain legal conditions and terms to control digital or physical modules. A smart contract does not depend on a state for its enforcement although it usually avoids issues related to human ambiguity. Therefore, it is a method to maintain performance on the deals of all involved parties [1]. A smart contract has a lot of benefits in comparison with traditional contracts:

- The digital nature of the smart contract guarantees that there is a final version of the contract which is written in detail and it will be executed accurately by computers once needed.
- A smart contract is paper-free, you do not have to print them. It is deployed to the blockchain and executed exclusively according to its encoded instructions and terms. unlike the paper contract, which could be destroyed or damaged or may exist in multiple versions and copies.
- By translating the contract into a computer-executable code, a smart contract is accurate. All terms and conditions are explicit.

- A smart contract is clear, there is no misinterpretation.
- Smart contracts are usually stored on the blockchain, which records all the necessary details. No need for extra storage or backup.
- Smart contracts provide trust for the involved parties.
- Smart contracts are executed quickly, no delay in their processing.
- Smart contracts are secure and transparent.

Blockchain-based smart contracts have the ability to affect a wide range of industries like the automotive industry. Blockchain-based smart contracts can be used to create new industries and reach new markets besides that they used to automate the existing processes. The blockchain guarantees an ideal platform for smart contracts by providing a digital, secure, and resilient environment for coding "if-then" statements and value transition and maintaining detailed and immutable transaction history.

4 Data Sources of Smart Contract in the Automotive Industry

Despite that smart contracts are stored on the blockchain, they receive data from external services called oracles that collect information from different sources [1]. Oracles are one or more external digital agents or sources trusted by the blockchain involved parties that monitor the external parameters defined in the smart contract and approve the execution of the contract if these parameters are satisfied. For example, an oracle can monitor the status of a car to determine its mileage, condition, usage statistics, etc. and changed its status on the blockchain. Then, the change in the status of the car could be detected by the smart contract, which can in turn triggers the payment activities related to the billing of its usage.

There are multiple types of oracles [1] depending on the collected data type and how they interact with their sources. The digital world needs to know about the physical world. The following categories can be defined:

- **Software oracles**: they usually handle information from an online system such as weather info, product and service prices, etc. The software oracle obtains the required information automatically by API calls and inserts the data into the smart contract.
- **Hardware oracles**: they are designed to collect data directly from the physical world sources, Internet of Things devices. For instance, movement sensors, GPS coordinates. The major challenging attribute for hardware oracles is to maintain data security and privacy while providing accurate and reliable data.
- **Consensus-based oracles**: some pieces of information can be unreliable, uncertain, tentative. Smart contracts may require reaching consensus on them before getting the data.

By the direction of data flow, we can distinguish:

- **Inbound oracles**: inbound oracles usually push information from the external world sources into the blockchain. For example, an automatic event triggers if some predefined property reaches a certain predetermined value.

- **Outbound oracles**: they allow smart contracts to send data to the blockchain network. For example, a smart lock in the physical world which receives the required payment on its blockchain address and unlocks automatically (Fig. 2).

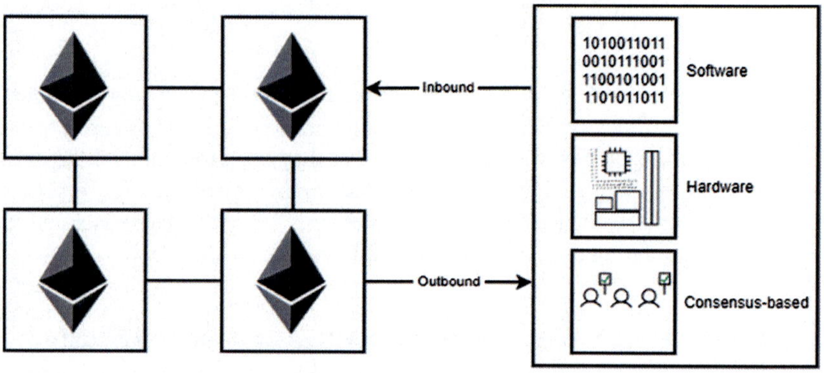

Fig. 2. Data sources of a blockchain

Practically, oracles are responsible for the correct execution of a smart contract since the insertion of wrong information will lead to an action that may not be rolled back easily (certain money transfers for example). So, because of this critical problem, many companies provide oracles with collected data verification features [1].

Recently, some blockchain-based applications have become more sophisticated and involved the use of oracles concepts, smart contracts and DAO (Decentralized Autonomous Organization), which is a distributed application implemented to make the interaction between multiple parties, humans or machines, effectively possible. This interaction is controlled by the blockchain application that is in turn controlled by a set of incorruptible and immutable rules defined in its source code [1].

5 Smart Contract Design

The smart contract design consists of two parts: the operational part and the non-operational part. The operational part usually simple and it is straightforward to encode. It consists of the automated part. Non-operational part however can be very complex: it may analyze various risk areas of the contract. In other words, the operational part describes the successful completion of the smart contract, the non-operational part is devoted to an imperfect completion or partial completion. [23] provides templates for smart contract design and urges standardization. [24] focuses on common mistakes and security considerations of smart contracts design. [25] investigated and collected some common smart contract design pattern, such as:

- Tokenization – for distributing transferable, countable goods represented by tokens,
- Authentication – to restrict unauthorized access,

- Safe transfer – to implement the business logic over the blockchain,
- Using oracles – as discussed earlier,
- Time constraint– to define the interval while the contact is alive,
- Termination – to deal with unresponsive or timed out contracts,
- Randomness – which is not a trivial thing as all nodes should obtain the same random number,

There are some other design patterns published, [26] lists some more patterns like:

- Pull payment – to deal with micropayment, bidirectional payment channels,
- State machine – to support finite-state automation, in which the system is in exactly one of a finite number of states,
- Commit and reveal – to hide information on the blockchain, e.g. when there is a competition. First, a hash of the data is committed and then everyone sends the real data (reveal),
- Ownership – to retain sensitive and maintenance operations to be executed by the owner of the contract only,
- Access restriction – to allow the execution of the contract when some restrictive conditions present,
- Data segregation – to separate contract logic from its data,
- Balance limit – to avoid the execution of unaffordable transactions,

A disadvantage of blockchain-based smart contracts is that they are public. [27] proposes splitting the smart contract into a public part containing the transfer between the parties and a private part containing the contact details. That would result in an on-chain and off-chain part of the contract.

6 Steps Toward Envisioned Automotive Industry Blockchain and Smart Contracts

According to [7] "Blockchains can be an extremely powerful technology for automotive organizations". Whether used to help secure information on car usage against hacking, enable financial payments at toll booths, provide traceability of parts in the supply chain or administer smart contracts for the various participants in finished vehicle distribution. The author mentioned that automotive companies can start by carefully evaluating where blockchains can provide the greatest gains. Three points were recommended for evaluation: manufacturer business network, network-wide standards [20, 21], and new revenue models.

The blockchain is considered as the latest buzzword in the automotive industry in [4]. It is now being widely described as a potential solution to some of the biggest challenges facing the automotive industry – secure data transactions, component provenance and location tracking – and the gateway to new products and services. A group of 30 major automotive giants launched the Mobility Open Blockchain Initiative (MOBI) [8], a research group aimed to accelerate the adoption of blockchain and the development of a digital mobility ecosystem. The new envisioned blockchain enables the smart contracts without the involvement of third parties when two parties reach an agreement (Smart Contract) and the execution of this agreement is contingent on a condition, for instance, time or account balance, the conditions will be validated based on the code included in the smart contract and when all conditions and terms are met the smart contract is executed the goods transferred between the parties automatically based on the agreement.

Finance, Securitization, and Smart Contracts (FSSC) is a working group under MOBI that aims to assess the potential value proposition of blockchain and working on interoperability standards that mobility stockholders ecosystem "including OEMs, auto financiers, dealerships, and others" [9].

Some focus areas of MOBI are as follows:

- digital vehicle identity,
- vehicle history and data tracking,
- supply chain tracking, transparency, and efficiency,
- autonomous machine and vehicle payments,
- secure mobile commerce,
- data markets for autonomous and human driving,
- car sharing and ride-hailing,
- usage-based mobility pricing, and payments for vehicles, insurance, energy, congestion, pollution, and infrastructure.

Many automotive executives believe that blockchain will be a disruptive force in the following years [5]. The mobility sector is changing from a product business to a mobility-as-service model. Successful services always focus on the user and new solutions. They must address both passenger and freight transport. The future of mobility is digital, and that's why today's automotive manufacturers need to completely rethink their business models to be successful in this data-driven era. Smart contracts will be used for automation and control purposes with intelligent grids and on distributed ledger make sense as they are resistant to manipulation. The implementation of smart contracts will enable the platform to act automatically in the vehicle registration ledger. The majority of the car registrations can be fully automated and this in turn shows the processes optimization potential in public administration. [6] suggest smart contracts as a record-keeping service: a secure, auditable, and immutable records of physical and digital assets and transfer of value service which issues new or exchanged of ownership without intermediaries.

Smart Contracts shall be extensively used to make the contract and payment processes more efficient and secure, overcoming the barriers of the transfer of vehicle data. It was chosen due to simplification and existing acceptance. [18] classifies automotive lifecycle into seven phases as follows:

1. Regulator Phase
2. Manufacturer Phase
3. Dealer Phase
4. Leading company Phase
5. User Phase
6. Maintenance Phase
7. Recycle Phase

The manufacturing phase includes supply chain management applications like [19, 22, 23, 30], which has big potential on its own. Recycle phase applications are listed in [28]. Swatchhcoin is a startup, it is a decentralized waste management system based on blockchain and smart contracts [31].

7 Conclusions

The development and migration to a data and value-driven world are boosted by the evolution speed of the technological Internet-enabled global world, increasing business competition, and the challenges of future mobility. In this complex environment, the use of smart contracts and blockchain can give to the automotive industry the required advanced platform which can distribute trusted and cyber-resilient information that challenges and contradicts the current non-collaborative organizational structures. In this paper we have overviewed the importance of blockchain and smart contracts as one of the most important blockchain key capabilities for cybersecurity in automotive industries, we have defined the smart contracts and listed its data sources overviewing the smart contracts design methodology and finally, we reviewed some real steps and opinions about the envisioned automotive industry blockchain and smart contracts.

Acknowledgements. This research was supported by the European Union and the Hungarian State, co-financed by the European Regional Development Fund in the framework of the GINOP-2.3.4-15-2016-00004 project, aimed to promote the cooperation between the higher education and the industry.

References

1. Fraga-Lamas, P., Fernández-Caramés, T.M.: A review on blockchain technologies for an advanced and cyber-resilient automotive industry. IEEE Access **7**, 45201–45218 (2019)
2. Cohn, A., West, T., Parker, C.: Smart after all: blockchain, smart contracts, parametric insurance, and smart energy grids. Georgetown Law Technol. Rev. **1**(2), 273–304 (2017)
3. Dolgui, A., Ivanov, D., Potryasaev, S., Sokolov, B., Ivanova, M., Werner, F.: Blockchain-oriented dynamic modelling of smart contract design and execution in the supply chain. Int. J. Prod. Res. **58**(7), 2184–2199 (2020)
4. Boyle, B., Brenner, A., Steger, S., Janssens, S., Rasamoela, M.: The Blockchain Bandwagon. Roland Berger Publications, Germany (2018)
5. Yannick, T., Schuetz, K., Preikschat, L., Ehrenhart, M.E., Moritz, M., Sebastian, B., Benjamin, S.: The Automotive Sector and Blockchain. A joint white paper by MHP and RIDDLE&CODE (2019)
6. Lawson, T.: Accelerating Technology Disruption in the Automotive Market. Blockchain in the Automotive Industry. Deloitte Publications, United Kingdom (2018)
7. Jones, M., Wollschlaeger, D., Stanley, B., Romeo, S., Ayres, G., Suzuki, N.: Daring to be First, How Auto Pioneers are Taking the Plunge into Blockchain. IBM Institute for Business Value, United States of America (2018)
8. Simon, K.: MOBI-the Mobility Open Blockchain Initiative, Launched. https://www.indrastra.com/2018/05/MOBI-Launched-004-05-2018-0002.html. Accessed 23 June 2020

9. MOBI Working group Homepage. https://dlt.mobi/mobi-working-groups/. Accessed 19 June 2020
10. Voshmgir, S.: Token Economy How Blockchains and Smart Contracts Revolutionize the Economy. Blockchain Hub, Berlin (2019)
11. Stalley, R.F.: An Introduction to Plato's Laws. Hackett Publishing, Indianapolis (1983)
12. Idelberger, F., Governatori, G., Riveret, R., Sartor, G.: Evaluation of logic-based smart contracts for blockchain systems. In: Alferes, J., Bertossi, L., Governatori, G., Fodor, P., Roman, D. (eds.) International Symposium on Rules and Rule Markup Languages for the Semantic Web, RuleML 2016: Rule Technologies. Research, Tools, and Applications, vol. 9718, pp. 167–183. Springer, Cham (2016)
13. Kosba, A., Miller, A., Shi, E., Wen, Z., Papamanthou, C.: Hawk: the blockchain model of cryptography and privacy-preserving smart contracts. In: 2016 IEEE Symposium on Security and Privacy (SP), pp. 839–858. San Jose, CA, USA (2016)
14. Christidis, K., Devetsikiotis, M.: Blockchains and smart contracts for the internet of things. IEEE Access **4**, 2292–2303 (2016)
15. Etherscan. https://etherscan.io/chart/address. Accessed 24 June 2020
16. Hu, Y., Liyanage, M., Mansoor, A., Thilakarathna, K., Jourjon, G., Seneviratne, A.: Blockchain-based Smart Contracts-Applications and Challenges. arXiv preprint arXiv:1810.04699 (2018)
17. Raskin, M.: The law and legality of smart contracts. SSRN (2016)
18. Sharma, P.K., Kumar, N., Park, J.H.: Blockchain-based distributed framework for automotive industry in a smart city. IEEE Trans. Ind. Inform. **15**(7), 4197–4205 (2019)
19. Supranee, S., Rotchanakitumnuai, S.: The acceptance of the application of blockchain technology in the supply chain process of the Thai automotive industry. In: Proceedings of the International Conference on Electronic Business (ICEB), Vol. 2017, pp. 252–257. ICEB, Dubai, UAE (2017)
20. Singh, M., Kim, S.: Blockchain based intelligent vehicle data sharing framework. arXiv preprint arXiv:1708.09721 (2017)
21. Ortega, V., Bouchmal, F., Monserrat, J.F.: Trusted 5G vehicular networks: blockchains and content-centric networking. IEEE Veh. Technol. Mag. **13**(2), 121–127 (2018)
22. Dujak, D., Domagoj, S.: Blockchain Applications in Supply Chain. Springer, Cham (2019)
23. Mondragon, A.E.C., Mondragon, C.E.C., Coronado, E.S.: Exploring the applicability of blockchain technology to enhance manufacturing supply chains in the composite materials industry. In: 2018 IEEE International Conference on Applied System Invention (ICASI), pp. 1300–1303. Chiba, Japan (2018)
24. Clack, C.D., Bakshi, V.A., Braine, L.: Smart contract templates: foundations, design landscape and research directions. arXiv preprint arXiv:1608.00771 (2016)
25. Delmolino, K., Arnett, M., Kosba, A., Miller, A., Shi, E.: Step by step towards creating a safe smart contract: lessons and insights from a cryptocurrency lab. International Conference on Financial Cryptography and Data Security. FC 2016: Financial Cryptography and Data Security, pp. 79–94. Springer, Berlin, Heidelberg (2016)
26. Bartoletti, M., Pompianu, L.: An empirical analysis of smart contracts: platforms, applications, and design patterns. International Conference on Financial Cryptography and Data Security. FC 2017: Financial Cryptography and Data Security, pp. 494–509. Springer, Cham (2017)
27. Wöhrer, M., Zdun, U.: Design patterns for smart contracts in the ethereum ecosystem. 2018 IEEE International Conference on Internet of Things (iThings) and IEEE Green Computing and Communications (GreenCom) and IEEE Cyber. Physical and Social Computing (CPSCom) and IEEE Smart Data (SmartData), pp. 1513–1520. Halifax, NS, Canada (2018)
28. Li, C., Palanisamy, B., Xu, R.: Scalable and privacy-preserving design of on/off-chain smart contracts. In: 2019 IEEE 35th International Conference on Data Engineering Workshops (ICDEW), pp. 7–12. Macao (2019)

29. Gupta, N., Bedi, P.: E-waste management using blockchain based smart contracts. 2018 International Conference on Advances in Computing. Communications and Informatics (ICACCI), pp. 915–921. Bangalore, India (2018)
30. Chidepatil, A., Bindra, P., Kulkarni, D., Qazi, M., Kshirsagar, M., Sankaran, K.: From trash to cash: how blockchain and multi-sensor-driven artificial intelligence can transform circular economy of plastic waste. Adm. Sci. **10**(2), 23 (2020)
31. Wang, S., Qu, X.: Blockchain applications in shipping, transportation, logistics, and supply chain. Smart Transportation Systems, pp. 225–231. Springer, Singapore (2019)
32. Swatchhcoin whitepaper. http://swachhcoin.com/overview.pdf. Accessed 21 June 2020
33. Blockchain-internet security. https://www.internetsociety.org/issues/blockchain/. Accessed 02 July 2020
34. The growing list of applications and use cases of blockchain technology in business and life. https://www.businessinsider.com/blockchain-technology-applications-use-cases. Accessed 03 July 2020

A Review on the Differences Between Particle Emission, Filtration and Regeneration of Particulate Filters of Diesel and Gasoline Engines

Péter Nagy(✉) and Ibolya Zsoldos

Széchenyi István University, Győr 9026, Hungary
nagy.peter@sze.hu

Abstract. The exhaust aftertreatment systems of diesel passenger cars have been equipped with diesel particulate filters since 2009 when the EURO 5 standard was introduced. This device is a trap for particulates, which originate from internal combustion engines in soot form. These particles must be filtered because they are able to enter the bloodstream through the human respiratory system and can cause an adverse effect on health. Investigations conducted after the 2016 diesel scandal concluded that gasoline engines with direct injection systems commonly in use today, also produce a high number of particles, therefore they must be equipped with gasoline particulate filters. The current EURO 6d TEMP regulation introduced in September 2017 specifies that the NEDC-regulation has to be replaced gradually with the WLTP investigation and the RDE realistic vehicle tests, thus making the real results of the measurements impossible to avoid. Also the particulate emission per one hundred kilometers will be observable in case of direct injection gasoline engines. This paper presents the formation of soot particles during the combustion processes of diesel and gasoline engines, the deposition of particles in particulate filters and explores and analyses the differences between the regeneration processes due to the deposition of particles.

Keywords: Particle Emission · Particulate Filter · Deposition · Regeneration Process · GPF · DFP

Definitions/abbreviations: NEDC - New European Driving Cycle · WLTP - Worldwide Harmonized Light-Duty Vehicles Test Procedure · RDE - Real Driving Emissions · PM - Particle Mass · WHO - World Health Organization · WLTC - Worldwide harmonized Light vehicles Test Cycles · PEMS - Portable Emissions Measurement System · PN - Particle Number · ECU - Electronic Control Unit · PFI - Port Fuel Injection · DI - Direct Injection. · EGR - Exhaust Gas Recirculation · AFR - Engine Fuel Ration. · GDI - Gasoline Direct Injection · GPF - Gasoline Particulate Filter · DPF - Diesel Particulate Filter · TWC - Three-way catalysts · EDX - Energy Dispersive X-Ray Analysis · SEM - Scanning Electron Microscopy

K. Jármai and K. Voith (Eds.): VAE 2020, LNME, pp. 158–173, 2021.
https://doi.org/10.1007/978-981-15-9529-5_14

1 The Effect of Particles on Human Health

There are several types of pollutants from the incomplete combustion of an internal combustion engine. Carbon dioxide, which is a greenhouse gas, or nitrogen oxides, which can cause acid rain. In addition to their impact on the environment, they also have a negative impact on the human body. Another by-product is soot particles, which need to be measured since Carbon-based soot particles are relatively small in size with a large surface and with a high potential for adherence. So it is not necessarily the soot particles themselves that are dangerous, but the adhering substances may have adverse health effects. These include the risk of developing asthma and cancer [1–3]. Table 1 shows, how the different size of inhaled particles pass through the body and get trapped.

Table 1. The effect of particle size on the human body [4]

Particulate	Particle Size ($\leq\mu m$)	Influences on Human Health
PM_{100}	100	Persist in the air and no evidence of adverse effects on human health
PM_{10}	10	Enter the respiratory system, deposit in the respiratory tract and cause respiratory diseases
$PM_{2.5}$	2.5	Get into the alveoli through the respiratory tract and then enter into the blood circulation, causing various diseases
$PM_{0.1}$	0.1	

It is evident that the human lung is unable to filter particles below the size of $PM_{2.5}$ (called ultra-fine particles), which are admitted into the bloodstream, adhered to contaminants, where they can exert their harmful effects. Among other things, they cause oxidative stress (an unfavorable effect on the antioxidant-prooxidant balance, which is an important means of maintaining health) on a type of white blood cell called macrophages, whose primary function is to kill bacteria and cancer cells in the body. Experiments were carried out with animals and with children living closer to high traffic roads. Allergic sensitization and asthma were more common in those cases [5, 6]. According to WHO data, nearly 4.2 Mio. People worldwide die early each year from air pollution, where the illness is mainly caused by $PM_{2.5}$ or smaller particles [7]. In today's modern cars, both diesel and Otto engines, the formed particle size is well below the lung's filtration ability.

2 Test Methods

Because of their negative impact on the environment and health, various measurement procedures were developed for internal combustion engines to measure their emission values. Ruing the emission tests the composition of the exhaust gas is measured and analyzed. Starting from July 1992 limits were set on internal combustion engine emissions in Europe. These are called Euro norms. As a result of the increasingly stringent regulations to date, emissions from internal combustion engines are now a fraction of the initial state. The norms regulate, among other things, the NOx, CO, HC emissions of diesel and gasoline engines, the number and mass of nanoparticles emitted, as well as the test conditions and the type of measurement cycle used.

Test cycles are internationally uniform measurement processes, where vehicles are tested on a chassis dynamometer. Reproducing the load of the vehicle and environment adapted to real traffic. The cycles are carried out with different time-frame, load, temperature, etc. parameters according to the standard under laboratory conditions. Vehicle manufacturers among enforced emission and fuel consumption limits can also inform customers, for example, about the fuel consumption of a particular vehicle, so the customer can select the optimal for their purpose. For example, in the United States, the FTP-75 cycle is used due to different requirements and standards.

The NEDC test method was introduced in 1970, which was later expanded with consideration of the city traffic. As a result of stricter environmental regulations, it was necessary to develop a new, more close to real life test cycle, which gives a more overall test that accommodates to modern car operating conditions. As of 1 September 2017, the WLTP has been progressively introduced, which is compared with the previous NEDC in Table 2.

Table 2. Differences between NEDC and WLTC class 3 [8]

	NEDC	WLTC Class 3
Distance [km]	11.023	23.262
Duration [s]	1180	1800
Idle time [s]	280	235
Phases [#]	2	4
Average speed/w idle (w/o idle) [km/h]	33.6 (44.7)	46.5 (53.5)
Mas speed [km/h]	120.0	131.3
Max acceleration [m/s^2]	~ 1.0	~ 1.7

Within the WLTP, three different cycles have been created, depending on the power to weight ratio of the vehicles, which are WLTC class 1–3. As shown in Fig. 1, the WLTP test lasts longer, with longer distances, and has more intense deceleration and acceleration stages at a higher average and top speeds.

WLTP takes into account the impact of built-in consumers (such as air conditioning, seat heating, etc.) on vehicle operation and includes predetermined gear shift points in the cycle program. As of September 2017, the Euro 6d-TEMP standard has made it mandatory to obtain RDE measurements for all newly acquired type certificates, which measure NOx, CO and particulate emissions on the road using the PEMS. The measured emission and fuel consumption results are real, and therefore satisfactory.

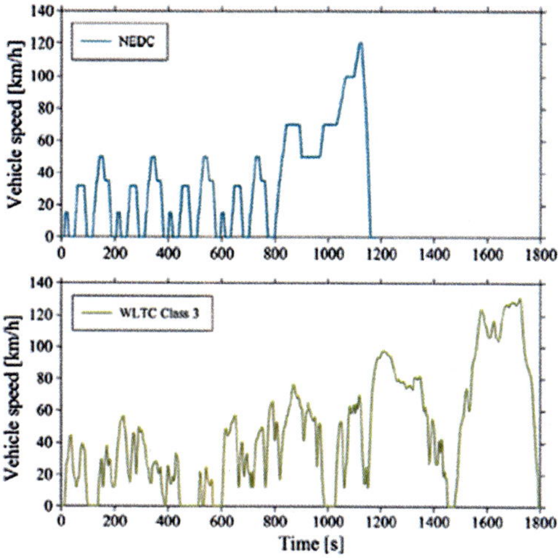

Fig. 1. NEDC and WLTP cycle diagrams [8]

3 Soot Particle Formation

Basically, it is true for each engine type, that PM and PN depend on their origin: the amount and composition of the lubricant consumed, the type of fuel and injection pressure, the location of the injection, and the wear particles. In diesel engines besides the lubricating oil, the fuel has a lubricating role as well. In gasoline engines, only the lubricating oil has this function. The full synthetic oils used in today's modern engines are equipped with additives based on application (for example passenger or race car) and engine type. The components are mostly inorganic substances: magnesium, zinc, calcium, phosphorous, and sulfur, which slow down the oil degradation, block foaming, keep contaminants floating, reduce friction, but reaching the combustion chamber they promote soot forming [9]. As low-viscosity oils are used in passenger cars, the oil scraper ring cannot fully remove the oil from neither the coated nor the honed cylinder wall, since the oil stays in the honing grooves and slightly gets into the combustion chamber, being present during combustion. Previous research has shown that oil consumption (less than 0.1% in today's engines) promotes diffusion combustion, so it is responsible for about 25% of particulate emissions [10]. Otto and Diesel engines have different emission rates for particles.

3.1 Particle Emissions of Gasoline Engines

Gasoline evaporates well, the vapor of gasoline mixed with air forms a highly flammable mixture. In gasoline engine mixture forming the aim is to create a homogeneous air-fuel mixture. Due to the relatively higher engine speed of gasoline engines (compared to diesel engines), less time is left per cycle to maintain a homogeneous air-fuel mixture. These engines aim for the so-called stoichiometric operation with a mixture of $\lambda = 1$, when exactly enough fuel is injected into the intake air for combustion, resulting in complete combustion of both components. In practice, however, the lambda value is constantly changing and is controlled by the ECU during vehicle operation.

During real operation, there is no perfect homogeneous mixture formation, which can result in fuel droplets being left without oxygen, resulting in diffusion combustion over a well-defined surface. In the case of unmixed fuel droplets, actually $\lambda = 0$, so the fuel cannot oxidize, resulting in soot forming.

3.2 PN Reduction on Gasoline Engines

There are several factors concerning air-fuel mixing. Fuel can be injected in the intake port, or in the combustion chamber, although both have advantages and disadvantages. Previously port fuel injection was mainly used, then with direct injection in the combustion chamber, the PN number of the engines has risen. The simple explanation is that when PFI technology was used, gasoline and air already got mixed in the intake manifold, allowing more time for approximate homogeneity. With DI technology, for optimal power and fuel consumption, the injections take place in a short period of time at a specific pressure and location compared to the piston's top dead center position, allowing less time for mixture forming. Modern engines use both injection technology at the same time. To compare these technologies several studies were carried out.

Injection pressure is another important factor. The increased pressure provides better air fuel mixture. Choi et al. [11] studies show that by increasing the injection pressure of a petrol engine with the same parameters, the PM is slightly reduced, but the PN is reduced to a fraction. Cylinder pressure increase and mixture formation can also be enhanced with supercharging systems (compressor, turbocharger) used by car manufacturers in the downsizing trend. By increasing the charge air pressure, not only the power and torque will increase, but the mixture forming will be also enhanced, also with increased air temperature caused by the compression, particle emissions will be reduced [12]. PN is also influenced by the intake air and combustion temperature, their increase has a positive effect on reducing the amount of soot, as opposed to NOx, where emissions will increase [13, 14].

Based on Chan et al. [15] investigations the cold start phase gives the main part of the PN emission with both PFI and DI engines. During a cold start, the combustion process gradually eases close to the cylinder wall and the piston, resulting in flame extinction, which highly affects the emissions of unburnt HC and particulates [16]. For PFI engines, cold running causes up to 10 times more PN than warm running due to gasoline enrichment in the mixture, when the stoichiometric mixture ratio deteriorates, $\lambda < 1$. Starting and operating in cold ambient air produces more ultra-fine particles than solid, larger-sized particles [17]. The maximum point of their formation is around $\lambda \approx 1{,}1$, which is close to the optimum air ratio for gasoline engines. According to the cycle describing the operation of an internal combustion engine, the higher the temperature step between combustion and intake air temperature is, the greater the work produced by the engine. The upper limit of the combustion temperature is due to the formation of nitrogen oxides, so it is necessary to lower the lower limit, i.e. the intake air temperature, which can be achieved effectively by EGR (about 15–20% for gasoline engines) and charge air cooling. However, these again have a negative effect on PN. Like diesel engines, PN decreases in nucleation mode but increases in accumulation mode due to lack of afterburning [18–20].

Hergueta et al. [21] and Myung et al. [22] have shown that fewer soot particles are produced compared to normal gasoline when using butanol gasoline (B33) and ethanol gasoline (E25), which can be explained by faster and higher temperature combustion, lighter hydrocarbon chains and higher oxygen content. With cold intake air and E10 fuel, DI engines showed increasing particle emissions, whereas with PFI it increased only under extreme cold conditions [15].

In certain operating conditions (cold start, acceleration phase, uphill climb or higher speed) the air-fuel mixture must be enriched with fuel. However, for DI, excess fuel may wash the oil film off the cylinder wall, so multi-stage injection is recommended. Rapid warming to operating temperature is required for optimal catalyst performance. In order to allow the catalyst to warm up faster, a design tendency to reduce the distance between the combustion chamber and the catalyst can be observed. With new engines, the use of integrated exhaust manifolds is becoming common, so that the exhaust gases, after leaving the cylinder head, actually enter the turbine housing, to which the catalyst is directly connected. In addition to increased power, the exhaust gas temperature increases and the exhaust gas flow distance is shorter to the catalyst, thus reducing heat loss. As a consequence, excess fuel is injected, so the evaporative heat of the fuel cools the combustion chamber and the catalytic converter, thereby preventing its damage [23]. Fuel particles that do not oxidize produce excess soot particles and result in higher CO_2 emissions. Table 3 shows the effects of various engine parameter changes in particle emissions.

Table 3. Effects of various engine parameters on particulate number [13]

Variable	Effect	Particulates
Engine load	increase ↑	increase ↑
Fuel injection pressure	↑	↓
EGR	↑	↑
Inlet air temperature	↑	↓
Exhaust back pressure	↑	↓
λ (AFR)	↑	↑
Spark timing	advanced	↓
Fuel injection timing	delayed	↓

3.3 Diesel Particulate Emissions

Compared to a gasoline engine, the fundamental difference is in the combustion process. Compressing the constant amount of intake air raises its temperature and the diesel fuel injected into it ignites. The thicker diesel fuel is injected up to ten times the pressure of gasoline for higher power and reduced pollutant emissions. At the moment of injection, as the diesel fuel mixes with the hot oxygen, the combustion process begins immediately. Compared to GDI, the lower speed range allows more time for mixture forming, but it is not easy to create a homogeneous mixture due to the heavier diesel fuel dissolution. In order to ignite the inner part of the fuel jet, an excess of air is required: $\lambda \geq 1.3$. However, even this does not ensure proper mixing, as fuel molecules remain which are not oxidized, resulting in diffusion combustion. Diesel engines operate at a lower temperature than gasoline engines, partly due to the extended expansion process and partly due to excessive exhaust gas recirculation of up to 50%. Shi et al. [24] and Hussain et al. [25] studies show that by increasing the amount of exhaust gas recirculation, the emissions of soot particles increase, while the NOx emission are reduced, this can be seen in Fig. 2.

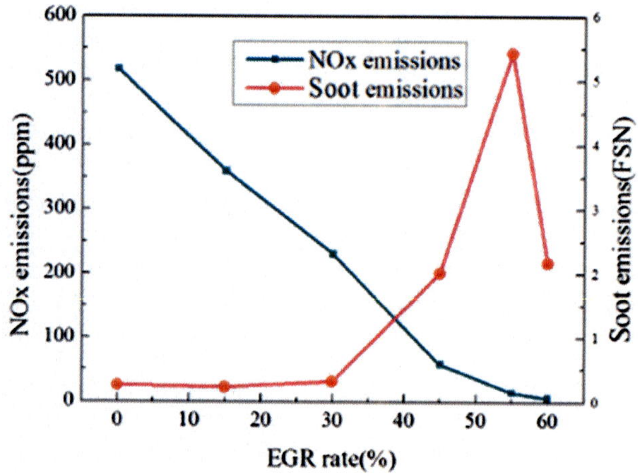

Fig. 2. The effect of EGR rate on NOx and soot emissions [24]

Although increased injection pressure provides better mixture forming, the emitted particle number is reduced as well as their size [26, 27]. While lungs were able to filter particles produced by older engines due to their larger size, they may not filter smaller particles created by higher injection pressures, allowing them to enter the bloodstream more easily.

3.4 Comparison

It is important to emphasize that what has been described so far concerns light-duty vehicles that have been optimized for normal, everyday use. Compared to this, the design and operating conditions used in motor sports and heavy-duty vehicles highly differ, which also affects the emission of soot particles. The main differences between the two engine types in terms of particle emissions are summarized in Table 4.

By comparing the processes described above, conclusions can be made for each point. With an ideal stoichiometric and homogeneous mixture, gasoline vapor burns faster and under better conditions than diesel fuel droplets sprayed into the combustion chamber, which is more difficult to mix with air when injected with excess oxygen and

Table 4. Differences between engine types in terms of particle emissions

Heading level	Gasoline engine	Diesel engine
Fuel miscibility	Vapor mixable	Not vaporizing
Air - fuel ratio	$\lambda \approx 1$	$1{,}3 \leq \lambda$
Injection pressure	150–200 bar	1000–2000 bar
Operational temperature	Higher	Lower
Engine speed	Higher	Lower
Surface/volume ratio	Varies by model	Varies by model

at higher pressures. The lower temperature during diesel expansion is not enough to burn the soot. As a result, particle size is essentially smaller for gasoline engines than for diesel engines [28].

According to Banerjee et al. [29], in the case of a diesel engine operating with excess power need (e.g. acceleration phase), only a negligible change in PN occurs, but the particle size increases, whereas in gasoline engines there is a sharp increase in PN with smaller particles. Basically, GDI engines without GPF have higher particle emissions than modern diesel engines with DPF, so gasoline engines need to be fitted with a particulate filter as well [30].

4 Particulate Filter

In the previous section, it was described what factors influence the formation of nano-sized soot particles that are harmful to health in the case of the two main engine types, Otto, and diesel engines. It is clear from the processes presented that there are trade-offs in engine design when it comes to emissions. As stated earlier, for example, increasing engine operating temperature can reduce particle emissions, but on the other hand NOx emissions will increase. In order to keep the number of particles emitted within the emission limits in all operating conditions, manufacturers have developed several methods.

The main task of ceramic monolithic catalysts used in vehicles was to convert HC and carbon monoxides into non-harmful substances, followed by TWC, which also limited the amount of NOx released into the atmosphere. Exhaust after treatment systems is built in several ways. The most common design to reduce PN is when diesel or gasoline particulate filters (depending on engine type) are installed after the TWC. There's also a design used where two catalysts are connected in series in the exhaust system. As the exhaust gas flows, some of the soot particles adhere to the catalyst channel wall, thereby reducing PN. There is also a solution by applying a washcoat layer to the particle filter wall, thus reducing the role of the TWC or totally leaving it out from the system. According to Tanaka et al. [31] research, the latter solution is although effective in reducing pollutant emissions but has resulted in a significant increase in exhaust gas pressure drop with cross-wall flow, which degrades engine operation. The constructions used are shown in Fig. 3.

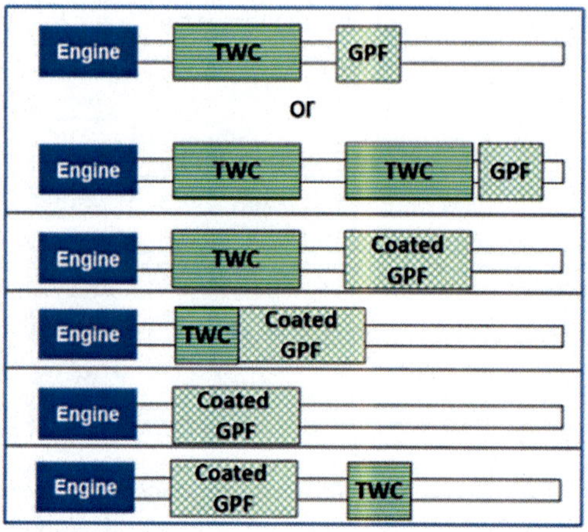

Fig. 3. Possible exhaust gas after treatment system structures [32]

Under current regulations, the emission of $6x10^{11}$ particles per kilometer is permissible, but this limit can only be effectively met with the use of particulate filters. The efficiency of particle filters is close to 100% filtration efficiency [33]. Demuynck et al. [34] studied particle emissions during standard measurement cycles (NEDC, WLTP), using different fuels, with and without particulate filters. The results prove that using a filter can achieve 50–70% reduction in particulate emissions so that the values are well below the EURO 6d TEMP specification.

4.1 Comparison of Differences in Particulate Filter of Otto- and Diesel-Engines

Depending on the type of engine, there are basically two main groups. DPF used on diesel engines and GPF used on gasoline engines. With the introduction of particle limitations on Otto engines, there were experiments with integrating the previously proven DPF into the exhaust after-treatment system of the Otto engines. However, based on test and application experience, DPF filters were not suitable for gasoline engines because of the different operating parameters. For example, the GDI engines have higher exhaust flow rate and temperature than the diesel engines, so the pressure drop is higher too. For this reason, wall thickness, cell density and permeability need to be adjusted, taking into account, however, that GDI engines have less soot emission and deposition than diesel and thus have different filtration efficiencies [35]. Based on the results so far, ceramics with different compositions have proven to be the most effective solution. DPF and GPF material are similar, which are mostly cordierite, silicon carbide (SiC), or aluminum titanate (Al2TiO5), but Otto-engine's exhaust gas temperature is higher than Diesel, so the filter cartridge of GPF has to be more thermal resistant than DPF [36].

For example, their shape may be circular or oval in and vary in size. There are straight flow and wall through flow filters, the difference between them is shown the Fig. 4.

The design of straight flow filters is similar to catalysts in which the exhaust gas enters and exits the same pipe. With wall through flow filters, the end of the pipe in which the exhaust gas enters is closed, forcing the gases to flow through the porous

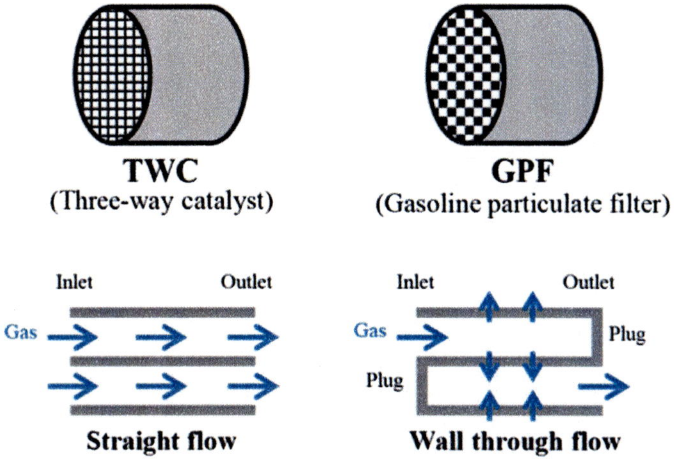

Fig. 4. The structure of straight flow and wall through flow filters [31]

wall of the channel [31]. The porosity of the material usually varies between 45–65%. Depending on the manufacturer and the particular engine specification [35]. Partly the porosity determines what size of the soot particles are released to the atmosphere or get trapped.

5 Soot and Ash Deposits

The wall through flow particulate filter introduced in the previous chapter is the most widespread and effective solution for trapping soot particles. When entering the filter inlet pipe, the particles show multiple deposition patterns based on a variety of influencing factors. Conditions for soot formation are the following: percentage composition of oil additives, engine operating parameters, exhaust gas velocity and temperature, filter saturation, and regeneration processes which are discussed in more detail in the following all have an effect on the shape, the image, and the size of the particles [37].

5.1 Exhaust Gas Velocity

As a result of several research work, it can be observed that initially, the deposition takes place along the wall and then, with the cross-section narrowing, the velocity of the exhaust gas increases according to Bernoulli's law due to the smaller cross-section. The higher exhaust gas velocity transported the particles mostly towards the end of the filter [37, 38]. Among other things, it was noticeable at higher exhaust gas speeds that the density of the ash deposit was higher.

5.2 The Effect of Oil Composition

One of the simplest ways to test the filter saturation without dismounting is to place pressure sensors before and after the filter. The difference between the values measured here gives the so-called pressure drop from which the vehicle's ECU calculates and indicates the state of filter saturation. The measured pressure drop gives an estimate of the degree of deposits. Sappok et al. [38] carried out research on ash from burning soot particles using EDX and SEM equipment to determine the composition of the ash. As seen in Fig. 5 ash from Ca and Mg oil additives has a higher sealing effect, so the measured pressure difference is also higher, while Zn has a smaller effect, although the amount of ash from Ca and Mg additivated oils was half of Zn additivated oils, as shown in Fig. 6.

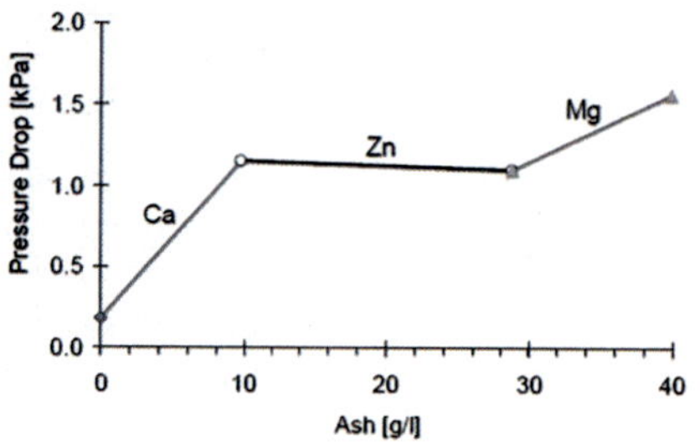

Fig. 5. Effects of different oil additives on pressure drop [38]

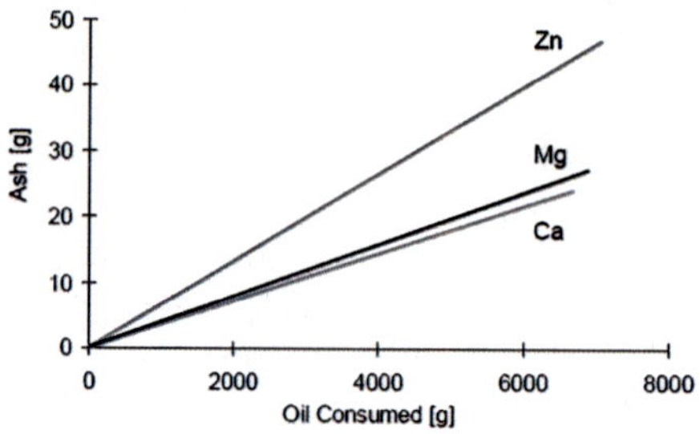

Fig. 6. Main components of ash after regeneration [38]

5.3 Porosity

An important parameter of filters is the degree of porosity, which affects, among other things, the efficiency of the filter and the degree of backpressure. According to C. Lambert et al. [39] studies, the filtration efficiency of a clean GPF is about 60%. During use, the soot particles penetrate the porous material and then gradually fill it, reducing the porosity, providing a smaller surface area for the free flow of the exhaust gases, thereby increasing the filtration efficiency up to 80–90% shown in Fig. 7, in contrast to DPF, where the efficiency is nearly 99% with proper operation.

Thus, the greater the porosity, the lower the filtering efficiency and backpressure will be. These values increase in direct proportion to the use of a washcoat and deposition build up over time. T. Ishizawa et al. [40] confirmed by measurements that, as shown in Fig. 8 the pattern of soot/ash also changes as a function of porosity.

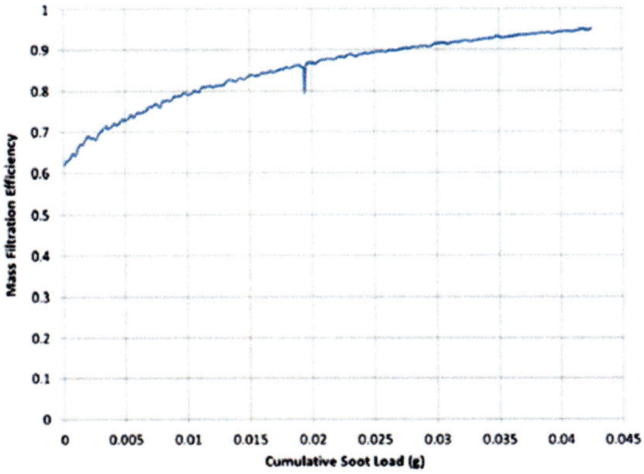

Fig. 7. The effect of soot filling on filtration efficiency [39]

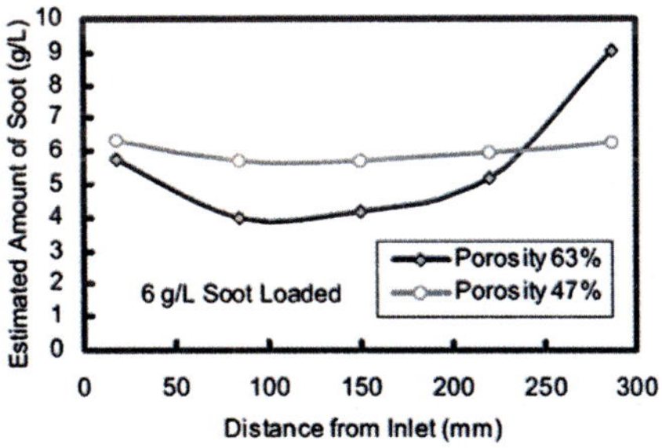

Fig. 8. Amount of ash deposits at different distances as a function of porosity [40]

It can be seen that relative to the inlet pipe a quite uniform amount of soot/ash has built-up over the wall of the smaller porosity filter. In contrast, a higher porosity filter showed greater deposition toward the end of the pipe. This can also be explained by the fact that ash agglomerates may be more easily separated from the wall of the higher porosity filter and accumulate towards the end of the filter during the regeneration process.

6 Particulate Filter Regeneration

As described earlier, the soot from the combustion process causes a pressure drop in the filter. Depending on the type of engine and the current operating condition, the size and amount of soot particles may vary. As the soot particles begin to fill the porous material of the filter (through which the exhaust gas flows), the measured pressure difference shows an increasing tendency. Then, as said before, the filtration efficiency increases. Filter saturation reduces engine performance and can increase fuel consumption due to the increasing exhaust backpressure. Reaching a specific pressure drop, the engine controller indicates the need for a regeneration process, which means that the ash deposits in the filter will be burned. After regeneration, the organic components are burnt, but the inorganic ones become ash, which remains in the filter.

The regeneration process is the self-cleaning of the filter without mechanical disassembly, which has two types: passive and active regeneration. To start the regeneration process externally excess heat supply might be required, therefore, detailed, intelligent thermal management is needed. Different types of regeneration show different deposition patterns in the filter channel (Fig. 9).

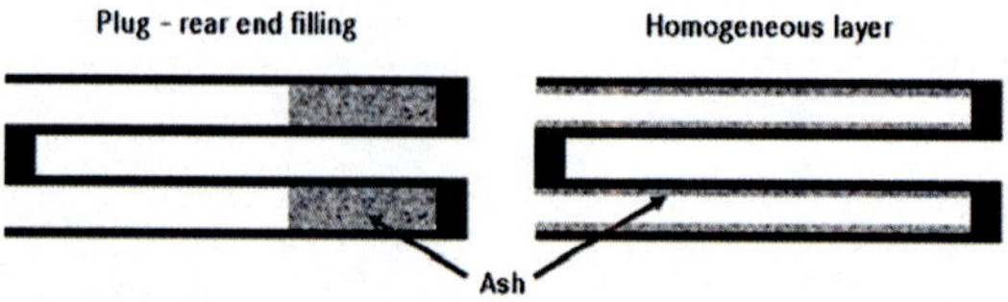

Fig. 9. Deposit pattern of active regeneration can be seen on the left side and passive regeneration on the right side [37]

During active regeneration on the left, the soot deposited along the wall is cyclically heated up (when the engine controller detects a predetermined drop in pressure) and burned. The engine controller adjusts the engine operating parameters to raise the exhaust gas temperature to start the regeneration process, which results in enough soot and inorganic components to be burned into ash. The ash is separated from the filter wall by repetition of regeneration processes and the shear stress of the exhaust gas, will be transported to the end of the pipe where it starts to build-up. SEM and EDX examinations show that the plug-like deposition is a result of the most easily detaching ash from Ca additives from the filter wall. This form of deposition results in less pressure drop and lower filtration efficiency on the filter. The cyclical process is carried out on a higher temperature than passive regeneration, and the ash agglomerates are larger but have a higher permeability, which increases with repetition.

On the right is the pattern of passive regeneration deposits, where the regeneration process happens automatically when the temperature of the exhaust gas rises (for example, uphill, highway or high rpm driving) [37]. Passive regeneration does not require any special intervention from the system, it only gives warning and information about the current filter saturation and condition. Passive regeneration takes place at a lower temperature than the active one, so the resulting ash particles, agglomerates are smaller, which, due to Van der Waals bonding, do not separate from the filter wall but spread evenly on its surface. This results in higher pressure drop, but better filtration efficiency [40].

The ash from burning the inorganic components of the soot accumulates in the filter. Not directly depending on mileage, but rather on operating parameters, the remaining ash gradually chokes the engine, limiting the exhaust gas flow.

7 Summary

Processes occurring in internal combustion engines and exhaust gas aftertreatment systems include complex, multi-stage phenomenon from the exhaust gas forming to particle filtration. From time to time, as a result of intensive research better and better solutions are found for particle filtration methods and processes. Certain factors of the process require research and development for better solutions. The author has started a PhD research work to develop accurate, reliable measurement methods for particle filtration and to detect adverse effects on the engine testbench in laboratory conditions. The first part of the PhD research work is the literature review presented in this article. I find it important to study in more detail the interaction of the material of the particulate filters with the individual inorganic materials, which may provide an opportunity to positively influence the life of the filter.

References

1. Riedl, M., Diaz-Sanchez, D.: Biology of diesel exhaust effects on respiratory function. J. Allergy Clin. Immunol. **115**(2), 221–228 (2005). https://doi.org/10.1016/j.jaci.2004.11.047
2. Ristovski, Z.D., Miljevic, B., Surawski, N.C., Morawska, L., Fong, K.M., Goh, F., Yang, I.A.: Respiratory health effects of diesel particulate matter. Respirology **17**(2), 201–212 (2012). https://doi.org/10.1111/j.1440-1843.2011.02109.x
3. Chio, C.P., Liao, C.M., Tsai, Y.I., Cheng, M.T., Chou, W.C.: Health risk assessment for residents exposed to atmospheric diesel exhaust particles in southern region of Taiwan. Atmos. Environ. **85**, 64–72 (2014). https://doi.org/10.1016/j.atmosenv.2013.11.072
4. Wang, C., Tu, Y., Yu, Z., Lu, R.: PM2.5 and cardiovascular diseases in the elderly: an overview. Int. J. Environ. Res. Public Health **12**(7), 8187–8197 (2015). https://doi.org/10.3390/ijerph120708187
5. Donaldson, K., Li, X.Y., MacNee, W.: Ultrafine (nanometre) particle mediated lung injury. J. Aerosol Sci. **29**, 553–560 (1998). https://doi.org/10.1016/S0021-8502(97)00464-3
6. Van Vliet, P., Knape, M., de Hartog, J., Janssen, N., Harssema, H., Brunekreef, B.: Motor vehicle exhaust and chronic respiratory symptoms in children living near freeways. Environ. Res. **74**(2), 122–132 (1997). https://doi.org/10.1006/enrs.1997.3757
7. WHO, Ambient (outdoor) air pollution. https://www.who.int/news-room/fact-sheets/detail/ambient-(outdoor)-air-quality-and-health, Accessed 17 Feb 2020
8. Tsokolis, D., Tsiakmakis, S., Dimaratos, A., Fontaras, G., Pistikopoulos, P., Ciuffo, B., Samaras, Z.: Fuel consumption and CO2 emissions of passenger cars over the New Worldwide Harmonized Test Protocol. Appl. Energy **179**, 1152–1165 (2016). https://doi.org/10.1016/j.apenergy.2016.07.091
9. Sappok, A., Rodriguez, R., Wong, V.: Characteristics and effects of lubricant additive chemistry on ash properties impacting diesel particulate filter service life. SAE Int. J. Fuels Lubr. **3**, 705–722 (2010). https://doi.org/10.4271/2010-01-1213
10. Sonntag, D.B., Bailey, C.R., Fulper, C.R., Baldauf, R.W.: Contribution of lubricating oil to particulate matter emissions from light-duty gasoline vehicles in Kansas City. Environ. Sci. Technol. **46**, 4191–4199 (2012). https://doi.org/10.1021/es203747f
11. Choi, Y., Lee, J., Jang, J., Park, S.: Effects of fuel-injection systems on particle emission characteristics of gasoline vehicles. Atmos. Environ. (2019). https://doi.org/10.1016/j.atmosenv.2019.116941
12. Tanin, K. V., Wickman, D. D., Montgomery, D.T., Das, S., Reitz, R.D.: The influence of boost pressure on emissions and fuel consumption of a heavy-duty single-cylinder D.I. diesel engine. SAE Technical Paper Series (1999). https://doi.org/10.4271/1999-01-0840
13. Leach, F., Stone, R., Richardson, D., Lewis, A., Akehurst, S., Turner, J., Remmert, S., Campbell, S., Cracknell, R.F.: Particulate emissions from a highly boosted gasoline direct injection engine. Int. J. Engine Res. **19**(3), 347–359 (2017). https://doi.org/10.1177/1468087417710583

14. Weber, C., Sundvor, I., Figenbaum, E.: Comparison of regulated emission factors of Euro 6 LDV in Nordic temperatures and cold start conditions: Diesel- and gasoline direct-injection. Atmos. Environ. **206**, 208–217 (2019). https://doi.org/10.1016/j.atmosenv.2019.02.031
15. Chan, T., Meloche, E., Kubsh, J., Brezny, R.: Impact of ambient temperature on gaseous and particle emissions from a direct injection gasoline vehicle and its implications on particle filtration. SAE Int. J. Fuels Lubr. **6**(2), 350–371 (2013). https://doi.org/10.4271/2013-01-0527
16. Badshah, H., Kittelson, D., Northrop, W.: Particle Emissions from light-duty vehicles during cold-cold start. SAE Int. J. Engines **9**(3), 1775–1785 (2016). https://doi.org/10.4271/2016-01-0997
17. Yusufa, A.A., Inambao, F.L.: Effect of cold start emissions from gasoline-fueled engines of lightduty vehicles at low and high ambient temperatures: Recent trends. Case Stud. Thermal Eng. **14**, 100417 (2019). https://doi.org/10.1016/j.csite.2019.100417
18. Hedge, M., Weber, P., Gingrich, J., Alger, T., Khalek, I.: Effect of EGR on particle emissions from a GDI engine. SAE Int. J. Engines **4**(1), 650–666 (2011). https://doi.org/10.4271/2011-01-0636
19. Lattimore, T., Wang, C., Xu, H., Wyszynski, M.L., Shuai, S.: Investigation of EGR effect on combustion and PM emissions in a DISI engine. Appl. Energy **161**, 256–267 (2016). https://doi.org/10.1016/j.apenergy.2015.09.080
20. Wang, S., Zhu, X., Somers, L.M.T., de Goey, L.P.H.: Effects of exhaust gas recirculation at various loads on diesel engine performance and exhaust particle size distribution using four blends with a research octane number of 70 and diesel. Energy Conv. Manag. **149**, 918–927 (2017). https://doi.org/10.1016/j.enconman.2017.03.087
21. Hergueta, C., Tsolakis, A., Herreros, J.M., Bogarra, M., Price, E., Simmance, K., York, A.P.E., Thompsett, D.: Impact of bio-alcohol fuels combustion on particulate matter morphology from efficient gasoline direct injection engines. Appl. Energy **230**, 794–802 (2018). https://doi.org/10.1016/j.apenergy.2018.08.076
22. Myung, C.L., Lee, H., Choi, K., Lee, Y.J., Park, S.: Effects of gasoline, diesel, LPG, and low-carbon fuels and various certification modes on nanoparticle emission characteristics in light-duty vehicles. Int. J. Automot. Technol. **10**, 537–544 (2009). https://doi.org/10.1007/s12239-009-0062-9
23. Huang, Y., Hong, G., Cheng, X., Huang, R.: Investigation to charge cooling effect of evaporation of ethanol fuel directly injected in a gasoline port injection engine. SAE Technical Paper (2013). https://doi.org/10.4271/2013-01-2610
24. Shi, X., Liu, B., Chao, Z., Jingchao, H., Zeng, Q.: A study on combined effect of high EGR rate and biodiesel on combustion and emission performance of a diesel engine. Appl. Therm. Eng. **125**, 1272–1279 (2017). https://doi.org/10.1016/j.applthermaleng.2017.07.083
25. Hussain, J., Palaniradja, K., Alagumurthi, N., Manimaran, R.: Effect of exhaust gas recirculation (EGR) on performance and emission characteristics of a three cylinder direct injection compression ignition engine. Alexandria Eng. J. **51**, 241–247 (2012). https://doi.org/10.1016/j.aej.2012.09.004
26. Agarwal, A.K., Srivastava, D.K., Dhar, A., Maurya, R.K., Shukla, P.C., Singh, A.P.: Effect of fuel injection timing and pressure on combustion, emissions and performance characteristics of a single cylinder diesel engine. Fuel **111**, 374–383 (2013). https://doi.org/10.1016/j.fuel.2013.01.077
27. Xu, Z., Li, X., Guan, C., Huang, Z.: Effects of injection pressure on diesel engine particle physico-chemical properties. Aerosol Sci. Technol. **48**(2), 128–138 (2014). https://doi.org/10.1080/02786826.2013.862589

28. A study of the number, size & mass of exhaust particles emitted from European diesel and gasoline vehicles under steady-state and european driving cycle conditions. Prepared for the CONCAWE Automotive Emissions Management Group by its Special Task Force AE/STF-10 (1998). Accessed 04 Mar 2020
29. Banerjee, T., Christian, R.A.: Effect of operating conditions and speed on nanoparticle emission from diesel and gasoline driven light duty vehicles. Atmos. Poll. Res. **10**, 1852–1865 (2019). https://doi.org/10.1016/j.apr.2019.07.017
30. Platt, S.M., El Haddad, I., Pieber, S.M., Zardini, A.A., Suarez-Bertoa, R., Clairotte, M., Prévôt, A.S.H.: Gasoline cars produce more carbonaceous particulate matter than modern filter equipped diesel cars. Sci. Rep. **7**(1), 1–9 (2017). https://doi.org/10.1038/s41598-017-03714-9
31. Tanaka, A., Miyoshi, N., Sato, A.: Development of low pressure and high performance GPF catalyst. SAE Technical Paper 2018–01–1261 (2018). https://doi.org/10.4271/2018-01-1261
32. AECC - Technical Summaries –The Gasoline Particulate Filter (GPF) (2017) https://www.aecc.eu/wp-content/uploads/2017/11/2017-AECC-technical-summary-on-GPF-final.pd. Accessed 10 Mar 2020
33. Kattouah, P., Kato, K., Thier, D., Ohara, E., Vogt, C.D., Ito, Y., Shimoda, T., Aoki, T., Yuuki, K., Sakamoto, H.: Advanced Gasoline Particulate Filter for Effective Gasoline Emission Control Beyond Euro 6. Emission Control (2014)
34. Demuynck, J., Favre, C., Bosteels, D., Hamje, H., Andersson, J.: Real-world emissions measurements of a gasoline direct injection vehicle without and with a gasoline particulate filter. SAE Technical Paper (2017). https://doi.org/10.4271/2017-01-0985
35. Boger, T., Cutler, W.: Reducing Particulate Emissions in Gasoline Engines. SAE International, Warrendale (2018). (978-0-7680-9417-6)
36. Ito, Y., Shimoda, T., Aoki, T., Shibagaki, Y., Yuuki, K., Sakamoto, H., Kato, K.: Advanced ceramic wall flow filter for reduction of particulate number emission of direct injection gasoline engines. SAE Technical Paper Series (2013). https://doi.org/10.4271/2013-01-0836
37. Dittler, A.: Ash Transport in Diesel Particle Filters. SAE Technical Paper (2012). https://doi.org/10.4271/2012-01-1732
38. Sappok, A., Wong, V.W., Morrow, R., Zisholtz, E., Doustar, I., Govani, I.: Investigation of ash formation, accumulation, and distribution in diesel particulate filters using lubricant additive tracers. In: ASME 2011 Internal Combustion Engine Division Fall Technical Conference, ICEF 2011–60072, pp. 1–15 (2011). https://doi.org/10.1115/ICEF2011-60072
39. Lambert, C., Chanko, T., Dobson, D., Liu, X., Pakko, J.: Gasoline particle filter development. Emiss. Control Sci. Technol. **3**, 105–111 (2017). https://doi.org/10.1007/s40825-016-0055-x
40. Ishizawa, T., Yamane, H., Satoh, H., Sekiguchi, K.: Investigation into ash loading and its relationship to DPF regeneration method. SAE Int. J. Commer. Veh. **2**(2), 164–175 (2010). https://doi.org/10.4271/2009-01-2882

Electromagnetic Emission Rates Between 2-Phase and 3-Phase Motors

Dániel Erdősy(✉), István Bodnár, and Rafael Ruben Boros

University of Miskolc, Miskolc, Hungary
{elkedani,vegybod,elkruben}@uni-miskolc.hu

Abstract. In all departments of electronic design and manufacturing, electromagnetic compatibility (EMC) is one of the most important factors. As a part of a bigger project, we study BLDC motors. The main type has 2-phase stator winding, but another has 3-phase and this article is about comparing these two different motors with each other. The base of the comparison is EMC, even though we know in advance what results are possible. 3-phase windings are more symmetrical electrically, but we did not know in these cases exact numbers. In the first section, we introduce the project and BLDC motors briefly to understand what we work on. In the second section, our measuring method, arrangement, evaluation process will be written. Finally, the most important, our results and consequences will be also included, according to the title. Unless we do not have proper EMC measuring equipment, we are not possible to show proper emission numbers, but for comparing different versions of these motors, the method we used can work fine.

Keywords: BLDC motor · EMC · 2-phase BLDC

1 The Base Project and the BLDC Motors

In everyday life, all devices need EMC to work properly. Some of them can be interfered with others easily, while others radiate lots of electromagnetic (EM) waves. Nowadays EMC is becoming more important than ever, because of the growing number of electronic devices, gadgets, smart devices we use. In the automotive industry, due to the increasing number of comfort, safety and driver assistance electronic systems, EMC is a main property inside and outside the vehicle. Modern vehicles with internal combustion engines carry so much electronic equipment, that it is hard to say that they are powered by gas. Different hybrid types, full electric ones with much bigger battery packs involve advanced driving and charging circuits, which are not the easiest task to design from EMC perspective.

1.1 The Base Project

We work in a bigger project in which we must improve the EMC performance of a BLDC motor. This motor is the driver of an automotive cooling fan. It is electronically

K. Jármai and K. Voith (Eds.): VAE 2020, LNME, pp. 174–185, 2021.
https://doi.org/10.1007/978-981-15-9529-5_15

controlled, just 12 V DC and a PWM signal should be connected to the motor to supply power and adjust its rotation. The fan assembly is made of plastic, so it does not influence EMC, but it gives a good static load on the motor, so we have measured not only the motor but the whole assembly. This motor fulfills the required EMC standards, but in the future, these standards might be stricter especially in electric vehicles, so it must be improved.

Designing a whole new motor is way more expensive than inspecting the current one and find out what should be changed to reduce the radiated emission rates. We had and have different plans, for example changing the main bearing between the rotor and stator, changing the material of the stator, making a more precise magnetic field for the rotor and so on. The one which this article is about is changing the phase number of the stator. Of course, we cannot just simply make another winding, put it on and measure, because the iron core and the control electronics do not allow this. In the product range of the motor, there is another similar motor which has the needed stator winding, so we measured how different the two motors.

1.2 BLDC Motors

BLDC motors are one type of synchronous motor. This means the magnetic field generated by the stator and the magnetic field generated by the rotor, rotate at the same frequency. BLDC motors do not produce “slip” that is normally seen in induction motors. BLDC motors come in single-phase, 2-phase and 3-phase configurations. Corresponding to its type, the stator has the same number of windings. Out of these, 3-phase motors are the most popular and widely used. A brushless motor is constructed with a permanent magnet rotor and wire wound stator poles. Electric energy is converted to mechanical energy by the magnetic attractive forces between the permanent magnet rotor and the rotating magnetic field induced in the wound stator poles [1].

Figure 1 shows a three-phase, 12-pole BLDC motor construction. On the left side of Fig. 1 is the stator and on the right side is the rotor with the magnetic poles. This is the 3-phase motor we will measure in this study. The iron core material of the stator is made of M400-50A steel.

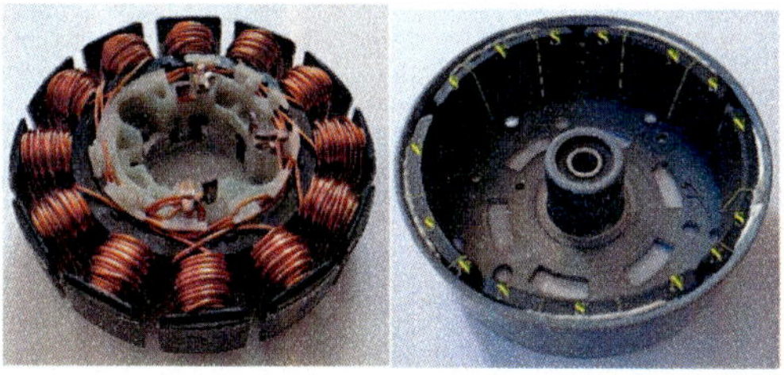

Fig. 1. Three-phase, 12-pole BLDC motor construction (provided by Bosch)

The three-phase stator winding is fed by a three-phase inverter and the electronic commutation is implemented also by the inverter. This motor has a delta-connected winding. The commutation of a BLDC motor is controlled electronically. To rotate the BLDC motor, the stator windings should be energized in a proper sequence. It is important to know the exact rotor position to understand which winding must be energized in sequence. Rotor position is sensed usually by Hall effect sensors embedded into the stator.

Most BLDC motors have three Hall sensors embedded into the stator on the non-driving end of the motor. Whenever magnetic poles of the rotor pass near the Hall sensors, they give a high or low signal, indicating the N or S pole passing next to the sensors. Based on the combination of these three Hall sensor signals, the exact sequence of commutation can be determined. Hall sensors increase the price of the motor. It is possible to leave these sensors if we measure the motors back EMF[1]. When the rotor rotates, each winding voltage is generated by the changing magnetic field. This voltage is known as the back EMF, which opposes the main voltage supplied to the windings according to Lenz's Law. The polarity of these back EMF is in the opposite direction of the energizing voltage [1–3].

The other motor we measured is shown in Fig. 2. This motor is 2-phase and the rotor magnetic poles are the same as the three-phase.

Fig. 2. 2-phase, 12-pole BLDC motor construction (provided by Bosch)

Three-phase motors are better than two-phase design because the rotating magnetic field is much more symmetrical. The three-phase windings make less EMC noise [4].

Both rotors are made of Strontium Oxide ferrite permanent magnets. Nowadays, ferrite magnets are frequently used because they are less expensive and can produce high magnetic flux densities. As a result, size-to-weight ratio is a good value with high torque output [5, 6].

2 Measuring Environment and the Evaluation Method

The type of EMC measurement must be measured in a proper EMC laboratory. There are outdoor measuring techniques, but it is hard to find the right place. Our measurements

[1] EMF – Electromotive Force.

were made in a shielded laboratory. This lab is owned by the University of Miskolc, Institute of Physics and Electrotechnics, Department of Electrical and Electronic Engineering. It is in building A3, on the basement floor. This lab is not an accredited, proper EMC laboratory, but it has a good electromagnetic shielding (with a maximum shielding frequency around 1 GHz), and we made our best to build proper test conditions. It must be mentioned that the later detailed measuring data is only for reference, it cannot be used for qualification purposes [7, 8].

2.1 Measuring Equipment and Arrangement

We used a handheld spectrum analyser, FSH-8 from Rohde & Schwarz for the measurements and a computer to collect the data with. The antenna was a HE300HF antenna (also from Rohde & Schwarz). This antenna is a direct antenna with a frequency range from 9 kHz to 20 MHz. The energy source for the BLDC motor was a 12 V lead acid battery. It was not being charged under testing, but for the short working periods, the battery voltage did not change significantly, and it also does not affect the emission rates. We used an analogue PWM generator to produce the control signal (100 Hz PWM with duty cycle between 10% and 90%). As the PWM generator uses a potentiometer to the duty cycle, we manually set it, checking with an oscilloscope [9].

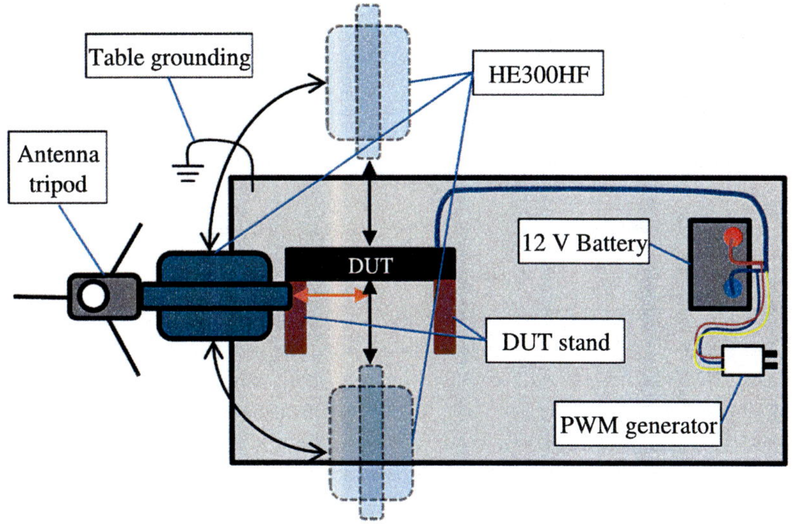

Fig. 3. The measuring environment (without instruments)

Figure 3 shows the measuring arrangement in the shielded laboratory. 3 antenna positions are shown in the figure, where the arrows mean 30 cm from the centre of the rotor. Horizontal and vertical antenna positions are not shown on the figure. DUT[2] means the fan assembly with the wooden holding frame (brown rectangles). The thick

[2] DUT – Device Under Test.

blue wire is the cable for drive. Thinner red and blue wires are connected to the battery + and – terminals, the thinner yellow wire is the PWM signal from the PWM generator (white box with 2 potentiometers). The 2 × 1 m table is covered with steel and earthed to a grounding rod.

In most cases 3 people did the measurements, during testing they stay always on the same place. Operator 1 manages the computer, Operator 2 manages the DUT (when it must be changed) and the antenna (positioning and rotating), Operator 3 manages the PWM generator. With this test setup, we can ensure that background noise is always the same. There is just one radio station which is in our measuring frequency range, but thanks to our evaluation method it can be ignored [10].

Due to the direct antenna, we could make horizontal and vertical antenna arrangements in 3 places around the DUT. In Fig. 3 the antenna is mounted in front of the BLDC motor (axial measurements), exactly 30 cm from it and pointing to its center. Figure 4 shows the side measuring arrangement (radial setup) where we keep the 30 cm distance from the center of the motor and the height just as the previous setup. The last antenna position is behind the module (also axial measurements). Antenna height and distance from the center is the same as before.

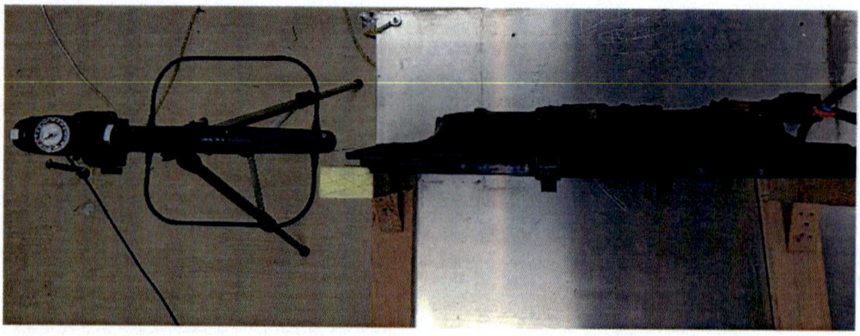

Fig. 4. Antenna arrangement for side measurements

The inspected motors are controlled by a PWM signal. The duty cycle is between 10 and 90%. According to the manufacture's database, commonly used PWM duty cycles are 25%, 50%, 70% and 90%. We made our measurements on these PWMs, in some cases, also on 78%, because here are the biggest electromagnetic emission rates at some modules (still in research why exactly here and why not at all modules). If there are no maximum values at 78%, they are around the 70% point. Later we will mention these as maximum points.

2.2 Spectrum Analyzer Settings and Evaluation Method

Evaluation method:

- Radial and axial (in front of and behind the DUT) measurements
- Horizontal and vertical antenna position (Table 1)
- Duty cycle of the controlling PWM signal: 0%, 25%, 50%, 70%, 90%
- Measurement frequency range: 100–1500 kHz

Table 1. Spectrum analyzer settings (Imported from the program which records measured data)

Settings	Value
Middle Frequency	800 000 Hz
Frequency offset	0 Hz
RF dampening	manual
RF dampening	0 dB
Preamp	ON
RF input	50 Ω
RBW	10 000 Hz
VBW	100 000 Hz
Sweep time	200 ms
Type of measurement	Average
Detection method	RMS
Primary transducer	HE300A-HF
Number of averages	10

- Evaluated frequency range: 900–1100 kHz
- 4 distinct frequencies have been chosen to evaluate the amplitude: 900 kHz, 966,667 kHz, 1033,33 kHz, 1100 kHz
- In the examined frequencies RMS (Root Mean Square) value of the amplitude is calculated (1)
- We calculated the ratio between the RMS values of the original DUT and the modified DUT. It is marked with "H". If H < 1, then the modification made an improvement if H is about 1, there were no improvement and if H > 1, that means the modified version performed worse

$$RMS = \sqrt{\frac{1}{N} \cdot \sum_{i=1}^{N} x_i^2} \tag{1}$$

where RMS is Root Mean Square, N is the number of samples, i is the index number, xi is the i-th value of x.

3 Results

3.1 Emission Rates

Tables 2, 3, 4, 5, 6 and 7 show emission rates of both 2 and 3 phase motors in all measuring directions. We measured with 2 antenna positions (vertical and horizontal), but due to the results, in horizontal position emission rates were always less than in vertical position. That is why we only evaluate vertical position results in this paper. Tables show a similar trend according to duty cycle between the 2 and 3 phase motors.

Table 2. Emission rates in front of the 2-phase motor

[dBμV/m]	PWM			
Frequency [Hz]	25%	50%	70%	90%
900 000.00	7.758	13.054	15.559	1.901
966 666.67	8.550	14.334	13.748	2.246
1 033 333.33	7.989	13.744	14.274	3.282
1 100 000.00	9.035	12.510	11.944	2.303
RMS	7.900	12.021	13.042	2.382

Table 3. Emission rates next to the 2-phase motor

[dBμV/m]	PWM			
Frequency [Hz]	25%	50%	70%	90%
900 000.00	2.491	2.130	1.458	0.258
966 666.67	2.559	2.447	1.818	0.456
1 033 333.33	3.428	2.059	1.852	0.419
1 100 000.00	2.089	1.521	1.699	0.508
RMS	2.386	1.701	1.175	0.384

Table 4. Emission rates behind the 2-phase motor

[dBμV/m]	PWM			
Frequency [Hz]	25%	50%	70%	90%
900 000.00	3.969	7.500	14.344	2.409
966 666.67	5.910	6.577	13.434	2.000
1 033 333.33	5.874	6.057	10.973	2.513
1 100 000.00	3.709	5.984	8.815	2.030
RMS	4.764	6.771	10.657	2.098

Table 5. Emission rates in front of the 3-phase motor

[dBμV/m]	PWM			
Frequency [Hz]	25%	50%	70%	90%
900 000.00	0.037	0.287	0.200	0.087
966 666.67	0.289	0.180	1.614	0.269
1 033 333.33	0.224	0.017	1.577	0.396
1 100 000.00	0.117	0.147	0.496	0.099
RMS	0.400	0.433	1.229	0.021

Table 6. Emission rates next to the 3-phase motor

[dBμV/m]	PWM			
Frequency [Hz]	25%	50%	70%	90%
900 000.00	0.829	0.697	0.484	0.218
966 666.67	0.211	0.123	0.065	0.128
1 033 333.33	0.554	0.579	0.879	0.181
1 100 000.00	0.940	0.460	0.855	0.425
RMS	0.312	0.256	0.166	0.106

Table 7. Emission rates behind the 3-phase motor

[dBμV/m]	PWM			
Frequency [Hz]	25%	50%	70%	90%
900 000.00	1.744	1.181	3.372	2.081
966 666.67	1.891	1.933	3.351	1.784
1 033 333.33	1.704	1.929	3.860	1.865
1 100 000.00	1.341	1.813	3.723	1.872
RMS	1.500	1.606	3.339	1.468

If we check the RMS values, we can see that there is a maximum point in the amplitude in function of PWM duty cycle. The maximum points are also showed in Fig. 5. These points are at 70% duty cycle, which we mentioned earlier (end of Sect. 2.1).

Figure 5 shows the RMS value of emission rates in front of and behind the 2 motors. It can be clearly seen that the 2-phase motor emits more in front than behind. We could predict this, after inspecting the windings. At the 3-phase motor, this trend works opposite, backward emission rates are less than forward emission rates. Each motor has the control PCB in the back, so it can be stated that control logic of the 3-phase motor

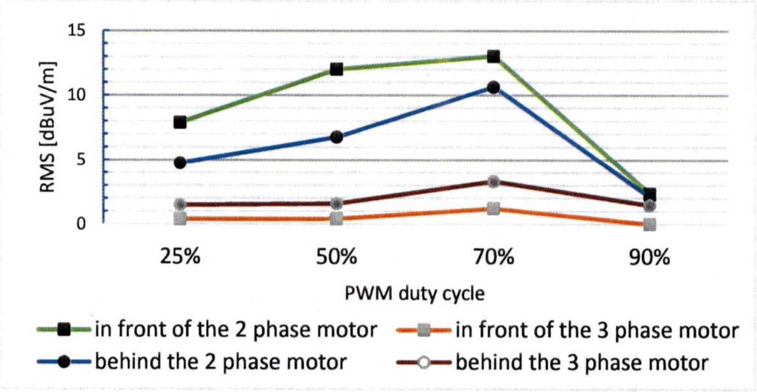

Fig. 5. Comparing the 3-phase and 2-phase motors in function of duty cycle

emits more electromagnetic waves than its windings do. In numbers: the 2-phase motor emits 43.24% more EM waves in front than behind, while 3-phase one emits 67.99% less in front than behind. Side directional rates at the 2-phase motor are 83.74% less, at the 3-phase motor are 97.06% less than forward radiation.

Since the 3-phase motor will replace the 2-phase one on the market, we had to compare them with each other. Table 8 shows the ratio between the measured emission rates in front of the motors. The new generation 3-phase motor emits much less EM noise than the older 2-phase design. On average, the 3-phase one emits 94.34% less EM noise than the 2-phase one.

Table 8. Ratio between front emission rates of 3-phase and 2-phase motors

3-phase/2-phase [%]	PWM			
Frequency [Hz]	25%	50%	70%	90%
900 000.00	0.48	2.20	1.29	4.58
966 666.67	3.38	1.26	11.74	11.98
1 033 333.33	2.80	0.12	11.05	12.07
1 100 000.00	1.29	1.18	4.15	4.30
RMS	5.06	3.60	9.42	0.88

Table 9 shows the same ratio, but with the side emission values. There are still significant differences between the motors like the previous table showed. Here the 3-phase motor emits an average 81.57% less noise than the 2-phase version.

Table 10 shows the ratio between the backwards radiation of the 2 motors. These values due to the construction design belong to the control board. According to the fact that each motor has similar control electronics, so there is a closer connection between the measured values. On average, the difference is just 56.90% between the motors (for the benefit of the 3-phase motor).

Table 9. Ratio between side emission rates of 3-phase and 2-phase motors

3-phase/2-phase [%]	PWM			
Frequency [Hz]	25%	50%	70%	90%
900 000.00	33.28%	32.72%	33.20%	84.50%
966 666.67	8.25%	5.03%	3.58%	28.07%
1 033 333.33	16.16%	28.12%	47.46%	43.20%
1 100 000.00	45.00%	30.24%	50.32%	83.66%
RMS	13.08%	15.05%	14.13%	27.60%

Table 10. Ratio between backwards emission rates of 3-phase and 2-phase motors

	PWM			
Frequency [Hz]	25%	50%	70%	90%
900 000.00	43.94%	15.75%	23.51%	86.38%
966 666.67	32.00%	29.39%	24.94%	89.20%
1 033 333.33	29.01%	31.85%	35.18%	74.21%
1 100 000.00	36.16%	30.30%	42.23%	92.22%
RMS	31.49%	23.72%	31.33%	69.97%

The final fan assembly is built in vehicles with a radiator in front of it, which highly reduces the front emission rates. The side and backwards emissions are not reduced by the radiator, so they can cause unwanted side-effects in other electrical components. Electromagnetic shielding must be applied to reduce or eliminate these emissions.

Figure 6 shows the distribution of EM noise on different PWM duty cycles. The bigger area under the curve means bigger EM noise on that duty cycle in that direction. You can see that the biggest noise emission forward and backward is at 70% duty cycle. Sideway emission is the biggest at 25% duty cycle. Increasing duty cycle causes decreasing sideway radiation.

Figure 7 shows also the measured radiations but as a function of measuring direction. The bigger area the circle has means bigger noise emission in that way (at that duty cycle). It can be clearly seen, while at the 3-phase motor backward radiation, while at the 2-phase motor forward radiation is authoritative. These conclusions can be drawn also from the tables above.

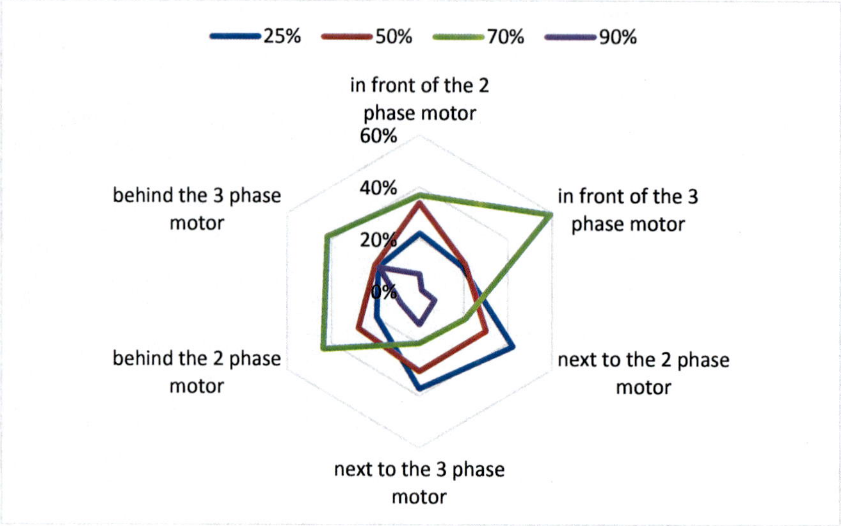

Fig. 6. EM noise distribution at different duty cycles

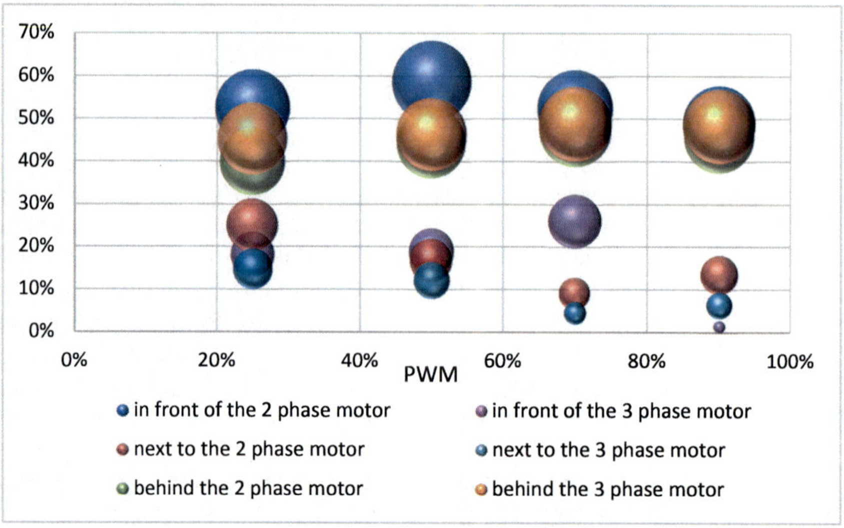

Fig. 7. EM noise distribution in different ways of measurements

4 Summary

Our goal was to measure and compare how much EMC noise is radiated by 2 different but also similar BLDC motors. We had a 2-phase and a 3-phase motor to measure. According to literature, a 3-phase system is more symmetrical therefore lower EM noise must be emitted. We could build the required measurement environment in a shielded laboratory, with which we can measure and compare the mentioned motors. In our case results prove the literature, 2-phase motor performed worse than 3-phase one [11].

Acknowledgement. "This research was supported by the European Union and the Hungarian State, co-financed by the European Regional Development Fund in the framework of the GINOP-2.2.1-15-2017-00090 project, titled "E-mobility from Miskolc: Improvement of Coolant Pump and Engine Cooling Fan Taking into Account the Higher Quality Requirements in Electric Vehicles"."

References

1. Yedamale, P.: Brushless DC (BLDC) Motor Fundamentals (AN885), p. 20. Microchip Technology Inc. (2003)
2. Krall, F., Gruebler, H., Muetze, A.: Angle modulated switching strategy for fractional horsepower BLDC motors for improved electromagnetic compatibility. In: 2019 21st European Conference on Power Electronics and Applications (EPE '19 ECCE Europe), pp. 1–9. IEEE, Genova, Italy (2019)
3. Park, J., Wellawatta, T.R., Ullah, Z., Hur, J.: New equivalent circuit of the IPM-Type BLDC motor for calculation of shaft voltage by considering electric and magnetic fields. IEEE Trans. Ind. Appl. **52**(5), 3763–3771 (2016)
4. Perotti, M., Fiori, F.: Software based control of the EMI generated in BLDC motor drives, In: 2016 International Symposium on Electromagnetic Compatibility - EMC EUROPE, pp. 417–421. IEEE, Wroclaw, Poland (2016)
5. Lee, S., Hur, J.: Detection technique for stator inter-turn faults in BLDC motors based on third-harmonic components of line currents. IEEE Trans. Ind. Appl. **53**(1), 143–150 (2017)
6. Kim, H., Hur, J.: Dynamic characteristic analysis of irreversible demagnetization in SPM- and IPM-Type BLDC motors. IEEE Trans. Ind. Appl. **53**(2), 982–990 (2017)
7. Automotive EMC Testing: CISPR 25, ISO 11452-2 and Equivalent Standards
8. ISO-11452: Road vehicles – Component test methods for electrical disturbances from narrowband radiated electromagnetic energy – Part 2: Absorber-lined shielded enclosure Second Edition 2011-11-01
9. CISPR 16-1-4: Specification for radio disturbance and immunity measurement apparatus and methods Part 1-4 radio disturbance and immunity measuring apparatus – Antennas and test sites for radiated disturbance measurements. 3rd Ed. IEC, Geneva, Switzerland (2010)
10. Vidmar, G., Štibelj, B., Rihtarsic, B., Zagirnyak, M., Miljavec, D.: Evaluation of different mitigation techniques for electromagnetic interference caused by common mode voltage in BLDC outer-rotor motor, In: 2012 15th International Power Electronics and Motion Control Conference (EPE/PEMC), pp. DS1c.3-1–DS1c.3-5. IEEE, Novi Sad, Serbia (2012)
11. Baek, S.: Optimum shape design of a BLDC motor for electric continuous variable valve timing system considering efficiency and torque characteristics. Microsyst. Technol. **24**, 4441–4452 (2018)

Life Cycle Assessment of Traditional and Electric Vehicles

Rafael Ruben Boros(✉), István Bodnár, and Dávid Matusz-Kalász

University of Miskolc, Egyetemváros út 1, Miskolc, Hungary
{elkruben,vegybod,elkmkd}@uni-miskolc.hu

Abstract. This paper presents the LCA (Life Cycle Assessment) of traditional and electric vehicles. Among the traditional cars, we examined diesel, petrol, LPG, and CNG, as well as petrol hybrid and diesel hybrid cars. In case of analysis took into production (indirect) and in-use (direct) environmental emissions from a vehicle. While using traditional vehicles, direct emissions are higher than for electric vehicles. In contrast, the environmental impact categories from vehicle production show reverse proportionality. Among the cumulative impact categories, there is no large difference. Depending on the design life and mileage of the vehicles and how they are used, an electric car can pose an even greater environmental risk over its lifetime than a conventional car.

Keywords: LCA · Electrical vehicles · Diesel and petrol vehicles

1 Introduction

The accelerated extraction of crude oil and the gradual depletion of oil fields, as well as their environmental impacts, are one of the most controversial topics in modern society. Crude oil consumption is concentrated in two areas. One area is plastics production, the other is fuel production. In the 20th century, vehicles equipped with internal combustion engines spread explosively and replaced conventional, animal-powered, or steam-powered machines. While in 1900 barely 100 million barrels of crude oil were produced annually, in 2019 this value had already exceeded 3,000 million barrels. This about thirty times increase also left a very significant mark on the environment. In addition to the carbon footprint, the depletion of fossil fuels as well as the effects of smog formation have also become key issue. However, we have to reckon with environmental damage not only for vehicles running on conventional fuels (petrol and diesel), but also for electric vehicles. While indirect and direct emissions can be quantified in the former, only indirect emissions can be expected in the latter. If we look at vehicles from a life cycle perspective, both indirect and direct emissions should be considered. For conventional vehicles, direct emission is authoritative, while for electric vehicles it is indirect. In the case of electric cars, the way in which the electricity used is generated plays a very decisive role. It does not matter whether the electricity comes from a fossil (e.g. coal-fired) or a renewable power plant (e.g. solar). For electric cars, the battery is another

K. Jármai and K. Voith (Eds.): VAE 2020, LNME, pp. 186–193, 2021.
https://doi.org/10.1007/978-981-15-9529-5_16

key factor. The production of batteries and their recycling and disposal have significant environmental effects. The question arises, is an electric car really more environmentally friendly than a traditional one?

2 The Methods of Life Cycle Assessment

Nowadays, Life Cycle Assessment (LCA) is one of the most popular environmental management system tools, the application of which is the most expedient, especially for substitute services, products, and technologies. Apropos of LCA, we quantify and estimate the environmental impact of a product, technology or service over its entire life cycle (production, distribution, use, disposal of waste) and what and how much natural resources it uses (including energy expenditure).

Based on ISO 14040, life cycle analysis can be defined as follows: "a method of assessing product-related environmental factors and potential impacts that takes stock of the inputs and outputs of a system of product-related processes; assess the potential environmental impacts associated with them; interprets the results of the inventory analysis and impact assessment phases taking into account the objectives of the study" [1].

In the classical sense, a product or technology can only be described as environmentally friendly if the product itself and the waste generated during its production and consumption do not have a direct and indirect impact on the environment. In this sense, the ecological balance of a given product in the "cradle to grave" approach is positive, but at least neutral to the environment. In practice, a production and consumption process that can be considered waste-free is inconceivable. Therefore, the use of the positive indicator itself, in this case, suggests that adverse environmental impacts on products and technologies are smaller. Recognizing the environmentally friendly nature of a product or technology is not an easy task and in practice, we may encounter many contradictions [1].

The interpretation phase: Life cycle interpretation is the final phase of the LCA procedure, in which the results of the LCI or an LCIA, or both, are summarized and discussed as a basis for conclusions, recommendations and decision-making in accordance with the goal and scope definition [1].

The Life Cycle Assessment study has four main phases. In the first stage, we define the purpose and scope of the analysis. In the second stage, we perform an inventory analysis of whether basic physical laws, such as the law of conservation of matter and energy, prevail. The third stage is the most interesting, as this is where the impact assessment takes place. Based on the inventory data, we assign an environmental impact to each material and energy flow. Environmental impacts are classified into so-called environmental impact categories. Finally, the fourth stage is the comparison, evaluation, and decision-making of LCA results [1, 13].

In the second stage of the analysis, a reference unit is defined for each impact category, against which other inventory data with the same impact are compared. For example, greenhouse gases are measured in kg CO_2 equivalent. The impact of 1 kg of CO_2 on global warming represents 1 kg of CO_2 equivalent. The contribution of methane impact categories to global warming is given in kg CO2 equivalent, which can be 21, 23 or 25 kg, depending on the method used. Inventory data can be linked to multiple impact categories [1, 10].

There are different methods for weighting impact categories that allow the expression of potential environmental impacts using an indicator after life-cycle assessment. There are different methods for weighting impact categories that allow the expression of potential environmental impacts using an indicator after life-cycle modelling. In our studies, we can also use a free trial version of commercially available life cycle analysis software called GaBi. The GaBi 8 software contains about 100 evaluation methods (e.g. CML 96, CML 2001, EDIP 97, EDIP 2003, EPFL 2002+, TRACI, Eco-indicator 95, Eco-indicator 99 etc.), thus practically covering the most widely used current methods. The environmental impact categories according to the CML are given in Table 1.

Table 1. The environmental impact categories according to the CML.

Environmental impact categories	Reference
Global Warming Potential (GWP)	kg CO_2- Equivalent
Acidiphication Potential (AP)	kg SO_2- Equivalent
Eutrophication Potential (EP)	kg phosphate- Equivalent
Human Toxicity Potential (HTP)	kg DCB- Equivalent
Photochemical Ozone Creation Potential (POCP)	kg ethylene- Equivalent
Ozon Layer Depletion Potential (ODP)	kg R11- Equivalent
Terrestric Ecotoxicity Potential (TETP)	kg DCB- Equivalent
Marine Ecotoxicity Potential (MAETP)	kg DCB- Equivalent
Freshwater Aquatic Ecotoxicity (FAETP)	kg DCB- Equivalent
Abiotic Depletion Potential fossil (ADP fossil)	MJ
Abiotic Depletion Potential elements (ADP elements)	kg Sb- Equivalent

3 Traditional and Electrical Vehicles

Internal combustion engines in vehicles require much more maintenance than electric motors used in electric cars. Electric cars are fitted with electric motors that do not contain carbon brushes. Electric cars are equipped with a permanent magnet synchronous machine (PMSM) that can deliver high torque even when starting the car at 0 km/h. These electric motors require large Li-ion battery packs. Batteries generate electricity electrochemically, so their production, use and recycling are not environmentally friendly. The Li-ion battery, on the other hand, is flammable, so its use is not safe. The materials that make up a Li-ion battery, when released into the environment, do more damage than carbon dioxide (2012/19/EU). Battery life is estimated at 20–22 years, so ideally no battery needs to be replaced during the life of the car. However, the reality does not reflect this, the capacity of batteries is reduced by about 3–4% per year. Because of this, the battery pack will be replaced prematurely because the range of the car will be greatly reduced at low capacity. Obsolete batteries can still be used for other purposes, but this does not cause a large reduction in impact categories. Due to the rapid spread of electric car use, there is a growing demand for electricity. For this reason, transmission lines also need to be scaled up and more installed [5, 6].

In internal combustion cars, lubricating oil changes and fuel filter replacements cause a lot of pollution. The production of lubricating oils is also associated with high carbon dioxide emissions. Internal combustion cars also have a battery, which also increases emissions, but these batteries are smaller in size and less flammable. Improperly maintained cars easily release lubricating oil or fuel into the environment, which gets into the ground [11].

4 The LCA Results

The Life Cycle Analysis was performed for 7 passenger cars with different fuels. The analysis in each case consists of two parts. On the one hand, we examined direct emissions from fuel consumption and, on the other hand, indirect emissions from car production. According to the EPD (Environmental Product Declaration) (ISO 14025, ISO 14044, ISO 15804), these so-called Life stages “A” and “B”. In both cases, a distance of 1 km was chosen as the functional unit [9]. For all vehicles, it was assumed that no major service was required for the first 150,000 km of service life. The total lifetime was taken as 300;000 km. Details of the vehicles and fuels tested are given in Table 2 [3, 4].

Table 2. Baseline data for LCA analysis.

Tested vehicle types	Tested fuel types
EURO 6 Diese 11400–2000 cm^3	Diesel EU-28
EURO 6 Petrol 1400–2000 cm^3	Petrol EU-28
EURO 6 LPG 1400–2000 cm^3	LPG EU-28
EURO 6 CNG 1400–2000 cm^3	CNG EU-28
EURO 6 Diesel-hybrid 1400–2000 cm^3	Diesel EU-28, Electricity EU-28
EURO 6 Petrol-hybrid 1400–2000 cm^3	Petrol EU-28, Electricity EU-28
Pure electric 100 kW electric motor (50 kWh battery bank)	Electricity EU-28

Table 3 compares the impact categories from the fuel used to cover 1 km. It can be observed that the largest direct greenhouse gas emissions are found in the petrol car. The smallest value is represented by the electric car. This value can be further reduced if the electricity used comes from renewable energy sources or a nuclear power plant. It can also be seen that the direct emissions of hybrid vehicles are lower than those of their pure diesel or gasoline-powered counterparts. This finding is only valid for the EURO 6 category. If we consider older hybrid vehicles of EURO 3 or EURO 4 environmental class, their emissions exceed those of conventional fuels of EURO 6 category [2, 8, 12].

If emissions are summed and compared to the electric car, the petrol vehicle is up 269.6%, the LPG up 265.6%, the CNG up 263.5%, and the gasoline hybrid up 235.7%, diesel has 207.0% higher diesel direct emissions and 185.6% higher direct emissions.

Table 3. Impact categories from (indirect) fuel consumption.

	GWP	AP	EP	ODP	ADP elements	ADP fossil	HTP
Diesel	0.131	1.10E-04	2.99E-05	2.38E-17	1.11E-08	1.950	2.98E-03
Petrol	0.158	1.44E-04	2.38E-05	6.32E-17	1.18E-08	2.550	6.16E-03
LPG	0.140	1.25E-04	9.60E-06	2.74E-17	5.13E-09	2.530	4.68E-03
CNG	0.133	6.77E-05	3.51E-06	2.94E-17	1.04E-10	2.520	1.41E-04
Diesel-hybrid	0.123	1.74E-04	3.05E-05	1.61E-16	1.19E-08	1.741	4.29E-03
Petrol-hybrid	0.144	2.01E-04	2.56E-05	1.92E-16	1.24E-08	2.222	6.83E-03
Electricity	0.089	4.28E-04	3.30E-05	7.08E-16	1.49E-08	0.908	9.52E-03

Table 4 shows the impact categories from vehicle production per 1 km. It can be observed that in all environmental impact categories, the electric car has the highest impact categories. This is understandable, as the battery technology currently used represents a very significant environmental risk. The electric car is followed by gasoline and diesel-powered hybrids. In terms of total impact categories, the petrol-powered car increased by 35.3%, the LPG-powered car by 34.7%, the CNG by 34.58%, the diesel by 29.23%, and the petrol hybrid by 25.67% and the diesel hybrid has 19.81% lower indirect emissions.

Table 4. Impact categories from vehicle production.

	GWP	AP	EP	ODP	ADP elements	ADP fossil	HTP
Diesel car	3.46E − 02	8.58E − 05	1.64E − 05	4.46E − 12	5.07E − 06	1.71E + 02	2.70E − 04
Petrol car	1.21E − 02	8.08E − 05	1.82E − 05	4.08E − 12	4.63E − 06	1.57E + 02	2.17E − 03
LPG car	3.58E − 02	2.67E − 05	7.55E − 06	4.12E − 12	4.68E − 06	1.58E + 02	5.85E − 06
CNG car	2.65E − 02	2.90E − 05	7.62E − 06	4.13E − 12	4.69E − 06	1.58E + 02	6.53E − 06
Diesel-hybrid	1.96E + 00	5.56E − 03	5.29E − 04	5.00E − 12	5.68E − 06	1.92E + 02	1.21E − 01
Petrol-hybrid	1.99E + 00	5.68E − 03	5.43E − 04	4.63E − 12	5.26E − 06	1.78E + 02	1.26E − 01
Electric car	2.05E + 01	5.81E − 02	5.45E − 03	5.74E − 12	6.52E − 06	2.20E + 02	1.29E + 00

We summarize direct and indirect emissions. The results shown in Table 5 are then obtained. It can be observed that the electric car has the highest environmental impact values. All this is illustrated in Fig. 1 based on the relative distribution of total impact categories. The larger the area under a curve, the greater the environmental impact of the car associated with it.

If we examine the percentage distribution of the environmental load of each car, we get Fig. 2 as a result. From a life cycle perspective, the most environmentally friendly car is gasoline-powered. This is followed in turn by LPG, CNG, diesel, gasoline hybrid, diesel hybrid and finally the electric car.

There is no clear significant difference between the results reported above. This is because, in addition to the mileage expressed in km, the operating hours and annual km

Table 5. Cumulative impact categories (direct and indirect).

	GWP	AP	EP	ODP	ADP elements	ADP fossil	HTP
Diesel car	1.66E – 01	1.96E – 04	4.63E – 05	4.46E – 12	5.08E – 06	1.73E + 02	3.25E – 03
Petrol car	1.70E – 01	2.25E – 04	4.20E – 05	4.08E – 12	4.65E – 06	1.59E + 02	8.33E – 03
LPG car	1.76E – 01	1.52E – 04	1.72E – 05	4.12E – 12	4.68E – 06	1.61E + 02	4.69E – 03
CNG car	1.60E – 01	9.67E – 05	1.11E – 05	4.13E – 12	4.69E – 06	1.61E + 02	1.48E – 04
Diesel-hybrid	2.09E + 00	5.73E – 03	5.60E – 04	5.00E – 12	5.69E – 06	1.94E + 02	1.26E – 01
Petrol-hybrid	2.13E + 00	5.88E – 03	5.68E – 04	4.63E – 12	5.27E – 06	1.80E + 02	1.33E – 01
Electric car	2.06E + 01	5.86E – 02	5.48E – 03	5.74E – 12	6.53E – 06	2.21E + 02	1.30E + 00

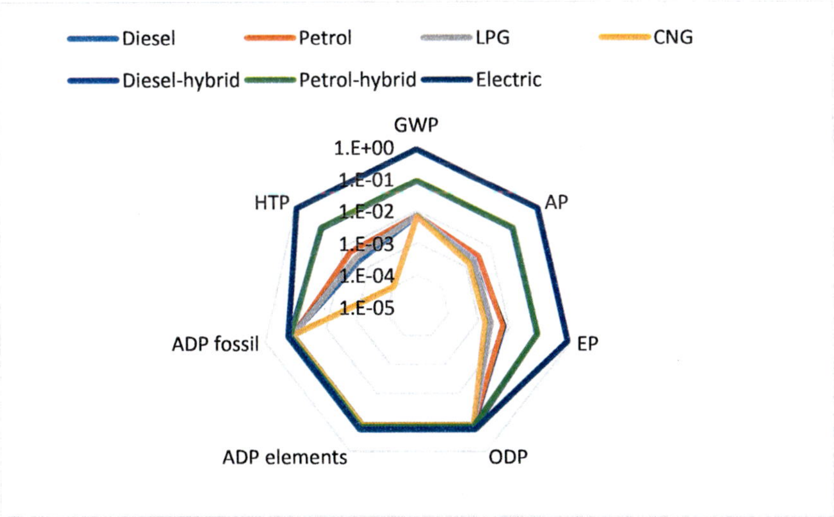

Fig. 1. Cumulative impact categories.

usage, as well as the mode of operation, must not be disregarded. In urban conditions, not only consumption but also wear and tear is more significant, so those cars may need more maintenance. It does not matter whether the vehicle reaches the planned service life of 10 or 20 years. Older cars are more prone to corrosion, so there is a greater environmental impact of maintenance due to the body being locked and repainted. For LPG and CNG vehicles, replacing the gas storage tank, while for an electric car, replacing the battery has an additional environmental impact. Battery replacement is more authoritative, so the life-cycle environmental impact categories of electric vehicles outweigh the same environmental impacts of conventional vehicles [7, 14].

Consider the case where each vehicle is used under the same conditions. Then operating expenses can be divided into two groups. One group includes factors that occur equally in all vehicles. Examples are the environmental effects of replacing wearing parts (such as brake pads and brake discs). The other group includes those that occur only for individual vehicles. For example, in cars with an internal combustion engine,

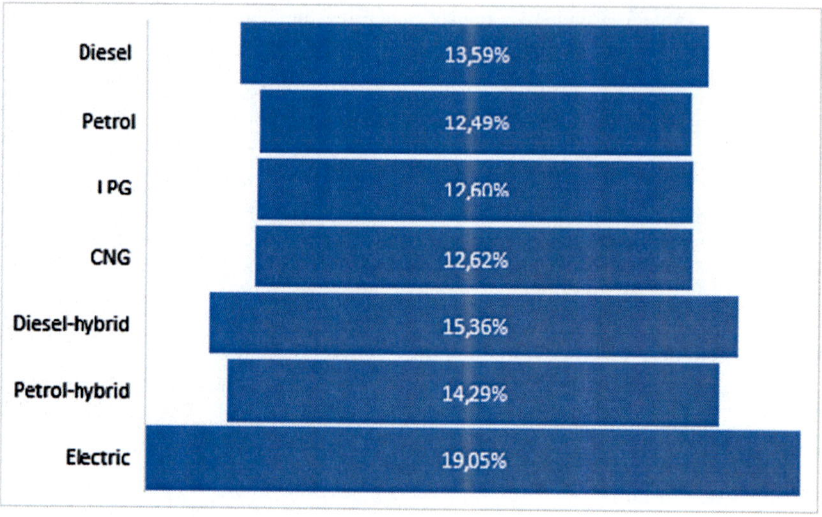

Fig. 2. Relative emission.

changing the engine oil or maintaining the engine itself (control change). Since the factors in the first group occur equally, they do not substantially change the proportions shown in the above price range. Those in the second group, on the other hand, do. Due to oil changes and maintenance of internal combustion engines, the emission rates of those cars are increasing. Since the electric motors of electric cars are practically maintenance-oriented, the impact categories there are unchanged, i.e. the proportions converge, equalize, or even the weights can be placed t. If the batteries of electric and hybrid cars can withstand the planned service life of 300,000 km, then the life-cycle impact categories of each car can be of the same order of magnitude, almost the same value.

At the end of the life cycle, we still must reckon with the impact categories from the disposal/recycling. Here, in the first place, the utilization of batteries may differ from that of cars without batteries. Overall, therefore, it can be said that there is no significant difference between the life-cycle environmental impacts of individual hairy vehicles, so it is appropriate to conduct further studies on specific cars.

5 Conclusions

Overall, there are no significant differences in the environmental impact of EURO 6 traditional cars and electric cars. While in the case of traditional cars, direct emissions are authoritative, in the case of electric cars, the production and recycling life cycle of the car is more environmentally burdensome. Depending on the design life and mileage of the vehicles and how they are used, an electric car can pose an even greater environmental risk over its lifetime than a traditional car. Of course, a number of other factors (e.g. maintenance and refurbishment) also depend on the life-cycle environmental impact categories. The environmental impact of electric cars can be reduced if the electricity they use comes from renewable energy sources or nuclear power plants. The continuous development of battery technology can also benefit electric cars. Based on all this, the effects caused by each vehicle can be balanced. Diesel cars older than EURO 6 that do not use a particulate filter are typically more polluting than their electric counterparts. The environmental impact of hybrid vehicles is heavily dependent on the type of battery and charge, but they are typically more environmentally burdensome than their traditional counterparts. The future goal is a full lifecycle comparison.

References

1. Hauschild, M.Z.: Introduction to LCA Methology, Life Cycle Assessment, pp. 59–66 (2017)
2. Donateo, T., Ingrosso, F., Bruno, D., Laforgia, D.: Effect of driving conditions and auxiliaries on mileage and CO2 emissions of a gasoline and an electric city car, SAE Technical Paper, 2014-01-1812, p. 13 (2014)
3. Danilecki, K., Mrozik, M., Smurawski, P.: Changes in the environmental profile of a popular passenger car over the last 30 years – Results of a simplified LCA study. J. Clean. Prod. **141**, 208–218 (2017)
4. Piotrowska, K., Kruszelnicka, W., Bałdowska-Witos, P., Kasner, R., Rudnicki, J., Tomporowski, A., Flizikowski, J., Opielak, M.: Assessment of the environmental impact of a car tire throughout its lifecycle using the LCA method. Materials **12**, 4177, p. 12 (2019)
5. Boureima, F.S., Messagie, M., Matheys, J., Wynen, V., Sergeant, N., Van Mierlo, J., De Vos, M., De Caevel, B.: Comparative LCA of electric, hybrid, LPG and gasoline cars in Belgian context. World Electr. Veh. J. **3**, 469–479 (2009)
6. Reinhardt, R., Pautzke, F., Schröter, M., Wiemers, M.: A case study of sustainable manufacturing strategy: comparative LCA of wheel hub engine for solar car application. In: 2017 International Conference on Research and Education in Mechatronics (REM), pp. 1–6. IEEE, Wolfenbuttel (2017)
7. Weymar, E., Finkbeiner, M.: Statistical analysis of empirical lifetime mileage data for automotive LCA. Int. J. Life Cycle Assess. **21**, 215–223 (2016)
8. Antonio, F., Rizzo, G., De Feo, G., Landolfil, S.: Converting a conventional car into a hybrid solar vehicle: a LCA approach. IFAC-PapersOnLine **51**(31), 188–194 (2018)
9. Schaubroeck, S., Schaubroeck, T., Baustert, P., Gibon, T., Benetto, E.: When to replace a product to decrease environmental impact?—A consequential LCA framework and case study on car replacement. Int. J. Life Cycle Assess. **25**, 22 (2020)
10. Ding, N., Pan, J., Zhang, Z., Yang, J.: Life cycle assessment of car sharing models and the effect on GWP of urban transportation: a case study of Beijing. Sci. Total Environ. **688**, 1137–1144 (219)
11. Del Pero, F., Delogu, M., Pierini, M.: The effect of lightweighting in automotive LCA perspective: Estimation of mass-induced fuel consumption reduction for gasoline turbocharged vehicles. J. Clean. Prod. **154**, 566–577 (2017)
12. Qiang, Z.Y., Quang, W.Y., Yuan, S.H.: Energy management of low voltage power supply of plug-in hybrid electric vehicle. IFAC-PapersOnLine **51**(31), 1–6 (2018)
13. Öztas, S.K.: The limitations of LCA methodology towards sustainable construction materials. In: International Sustainable Buildings Symposium ISBS 2017: Proceedings of 3rd International Sustainable Buildings Symposium (ISBS 2017), pp. 102–113. Springer, Dubau (2017)
14. Moro, A., Helmers, E.: A new hybrid method for reducing the gap between WTW and LCA in the carbon footprint assessment of electric vehicles. Int. J. Life Cycle Assess. **22**, 4–14 (2017)

Special Optimization Process for Warehouse Layout Design

György Kovács(✉)

Institute of Logistics, University of Miskolc, Miskolc, Hungary
altkovac@uni-miskolc.hu

Abstract. Warehousing activity is essential for all manufacturing companies (final assemblers and suppliers) of the global supply chains. Warehouses are also key elements of global automotive supply chains because the optimal formation and operation of warehouses play important role in manufacturing companies in order to maintain and increase their competitiveness. Warehousing is one of the most important and cost-intensive logistical processes in which the main goal is to reduce costs and improve efficiency. Determination of the ideal warehouse layout is a special optimization process, not a typical mathematical optimization. Therefore, the warehouse layout design has many special characteristics. In the article, the detailed procedure and characteristics of the special optimization process of the warehouse layout design are described. Furthermore, the study also introduces the most commonly used objective functions, their calculation methods, and a large number of design constraints and limitations. The author elaborated a three-step method for defining the ideal warehouse layout which is also introduced in the paper. The main added-value of the study is, that a detailed procedure of the special optimization process of warehouse layout design was elaborated and described by the author. There is a gap in the existing literature in the research topic of the special optimization process of warehouse layout design because there is not available any article which especially discusses the complex and detailed procedure of the special optimization process in any depth.

Keywords: Warehouse layout design · Special optimization process · Ideal warehouse layout

1 Introduction

Increasing global competition, rapidly changing customer demands, more complex networks of global supply chains resulted in significant changes in the production and logistics sector. In the production sector, the production conceptions have changed from the traditional mass ("Push" – "make to stock") production to the Industry 4.0 concept. These changes require efficient operation of logistical processes especially in the most important sectors, i.e. in transportation and warehousing [1, 2].

The before mentioned global tendencies resulted in significant changes also in the warehouse sector. Warehousing is an important and cost-intensive logistical process in

K. Jármai and K. Voith (Eds.): VAE 2020, LNME, pp. 194–205, 2021.
https://doi.org/10.1007/978-981-15-9529-5_17

global supply chains, also in automotive supply chains. The main goal of warehousing is to reduce costs and improve efficiency. Therefore, the topic of the article is relevant and up-to-date, because the main aims of the manufacturing companies are the optimal design and operation of the warehouses in order to maintain and increase their competitiveness [3–5].

The following global tendencies can be defined in the warehouse sector:

- The main goal in the warehouse sector is the reduction of inventories in order to reduce costs.
- The spatial concentration of stocking, because fewer warehouses in global supply chains result in significant cost saving.
- Implementation of the elements of the Industry 4.0 concept. Automation of warehousing processes, application of innovative technologies and machines.
- Application of Pull philosophy results in reduction of inventories. Furthermore, the establishment of new inventory strategies provides both enhance the speed of material flow and reduction of stocking time (e.g. Just in Time, Cross Docking and Vendor Managed Inventory strategies result in stock reduction and scheduled flow of goods in the supply chains).
- Formation of the optimal warehouse layout and efficient operation of the warehouse (e.g. application of identification systems and Warehouse Management Systems) [6, 7]

The selection of the ideal warehouse layout is a special optimization process, not a typical mathematical optimization. Since the number of layout alternatives is huge or infinite [8, 9]; therefore, the formation and the evaluation of all possible alternatives are impossible. Consequently, the global optimal warehouse alternative is very difficult or impossible to define, only the best solution, the "ideal" layout alternative can be determined.

Furthermore, the warehouse layout design is always a complex and unique task; in addition, it can be concluded, that a uniform and standard procedure for the warehouse layout design is not available either in practice or in literature. Therefore, each researcher and industrial expert elaborates and applies their own procedure differently. Consequently, the procedures relating to the warehouse layout design of each author are unique and can contribute to the recent state of the research field if they differ from other published procedures and at the same time lead to significant efficiency improvement.

In the article, the characteristics and the detailed procedure of the special optimization process of the warehouse layout design are described. Furthermore, the study also introduces the most important and commonly used objective functions, their calculation methods, and a large number of design constraints and limitations.

The author elaborated on the three-step method for defining the ideal warehouse layout which is also introduced in the paper. The three phases of the method are the following: 1) Alternative warehouse layouts have to be created considering the design constraints and limitations. 2) The huge number of alternative warehouse layouts has to be reduced by a heuristic method and by continuous iteration. 3) Then the reduced, smaller number of warehouse layout alternatives has to be compared. The ideal layout has to be selected based on the defined most important objective functions by the application of the systematic search method. This three-steps method can be applied in the practice in all cases of warehouse layout design.

The main added-value of the study is, that the detailed procedure of the special optimization process of warehouse layout design was elaborated and described by the author. It is novel because there is not available any publication in the existing literature which especially discusses the complex and detailed procedure of the special optimization process in any depth.

2 Special Optimization Procedure and Characteristics of the Warehouse Layout Design

Determination of the ideal warehouse layout is a special optimization process, not a typical mathematical optimization. **The reasons of the specialty of the optimization process are the following:**

- In the warehouse layout design procedure, the number of possible alternatives is huge or infinite [8]. Therefore, the formation and evaluation of all possible alternatives are impossible. This is one of the reasons that heuristic method and continuous iteration have to be used during the optimization process. On the contrary, in case of the classical optimization, the number of possible alternatives is finite. Consequently, the formation and evaluation of all possible alternatives are possible.
- Based on the before mentioned facts, in case of the warehouse layout design procedure, the global optimal layout is very difficult or impossible to define according to the opinion of most of the researchers [10, 11]. On the contrary, in case of the classical optimization, the global optimal solution always can be defined.
- Multi-objective optimization has to be applied in every case of warehouse layout design. The number of objective functions is large; therefore, their priority and weight must be determined.
- The number of design constraints and limitations is huge; among these, there are many interactions. The design constraints and limitations must be considered simultaneously from the beginning to the end of the optimization procedure.
- Most of the design constraints and limitations cannot be defined by mathematical formulas. Therefore, this is one of the reasons that heuristic method and continuous iteration have to be used during the optimization process.

3 General Procedure and Main Practical Steps of the Warehouse Layout Design

The main practical steps of the general procedure of the warehouse layout design are the following [6, 12, 13]:

At first design aims, management's demands, furthermore, the main and supporting activities of the warehousing process have to be defined. Main design aims are: maximal utilization of space, equipment and maximal efficiency of operations, etc.

Step 2: Preparing forecasts relating to design data.

Forecasts have to be prepared relating to the most important data needed for the design [e.g. types and volumes of goods to be stored; space requirement for storage and other warehouse operations (e.g. space for receiving, loading-in, loading-out, order picking, etc.) and the required warehouse infrastructure].

Step 3: Defining the most important objective functions of the warehouse layout design.

At first, the most important objective functions have to be determined relating to the given warehouse layout design. Then the priority (weight) of the different objective functions has to be defined.

Step 4: Defining the design constraints and limitations of the warehouse layout design.

The important design constraints and limitations have to be determined relating to the given warehouse layout design.

Step 5: Forming alternative warehouse layouts.

Alternative warehouse layouts have to be created considering design constraints and limitations.

Step 6: Reducing the number of potential warehouse layout alternatives.

The huge number of warehouse layout alternatives has to be reduced by heuristic method and by continuous iteration.

Step 7: Comparison of potential warehouse layout alternatives.

The reduced, smaller number of layout alternatives have to be compared based on the defined most important objective functions by the application of the systematic search method.

Step 8: Selecting the ideal warehouse layout.

Based on the comparison the ideal, the best warehouse layout has to be selected.

Step 9: Implementation.

Implementation of the ideal warehouse layout in the practice.

4 Optimization Procedure of the Warehouse Layout Design

4.1 Objective Functions and Their Calculation Methods

During the optimization, the most important objective functions have to be defined. After it, the importance (weight) of the different objective functions has to be determined.

The main goals of the warehouse design are the following:

- maximize resource (e.g. space, equipment, human, etc.) utilization according to customer requirements, and
- maximize service level for the customers taken into consideration the design constraints and limitations [14, 15].

The most important and common used objective functions are the following:

1) maximal storage capacity of the warehouse;
2) maximal utilization of floor area of the warehouse;
3) maximal utilization of space in the warehouse;
4) minimal material flow distances and total material workflow;
5) minimal investment cost;
6) minimal operating cost;
7) minimal time and cost of receiving, loading-in, loading-out, order picking, shipping and other activities;
8) maximal utilization of human and equipment;
9) maximal service level of the storage technology (flexibility of the storage system, etc.);
10) maximal service level of the material handling system (reliability and flexibility of machines, etc.).

Generally, most of the warehouse designers apply only the maximal storage capacity objective function for the selection of the ideal warehouse layout in the practice. On the contrary – in my opinion – the best warehouse layout can be defined if – at least – the before mentioned 1–6 objective functions are taken into consideration during the selection of the ideal warehouse layout. Consequently, the calculation methods of the 1–6 objective functions are described in the next part of the article.

1) Maximal storage capacity of the warehouse

One of the most important metrics is the maximum amount of unit loads (UL) that can be stored in the warehouse ($N_{\max i}$).

$$N_{\mathrm{max}i} = n_{xi} \cdot n_{yi} \cdot n_{zi} \,[\mathrm{UL}] \tag{1}$$

where:

n_{xi} – number of unit loads that can be stored in transversal direction of the warehouse,
n_{yi} – number of unit loads that can be stored in longitudinal direction of the warehouse,

n_{zi} – number of layers of unit loads that can be stored in vertical direction,
i – identifier of a given warehouse layout alternative.

2) Maximal utilization of the floor area of the warehouse

- Ratio of the area used for storage to the total floor area of the warehouse (α_i)

$$\alpha_i = \frac{A_{fi}}{A_t} \cdot 100\,[\%] \tag{2}$$

– A_{fi} – warehouse area used for storage in case of the i-th warehouse layout alternative:

$$A_{fi} = L_{fxi} \cdot L_{fyi}\left[\mathrm{m}^2\right] \tag{3}$$

where:

L_{fxi} – length of the area used for storage in x direction in case of the i-th alternative,
L_{fyi} – length of the area used for storage in y direction in case of the i-th alternative,
i – identifier of a given warehouse layout alternative.

– A_t – total floor area of the warehouse:

$$A_t = L_{tx} \cdot L_{ty}\left[\mathrm{m}^2\right] \tag{4}$$

where:

L_{tx} – total length of the warehouse in x direction,
L_{ty} – total length of the warehouse in y direction.

- Ratio of the useful storage floor area to the total floor area of the warehouse (α_i^{ht})

$$\alpha_i^{ht} = \frac{A_{hi}}{A_t} \cdot 100\,[\%] \tag{5}$$

– A_{hi} – useful storage floor area of the warehouse:

$$A_{hi} = A_{fi} - (T_{fi} + T_{ei})\left[\mathrm{m}^2\right] \tag{6}$$

where:

T_{fi} – area of aisles in the warehouse in case of the i-th layout alternative,
T_{ei} – areas required for receiving, sorting, shipping and other additional activities in case of the i-th layout alternative.

- Ratio of useful storage floor area to storage area of the warehouse (α_i^{hf})

$$\alpha_i^{hf} = \frac{A_{hi}}{A_{fi}} \cdot 100\,[\%] \tag{7}$$

3) Maximal utilization of space in the warehouse

- Ratio of the space used for storage to the total space of the warehouse (β_i)

$$\beta_i = \frac{V_{fi}}{V_t} \cdot 100\,[\%] \tag{8}$$

where:

V_{fi} – space used for storage in the warehouse in case of the *i*-th layout alternative,
V_t – total space of the warehouse in case of the *i*-th layout alternative.

- Ratio of useful storage space to the total storage space of the warehouse (β_i^{ht})

$$\beta_i^{ht} = \frac{V_{hi}}{V_t} \cdot 100\,[\%] \tag{9}$$

where:

V_{hi} – useful storage space in the warehouse in case of the *i*-th layout alternative.

- Ratio of useful storage space to the space used for storage in the warehouse (β_i^{hf})

$$\beta_i^{hf} = \frac{V_{hi}}{V_{fi}} \cdot 100\,[\%] \tag{10}$$

4) Minimal material flow distances and total material workflow

During the design of material flow paths, the main material flow directions, the connection points between the internal and external material flow ways and the characteristics of the warehouse building have to be considered which determine the directions of the main service aisles (longitudinal or transversal arrangement).

Data relating to the material flow intensity and turnover of the different unit loads are important for the optimal operation of the warehouse, to minimize material flow distances and total material workflow.

Material workflow (E_{MWF}) is a widely used objective function for the description of the amount of workflow that should be minimized [16, 17].

Material workflow can be calculated:

$$E_{MWFi} = \sum_{j=1}^{n} \sum_{k=1}^{n} q_{ijk} l_{ijk}\,[\mathrm{UL \cdot m}] \tag{11}$$

where:

q_{ijk} – elements of the $\mathbf{Q}_i$ material flow matrix, which means the material flow between the j-th point and k-th point of the warehouse in case of the i-th layout alternative,
l_{ijk} – elements of the $\mathbf{L}_i$ distance matrix, which means the distance between the j-th point and k-th point of the warehouse in case of the i-th layout alternative.

Matrix of material flow:

$$\mathbf{Q}_i = \begin{matrix} 1 \\ \vdots \\ j \\ \vdots \\ n \end{matrix} \overset{\begin{matrix} 1 & \dots & k & \dots & n \end{matrix}}{\begin{bmatrix} & & & & \\ & & & & \\ & & q_{ijk} & & \\ & & & & \\ & & & & \end{bmatrix}} [\mathrm{UL}]$$

Matrix of distances:

$$\mathbf{L}_i = \begin{matrix} 1 \\ \vdots \\ j \\ \vdots \\ n \end{matrix} \overset{\begin{matrix} 1 & \dots & k & \dots & n \end{matrix}}{\begin{bmatrix} & & & & \\ & & & & \\ & & l_{ijk} & & \\ & & & & \\ & & & & \end{bmatrix}} [\mathrm{m}]$$

The material handling cost is proportional to the quantity of material flow and the travel distances. This cost is linearly proportional to the material workflow.

5) Minimal investment cost

The construction of warehouses requires significant investment. Therefore, the minimization of investment costs is an important design aim. The investment cost can be calculated for each alternative:

$$K_{Inv_i} = K_{build_i} + K_{rack_i} + K_{mach_i} + K_{inf_i} + K_{oth_i} \text{ [euro]} \tag{12}$$

where:

K_{build_i} – cost of construction/reconstruction of the warehouse building,
K_{rack_i} – cost of storage racks,
K_{mach_i} – cost of material handling equipment,
K_{inf_i} – cost of the IT investment required for warehouse management,
K_{oth_i} – other additional costs in connection with construction/recon- struction of the warehouse building (e.g. planning, fire protection, security system, etc.),
i – identifier of a given warehouse layout alternative.

6) Minimal operating cost

The economic operation of warehouses in the long term is also an important design aim. The operating cost can be calculated as the sum of the following cost components for each alternative:

$$K_{Oper_i} = K_{handl_i} + K_{hum_i} + K_{ener_i} + K_{oth_i} \text{ [euro]} \tag{13}$$

where:

K_{handl_i} – cost of material handling inside the warehouse building,
K_{hum_i} – cost of human resource,
K_{ener_i} – cost of energy required for the operation of the warehouse,
K_{oth_i} – other costs in connection with the operation of the warehouse (maintenance, administration, etc.),
i – identifier of a given warehouse layout alternative.

4.2 The Most Important Design Constraints and Limitations

During the optimization design constraints and limitations have to be defined relating to the given warehouse layout design. The number of design constraints and limitations is huge; among these, there are many interactions. The design constraints and limitations must be considered simultaneously from the beginning to the end of the optimization procedure.

Types, variety, volume and turnover of goods to be stored in the warehouse or the area that can be used for storage is continuously changing over time. Therefore, the warehouse must be adapted flexibly to changes in the long term. Consequently, the warehouse must be easily adaptable and reconfigurable, which must be taken into account during the warehouse layout design.

Generally, the most important and common used design constraints and limitations are the following:

- Architectural characteristics of the warehouse building (e.g. floor area and height of the building; location of roof supporting pillars; location of warehouse dock doors; loading capacity of the floor, etc.).
- Types, number, geometrical dimensions, weight, stackability, turnover of the unit loads and unit load forming devices to be stored; furthermore, special storage requirements (e.g. cooling, heating, hazardous goods, etc.).
- Area used for main warehouse operations (e.g. receiving of goods; storage; loading in; loading out, etc.).
- Area used for further warehouse operations (e.g. order picking; checking; labeling; storage of packaging materials; waste storage, etc.).
- Possibilities of material flow paths, i.e. determining which are the typical material flow directions, which define the direction of the main service aisles (longitudinal or transversal aisles); furthermore, relations between internal and external material flow ways [18, 19].
- Types and characteristics of warehouse machines and equipment [20].
- Budget limitations of the warehouse layout design/redesign.

5 Three-Steps Method for Defining the Ideal Warehouse Layout

I elaborated a three-steps method for defining the ideal warehouse layout. **The three phases of the method are the following:**

1) **First phase:** Potential warehouse layout alternatives have to be created – consider design constraints and limitations – based on the existing literature (see Fig. 1) [3, 4, 14, 15].
   The potential adequate warehouse storage system, storage modes and further layout alternatives relating to the given warehouse layout design have to be selected.
2) **Second phase:** In the warehouse layout design procedure the formation and the evaluation of all possible alternatives are impossible because the number of possible alternatives is huge or infinite. (Consequently, in case of warehouse layout design

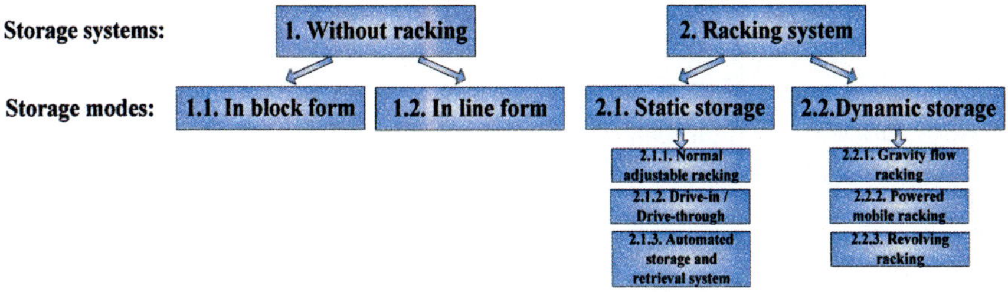

Fig. 1. General types of warehousing system.

classical optimization procedure cannot be completed, because in case of the classical optimization the number of possible alternatives is finite, so the formation and the evaluation of all possible alternatives are possible.)

Based on the before mentioned facts, the huge number of alternative warehouse layouts has to be reduced – consider the design constraints and limitations – by heuristic method and by continuous iteration.

3) **Third phase:** Then the reduced, smaller number of warehouse layout alternatives has to be compared based on the defined most important objective functions by the application of the systematic search method.

Based on the comparison the ideal, the best warehouse layout has to be selected. After it, the ideal warehouse layout has to be implemented in the practice.

This newly elaborated design method can be applied in the practice in every case of warehouse layout design.

6 Conclusion

Optimal formation and operation of warehouses are important tasks for manufacturing companies. Warehousing is an important and cost-intensive logistical process in global supply chains, also in automotive supply chains.

Determination of the ideal warehouse layout is a special optimization process, – not a typical mathematical optimization – which has many special characteristics. The reason of it is that the number of warehouse layout alternatives is huge or infinite. Therefore, the formation and evaluation of all possible alternatives are impossible. Consequently, the global optimal warehouse alternative cannot be defined, only the best solution, the "ideal" warehouse layout alternative can be determined.

In the article, the characteristics and the detailed procedure of the special optimization process of the warehouse layout design are described. Furthermore, the study also introduces the most important and commonly used objective functions, their calculation methods, and a large number of design constraints and limitations.

The author elaborated on a three-step method for defining the ideal warehouse layout which is also introduced in the paper. The three phases of the method are the following: 1) Alternative warehouse layouts have to be created considering the design constraints and limitations. 2) The huge number of alternative layouts has to be reduced by heuristic method and by continuous iteration. 3) After it, the reduced, smaller number of layout alternatives has to be compared based on the defined most important objective functions by the application of the systematic search method. The result of the comparison is the best, ideal warehouse layout, which has to be implemented in the practice.

The main added-value of the study is, that a detailed procedure of the special optimization process of warehouse layout design was elaborated and described by the author. It is novel because there is not available any publication in the existing literature which especially discusses the complex and detailed procedure of the special optimization process of warehouse layout design in any depth.

Acknowledgement. The research was supported by the Hungarian National Research, Development and Innovation Office - NKFIH under the project number K 134358.

References

1. Kovács, G.Y.: Changes in Production Philosophies – Efficiency Improvement of Production and Logistics Processes. University Press: University of Miskolc, Hungary (2017)
2. Yildirim, C., Oflaç, B.S., Yurt, O.: The doer effect of failure and recovery in multi-agent cases: service supply chain perspective. J. Serv. Theor. Pract. **28**(3), 274–297 (2018)
3. Hompel, M., Schmidt, T.: Warehouse Management. Springer, Berlin, Germany (2007)
4. Manzini, R. (ed.): Warehousing in the Global Supply Chain. Springer, New York (2012)
5. Telek, P., Landschützer, C.: Complexity analysis of material handling design processes. Adv. Logistic Syst. Theor. Pract. **11**(1), 61–76 (2017)
6. Kovács, G.Y.: Methodology and Procedure of Warehouse Layout Design. University Press: University of Miskolc, Hungary (2020). (in Hungarian)
7. Kovács, G.Y.: Process and Method for Warehouse Design. University Press: University of Miskolc, Hungary (2018). (In Hungarian)
8. Anjos, M.F., Vieira, M.V.C.: Mathematical optimization approaches for facility layout problems: the state-of-the-art and future research directions. Eur. J. Oper. Res. **261**, 1–16 (2017)
9. Hosseini-Nasab, H., Fereidouni, S., Ghomi, S.M.T.F., Fakhrzad, M.B.: Classification of facility layout problems: a review study. Int. J. Adv. Manuf. Technol. **94**, 957–977 (2018)
10. Heragu, S.S.: Facilities Design. CRC Press (2016)
11. Riyad, H., Kamruzzaman, R., Subrata, T.: Increasing productivity through facility layout improvement using systematic layout planning pattern theory. Glob. J. Res. Eng. **14**(7), 1–7 (2014)
12. Kovács, G.Y.: Warehouse design - determination of the optimal storage structure. Acta Technica Corviniensis Bull. Eng. **1**, 63–66 (2017)
13. Cselényi, J., Smid, L., Kovács, G.Y.: Evaluation methods of storage capacity between manufacturing levels of EEES at changing product structure. In: MicroCAD 2002 International Scientific Conference, Miskolc, Hungary, pp. 63–71 (2002)
14. Mulcahy, D.E.: Warehouse Distribution and Operations Handbook. McGraw-Hill, New York (1994)
15. Bartholdi, J.J., Hackman, S.T.: Warehouse & Distribution Science. elektronikus kiadás (2007). www.warehouse-science.com. Accessed 21 May 2020
16. Tompkins, J.A., White, J.A., Bozer, Y.A., Tanchoco, J.M.A.: Facilities Planning. Wiley, New York (2010)

17. Drira, A., Pierreval, H., Hajri-Gabouj, S.: Facility layout problems: a survey. Ann. Rev. Control **31**, 255–267 (2007)
18. Tompkins, J.A., Smith, J.D.: The Warehouse Management, Handbook. Tompkins Press, New York (1998)
19. Virág, Z.: Efficiency improvement of the warehousing system at Taurus Agricultural Tyre Ltd. Graduation thesis, University of Miskolc (2000). (in Hungarian)
20. Rushton, A., Croucher, P., Baker, P.: The Handbook of Logistics and Distribution Management. Kogan Page Limited, London (2006)

Towards Ammonia Free Retrofitting of Heavy-Duty Vehicles to Meet Euro VI Standards

I. I. Betsi-Argyropoulou, A. M. Moschovi, E. Polyzou, and I. Yakoumis(✉)

Monolithos Catalysts and Recycling Ltd., 83, Vrilissou Street, 11476 Polygono, Athens, Greece
yakoumis@monolithos-catalysts.gr

Abstract. Heavy-Duty Vehicles (HDVs) account only for 4% of the vehicle population. However, they are responsible for almost 33% of EU greenhouse gas emissions and have a high contribution to climate change. Diesel engine Heavy-Duty vehicles exhaust gas contains pollutants, hence HDV emissions regulations are becoming more and more stringent worldwide. Furthermore, in lean combustion conditions of the diesel engine, an increased concentration of oxygen is produced, making the reduction of NO_x challenging. Herein, a Catalytic Emission Control System (CECS) that is consisted of a Three-Way Catalyst (TWC), a Diesel Particulate Filter (DPF), an Oxygen Reduction System (ORS) and a Reduction Catalyst (RC) is presented. The aim of this CECS is to retrofit the exhaust system of various Pre-Euro up to Euro III HDVs. The Oxygen Reduction System (ORS) is a multi-tubular formation consisted of carbonate membranes that allow the permeation of oxygen, thus eliminating the use of the corrosive and dangerous ammonia (NH_3) to improve the performance of the RC. The permeated oxygen can be either released to the atmosphere or used onboard to improve the efficiency of the engine utilizing the high O_2 content. Lab experiments, on carbonate membranes, showed a permeation rate up to 75 μmol/sec, which is promising for utilization for ORS scale up applications.

Keywords: Heavy-duty vehicles · Diesel · Oxygen reduction system

Abbreviations

HDV	Heavy-Duty Vehicle
CECS	Catalytic Emission Control System
DPF	Diesel Particulate Filter
DOC	Diesel Oxidation Catalyst
TWC	Three-Way Catalyst
SCR	Selective Reduction Catalyst
RC	Reduction Catalyst
ORS	Oxygen Reduction System
IPCC	Intergovernmental Panel on Climate Change
GHG	Greenhouse Gases

K. Jármai and K. Voith (Eds.): VAE 2020, LNME, pp. 206–221, 2021.
https://doi.org/10.1007/978-981-15-9529-5_18

HC	Hydrocarbon
PM	Particulate Matter
PN	Particulate Number
VECTO	Vehicle Energy Consumption Calculation Tool
NSR	Non-Selective Reduction
PGM	Platinum Group Metal
SGB	Synthetic Gas Bench
GC	Gas Chromatograph
TCD	Thermal Conductivity Detector

1 Introduction

1.1 Diesel engine Vehicles

Diesel engines have high efficiency, durability and reliability together with their low-operating cost. These important features make them the most preferred engines, especially for Heavy-Duty Vehicles. The interest in diesel engines has risen substantially recently. In addition to the widespread use of these engines with many advantages, they play an important role in environmental pollution problems worldwide. Diesel engines are considered as one of the largest contributors to environmental pollution caused by exhaust emissions, and they are responsible for several health problems as well. Many policies have been imposed worldwide in recent years to reduce the negative effects of diesel engine emissions on human health and the environment. Extensive research has been carried out on both diesel exhaust pollutant emissions and after-treatment emission control technologies [1]. There are two categories of diesel engines: Open-chamber (direct-injection) and Divided-chamber (indirect-injection) engines. Open-chamber engines are preferred for Heavy-Duty applications because they offer the highest fuel economy. Divided-chamber (indirect-injection) engines have been preferred for light-duty applications because they are less sensitive to variations in fuels, have a wider range of speeds (and therefore greater power/weight ratio), run more quietly and emit fewer [2].

1.2 Climate Change and Heavy-Duty Vehicles

The transportation sector is one of the largest contributors to anthropogenic global greenhouse gas (GHG) and CO_2 emission. Additionally, Heavy-Duty Vehicles and buses account for 46.5% of global CO_2 emissions from road transport as presented in Fig. 1 [3].

The vast majority of Heavy-Duty Vehicles (HDVs) in the EU are powered by diesel engines. Although they account only for 4% of the vehicle fleet [4], they have been identified as important sources of both pollutant and Greenhouse Gas (GHG) emissions [5].

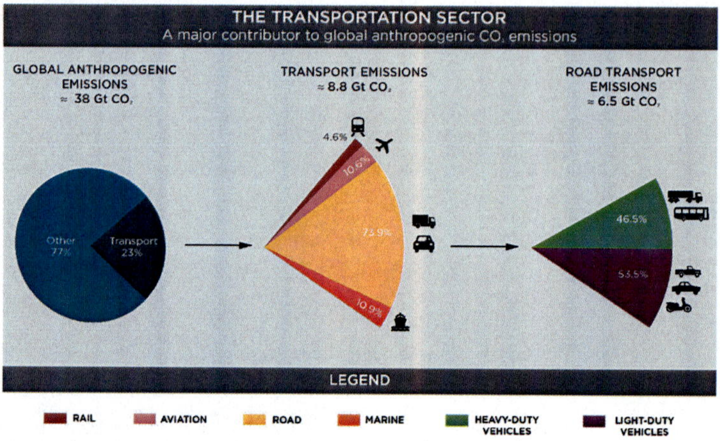

Fig. 1. Global anthropogenic CO_2 emissions [3].

1.3 The Emission from Diesel Engines of HDVs

The major products of the complete combustion of petroleum-based fuels in an internal combustion engine are CO (12%) and H_2O (11%). O_2 (9%) and the total pollutant gases (CO, HC, NO_x, SO_2 and PM) adding up to 1%, with N_2 from air comprising most (67%) of the remaining exhaust, as depicted in Fig. 2 [6]. A very small portion of the N_2 is converted to NO_x and some nitrated HCs. Excess O_2 is also emitted, depending on the operating conditions of the engine. Diesel engines operate with excess air (Air/Fuel ratio, ≃25–30/1) [7].

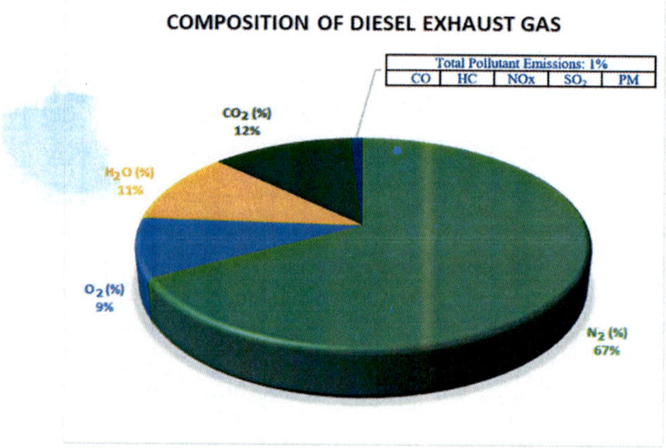

Fig. 2. The compositions of diesel exhaust gas.

Carbon Monoxide (CO) results from the incomplete combustion in the engine, where the oxidation process does not occur completely. This concentration is largely dependent on the air/fuel mixture ratio. The air/fuel mixture ratio is higher when excess-air factor

(λ) is < 1.0, which is classified as a rich mixture. Diesel engines are lean combustion engines, which have a consistently high air/fuel ratio ($\lambda > 1$). Thus, CO concentration is very low with respect to the rest of the gases in the gas exhaust stream [8]. Hydrocarbons (HC) emissions are composed of unburned fuels as a result of insufficient temperature, which occurs near the cylinder wall. HC consist of thousands of species, such as alkanes, alkenes, and aromatics. They are normally stated in terms of equivalent CH_4 content [6]. Particulate matter (PM) emissions in the exhaust gas are generated by the combustion process. They may be originated from the agglomeration of very small particles of partially burned fuel, lube oil, ash content and cylinder lube oil or sulfates and water. Finally, Nitrogen Oxides (NO_x) treatment is of significant importance. Diesel engines use highly compressed hot air to ignite the fuel. Air, mainly composed of O_2 and N_2, is initially drawn into the combustion chamber. Then, it is compressed, and the fuel is injected directly into this compressed air at about the top of the compression stroke in the combustion chamber. The fuel is burned and the heat is released. Normally in this process, the N_2 in the air does not react with O_2 in the combustion chamber and it is emitted identically out of the engine. However, high temperatures above 1600 °C in the cylinders cause the N_2 to react with O_2 and generate NO_x emissions. The amount of produced NO_x varies with respect to the maximum temperature in the cylinder, oxygen concentrations, and residence time [6].

1.4 Euro Standards and Catalytic converters introduction

As far as pollutant emissions are concerned, the EU adopted vehicle emission regulations, as described in Tables 1 and 2 and in Fig. 3. In recent times, these regulations are becoming more stringent by introducing the Euro VI standard, (EU Commission, 2011). Euro VI includes more stringent emission limits for HC, PM and NO_x, while for the first time introduced a limit for solid Particle Number (PN).

To comply with the stricter regulation of exhaust emissions, the use of catalytic converters on the exhaust system of the vehicles was introduced. Catalytic converters are washcoated ceramic monoliths with catalytic materials and have a honeycomb structure. Most common catalytic converters are: Diesel Oxidation Catalysts (DOC) and Selective Reduction Catalyst (SCR). In more detail, Diesel Oxidation Catalysts (DOC) are catalytic converters designed specifically for diesel engines and equipment to oxidise Carbon monoxide (CO), Hydrocarbons (HC) and Particulate matter (PM) emissions by reducing the Soluble Organic Fraction (SOF), which is the organic fraction of diesel particulates, according to Eqs. 1–3.

$$CO + O_2 \rightarrow 2\,CO_2 \tag{1}$$

$$\text{Hydrocarbons} + O_2 \rightarrow CO_2 + H_2O \tag{2}$$

$$\text{SOF} + O_2 \rightarrow CO_2 + H_2O \tag{3}$$

The oxidation reactions occur on the surface of a metal catalyst, whether it is either DOC or TWC or an SCR. Since the reaction depends on a surface where species can be adsorbed on and react, the greater the surface area provided, the higher the conversion will be. In other words, the reaction rate and conversion are directly proportional to the catalyst's surface area.

The role of the honeycomb-like structure of monoliths is to increase the exposed surface area covered by the catalyst layer. As the number of channels in the monolith increases, the surface area increases respectively. In SCR devices, the catalyst is employed to convert NO_x into N_2 and H_2O. Three-Way Converters (TWC) convert the three main pollutants (unburned HC, CO, and NO_x) simultaneously, by catalysing (enhancing) the following chemical reactions (Eq. 4–6) at the exhaust systems.

Reduction of nitrogen oxides to (N_2):

$$2H_2 + 2NO \rightarrow 2H_2O + N2 \tag{4}$$

Oxidation of carbon monoxide to carbon dioxide:

$$2CO + O_2 \rightarrow 2CO_2 \tag{5}$$

Oxidation of unburnt hydrocarbons (HC) to carbon dioxide and water, in addition to the above NO reaction:

$$\text{Hydrocarbons} + O_2 \rightarrow H_2O + CO_2 \tag{6}$$

Furthermore, in order for soot particles to be removed from the exhaust, the Diesel Particulate Filter (DPF) technology was introduced. DPF traps the particulate matter within a filter from the exhaust by passing the exhaust gases through a ceramic wall flow filter.

Table 1 and Table 2 contain a summary of the emission standards and their implementation dates. Dates in the tables refer to new type approvals—the dates for all vehicles are in most cases one year later. There are two sets of emission standards, with different type of testing requirements:

- Steady-State Testing: Table 1 lists emission standards applicable to diesel (compression ignition, CI) engines only, with steady-state emission testing requirements.
- Transient Testing: Table 2 lists standards applicable to both diesel and positive ignition (PI) engines, with transient testing requirements [9].

Throughout years, Euro Standards for HDVs introduced various catalysts and systems to reduce gas emissions [10]. In Fig. 3, Catalytic Systems of HDVs according to the year of manufacture are presented.

Euro I (1992) set the first limits for various gases, without any catalytic device.

Euro II (1996) first time of DOC appearance in some vehicles.

Euro III (2000) the use of DOCs at the exhaust system was compelled [10].

Euro IV (2005, 2006) DPFs were mounted on HDVs [11].

Euro V (2008) with an even more stringent NO_x emission limit value. In 2010 the requirement of the SCR technology and the use of the AdBlue system (use of Urea) that reduces NO_x rapidly matured.

Euro VI (2013) included more stringent emission limits for HC, NO_x, PM and for the first time included solid particle number (PN) and NH_3 emissions, that raised researchers to focus on for a different NO_x reduction procedure.

Table 1. EU emission standards for heavy-duty CI (diesel) engines: steady-state testing [9].

Stage	Date	Test	CO	HC	NO_x	PM	PN	Smoke
			g/kWh				1/kWh	1/m
Euro I	1992, ≤ 85kW	ECE R-49	4.5	1.1	8.0	0.612		
	1992, > 85kW		4.5	1.1	8.0	0.36		
Euro II	1996.10		4.0	1.1	7.0	0.25		
	1998.10		4.0	1.1	7.0	0.15		
Euro III	1999.10 EEV only	ESC & ELR	1.5	0.25	2.0	0.02		0.15
	2000.10		2.1	0.66	5.0	0.10[a]		0.8
Euro IV	2005.10		1.5	0.46	3.5	0.02		0.5
Euro V	2008.10		1.5	0.46	2.0	0.02		0.5
Euro VI	2013.01	WHSC	1.5	0.13	0.40	0.01	8.0×10^{11}	

[a] *PM = 0.13* g/kWh *for engines < 0.75* dm^3 *swept volume per cylinder and a rated power speed > 3000* min^{-1}

Table 2. EU emission standards for heavy-duty CI (diesel) and PI engines: Transient testing [9].

Stage	Date	Test	CO	HC	NO_x	PM	PN	Smoke
			g/kWh				1/kWh	1/m
Euro III	1999.10 EEV only	ETC	3.0	0.40	0.65	2.0	0.02	
	2000.10		5.45	0.78	1.6	5.0	0.16[c]	
Euro IV	2005.10		4.0	0.55	1.1	3.5	0.03	
Euro V	2008.10		4.0	0.55	1.1	2.0	0.03	
Euro VI	2013.01	WHTC	4.0	0.16[d]	0.5	0.46	0.01	6.0×10^{11}[e]

[a] *for gas engines only (Euro III-V: NG only; Euro VI: NG + LPG)*
[b] *not applicable for gas fueled engines at the Euro III-IV stages*
[c] *PM = 0.21* g/kWh *for engines < 0.75* dm^3 *swept volume per cylinder and a rated power speed > 300* min^{-1}
[d] *THC for diesel (CI) engines*
[e] *PN limit for PI engines applies for Euro VI-B and later*

1.5 Emission Regulation and Conventional Catalytic Systems of HDVs: Pre-euro (Euro 0)- Euro VI - Use of Ammonia

HDVs were launched without a DOC up to 1996, with respect to EURO 0 and EURO I legislation. In 1996, when Euro II was implemented, the first DOCs were introduced to

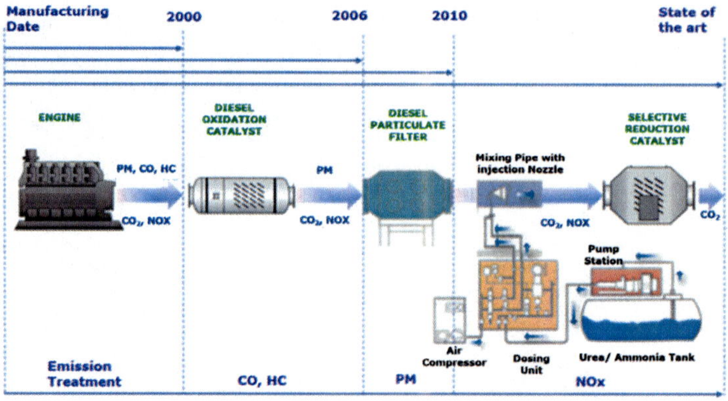

Fig. 3. Catalytic Systems of HDVs year of manufacture.

the exhaust systems, depending on the vehicles, but the use was very limited. In 2000, Euro III HDVs were obliged to have DOC [11].

A Conventional Catalytic System of Euro V-Euro VI (and some Euro IV) HDVs, like shown in Fig. 4, is consisted of two catalysts, a DOC for oxidization of CO and HC and an SCR for oxidization of NO_x. Furthermore, it is consisted of an AdBlue system (use of Urea) that reduces NO_x and a DPF filter for removal of soot [4].

Euro VI legislation (EU, 2011) for HDV included more stringent emission limits for HC, NO_x, PM and for the first time included solid particle number (PN) and NH_3 emissions [12].

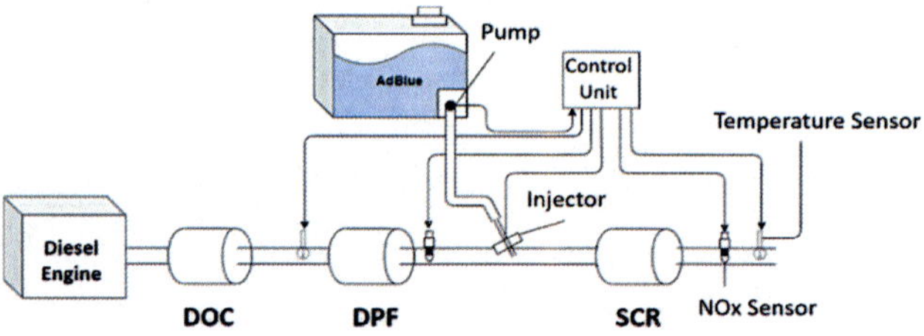

Fig. 4. Conventional Euro VI Catalytic System of HDVs.

Concerning the use of AdBlue system, a high purity urea/NH_3 solution is injected into the exhaust system to transform NO_x into N_2 and H_2O. NO_2 emissions are the highest at low driving speeds and low exhaust gas temperatures. This means that NO_2 emissions constitute a significant environmental problem in urban traffic. However, AdBlue functions properly only at exhaust temperatures, typically >250 °C which is not feasible in urban traffic. In such low temperature, severe exhaust blockages and subsequent engine damage is reported [13]. In that respect, research to develop different innovative methods, without the use of AdBlue, for controlling NO_x emissions should be expanded.

1.6 The Problem of Excess Oxygen at Diesel Heavy-Duty Vehicles and the Obstacles of Ammonia Utilization

As already mentioned, diesel engines are lean combustion engines with $\lambda > 1$. Diesel vehicles exhaust gases contain high O_2 concentrations up to 10% [11]. AdBlue system is not capable of removing oxygen from the exhaust gas stream. Removal of NO_x in an oxygen rich exhaust is extremely difficult for conventional TWC and SCR. TWC and SCR catalysts have several serious issues such as low catalytic activities, narrow temperature windows, and insufficient durability when operating in excess oxygen. Furthermore, conventional diffusion-based separation processes cannot be used because O_2 concentration gradient, between the exhaust gases and the atmosphere, allows only for inwards O_2 diffusion. As a result, the engine has to operate in a very narrow lean combustion range and this is the major obstacle for the improvement of fuel efficiency. This situation has prompted research on the development of a new catalyst technology that is capable of reducing NO_x in excess oxygen, that is, a NO_x storage–reduction (NSR) catalyst. Recently, catalysts for selective NO_x reduction by hydrocarbons under an oxidizing atmosphere have been extensively studied [14].

Ammonia (NH_3) is a toxic compound and is a precursor in the formation of atmospheric secondary aerosols. The particulate matter that is formed, namely ammonium nitrate and ammonium sulphate is also associated with other adverse health effects [15–18]. The SCR is an after-treatment system whose purpose is to reduce NO_x emissions by reacting to the NO and NO_2 with NH_3, formed by the reduction of the urea (AdBlue) injected into the system, on a catalyst surface. The over-doping of urea, low temperature in the system, and/or the catalyst degradation may lead to NH_3 emissions. Eventually, that concern led to the introduction of an ammonia emission limit for heavy duty vehicles (HDV) in the Euro VI standards. However, the NH_3 produced and emitted, by other in-use technologies has been neglected [19].

In this regard, this study is working towards replacing AdBlue technology, with an ORS that aims to reduce the O_2 concentration at the exhaust system, allowing the Reduction Catalyst to reduce NO_x.

2 Catalyst-Based Emission Control System Proposed in This Study

The Catalyst-based Emission Control System (CECS), which is studied in this article, focuses on retrofitting Heavy-Duty Vehicles, allowing them to meet emission and air quality EURO VI standards. Furthermore, this new CECS eliminates the use of the corrosive and dangerous NH_3.

According to the technology proposed here, retrofitting constitutes a solution for reducing vehicle emissions, specifically for the public HDV fleets of Greece, which are usually large, and renewal can be expensive. The rate of renewal of the HDV fleet is slow (average lifetime of HDV > 11 yrs). Most of the public HDVs run for small mileage per year (e.g. Galatsi's Municipality average truck mileage is 15,000 Km per year), operating within the limits of the municipality/port/airport. Thus, even if the HDVs are aged, they may operate effectively for a long time (in Greece the limit is 22 year) [11]. As a result, retrofitting these fleets constitute a reasonable solution.

The CECS, which is presented in Fig. 5, is a temporal demonstration. The CECS proposed in this study is consisted of a patented TWC containing Cu/Pd/Rh or a DOC containing Cu/Pd and a Reduction Catalyst (RC) containing Cu/Rh, reducing the use of rare and expensive Platinum Group Metals (PGMs) [20]. In addition, this CECS contains a DPF and finally, the patented ORS system [21]. The Layout of these catalysts and the DPF will be placed at the exhaust system of HDVs as depicted in Fig. 5, and subsequently, in a layout that will be further studied in order to achieve the highest NO_x reduction.

As mentioned above, the CECS which is presented in Fig. 5 is a temporal demonstration and the final layout will be consisted of an ORS, a TWC and a DPF as shown in Fig. 6.

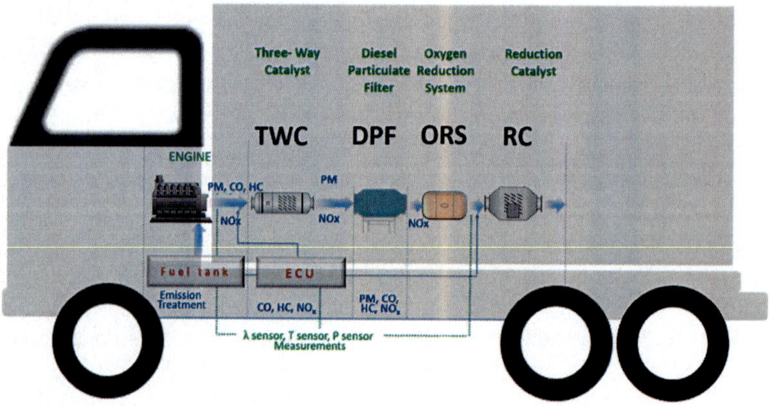

Fig. 5. Initial layout of Catalyst-based Emission Control System (CECS) based on the technology proposed by this study [11, 21].

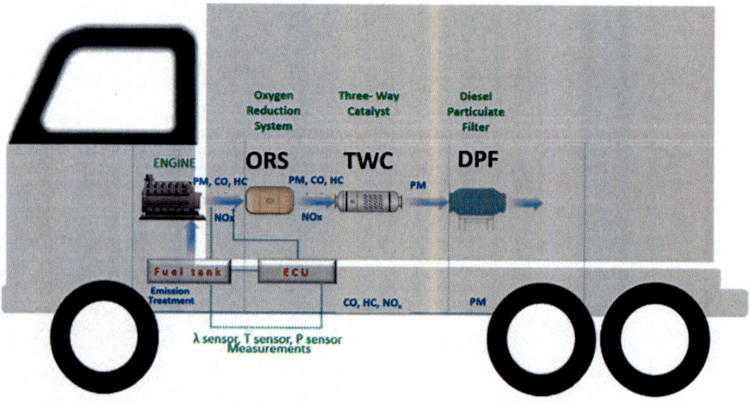

Fig. 6. Final CECS based on the technology proposed by this study [11, 21].

2.1 Oxygen Reduction System Operation

The Oxygen Reduction System (ORS) proposed here would be in a multi-tubular module formation as presented in Fig. 7, where the exhaust gas is fed through the lumen side and CO_2 and O_2 permeate across the membrane to the shell side, while the total surface area should be adjusted to obtain adequate permeation rate. In the produced shell side mixture, CO_2 and O_2 are present in a molar ratio of 2/1 and it can be either released to the atmosphere or to be used on-board to improve the efficiency of the engine, utilising the high O_2 content [21].

For reasons of space saving and minimization of energy losses, various layouts are proposed. Fig. 7a shows an example of a layout of multiple tubular membranes of circular cross-section. In addition, compact stacks of the described herein membranes can be used to limit energy losses. An example of a stack unit utilizing tubular membranes is shown in Fig. 7a, where the lumen side of all tubular membranes (1) is in gas contact with the gas inlet stream (4), through the gas distribution unit (2), and with the gas outlet stream (5) through the gas collection unit (3). The produced gas steam of CO_2 and O_2 that permeates through the membranes is released to the atmosphere (6).

In another embodiment, a metallic or ceramic or other refractory container is used to engage the above described membrane stack and minimize heat losses, as shown in Fig. 7b. The stack of membranes can be hermetically enclosed in a metal container (7), exhibiting a gas inlet (8) and gas outlet (6), allowing the management of the produced CO_2 and O_2 gas mixture that will be purged through the metal container (7). Furthermore, in a different layout, the exhaust gas is fed to the membrane from the gas inlet (8) and exits the membrane from the gas outlet (6), while the produced gas stream containing O_2 and CO_2 exits the membrane system from the gas outlet stream (5) with or without carrier gas in the gas inlet stream (4). The carrier gas can be either air or exhaust gas, while it can be in the temperature range between 25 and 800 °C, or preferably at the membrane operation temperature [21].

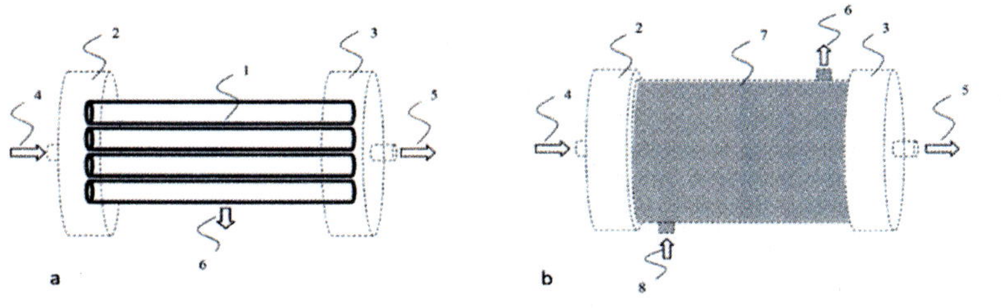

Fig. 7. Multi-tubular formation of ORS for scale-up applications to be tested on HDVs: a) Membrane stack without metallic cannister, b) Membrane stack, canned to manage of the produced O_2 and CO_2 gas mixture [21].

Initially, this system will be tested in an HDV, replacing the AdBlue technology, to procure the proper installation procedure on the HDVs and optimize the operational conditions of the ORS. Subsequently, the whole CECS will be tested as presented in Fig. 5 [21].

2.2 Dual Phase Membrane Operation

These multi-tubular module formations (Fig. 7) will be comprised by various dual phase membranes, consisting of a porous metal phase as substrate and a molten carbonate phase, conducting, respectively, CO_3^{2-} and electrons for selective permeation of CO_2 and O_2 at high temperatures, as shown in Fig. 8. The metal phase not only serves as a support, but also as an electron carrier. This type of membranes is typically used for CO_2 capturing. Such technologies have been mentioned in literature on separation and capture of carbon dioxide (CO_2) from gas streams [22]; however, they fail to describe the utilization for oxygen (O_2) removal and the application to lean-burn engine exhaust gas, where excess oxygen (O_2) conditions exist [21].

The rest of the exhaust gases are not able to permeate the membranes and are directly forwarded to the exhaust system as shown in Fig. 7. As mentioned in Sect. 1.6, TWC and SCR catalysts have several serious problems when operating in excess oxygen, thus, the use of his type of membranes can be a suitable solution.

The theoretical mechanism of the CO_2 and O_2 separation (Fig. 8) for each membrane is described in this section. CO_2 separation can be accomplished using a CO_2 partial pressure gradient as the driving force. On the upper stream membrane surface, CO_2, which is combining with electrons and oxygen, forms CO_3^{2-}, which is transported through the molten carbonate phase. The CO_3^{2-} releases electrons to form CO_2 and O_2 on the downstream membrane surface. The electron transports back, through the metal phase, toward the upstream membrane surface. No external electrodes and connectors are required in this dual-phase membrane. This dual-phase membrane allows permeation of CO_2 and O_2 in the 2/1 ratio for the production of O_2 enriched CO_2 stream, which is useful as the oxidant for oxyfuel processes [22]. This procedure is described in Fig. 8.

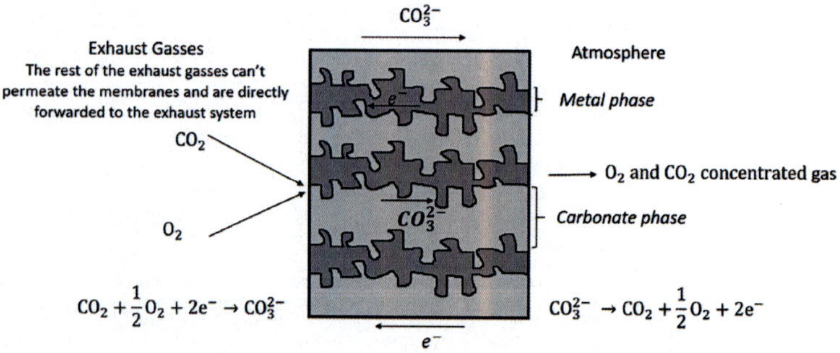

Fig. 8. Schematic representation of the proposed electrochemical membrane for O_2 removal from a diesel engine exhaust gas mixture.

3 Dual Phase Membranes Synthesis

Membranes with a molten carbonate phase in a porous stainless-steel (SS) support were prepared by a direct impregnation method. Membrane preparation conditions were optimized to obtain stable, gas-tight dual-phase membranes at a temperature window (400–550 °C). As porous SS support, closed-end SS316 sintered mesh and powder porous

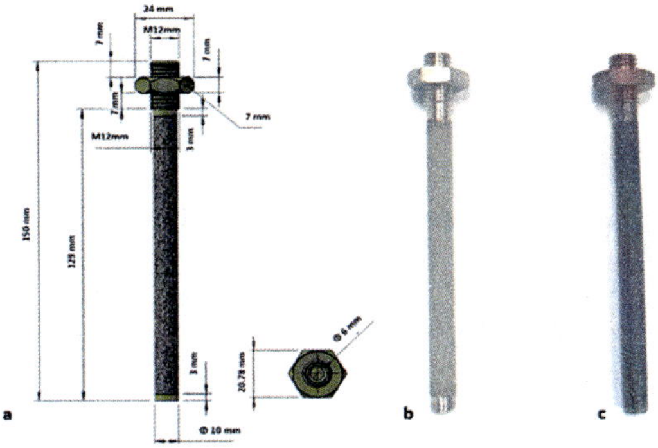

Fig. 9. a) Closed-end SS316 sintered mesh and sintered porous tubes, b) Stainless-steel filter-substrate and c) impregnated dual-phase membrane.

tubes with porosity size 1-10 μm were used, as shown in Fig. 9a–b. The carbonate mixture $Li_2CO_3/Na_2CO_3/K_2CO_3 = 43.5/31.5/25$ was selected, since it is the carbonate mixture with the lowest eutectic point. The operating temperature of these membranes is above the eutectic point of the mixture. The eutectic point of this carbonate mixture is 397 °C [23]. Carbonate mixture was melted at T > 400 °C, to assure that the whole synthesis layout temperature is above the mixture's eutectic point. The SS tubes were impregnated to create the dual phase membrane (Fig. 9c). So as for the dual phase membrane to be impregnated, pretreatment of the SS tubes at 500 °C was performed to avoid oxidization due to the temperature difference between molten carbonate mixture and SS filter, in order to achieve higher infiltration of carbonates at SS tube porous. The impregnated membrane was left to dry for 24 h and then calcinated at T > 400 °C for 1 h so as for the excess carbonates to be removed.

4 Experimental Procedure

A Synthetic Gas Bench (SGB) was built to study and control the reduction of O_2 content from HDVs exhaust in a lab scale simulation (measuring the permeability of a single membrane). A Gas Chromatograph (GC) was used for analysis like described in Fig. 10. The SGB of the ORS provides the ability to validate membranes permeability rate under various temperatures, humidity and gas mixture concentration conditions. Membranes were tested at the temperature range 300–550 °C, and the inlet gas concentration was: 12% CO_2, 8% O_2, 80% N_2. The impregnated SS tube with the carbonate material (dual phase membrane) is placed in the reactor (Fig. 11). In order for the temperature to be elevated and simulate automotive conditions, a custom-made furnace is used, and the reactor is placed in. The gas conditions of the experiment are controlled and monitored by the SGB, where the gas mixture is obtained. The gas mixture obtained from the SGB is connected to the reactor as inlet gas through the lumen side, where CO_2 and O_2 will permeate across the membrane to the shell side and released to the vent hood like shown in Fig. 11.

The outlet gas is sent to the GC for analysis. The SGB allows the use of air as a sweep gas at the shell side, to achieve the necessary partial pressure gradient to operate as the driving force for the enhancement of the permeation.

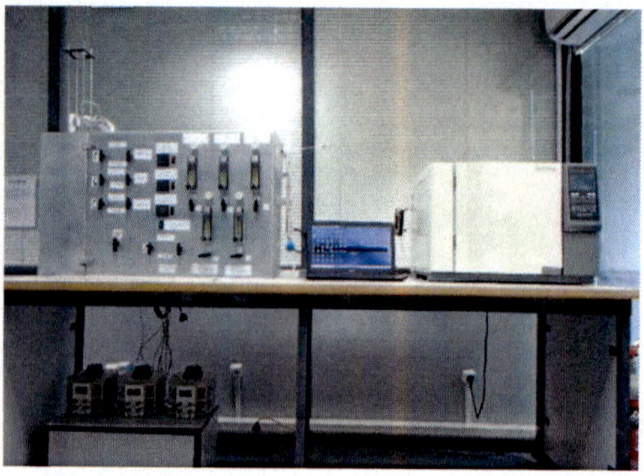

Fig. 10. Synthetic gas bench of oxygen reduction system and gas chromatography analyser.

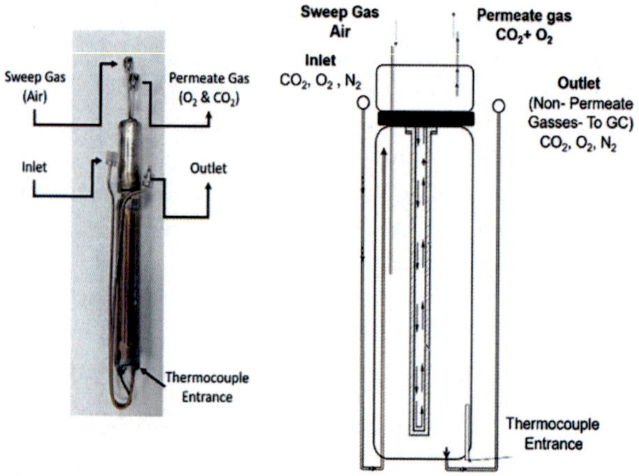

Fig. 11. Picture (left) and schematic figure (right) of the reactor of ORS, which is heated in a furnace, connected SGB of Fig. 10.

5 Results of Gas Permeation (O_2 Permeation Rate)

The permeation rate ($-r_{O_2}$) of gas O_2 in μmol/sec for a dual-phase membrane, as a function of temperature in the range of 300–550 °C, is presented in Fig. 12. The gas permeance of O_2 is relatively low and increases with the rising temperature almost monotonically. At the temperature window that the carbonate mixture changes state from solid state to complete liquification, a shoulder is noticed at the raising rate and finally the rate increases after the eutectic point of the carbonate mixture. After the carbonate mixture is liquified, $-r_{O_2}$ increases, with respect to the permeation rate in the solid state of the mixture. According to Fig. 12, dissolution and diffusion in molten carbonate phase of gas molecules are facilitated and enhanced by the temperature elevation.

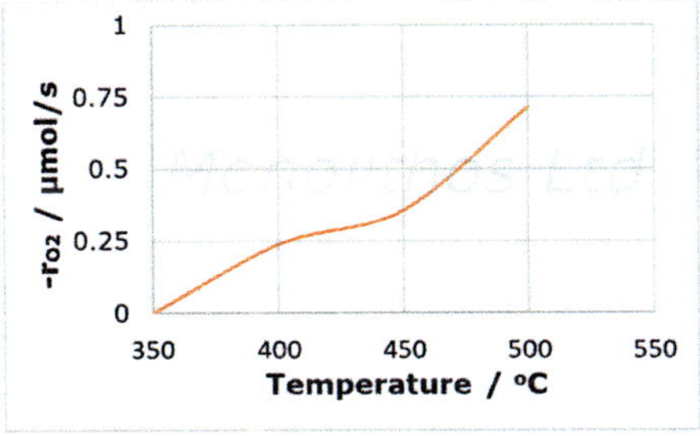

Fig. 12. O_2 permeation rate (μmol/sec) in temperature range 350–500 °C.

6 Conclusions

The purpose of this study is to reduce O_2 concentration of HDVs' exhaust emission, in order for TWC and RC to effectively reduce NO_x. Dual phase membranes that allow the CO_2 and O_2 permeance were studied to reduce the in the exhaust system. An Oxygen Reduction System will be used on diesel engine vehicles, utilizing these membranes in a multi-tubular formation.

The main objective of the HDVs' retrofit, with the proposed CECS, is for public HDVs fleets to meet Euro VI Standards, thus postponing their replacement, allowing them to operate within the emission limitations of EU for their remaining lifetime. The final version of this CECS will allow each and every Heavy-Duty Vehicle from Euro 0-Euro III to meet Euro VI standards, eliminating the use of Ammonium and PGMs use, thus, minimizing pollutant gas emissions and costs.

This CECS introduces an ORS to effectively reduce O_2 concentration of HDVs' exhaust emission, allowing TWC and RC to effectively reduce NO_x as mentioned.

Metal-carbonate dual phase membranes with a porous metal support and carbonate mixture were effectively prepared by the direct impregnation. Membrane preparation conditions were optimized to obtain a gas-tight dual phase membrane. Once the most permeative dual phase membrane synthesis composition is selected, the scale-up should be studied and optimized as well. The ORS will be tested on an HDV, replacing the AdBlue technology, to procure the proper installation procedure on the HDVs and optimise the operational conditions of the ORS.

An SGB was used to study and control the reduction of O_2 content from Heavy-Duty vehicles exhaust. For dual phase membranes preparation, carbonate mixture $Li_2CO_3/Na_2CO_3/K_2CO_3 = 43.5/31.5/25$ was selected, since it is the mixture with the lowest eutectic point. The membrane was tested at the temperature window 300–550 °C, and inlet gas concentration was: 12% CO_2, 8% O_2, 80% N_2. The permeation rate $-r_{O_2}$ was studied with respect to temperature. The results seem promising, concerning the

O_2 removal from the exhaust gas stream. Lab experiments showed a permeation rate of up to 75 μmol/sec. The permeance of CO_2 in the presence of oxygen increases with temperature and reaches a maximum at 500 °C, which confirms the workability of the dual phase membrane concept that was described in Sect. 2.2.

The shoulder, that is noticed at the raising rate, is representing of carbonate mixture phase change. After the carbonate mixture is transformed into liquid, the permeation rate $-r_{O_2}$ increases, with respect to the permeation rate in the solid state of the mixture. The operating temperature range of these membranes coincides with the mean temperature of HDVs' exhaust system, which makes it suitable for retrofit on HDVs applications.

Membrane's permeability will be further studied under various parameters that simulate the Heavy-Duty Vehicle diesel exhaust gas system, like the addition of humidity and utilization of air as sweep gas at the shell side of the membrane (ambient air) and wider temperature window. Synthetic Gas Bench inlet emissions will also simulate the emission concentration of a diesel engine. Finally, as an upscale (multi tubular formation), the total surface area will be adjusted to obtain an adequate permeation rate to change emission conditions from rich in oxygen ($\lambda >> 1$) to stoichiometric ($\lambda = 1$ or lower). The ORS will be firstly tested on an HDV, replacing the AdBlue technology, to assure proper function of the system. Subsequently, the whole CECS (TWC, DPF, ORS, RC) will be tested, in order to conclude on the final layout that will be consisted of an ORS, a TWC and a DPF.

Acknowledgements. This study was supported by CAT4HEAVY LIFE Environment and Resource Efficiency project (Contract No: LIFE17/ENV/GR/000352) and EIT Raw Materials INNOCAT project (Project Number: 18344).

References

1. Reşitoğlu, İ., Altinisik, K., Keskin, A.: The pollutant emissions from diesel-engine vehicles and exhaust aftertreatment systems. Clean Technol. Environ. Policy. **17**, 15–27 (2014)
2. National Research Council: Diesel Technology—Impacts of Diesel-powered Light-duty Vehicles. National Academy of Sciences, Washington DC (1982)
3. ICCT Website. https://theicct.org/blogs/staff/a-world-of-thoughts-on-phase-2. Accessed 25 Aug 2020
4. Grigoratos, T., Fontaras, G., Giechaskiel, B., Zacharof, N.: Real world emissions performance of heavy-duty Euro VI diesel vehicles. Atmos. Environ. **201**, 348–359 (2019)
5. Muncrief, R., Sharpe, B.: Overview of the heavy-duty vehicle market and CO_2 emissions in the European Union, ICCT (2015)
6. Stocker, T.F.: Climate Change 2013 - The Physical Science Basis, ICCT Report (2014)
7. Lassiter, D.V., Milby, T.H.: Washington DC, American Mining Congress: Health Effects of Diesel Exhaust Emissions: A Comprehensive Literature Review, Evaluation and Research Gaps Analysis (US NTIS PB-282–795) (1978)
8. Wu, C.W., Chen, R.H., Pu, J.Y., Lin, T.H.: The influence of air–fuel ratio on engine performance and pollutant emission of an SI engine using ethanol–gasoline-blended fuels. Atmos. Environ. **38**(40), 7093–7100 (2004)
9. DieselNet FAQ Website: Heavy-Duty Truck and Bus Engines, Regulatory Framework. https://dieselnet.com/standards/eu/hd.php. Accessed 25 Aug 2020

10. Can Cars Come Clean? OECD (2004)
11. Yakoumis, I.: European project CAT4HEAVY of LIFE environment and resource efficiency project application, contract No: LIFE17 ENV/GR/000352 (2017)
12. Mendoza-Villafuerte, P., Suarez-Bertoa, R., Giechaskiel, B., Riccobono, F., Bulgheroni, C., Astorga, C., Perujo, A.: NO_x, NH_3, N_2O and PN real driving emissions from a Euro VI heavy-duty vehicle. Impact of regulatory on-road test conditions on emissions. Sci. Total Environ. **609**, 546–555 (2017)
13. IEA: Technology Roadmap - Energy and GHG Reductions in the Chemical Industry via Catalytic Processes (2013)
14. Matsumoto, S.: Catalytic reduction of nitrogen oxides in automotive exhaust containing excess oxygen by NOx storage-reduction catalyst. CATTECH. **4**(2), 102–109 (2000)
15. Urban, C.M., Garbe, R.J.: Regulated and unregulated exhaust emissions from malfunctioning automobiles (1979). https://doi.org/10.4271/790696
16. Pope, C.A., III., Burnett, R.T., Thun, M.J., Calle, E.E., Krewski, D., Ito, K., Thurston, G.D.: Lung Cancer, Cardiopulmonary Mortality, and Long-term Exposure to Fine Particulate Air Pollution. JAMA **287**(9), 1132–1141 (2002)
17. Asman, W.A.H.: Factors influencing local dry deposition of gases with special reference to ammonia. Atmos. Environ. **32**(3), 415–421 (1998)
18. Verma, V., Ning, Z., Cho, A.K., Schauer, J.J., Shafer, M.M., Sioutas, C.: Redox activity of urban quasi-ultrafine particles from primary and secondary sources (2009)
19. Suarez-Bertoa, R., Zardini, A.A., Astorga, C.: Ammonia exhaust emissions from spark ignition vehicles over the New European Driving Cycle. Atmos. Environ. **97**, 43–53 (2014)
20. Yakoumis, I.: Copper and noble metal polymetallic catalysts for engine exhaust gas treatment, EP3569309A1 (2019)
21. Yakoumis, I., Souentie, S.: Device and process for the treatment of engine flue gases with high oxygen excess, EP3542887A1 (2018)
22. Chung, S.J., Park, J.H., Li, D., Ida, J.I., Kumakiri, I., Lin, J.Y.S.: Dual-phase metal–carbonate membrane for high-temperature carbon dioxide separation. Ind. Eng. Chem. Res. **44**(21), 7999–8006 (2005)
23. Mutch, G.A., Qu, L., Triantafyllou, G., Xing, W., Fontaine, M.L., Metcalfe, I.S.: Supported molten-salt membranes for carbon dioxide permeation. J. Mater. Chem. A. **7**(21), 12951–12973 (2019)

Applying Sustainable Logistics in Industry 4.0 Era

Mohammad Zaher Akkad(✉) and Tamás Bányai

Institute of Logistics, University of Miskolc, Miskolc, Hungary
zaherakkad91@gmail.com, alttamas@uni-miskolc.hu

Abstract. Sustainability and Industry 4.0 are very common terms nowadays that are used in almost all areas. Industry 4.0 has wide influence and effect because of its numerous new applications that come from the recent technological revolution. That led to the intelligent technologies that were built on the internet and interconnecting several areas, linking the fields of information technology, artificial intelligence, logistics systems, and environmental engineering to each other. On the other side, sustainability is an inclusive term that includes many dimensions, covers many areas like economy, industry, human, environment, and energy. Therefore, there increasing interest in applying sustainability to reach a better world. Logistics area is significantly affected by Industry 4.0 in an accelerated way and there is persistent need to use these new technologies to support sustainability by increasing the efficiency, reliability, and flexibility all along with saving energy and time with protecting the environment. Within the frame of this paper, the authors present an approach that combines sustainability and Industry 4.0 in the logistics area. After an introduction and theoretical background talk about the circular economy, reverse logistics, sustainability, and industry 4.0, the authors show modern applications aiming to apply circular economy and reverse logistics to save data, collect, move and treat waste in order to reuse, recycle and regenerate materials and energy. Then, it is discussed the expected results and outcomes regarding those applications on different aspects like sustainability, environment, and economics.

Keywords: Sustainability · Industry 4.0 · Logistics · Circular economy · Reverse logistics · Waste management

1 Introduction

The modern digital technologies, such as cyber-physical systems (CPS), the internet of things (IoT), and the internet of services (IoS) represent new models that are fast gaining ground in industrial transformation in the last few years [1]. On the other hand, the term 'Industry 4.0' was presented by Germany in 2011 [2]. The manufacturers are looking for a high standard of operational excellence through the developments of Industry 4.0 technologies [3]. However, to secure the advantages concerning Industry 4.0, there are requirements for building up a strategy for its useful applications, particularly while

K. Jármai and K. Voith (Eds.): VAE 2020, LNME, pp. 222–234, 2021.
https://doi.org/10.1007/978-981-15-9529-5_19

researching its effect on operations management [4]. In addition, executing these technologies requests spending big capital and controlling them to be put into practice [5]. Considering all of that, operational excellence should be looked at, which are economically functional and simultaneously to offset the high cost that might be found and could prevent the adoption of Industry 4.0 principles that aiming for sustainable developments. Beside Industry 4.0, other technological-based industrial activities under the premise of operational excellence include reverse logistics, lean operations, six-sigma, and many other smart manufacturing systems [3]. Further, it was specified [3] a few key performance dimensions that should be addressed in the context of operational excellence to improve sustainable supply chain including flexibility, collaboration, transparency, innovation, and relational capabilities of the supply chain [6].

On the other hand, the circular economy is an economic system that aims to eliminate waste and the continuous use of resources. Circular systems apply reuse, sharing, repair, refurbishment, and recycling approaches to create a closed-loop system, which minimizes the use of resource inputs and reduces waste scraps, pollution, and carbon emissions. It is possible to be argued that Industry 4.0 technologies can pave the way for circular economy principles, for example by tracking products post-consumption in order to recover components. Nevertheless, due to the very recent development of these ideas, the relationship between the circular economy and Industry 4.0 technologies has not been widely evaluated in the literature or practice, even though the two topics have largely been analyzed separately [7]. In the following chapter, the theoretical background is discussed tackling the circular economy, reverse logistics, Industry 4.0, and sustainability. The third chapter shows applications in the waste logistics area that use Industry 4.0 technologies and aim to raise sustainability. Within the fourth chapter, it is discussed the expected results of those applications and their effects on the different aspects. The last chapter contains a summary of this work.

2 Theoretical Background

2.1 Circular Economy

Circular economy (CE) is a new business mentality that can support associations and society to move towards sustainability [8]. The CE offers another and alternate point of view on the operational and official frameworks of production and consumption, this other perspective is centered on re-establishing the estimation of utilized assets. The CE recommends that a roundabout way to deal with vitality and materials can give financial, ecological, and social advantages [9] to organizations in order to replace the traditional attitude of 'take, make, use and dispose' that is also known as the linear economy with the CE attitude. Moreover, there have been obstructions to the fully applying of CE standards inside associations and supply chains in general [10]. It has been distinguished, for example, that an absence of data on the existing pattern of items, just as a lack of cutting-edge innovations for cleaner production, has decreased the reach of CE standards [11].

Concerning maintainable creation and utilization [12], the circular economy is a developing methodology focused on reaching the sustainable use of natural resources [8]. CE focuses on maximizing the circularity of resources and energy within production systems, because natural resources are scarce, and that waste at the end of its life may retain some value [13]. CE is based mainly on two cycles, one technical and one biological [14]. The technical cycle stresses the expansion of an item's life expectancy through a rapid order of circularity systems, which incorporate with reuse, repair, refurbishment, remanufacturing, and recycling [15]; technical cycles look to transform what is viewed as waste into assets for other creation frameworks. The biological cycle recovers environments by lessening exorbitant extraction of regular assets, utilizing sustainable materials and reusing vitality and natural waste by methods for anaerobic assimilation [16, 17].

CE cycles are ruled by three principles. These principles are, (1) protection of regular capital, which means making a harmony of utilization among sustainable and non-renewable assets; (2) expanding the life expectancy of assets through both natural and specialized cycles and (3) decrease of the negative impacts of production frameworks [14]. The Ellen MacArthur Foundation, a leading global charity in applying the CE's position on the list of decision-makers within the business, government, and academia [14], proposed the next six business actions that are called by the ReSOLVE framework to guide organizations through achieving the principles of the CE:

- Regenerate. This depends on a move to a renewable power source and materials. Biological cycles are utilized to empower the circulation of energy and materials and to change over organic waste into wellsprings of energy and raw material for other chains.
- Share. This is an integral part of a common economic perspective, where goods and assets are shared between individuals; thus, ownership loses its importance. Therefore, the products are designed to last longer, and maintenance must be provided to allow their reuse in order to extend the life of the product.
- Optimize. It is a central strategy technology. This model requires that organizations use digital manufacturing technologies, such as sensors, automation, radio-frequency identification, big data, and remote steering to reduce waste in production systems across supply chains. Organizations benefit from increased performance; for example, a predictive maintenance scheme can be planned based on real-time data reporting the conditions of machines [18].
- Loop. This is based on biological and technical cycles. Biological cycles, for instance, anaerobic digestion, are important to recapture the value of organic waste; technical cycles can restore the value of post-consumption products and packaging by utilizing repair, reuse, remanufacture, and recycling. Operations research approaches have been used to study these options [19].
- Virtualize. A service-focused strategy, which replaces physical with virtual and dematerialized products.
- Exchange. This includes replacing old and non-renewable goods with advanced and renewable ones.

It was carried out a comprehensive systematic literature review [20] on the CE in order to identify the research areas that have been studied so far. The finding was that three main CE related topics were studied: resource scarcity, environmental impact, and economic benefits. It was identified [21] the exchange of information as one of the big restrictions on the effectiveness of CE. Furthermore, it was additionally featured that it is basic to know the nature of materials flowing inside production frameworks after their assortment. Moreover, the apparent vulnerability with respect to costs, return on ventures, and course of events for usage regularly brings about introductory hesitance from companies to embrace such a driven objective. Overall, since rising advances dependent on the standards of Industry 4.0 have spread, it might now be able to overcome barriers of the CE by embracing rising advances identified with keen assembling. Mostly, it seems that new research is needed that is capable of shedding light on how organizations can gain competitive advantages by mitigating constraints on the effectiveness of the CE [7].

2.2 Reverse Logistics

Reverse logistics (RL) was defined [22] as "the process of planning, application, control of the operation, cost and flow of raw materials, the inventory process, finished products, the information related, from the point of consumption to the point of origin, in order to recover or create value or proper disposal". The comprehensive view detailed [23], defines RL as product return management, real-time inventory and workflow, tracking warranties, ordering and parts exchange, the flow of materials and information with suppliers, data analysis, execution of repairs, customer notification, and all the logistics flow and return. It was presented [24] distinct characteristics of RL, such as high uncertainties of supply in time, quantity and quality, and complexities of the operations, among other things. It was proposed [25] a framework based on the reverse flow of distribution from the producer to the user and back to the producer, with definite types of motivations (governance legislation, economic value, and ecological image) and types of disposal (reuse, repair, recycling, and remanufacturing), and discussed the issue of management separator counter network flows. It was proposed [26] a framework for RL that depends on five dimensions: (1) the return reasons; (2) reception structure; (3) the type of products and their characteristics; (4) recovery processes and options, and (5) the actors involved and their roles. Taking into consideration the characteristics of reverse logistics systems according to drivers (economics, legislation, and green citizenship), reasons (manufacturing returns, distribution returns, and customer returns), recovery processes (collection, inspection, classification, selection and types of products returned, composition, deterioration, and use) and agents (to the front of the supply chain, logistics reverse, and central and specialized). It was considered [27] several key reverse logistics problems by modeling techniques that can be helpful in understanding problems and developing solutions, considering a high number of variables. It was cited [28] five strategic factors that are important for the RL: costs, quality, customer service, environmental concerns, and political/legal concerns. It was presented [29] a framework composed of environmental factors (regulation and respect for the environment) and business factors (returns and customer satisfaction) and observed that there were a large number of models generalized in the literature [30]. It was confirmed [31] the need for further research on strategic aspects and organizational frameworks.

2.3 Industry 4.0

The term industrial revolution represents a quantum leap in the industry, which means raising the quantity, quality, or both in the industry and adopting innovative industrial methods through new technologies [32]. So far, there have been three industrial revolutions. We are now in the midst of the Fourth Industrial Revolution, or briefly called "Industry 4.0", which is now being developed and dominated by the different industrial sectors comprehensively [33]. The most prominent feature of Industry 4.0 is the adoption of intelligent technologies that rely on the Internet of Things and remove the lines that separate the physical, digital, and biological areas [34]. Industry 4.0 applications include the most recent technologies, especially in telecommunications, internet, and nanotechnology, which allowed us to use small devices with great efficiency [35]. This combination of advanced technologies empowered us to obtain various applications that have revolutionized the world of industry and changed the traditional concept of communication between machine and human into having the concept of communication between machine and machine [36, 37]. It is easy to observe the rapid pace of development of the industry, which makes it imperative for us to follow up the new applications of Industry 4.0 eagerly, so we can keep abreast of this development and benefit from it in our specialization, considering that these applications have moved logistics field to the next level in the future direction [38]. The pace of industrial development is constantly increasing. The results of the technological revolution that we are living, in addition to the intelligent technologies built on the internet and resulted from Industry 4.0 make it imperative to pursue these techniques in different software fields from CAD modeling [39] to digital twinning solutions [40].

2.4 Sustainability

A system might be called sustainable if it satisfies the sustainability requirements, which cover the related sustainability dimensions that are divided as following [41–44]: (1) Human sustainability: individual needs should be protected and supported with dignity and in a way that developments should improve the quality of human life and not threaten human beings. (2) Social sustainability: relationships of people within a society should be equitable, diverse, connected, and democratic. (3) Technical sustainability: technology must cope with changes and evolution reasonably, respecting natural resources. (4) Environmental sustainability: natural resources have to be protected from human needs and wastes. (5) Economic sustainability: a positive economic value and capital should be ensured and preserved. Therefore, sustainability covers comprehensively firstly the concept of "green in" systems, which means reducing energy and resource consumption and wastage in processes, and secondly the context of "green by" systems, which covers human and economic sustainability are improvements [42, 45, 46]. One dimension of sustainability cannot be reinforced without considering others. The five dimensions are interconnected with each other. For example, supporting more features related to individual and social dimensions would increase energy consumption, like influence economic and environmental dimensions. Moreover, 17 Goals were adopted by all United Nations Member States in 2015, as part of the 2030 agenda for sustainable development that set out a 15-year plan to achieve the Goals [47]. Those goals represent the inclusive aim of sustainability in different areas.

3 Logistic Applications Within the Industry 4.0 Environment

In order to show the results of using Industry 4.0 technologies in the logistics area, logistic applications that are related to waste containing, moving and separating are discussed in this chapter, next to other operations that aim to apply circular economy and reach the most sustainable results regarding saving time, material and energy. Each application is discussed separately to have clear details without overlap.

3.1 Waste Containers

Many Industry 4.0 technologies could be applied to the containers. A sensor that defines the amount of waste inside the container is available to be installed. The sensor can give more than one notification depending on the amount of waste, for instance, when it is 50% full, 75% full, and 90% full. Either light alert, sound alert, or both can be installed to work with those notifications. Radio Frequency Identification (RFID) technology can be defined as a powerful innovative gadget, which has been adopted in the development of IoT. However, there are two types of this technology, passive and active RFID systems. Passive RFID waits for a signal from an RFID reader. The reader sends energy, which converts that energy into an RF wave. Active RFID systems use battery-powered RFID tags that continuously transmit their own signal. By using active RFID systems, it can be connected to the live data management that receives the information directly. On the other hand, active RFID systems are much more expensive than passive ones. It should be considered that there are different types of containers as well. The waste containers might be defined depending on the waste types as organic and inorganic or as more specific targets, for instance, paper, glass, metal, or plastic. However, it is possible to use a smart container that can separate the inorganic waste by itself. This separation mechanism uses the relative relationship between the size and weight of the received waste, for each material, there is specific density, next to the sound that appears when the waste is received. The sensors define the data (sound, size, and weight) of this received waste. Other mechanisms can be used as well, like the ones that count on the IR (Infrared Radiation) or X-rays.

The containers are the first step of the waste collecting chain, which are in contact with users (the inhabitants) who would through the waste inside them. It is important to keep in mind that any change of the waste collecting procedures that should be done by the users should be essential, really needed and would not be changed again (at least for a long time relatively) because it costs a lot to make it clear and spread the right instruction about any new procedures to the users, furthermore, making different changes might confuse the users.

3.2 Collection and Transfer Trucks

To collect and transfer waste, special trucks should be used. Those trucks are defined either by the type of the transferred waste or by the needed size. The size can be important when the truck needs to collect the waste from narrow alleys. Active RFID systems should be used for these trucks to track their locations and movements. One of the most modern applications that can be used, is defining the truck's route by special algorithms.

The algorithms that create the routes of waste collection trucks target the containers that give notification that the container is almost full until it is enough to fill the truck capacity. If additional containers are needed, containers that give less full notification are added to the route. If there is a container with a full notification, it means there is an urgent condition and this container should be emptied within a short time. That reflects the flexibility that the system has by creating the routes depending on the needed waste amount and there is no need to have the same routes every day in order to choose the shortest way for the truck next to not having fully containers that could stay for a long time, which offers protection of possible biological environment pollution. Using a connection between the traffic lights and the trucks is possible to change the traffic lights into green when the trucks are near. This connection reflects positively on reducing carbon emissions.

3.3 Data Management

All the parts among any cyber-physical system can be directly connected to data management using the internet. This management stores data and deals directly with all the system parts. All data about the transportation, delivery of waste, collection trucks and waste quantities in each part of the system, as well as the records of surveillance cameras (if they are used), are saved. The algorithms that are used to create routes of waste collection trucks according to the waste type and quantities within the containers are counted within this management as well. Internet of Things, which is a newly emerging term, meaning the new generation of the internet that allows understanding between interconnected devices is used to make all the parts defined within the system to the management. These devices that use IoT technology include instruments, sensors, various artificial intelligence tools, and others. IoT technology makes the interconnecting idea goes beyond the traditional concept of connecting people with computers and smartphones over a single global network through traditional IP with the possibility to be free from the place. This data management can use the cloud computing technology that refers to the computer resources available on-demand via the network, which can provide several integrated computer services without being restricted to local resources for making it easier for the user. These resources include space for data storage, backup, and self-synchronization, as well as programming processing capabilities, task scheduling, and remote printing. When the user (or administrator in managing case) is connected to the network, he/she can control these resources through an easy programming interface that facilitates using and ignores internal processes. In cases that require more privacy, private servers can be used to store the received and processed data. If there is not enough more space in the servers, the new data will overwrite the oldest one.

On the other hand, data management gives the possibility to make direct online applications for raising sustainability and saving materials by the users, considering the users' types as individuals, workshops, factories, or companies. For instance, a user (textile factory) has excess materials like small textile pieces. This user can enter all the needed information on the platform (amount, type, and color). Any other user can find the available materials and can contact the offering user directly to get them. This application provides the possibility to use the excessed materials directly between the producers and reproducers. Moreover, it gives a space for creativity by finding out the available materials for upcycling and recycling that support sustainability without the need for transporting and treating steps.

3.4 Treatment Facilities

This is the final stage for the waste to be used or treated in the best manner. These facilities are divided according to the type of waste that it deals with into four sections mainly. Firstly, the renewal facility. In this section, waste is reused directly when it is possible or disassembled for other useful parts. For instance, about the targeted waste here, clothes, and electrical and mechanical equipment. After completion of the dismantling and evaluation phase, any excess material is transferred to one of the other sections. Secondly, recycle facility. In this section, the waste's raw materials are obtained for reuse, which means getting raw materials resources that can be used again by saving them from the waste that would be disposed of. Examples of the targeted waste here are paper, glass, plastic, and metal material. Thirdly, incineration facility. In this section, unusable and non-recyclable materials are collected to be burned; the obtained heat is used to produce energy. Fourthly, landfill facility. In this section, the waste (mainly the organic one) is buried after treatment to have faster biodegradation. The gases that are produced by the biodegradation of organic waste after burying can be collected and utilized. These facilities should be connected directly to the data management to collect all the information regarding the amounts of waste that were treated and ready to be transferred to the next stage or the excess materials that should be done by another facility and arrange the transferring trucks' time and route.

3.5 Local Composting

Composting is an effective way of the organic waste upcycling into compost that is used to fertilize the plants. Local composting units should be disrupted in appropriate locations and should be easy to access by local people. Those units provide the possibility to the local people to throw the organic waste directly that can be fertilized like fruits, vegetables, dairy products, cereals, bread, coffee filters, eggshells, and meat. Generally, if the waste can be eaten or grown in a field or garden, it can be composted. Items that cannot be fertilized include plastics, grease, glass, and minerals, spice packaging, plastic caps, plastic bags, silverware, drinking straws, bottles, polystyrene, or chemicals. Items such as red flesh, bones, and small amounts of paper are acceptable but take longer to degrade. Those local composting are connected to the data management so it can show the available space for waste until it is full and the time that is needed until completing the transformation into compost. When the compost is ready, it is used by the farms. However, there is a possibility to give back part or all of this compost to the local community if they wanted. The user who is participating in filling the units can have ID card that allows them to open the units and add their waste inside. Data management gives the possibility to the users to register and make accounts so they can save their share of waste in filling the units and the required compost they want. When the compost is ready, the data management will inform the selected users about the possible amount and time for them to collect their compost. The user should confirm his wanted share of the compost and he/she can use their ID card to collect the compost form the unit. The data management decides each person's share depending on the provided waste amount, number of participants, and the priority of spreading the compost to the farms depending on the season. By using local composting units, many advantages can be achieved. For instance, upcycling the waste into a useful item and saving time & efforts.

3.6 Ecological Footprint

The ecological footprint measures human demand for nature, that is, the amount necessary by nature to support humans or the economy. The calculations compare the required region for a biologically productive for human consumption with a biologically productive region in the world. In short, it is a measure of the human impact on the ecosystem of the planet, and it reveals the dependence of the human economy on natural capital. The available data that is stored within the data management allows making a direct application for the users reflects their attitude and lifestyle consequences. The users can choose the data types that are used to measure this footprint depending on their preferences. The used data can come from different resources such as the waste type and amount that the user throws since there is a special ID card for using the waste containers, resources spending like water and electricity, the daily purchased items, and the used transportation type. In addition, there is a possibility to allow the data management to send notifications to the users that suggesting improvements on the taken options that rise sustainability and minimize the negative impact on the environment.

4 Expected Results

The previous applications in waste management and sustainability areas have many advantages that encourage adopting and embracing them. These advantages can be outlined in the following points.

4.1 Structured System

Those applications support controlling the process of waste transporting in an ideal way. No excess waste in containers without collection nor the need to store waste above the required limit. In addition, data management & automation permit knowing the performance and efficiency of the system easily, allowing logistics management to fully regulate the operations of the collection system, which gives absolute flexibility in controlling the system. This flexibility is due to the connection of all parts to a single information-processing center within the data management, which allows the complete management with any modifications easily. Containers unloading functions can be modified as needed. For instance, it is possible to stop the containers unloading for one day or more. Instead of that, any container contains more than 50% on the previous day would be emptied. This flexibility is not affected by the size or population of the targeted city because of management automation.

4.2 Protecting Environment

By applying JIT (Just In Time) philosophy within those modern methods of waste's transfer and treatment, which means transferring the waste at the right time and amount without delaying, which offers protection of possible biological environment pollution. In addition to applying the applications that raise sustainability and encourage personal performance in reducing the environment negative impact.

4.3 Conducting Studies

The saved data through this system gives details about the collected and treated produced waste and its exact types and quantities continuously, allowing studies to be conducted through them. For instance, the waste's types and its increase can be identified throughout the year. In addition to the possibility of saving the number of the population associated with each container, which means studying the changes in waste quantities over time and expecting the resulting increase from the increase in population. This means it is possible to reach accurate results on the required areas and energies for treatment facilities to manage waste in the future.

4.4 Effectiveness

The direct connection of all parts to the data management means that it is easy to identify any problem that may occur immediately. Over time, a problem or malfunction would occur in any part of the waste collecting and treatment system. The system can identify this problem automatically with finding the best solution and send it directly to the specialist team depending on the problem type without the need of the human intervene, which means saving time and raising efficiency.

4.5 Upgrading

Although all the system parts and applications are linked to each other directly through the data management, it is very easy to adopt new technologies or develop one or more of the system parts in order to replace them with more modern ones. Because the linking in this system depends on the internet and smart technologies that do not require many physical connections, developing the system does not require big changes in general. Moreover, transforming waste processing into an investment project with an economic return supports the system's upgrading and development. This system focuses on both the waste collection part and the processing part, which means eventually obtaining the maximum amounts of reusable materials, recycled materials, and raw materials. In addition, the energy that would be produced while the waste' process treatment raises the economical return.

5 Summary

Even both sustainability and Industry 4.0 are famous terms were tackled in many studies especially the last few years in the different areas but studying the combination of these terms was not tackled directly enough. Logistics area is significantly affected by Industry 4.0 in an accelerated way and there is persistent need to use these new technologies to support sustainability by increasing the efficiency, reliability, and flexibility all along with saving energy and time with protecting the environment. Within the frame of this paper, the authors represented an approach that combines sustainability and Industry 4.0 in the logistics area. After an introduction and theoretical background talked about the circular economy, reverse logistics, sustainability, and industry 4.0, the authors showed modern applications within the waste management and sustainability areas aiming to apply circular economy and reverse logistics. Those applications aim to save data, collect, move, and treat waste to reuse, recycle, upcycle, and generate materials and energy while preserving time and effort and reducing environmental damage. Then it was discussed the expected results and outcomes regarding those applications in different aspects.

Acknowledgment. The described research was carried out as part of the EFOP-3.6.1-16-2016-00011 "Younger and Renewing University – Innovative Knowledge City – institutional development of the University of Miskolc aiming at intelligent specialisation" project implemented in the framework of the Szechenyi 2020 program. The realization of this project is supported by the European Union, co-financed by the European Social Fund.

References

1. Sun, Y., Yan, H., Lu, C., Bie, R., Thomas, P.: A holistic approach to visualizing business models for the internet of things. Commun. Mob. Comput. **1**(1), 1–7 (2012). https://doi.org/10.1186/2192-1121-1-4
2. Kagermann, H., Lukas, W., Wahlster, W.: Industrie 4.0 – Mitdem Internet er Dinge auf dem Wegzur 4. Industriellen Revolution [Industry 4.0: with the Internet of Things towards 4th industrial revolution], VDI Nachrichte (2011)
3. Mangla, S.K., Kusi-Sarpong, S., Luthra, S., Bai, C., Jakhar, S.K., Khan, S.A.: Operational excellence for improving sustainable supply chain performance. Resour. Conserv. Recycl. **142**, 277–278 (2019)
4. Holmström, J., Romme, A.G.L.: Five steps towards exploring the future of operations management. Oper. Manag. Res. **5**, 37–42 (2012)
5. Tortorella, G.L., Fettermann, D.: Implementation of industry 4.0 and lean production in Brazilian manufacturing companies. Int. J. Prod. Res. **56**(8), 2975–2987 (2018)
6. Deva, N.K., Shankarb, R., Qaiserc, F.H.: Industry 4.0 and circular economy: operational excellence for sustainable reverse supply chain performance. Resour. Conserv. Recycl. **153**(104583) (2020)
7. Jabbour, A.B.L.D.S., Jabbour, C.J.C., Filho, M.G., Roubaud, D.: Industry 4.0 and the circular economy: a proposed research agenda and original roadmap for sustainable operations. Ann. Oper. Res. **270**, 273–286 (2018)
8. McDowall, W., et al.: Circular economy policies in China and Europe. J. Ind. Ecol. **21**(3), 651–661 (2017)
9. Geissdoerfer, M., Savaget, P., Bocken, N.M., Hultink, E.J.: The circular economy: a new sustainability paradigm? J. Clean. Prod. **143**, 757–768 (2017)
10. Geng, Y., Doberstein, B.: Developing the circular economy in China: challenges and opportunities for achieving 'leapfrog development'. Int. J. Sustain. Dev. World Ecol. **15**(3), 231–239 (2008)

11. Su, B., Heshmati, A., Geng, Y., Yu, X.: A review of the circular economy in China: moving from rhetoric to implementation. J. Clean. Prod. **42**, 215–227 (2013)
12. Fahimnia, B., Sarkis, J., Gunasekaran, A., Farahani, R.: Decision models for sustainable supply chain design and management. Ann. Oper. Res. **250**(2), 277–278 (2017). https://doi.org/10.1007/s10479-017-2428-0
13. Ghisellini, P., Cialani, C., Ulgiati, S.: A review on circular economy: The expected transition to a balanced interplay of environmental and economic systems. J. Clean. Prod. **114**, 11–32 (2016)
14. MacArthur, D.E., Zumwinkel, K., Stuchtey, M.R.: Growth within: a circular economy vision for a competitive Europe. Report of Ellen MacArthur Foundation (2015)
15. Zhao, S., Zhu, Q.: Remanufacturing supply chain coordination under the stochastic remanufacturability rate and the random demand. Ann. Oper. Res. **257**(2), 661–695 (2015). https://doi.org/10.1007/s10479-015-2021-3
16. Bocken, N.M., Olivetti, E.A., Cullen, J.M., Potting, J., Lifset, R.: Taking the circularity to the next level: a special issue on the circular economy. J. Ind. Ecol. **21**(3), 476–482 (2017)
17. Murray, A., Skene, K., Haynes, K.: The circular economy: an interdisciplinary exploration of the concept and application in a global context. J. Bus. Ethics **140**(3), 369–380 (2017)
18. MacArthur, D.E., Waughray, D.: Intelligent assets. Unlocking the circular economy potential. Report of Ellen MacArthur Foundation (2016)
19. Loomba, A.P.S., Nakashima, K.: Enhancing value in reverse supply chains by sorting before product recovery. Prod. Plann. Control **23**, 205–215 (2012)
20. Lieder, M., Rashid, A.: Towards circular economy implementation: a comprehensive review in context of manufacturing industry. J. Clean. Prod. **115**, 36–51 (2016)
21. Winans, K., Kendall, A., Deng, H.: The history and current applications of the circular economy concept. Renew. Sustain. Energy Rev. **68**, 825–833 (2017)
22. Rogers, D.S., Tibben-Lembke, R.S.: Going Backwards: Reverse Logistics Trends and Practices. Logistics Executive Council, Pittsburgh (1999)
23. Shaik, M., Abdul-Kader, W.: A comprehensive performance measurement framework for reverse logistics enterprise. In: 61st Annual IIE Conference and Expo Proceedings (2011)
24. Wang, B., Sun, L.: A review of reverse logistics. Appl. Sci. **7**, 16–29 (2005)
25. Fleischmann, M., Bloemhof-Ruwaard, J.M., Dekker, R.D., Laan, A.E., Nunen, J.A.E.E., Wassenhove, L.N.: Quantitative models for reverse logistics: a review. Eur. J. Oper. Res. **103**, 1–17 (1997)
26. De Brito, M.P., Dekker, R.: A Framework for Reverse Logistics. ERIM Report Series Research in Management; Erasmus University Rotterdam: Rotterdam, The Netherlands (2003)
27. Rogers, D.S., Melamed, B., Lembke, R.S.: Modeling and analysis of reverse logistics. J. Bus. Logistics **33**, 107–117 (2012)
28. Dowlatshahi, S.A.: Strategic framework for the design and implementation of remanufacturing operations in reverse logistics. Int. J. Prod. Res. **43**, 3455–3480 (2005)
29. Meade, L., Sarkis, J., Presley, A.: The theory and practice of reverse logistics. Int. J. Logistics Syst. Manag. **3**, 56–84 (2007)
30. Rubio, S., Chamorro, A., Miranda, F.J.: Characteristics of the research on reverse logistics (1995–2005). Int. J. Prod. Res. **46**, 1099–1120 (2008)
31. Voigt, D., Filho, N.C., Macedo, M.A., Braga, T.G., Rocha, R.U.G.D.: Performance evaluation of reverse logistics: opportunities for future research. Sustainability **11**(5291) (2019)
32. Roblek, V., Mesko, M., Krapez, A.: A complex view of industry 4.0. Sage Open **6**(2) (2016)
33. Liao, Y.X., Deschamps, F., Loures, E.D.R., Ramos, L.F.P.: Past, present and future of Industry 4.0-a systematic literature review and research agenda proposal. Int. J. Prod. Res. **55**(12), 3609–3629 (2017)
34. Illés, B., Varga, K.A., Czap, L.: Logistics and Digitization. Lecture Notes in Mechanical Engineering, PP. 220–225 (2018)

35. Akkad, M.Z., Bányai, T.: Cyber-physical waste collection system: a logistics approach. Solutions for sustainable development. In: Proceedings of the 1st International Conference on Engineering Solutions for Sustainable Development, (ICESSD 2019), pp. 160–168. CRC Press, London (2019)
36. Dobos, P., Tamás, P., Illés, B., Balogh, R.: Application possibilities of the big data concept in Industry 4.0. IOP Conf. Ser. Mater. Sci. Eng. **448**(012011) (2018)
37. Zhong, R.Y., Xu, X., Klotz, E., Newman, S.T.: Intelligent manufacturing in the context of industry 4.0: a review. Engineering **3**(5), 616–630 (2017)
38. Glistau, E., Machado, N.I.C.: Industry 4.0, Logistics 4.0 and materials - chances and solutions. Mater. Sci. Forum **919**, 307–314 (2018)
39. Felhő, C.S., Kundrák, J.: CAD-based modelling of surface roughness in face milling. World Acad. Sci. Eng. Technol. **8**(5), 71–75 (2014)
40. Lu, Y.: Industry 4.0: a survey on technologies, applications and open research issues. J. Ind. Inf. Integr. **6**, 1–10 (2017)
41. Goodland, R.: The concept of environmental sustainability. Annu. Rev. Ecol. Syst. **26**(1), 1–24 (2002)
42. Penzenstadler, B.: Infusing green: requirements engineering for green in and through software systems. In: Proceedings of the Third International Workshop on Requirements Engineering for Sustainable Systems, (RE4SuSy) (2014)
43. Penzenstadler, B. Femmer, H.: A generic model for sustainability with process- and product-specific instances. In: Proceedings of the Workshop on Green In/By Software Engineering, pp. 3–8 (2013)
44. Razavian, M., Procaccianti, G., Tamburri, D.A.: Four-dimensional sustainable e-services. In: Proceedings of the 28th EnviroInfo, Shaker-Verlag AG (2014)
45. Calero, C., Piattini, M.: Puzzling out software sustainability. Sustain. Comput. Inf. Syst. **16**, 117–124 (2017)
46. Lago, P., Koçak, S.A., Crnkovic, I., Penzenstadler, B.: Framing sustainability as a property of software quality. Commun. ACM **58**(10), 70–78 (2015)
47. The Sustainable Development Agenda, the United Nations Homepage. https://www.un.org/sustainabledevelopment/development-agenda/. Accessed 23 July 2020

Design and Noise

Formulation of a Mathematical Model for the Prediction of Fatigue Life in the High Cycle Regime

Barna Szabó(✉) and Ricardo Actis

Engineering Software Research and Development, Inc., 111 West Port Plaza, Suite 825, St. Louis, MO, USA
barna.szabo@esrd.com

Abstract. Application of the concepts and procedures of verification, validation and uncertainty quantification with reference to a mathematical model formulated for the prediction of the fatigue life of structural and mechanical components in the high cycle regime are described. Such models are used for the formulation of design rules and estimating the probability of fatigue life remaining in support of condition-based maintenance (CBM) decisions.

Keywords: Design rules · Condition-based maintenance · Simulation governance

1 Introduction

The work described herein was motivated by problems associated with condition-based maintenance of high-value assets where the influence of small surface defects on fatigue life is of interest. The defects considered cannot be characterized in general by a single geometric parameter, such as a notch radius.

We consider a generalization of the results of fatigue tests performed on notch-free and notched coupons under constant cycle uniaxial loading conditions to variable cycle biaxial conditions. Our mathematical model comprises four sub-models:

(i) A deterministic model of linear elasticity for estimating the elastic stress field,
(ii) a deterministic predictor of fatigue failure defined on the elastic stress field that generalizes the results of experiments performed on notched coupons to arbitrary notches,
(iii) a statistical model for the generalization of fatigue data obtained from notch-free coupons to notched coupons under constant-cycle loading,
(iv) a model for the generalization of a constant-cycle fatigue model to arbitrary load spectra.

K. Jármai and K. Voith (Eds.): VAE 2020, LNME, pp. 237–244, 2021.
https://doi.org/10.1007/978-981-15-9529-5_20

Many plausible formulations can be proposed for the sub-models (ii) to (iv). These formulations are based on subjective choices influenced by intuition, experience and personal preferences. Therefore, it is necessary to have a process for objective ranking of candidate models based on their predictive performance. We employ such a process based on the principles and procedures of Bayesian statistics with emphasis on items (ii) and (iii).

The choice of a mathematical model from a set of competing models is conditioned on the available experimental data. It is expected that new ideas will be proposed in the future and the available experimental data will increase over time. Therefore ranking and selection of mathematical models is an open-ended problem: It is necessary to establish a process for systematic revision and updating mathematical models. In industrial and research organizations this falls under the administration of simulation governance and simulation project management [11].

It is assumed that the propensity of a metallic material to fail in fatigue is characterized by fatigue data collected from notch-free coupons (the S-N data) under constant cycle loading at various mean stress levels. The S-N data are then transformed to an equivalent stress vs. number of cycles to failure relationship which is used for calibrating a statistical model. The statistical model provides the probability density function for a fixed equivalent stress level from which probabilities of fatigue life can be calculated.

Generalization of notch-free fatigue data to notched specimens involves the formulation of a deterministic predictor. Traditionally this predictor is the fatigue stress concentration factor, denoted by K_f. The most commonly used definitions for K_f were proposed by Neuber and Peterson. They assumed that K_f depends on the geometric stress concentration factor, the notch radius and a material-dependent length parameter. We found that neither Neuber's nor Peterson's definition of K_f can be validated beyond a very narrow interval of notch radii and, furthermore, the same is true for the method of critical distances [9].

We have investigated an alternative generalization, which is described in this paper. We defined a predictor, denoted by G_α, which was calibrated against fatigue data of notched 24S-T3 (current designation: 2024-T3) aluminum test coupons under uniaxial loading conditions. The calibration data are available in references [3–7]. The predictor was then validated against independently obtained biaxial test records, also in the public domain [2].

2 Statistical Model

An investigation reported in [1] indicated that the random fatigue limit model proposed by Pascual and Meeker [8] is well suited for the statistical representation of the S-N data reported in [4]. This statistical model is based on the assumption that the random variable defined by $W = \log_{10} N$, where N is the number of cycles to failure, is normally distributed with mean $\mu(\sigma_{eq})$ and standard deviation s. The symbol σ_{eq} represents the equivalent stress that depends on the maximum stress σ_{max} and the cycle ratio R:

$$\sigma_{eq} = \sigma_{max} \left(\frac{1 - R}{2} \right)^{1/2} \quad (1)$$

The equivalent stress establishes equivalence between a constant cycle loading with arbitrary mean stress and constant cycle loading with zero mean stress. This definition is one of the frequently used definitions. In reference [1] the exponent was selected along with the other statistical parameters by maximizing the likelihood function.

In fatigue tests $\sigma_{eq}^{(i)}$ is fixed and the number of cycles (n_i) at which failure occurs is recorded. The set of data (n_i, $\sigma_{eq}^{(i)}$), i = 1,2,... recorded in fatigue tests where the first principal stress in the test section is either constant or varies with a small gradient is the S-N data.

The assumed functional form of the mean is

$$\mu(\sigma_{eq}) = A_1 - A_2 \log_{10}(\sigma_{eq} - A_3), \sigma_{eq} - A_3 > 0 \tag{2}$$

where the parameter A_3 is called the fatigue limit or endurance limit. In the random fatigue limit model A_3 is a random variable and $v = \log_{10} A_3$ is assumed to have normal distribution with mean μ_f and standard deviation s_f. Thus the statistical model has five parameters that were calibrated to the S-N data found in [3] by maximizing the likelihood function. The results of calibration are displayed in Table 1 below.

Table 1. Parameters of the random fatigue limit model

A_1	A_2	s	μ_f	s_f
7.191	1.991	0.1255	1.3438	0.0488

This is one of countless conceivable statistical models. Those models can be objectively ranked with reference to a fixed set of data by means of the Bayes factor. Details and examples are available in [1].

3 Predictor of Fatigue Failure

A predictor of fatigue failure, proposed in [10], is described in the following. This predictor is based on the assumption that the onset of fatigue failure can be correlated with the averaged volume integral of a linear combination of two stress invariants. It is defined as follows:

$$G_\alpha(\sigma_{ij}, R) = \frac{1}{V_c} \int_{\Omega_c} (\alpha I_1 + (1-\alpha)\bar{\sigma}) dV \left(\frac{1-R}{2}\right)^{1/2}, 0 \le \alpha \le 1 \tag{3}$$

where σ_{ij} is the stress tensor field, $I_1 = \sigma_{kk}$ is the first stress invariant, $\bar{\sigma}$ is the von Mises stress. The domain of integration is defined by:

$$\Omega_c = \{\boldsymbol{x} | \sigma_1 > \beta \sigma_{max} > 0\} \tag{4}$$

where $\boldsymbol{x}$ is the position vector, σ_1 is the first principal stress, σ_{max} is the maximum principal stress at a stress raiser such as notch root, and $\beta = \beta(V)$ is a material-dependent parameter, determined by calibration. It is a function of the highly stressed volume V, defined as follows:

$$V = \int_{\Omega_c} y(x)dV, \text{ where } y = \begin{cases} 1 & \text{when } \sigma_1(x) > \gamma\sigma_{max} > 0 \\ 0 & \text{otherwise} \end{cases} \quad (5)$$

The highly stressed volume depends on the geometric features and the loading condition. It is independent of material properties. Prior investigation indicated that the performance of the predictor is not sensitive to the choice of γ and, furthermore, $\gamma = 0.85$ is a reasonable value. For that reason γ was fixed at 0.85.

4 Calibration of the Predictor

The parameter β depends on the material properties and the elastic stress distribution in the vicinity of notches. A unique algorithmic procedure was developed for the computation of β and verification that the relative error in the computed data is not greater than 1 percent. For calibration the test records published in references [3, 7] were used. Details are available in [10]. Since the source data is available in US customary units, we will use the same units.

The parameter $\beta = \beta(\alpha)$ and the predictor G_α were computed for six equally spaced values of α in the interval [0,1] and the corresponding likelihood functions were computed for the calibration set. It was found that the maximum likelihood occurs at $\alpha = 0$. Therefore the predictor G_α is the von Mises stress averaged over a volume characterized by β, subject to the condition that the first principal stress is positive.

The calibration curve corresponding to $\alpha = 0$ is shown in Fig. 1. The domain of calibration is defined by the interval in which the highly stressed volumes of the calibration data lie. As seen in the figure, that interval is (2.0E−8, 8.0E−3) in^3.

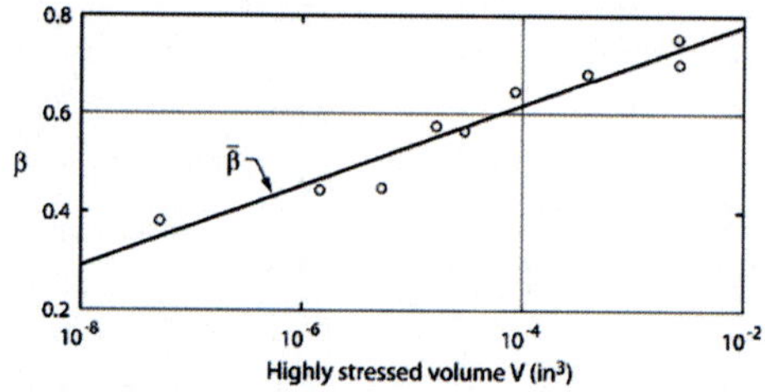

Fig. 1. The computed and fitted values of β corresponding to $\alpha = 0$.

The computed values of β were fitted by linear regression to obtain

$$\bar{\beta} = 0.9422 + 0.08184 \log_{10} V. \quad (6)$$

5 Validation

Records of experiments published in [2] were used for the purpose of testing the mathematical model in validation experiments. The specimens were fabricated from drawn tubing in accordance with ASTM Standard E2207. The specimens featured a 30 mm long gage section with an outside diameter of 29 mm and an inside diameter of 25.4 mm. A circular hole of 3.2 mm diameter was cut into the center of the test section by a drilling and reaming procedure. Remarkably, these experiments were performed approximately 60 years after the experiments on which the calibration of the predictor is based. The specimen is shown in Fig. 2.

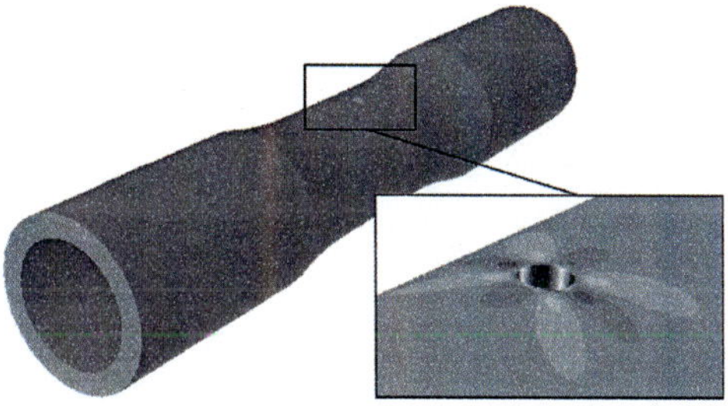

Fig. 2. (a) Specimen used in validation experiments. (b) Contours of von Mises stress in torsion.

Whereas the designations of the materials used for calibration and validation were equivalent, the material properties were not the same: The material properties of the specimens used for calibration (resp. validation) were: Tensile yield strength (0.2%): 372 MPa (resp. 330 MPa), ultimate tensile strength: 503 MPa (resp. 495 MPa), elongation: 18.2 (resp. 19.5). The fabrication processes were different as well. Such variabilities introduce uncertainties, however similar uncertainties are typically present in aluminum stock.

The prediction tested in the validation experiments was formulated as follows: "For a given value of G_α the probability is 90% that the number of cycles at which failure will occur (N_f) lies in the interval $N_1 < N_f < N_2$". This is equivalent to saying that if the number of specimens that fail outside of the specified interval is 10% or less of the total number of specimens tested then there is no reason to reject the model. We must bear in mind that in validation experiments we look for reasons that would justify rejection of a model, not to affirm its validity.

For example, for two in-phase axial-torsion fatigue experiments performed under the identical loading we found $Ga = 31.6$ ksi. The predicted interval from the survival function defined by the statistical model was between 77 and 671 thousand cycles. The outcomes were 130 and 302 thousand cycles which fall within the predicted interval, The survival function, the predicted interval and the realized outcomes are shown in Fig. 3.

The results of validation experiments, taken from reference [2], are listed in Table 2. The source data used for calibration of β were in US customary units therefore $G\alpha$ is reported in psi (pounds per square inch) units.

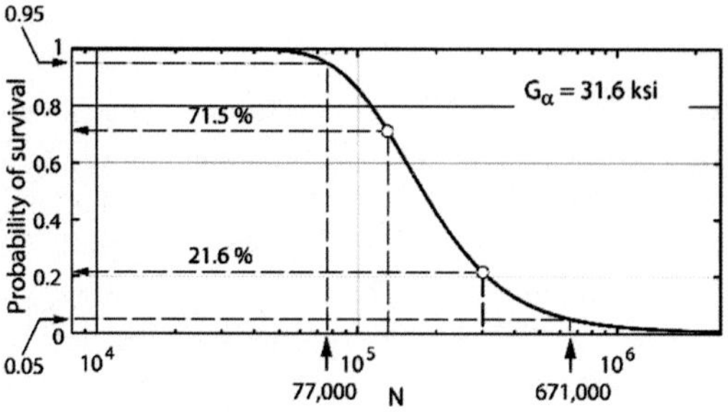

Fig. 3. The survival function and the outcomes of two in-phase axial-torsion fatigue experiments performed on notched 2024-T3 aluminum specimens.

Table 2. The results of validation experiments. Source: Ref. [2].

Load	σ_{max}	τ_{max}	G_α	N	Prob	Outcome	η	Remarks
	MPa	MPa	Psi	Cycles	%		%	
Axial	145	0	44030	9500	100.0	Fail	17.1	$\eta > \eta_{lim}$
Axial	130	0	39470	21670	99.2	Fail	8.0	
Axial	130	0	39470	31000	92.1	Pass	8.0	
Axial	115	0	34920	135450	26.5	Pass	1.6	
Axial	115	0	34920	145600	22.4	Pass	1.6	
Axial	98	0	29760	735000	13.5	Pass	0.0	
Torsion	0	108	42944	18500	97.4	Fail	34.4	$\eta > \eta_{lim}$
Torsion	0	91	36185	71890	61.0	Pass	6.9	
Torsion	0	91	36185	60140	75.6	Pass	6.0	
Torsion	0	76	30220	215000	57.6	Pass	0.0	
Torsion	0	64	25449	1800000	43.8	Pass	0.0	runout
Combined	81	50	31600	302000	6.3	Pass	0.7	In-phase
Combined	81	50	31600	130000	42.1	Pass	0.7	In-phase

The column labeled "Prob" indicates the probability of survival to the number of cycles at which the specimen fails. If this probability is between 5 and 95 percent then the outcome falls within the predicted interval. It is seen that 3 out of 13 experiments fall outside of the predicted interval. On closer examination, we find however that these specimens had larger amounts of plasticity than the others. Introducing the measure.

$$\eta = V_{yld}/V_{\beta} \tag{7}$$

where V_{yld} is the volume of material where the von Mises stress $\bar{\sigma}$, calculated from the solution of the problem of linear elasticity, is greater than the yield stress and V_β is the integration volume where $\sigma_1 > \beta\sigma_{max}$.

The predictor was formulated with the assumption that fatigue damage can be correlated with a linearly elastic stress field. This is known as a stress-life approach. It would be overly restrictive to limit the predictor to purely elastic stress fields however because the peak elastic stress can be very high at notch roots which would severely limit the applicability of the predictor. It is possible to admit small amounts for plastic deformation, provided that the plastic zone is sufficiently small so that plastic deformation is controlled by the surrounding elastic stress field.

The effects of plasticity, controlled by the elastic stress field, are taken into account through calibration of the parameter β. Naturally, this leads to the question: How large may the plastic zone be? Equivalently, what is the limiting value of η, denoted by η_{lim}. The choice of η_{lim} is one of the parameters that define the domain of calibration. In Table 2 we set $\eta_{lim} < 15.0\%$. Out of 11 qualified specimens, 1 failed outside of the predicted interval. Alternatively, if we set $\eta_{lim} < 8.0\%$ then there are 9 qualified specimens and the number of cycles at which failure occurred all fall within the predicted interval. In either case, we find no reason to reject the model.

6 Summary and Conclusions

We have formulated, calibrated and tested a mathematical model for the prediction of the probability of survival of mechanical and structural components made of 2024-T3 aluminum alloy subjected to constant cycle periodic loading at arbitrary mean stress levels in the high cycle regime. The model comprises four sub-models: A statistical model, a predictor of failure initiation caused by metal fatigue, a solid continuum model based on the assumptions of the linear theory of elasticity and a model for the generalization of constant cycle loading to arbitrary load spectra. In this paper we focused on the predictor and the statistical models while maintaining a strict protocol for solution verification: All computed data were verified to have not greater than 1% relative error.

The model was first calibrated on the basis of test records published in the 1950's. Calibration was in two parts; (i) a statistical model, called random fatigue limit model, was calibrated against test records of notch-free coupons cut from commercial aluminum sheets and (ii) the predictor calibrated against test records of notched coupons under uniaxial periodic loading conditions. Calibration of the predictor involves selection of the model parameters α and β.

Based on the calibrated model, predictions were made for the outcomes of fatigue tests performed on notched specimens fabricated from drawn tubing.

The domain of the predictor is defined by the interval of the highly stressed volume for which $\beta = \beta(V)$ was calibrated and the volume ratio η. In the present case the interval of highly stress volume for β is $(2.0E - 8 \leq V \leq 8.0E - 3)$ in^3 and the upper bound of η can be at least as high as 15%. Within this domain of calibration the model is considered to have been validated.

Of course, it is possible, even likely, that a better model will be proposed in the future. That model will have to be shown to produce a greater number of successful predictions in the same number trials and/or to have a larger domain of calibration for the same or a larger set of experimental data.

Industrial and research organizations collect experimental data over long periods of time. Interpretation and generalization of data through mathematical models must follow procedures similar to those outlined in this paper.

References

1. Babuška, I., Sawlan, Z., Scavino, M., Szabó, B., Tempone, R.: Bayesian inference and model comparison for metallic fatigue data. Comput. Methods Appl. Mech. Eng. **304**, 171–196 (2016)
2. Gates, N., Fatemi, A.: Notch deformation and stress gradient effects in multiaxial fatigue. Theor. Appl. Fract. Mech. **84**, 3–25 (2016)
3. Grover, H., Bishop, S., Jackson, L.: Fatigue Strengths of Aircraft Materials. Axial-load Fatigue Tests on Unnotched Sheet Specimens of 24S-T3 and 75S-T6 Aluminum Alloys and of SAE 4130 Steel. NACA Technical Note 2324, March 1951
4. Grover, H., Bishop, S., Jackson, L.: Fatigue Strengths of Aircraft Materials. Axial-load Fatigue Tests on Notched Sheet Specimens of 24S-T3 and 75S-T6 Aluminum Alloys and of SAE 4130 Steel with Stress Concentrations Factors of 2.0 and 4.0. NACA Technical Note 2389, June 1951
5. Grover, H., Bishop, S., Jackson, L.: Fatigue Strengths of Aircraft Materials. Axial-load Fatigue Tests on Notched Sheet Specimens of 24S-T3 and 75S-T6 Aluminum Alloys and of SAE 4130 Steel with Stress Concentrations Factors of 5.0. NACA Technical Note 2390, June 1951
6. Grover, H., Hyler, W., Jackson, L.: Fatigue Strengths of Aircraft Materials. Axial-load Fatigue Tests on Notched Sheet Specimens of 24S-T3 and 75S-T6 Aluminum Alloys and of SAE 4130 Steel with Stress Concentrations Factor of 1.5. NACA Technical Note 2639, February 1952
7. Grover, H., Hyler, W., Jackson, L.: Fatigue Strengths of Aircraft Materials. Axial-Load Fatigue Tests on Notched Sheet Specimens of 2024-T3 and 7075-T6 Aluminum Alloys and of SAE 4130 Steel with Notched Radii of 0.004 and 0.070 inch. NASA Technical Note D-111, September 1959
8. Pascual, F., Meeker, W.: Estimating fatigue curves with the random fatigue-limit model. Technometrics **41**(4), 277–289 (1999)
9. Szabó, B., Actis, R., Rusk, D.: Validation of notch sensitivity factors. J. Verification Validation Uncertainty Quantification **4**(1), 1–8 (2019)
10. Szabó, B., Actis, R., Rusk, D.: Predictors of fatigue damage accumulation in the neighborhood of small notches. Int. J. Fatigue **92**, 52–60 (2016)
11. Szabó, B., Actis, R.: Simulation governance: technical requirements for mechanical design. Comput. Methods Appl. Mech. Eng. **249–252**, 158–168 (2012)

Experimental Investigation of Vibroacoustic Behaviour of an Automotive Turbocharger with Semi-floating Bearing

Balázs Rácz(✉), Márk Pesthy, Péter Sass, and Jan Rohde-Brandenburger

Department of Internal Combustion Engines and Propulsion Technology, Széchenyi István University, Egyetem tér 1, Győr 9026, Hungary
racz.balazs@ga.sze.hu

Abstract. Due to the strict European emission standards and the constant aspiration for the higher power density, turbochargers became essential components of the modern internal combustion engines. Turbochargers are high-speed operating machines thus the design of the rotor and the bearing system requires special attention. The motions of the rotor are affected by several parameters, such as bearing design, clearances, structure of the surface and also the quality and the physical properties of the used lubricant. If the motions of the rotor are intensive in a wide rotational speed range, the bearing load increases, resulting in a reduced lifespan. The motion of the rotor induces vibrations, which leads to audible noise emission to the environment.

In this article, the vibrations of a four-cylinder spark ignition engine's turbocharger are presented, based on component test-bench experiments. Furthermore, the main vibration components and their influencing factors are briefly introduced. During the experiments, the noise and vibrations of the turbocharger have been measured with different viscosity grade oils from 20 °C to 140 °C inlet temperature. The results showed that the amplitudes of both the synchronous and subsynchronous vibrations changed significantly and the volumetric flow is highly dependent on the temperature. The effect of the changing oil temperature will be analyzed with an emphasis on the subsynchronous vibrations and the possible cause of the phenomenon will be presented. Finding the optimal parameters with the lowest possible vibration response could result in an extended lifetime and provides important information for the balancing process during production.

Keywords: Turbocharger · Rotordynamics · Tribology · NVH analysis · Lubricants

1 Introduction

Despite the current initiatives in the automotive industry, based on the predictions and the sales data, the internal combustion engine (ICE) will be dominantly present in the newly produced vehicles in the next decades. The main reasons for the presence of ICE vehicles on the market are the decent efficiency, high driving range, the well-established

K. Jármai and K. Voith (Eds.): VAE 2020, LNME, pp. 245–255, 2021.
https://doi.org/10.1007/978-981-15-9529-5_21

filling station infrastructure, moreover, the globally caused "cradle to grave" CO_2 emission calculation. During the testing and development of the engines, a high number of parameters should be considered from various sides such as the manufacturing, engineering challenges, and customer needs. Due to the strict European emission regulations, the engineers are required to develop and improve technology to reduce emissions on test procedures. The turbocharger is one of the engine components, which can provide specific efficiency increase, CO_2 emission reduction and observing future emission regulations even with the current technical level, which is indispensable to fulfill customer needs [1]. The feasible efficiency and specific performance of the engine, moreover, the quantity of harmful emission during RDE cycle (real driving emissions) can be improved on a relevant scale with the development of the actual charging systems [2]. The mentioned development trends originate new requirements upon the turbocharger itself. For example, increasing turbine inlet temperature of the exhaust gas, which leads to high thermal load of the turbine housing or higher achievable rotational speeds due to the need for larger pressure ratios and air mass flow into the engine, also operation during extreme conditions is expected. These conditions could be insufficient lubrication, extreme cold or high temperatures or limited cooling capacity [3].

Decreasing the viscosity level of the engine lubricants is a general development trend in the automotive industry as well [4]. The lower viscosity grade oils provide several possibilities to reach preferable exhaust gas emission values by decreasing the friction between the engine components and improving the cold-start properties of the engine. The change of the oil viscosity requires further developments, such as the research of the necessary oil additives and the investigation of the effect of the new lubricant consistency to the contact surfaces [5]. The change of the viscosity grade of the lubricants affects the operation of the turbocharger as well. All the components of a turbocharger are exposed to significant mechanical and thermal loads due to different operating conditions. To fulfil customer needs by achieving better throttle response and high performance of the engine, the turbocharger rotational speed can reach extreme values, even 300.000 RPM. The bending modes of the rotor are changing constantly during run-up at these high speeds; therefore, the loads of the bearings are changing dynamically during operation. The inlet oil parameters (viscosity, pressure, temperature) affect the behaviour of the rotor during operation and the occurring vibrations [6]. The bending of the rotor and the occurring vibrations are related to each other, hence with the investigation of their relations, the measured vibrations can refer to the operation and condition of the turbocharger without disassembling. Moreover, the lubricant acts as a damping medium in the bearing system, thus, applying vibration analysis methods to the turbocharger provides information regarding the effect of the used properties on the balancing process during production.

2 Measuring System

The experiments were implemented on a cold-gas operating turbocharger component test-bench, specially developed for vibration diagnostic tests (Fig. 1). The turbocharger is driven by pressurized air, produced by a mechanically driven centrifugal compressor. The control system of the test-bench allows normalized tests, with a pre-defined turbocharger run-up gradient. An additional oil conditioning unit supplies the turbocharger with lubricant. The extensive controllability of the mobile oil conditioning unit (MOCU) is highly important for the investigation of the effect of the changing oil properties to the turbocharger's vibrations. With the use of the MOCU, the oil pressure can be controlled from 0.1 to 6 bar and the oil inlet temperature from 10 °C to 150 °C, which is enough to completely cover the occurring values during real environment operation.

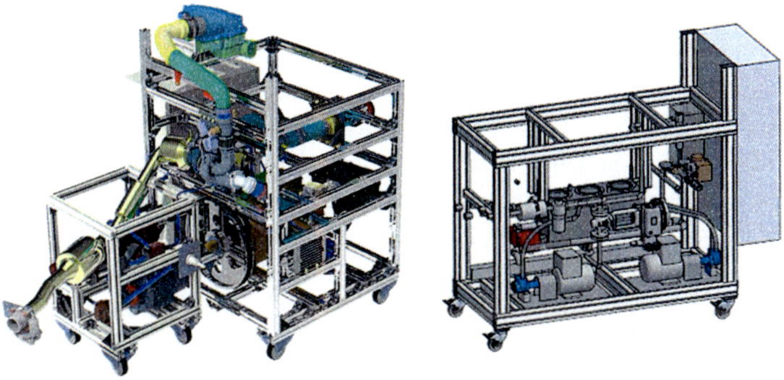

Fig. 1. Turbocharger component test-bench (left) and mobile oil conditioning unit (right)

To measure the vibrations of the turbocharger, acceleration sensors have been used. The acceleration sensors are mounted on an adapter in three directions, which is directly fitted to the bearing housing of the turbocharger (Fig. 2). To measure the rotational speed of the rotor, Picoturn eddy-current type rotational speed sensor was used, which is capable of measuring the rotor speed up to 320.000 RPM. The signal processing for both the speed sensor and the acceleration sensors was solved with an IMC Cronosflex data acquisition system. The Cronosflex has 100 kHz sampling frequency per channel, therefore, the system is suitable for the vibration measurements [7].

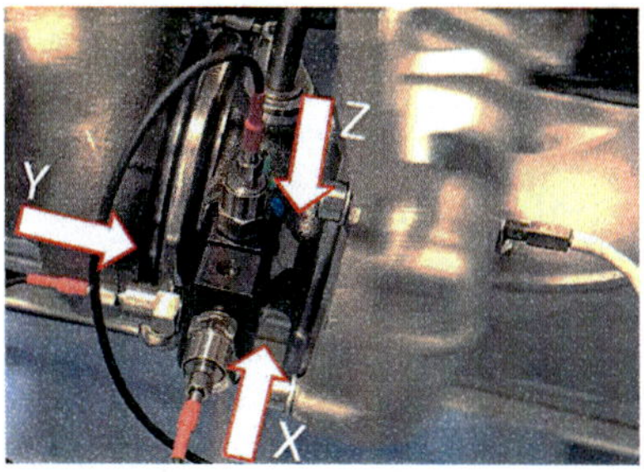

Fig. 2. Acceleration sensors mounted on the turbocharger bearing housing in 3 directions

During the experiment, normalized turbocharger run-up tests were implemented. The scope of the measurements is to determine the effect of the viscosity of the used lubricant to the vibrations of the turbocharger. Thus, three different viscosity grade oils were used, SAE 0W20, SAE 0W30 and SAE 5W40, which are widely used lubricants in the automotive segment. The oil inlet pressure was constant 3 bar and the oil inlet temperature varied between 20 °C and 140 °C with a step of 20 °C. At every temperature step, the turbocharger had a run-up phase from 0 to 130.000 RPM in 60 s and the occurring accelerations, the rotor speed and the oil parameters (inlet and outlet temperature, pressure, volumetric flow) were recorded.

3 Vibrations of the Turbocharger

To investigate the effect of the changing lubricant parameters to the turbocharger's vibration, extensive vibration analysis is necessary to determine the vibration components, which are affected by the oil properties. In this chapter, the main vibration components of the experimental turbocharger are analyzed and presented based on preliminary measurements. Fourier spectra are the most suitable for the spectral analysis of the acceleration data in stationary working points, on the other hand, during a dynamic test, the turbocharger's rotational speed changes as well as the frequency of the vibration components, thus the use of Campbell diagrams is necessary [8]. The Campbell diagram is a 3-dimensional diagram, which shows the amplitudes of the turbocharger's vibrations depending on rotational speed and frequency (Fig. 3).

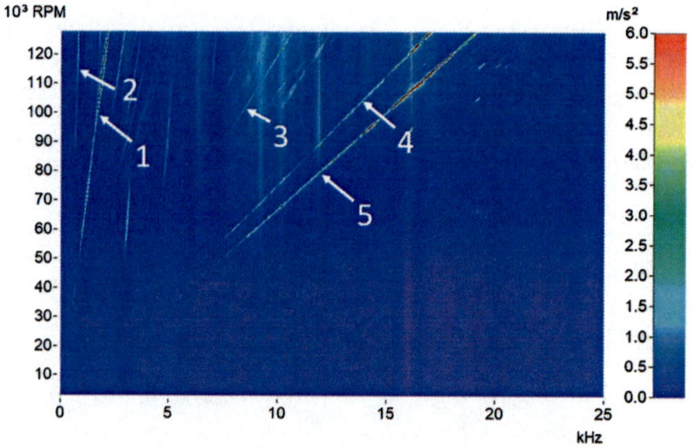

Fig. 3. Campbell diagram of the investigated turbocharger

Figure 3 shows the already determined vibration components of the turbocharger based on experiments and literature research. The simplest vibration component to find on this spectrum is the first order vibration (marked with 1), which equals the frequency of the rotor speed. The cause of this vibration is mainly the imbalance of the rotor assembly of the turbocharger (turbine wheel, shaft, compressor wheel and compressor wheel nut) [6]. The implementation of a perfect balancing method is impossible, so the synchronous vibration component is always visible on the Campbell diagram under real circumstances. The vibration component marked with 2 shows the oil whirling noise (subsynchronous vibration) of the turbocharger. This paper is mainly focusing on the investigation of the changes and influencing factors of this phenomenon. The frequency of the oil whirling vibration is lower than the frequency of the synchronous vibration and usually found on the spectra between 0.2 and 0.6 times the synchronous vibration. In this spectrum, a jump can be seen in the frequency of the oil whirling vibration at 104.000 RPM. The cause of this frequency jump is related to rotordynamic phenomena. It is presumed, that the vibration mode of the rotor system is changing at this point. The frequency and the magnitude of the oil whirl are highly dependent on the damping of the bearing system, which is affected by oil temperature, viscosity and the inner and outer radial bearing clearances. It is important to note that the design of the bearing system influences the oil whirling vibration as well. The used turbocharger for the experiment is supported by a semi-floating radial bearing (Fig. 4), which is not rotating, contrary to the full floating bearing, which rotates during the operation with a speed presumably 0.2 to 0.7 times the rotor speed (depending on the actual rotational speed and operating temperature) [9, 10].

Fig. 4. Semi-floating bearing (left) and full floating bearings (right)

The vibration marked with number 3 in Fig. 3 shows an undesired noise, which was detected during the vibration analysis of the test-bench itself. The necessary amount of air pressure and mass flow to spin up the turbocharger on the test-bench is created by a Rotrex type centrifugal compressor [11]. The indicated vibration component is the 7th order superharmonic vibration of the centrifugal compressor. The impeller of the Rotrex compressor emits air-borne noises with a frequency of the number of vanes times the rotational speed. This vibration frequency, in case of turbochargers and compressors, is called the vane pass frequency, which occurs due to the pressure differences at the edge of the impeller during rotation, which results in pulsation. The vibrations marked with 4 and 5 on the spectrum are formed due to the same phenomenon, which indicates the 8th and 9th order synchronous vibrations of the turbocharger. The turbocharger's compressor wheel has 8 vanes and the turbine wheel has 9 vanes; hence, the mentioned vibrations are also occurring with vane pass frequency [12].

To investigate the effect of the properties of the used oil to the occurring vibrations, the detailed analyzation of the subsynchronous and first order synchronous vibrations is necessary, since the lubricant's physical properties affect the oil whirling phenomenon and an assumption of the experiment is that the changing damping coefficient of the bearing system will influence the magnitude of the first order harmonic vibrations as well [13].

4 Effect of the Lubricant Viscosity to NVH (Noise, Vibration and Harshness) Level Based on Experiments

In this chapter, the experiments with different viscosity grade lubricants are presented. Based on the vibration analysis of the turbocharger, the measurements are aiming to investigate the first order synchronous vibrations and the oil whirling vibrations. The oil volumetric flow through the turbocharger's bearing system has been recorded in the oil conditioning unit (Table 1). The lubricant volumetric flow is highly dependent on the temperature and the viscosity of the used oil as well as the turbocharger rotational speed. By increasing the oil inlet temperature from 20 °C to 140 °C, the quantity of the volumetric flow increased by 271% (0W20), 260% (0W30) and 320% (5W40). In each measurement point, using the lower viscosity grade oil resulted in higher volumetric flow, thus both the oil temperature change and the oil viscosity affected the amount of lubricant flowing through the bearing system. The structure of the experiment is defined in Sect. 2.

Table 1. Measured maximum oil volumetric flow rates at different temperatures

Oil temperature (°C)	Oil volumetric flow (l/min)		
	0W20	0W30	5W40
20	0.98	0.88	0.69
40	1.31	1.17	1.02
60	1.57	1.42	1.25
80	1.89	1.65	1.49
100	2.17	1.91	1.77
120	2.44	2.11	2.01
140	2.66	2.29	2.21

During the measurements, evaluation method development was necessary, to analyze the emphasized vibration components in the most efficient way. The synchronous vibrations of the turbocharger were evaluated with vibration order tracking based on the rotational speed signal. The evaluation of the subsynchronous vibration is more complex since its frequency is not harmonic with the rotational speed and as mentioned in Sect. 3, a frequency jump occurs in the oil whirling vibration at high rotor speeds. Special band pass filters have been used to precisely show the amplitudes of the oil whirling vibration on two-dimensional diagrams.

Figure 5 shows the amplitudes of the oil whirling vibrations with 80 °C oil inlet temperature. It is clearly visible, that with the SAE 0W20 viscosity grade lubricant, the oil whirling vibrations have significantly higher amplitudes than with the higher viscosity lubricants. The magnitude of the oil whirling vibration starts to increase at higher rotor speeds, but the start of this increase is influenced by both the oil temperature and the

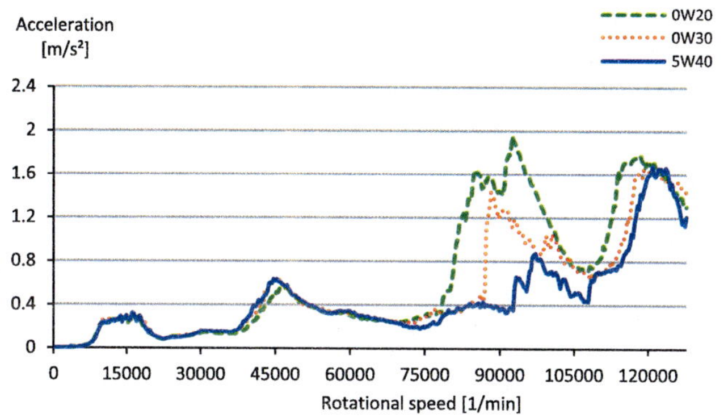

Fig. 5. Amplitudes of the subsynchronous vibrations in function of the rotational speed with 80 °C oil inlet temperature

viscosity grade according to the measured values. Based on the data shown in Fig. 5, by using lower viscosity grade lubricant, the start of the magnitude increase of the oil whirling vibration occurs at lower rotor speed. In case of 0W20 lubricant, the amplitude increase happens at 79.000 RPM, while with 0W30 its 87.000 RPM and with the use of 5W40 oil, the amplitude increase begins at 93.000 RPM.

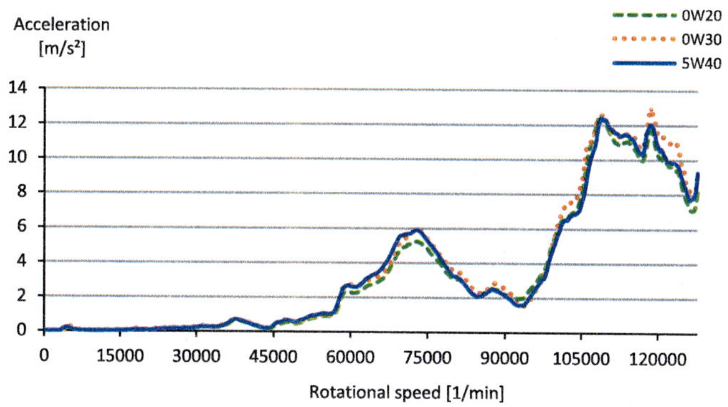

Fig. 6. Amplitudes of the first order vibrations in function of the rotational speed with 80 °C oil inlet temperature

The viscosity of the used lubricant influences the first order vibrations as well. Figure 6 shows the amplitudes of the first order synchronous vibrations with 0W20, 0W30 and 5W40 viscosity grade oils with 80 °C oil inlet temperature. The hypothesis of the experiment is that a correlation can be found between the magnitude of the synchronous and the subsynchronous vibrations. The higher the magnitude of the oil whirling vibration, the lower amplitudes can be measured in case of the first order vibrations. As Fig. 6 shows, slightly lower amplitudes can be measured with the use of 0W20 viscosity grade oil, than in the case of 0W30 and 5W40. On the other hand, with 0W30 oil the amplitudes of the vibration are not obviously lower than with 5W40

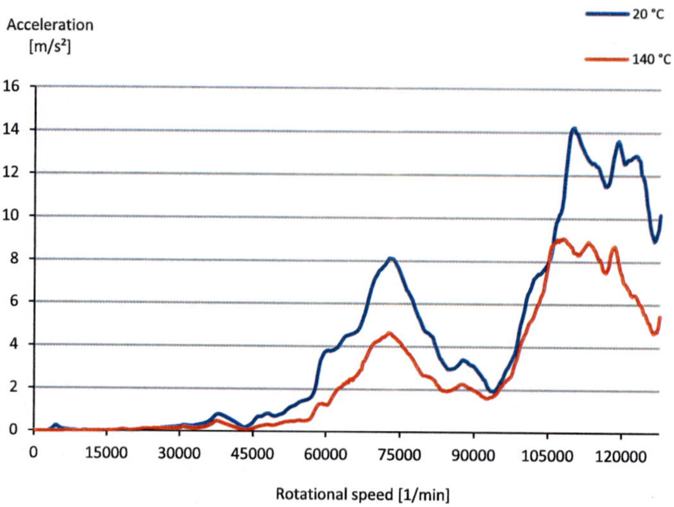

Fig. 7. Amplitudes of the first order vibrations in function of the rotational speed with 0W20 viscosity grade lubricant at different temperatures

viscosity grade oil. Hence, based on the diagram of Fig. 6 the effect of the oil viscosity to the magnitude of the synchronous vibration cannot be specifically classified. The measured oil volumetric flow values in case of 80 °C oil inlet temperature were 1.89 l/min (0W20), 1.65 l/min (0W30) and 1.49 l/min (5W40). It is important to note, that the oil volumetric flow through the bearing system varied during the run-up phase. The shown values in Table 1 are the maximum measured oil volumetric flow values during the run-up phase of the turbocharger. The slight difference between the measured first order vibrations with 80 °C oil inlet temperature, can be explained by the relatively small deviation in the volumetric flow values of the different viscosity lubricants. To clearly show the damping effect of the oil quantity in the bearing system, the comparison of measured points with a bigger oil volumetric flow difference is necessary.

Figure 7 shows the amplitude of the first order synchronous vibrations with 0W20 viscosity grade oil in case of 20 °C and 140 °C oil inlet temperature. The measured oil volumetric flow values are 0.98 l/min at 20 °C and 2.66 l/min at 140 °C, which is a significant difference. This time, the amplitudes in case of the experiment with 140 °C oil inlet temperature, are visibly lower, than with 20 °C oil inlet temperature at both high and low rotational speed range.

Figure 8 shows the amplitude changes between the experiments with 20 °C and 140 °C oil inlet temperature. From 30.000 to 130.000 RPM, the achieved average amplitude decrease is 40%. Hence, it can be stated that the temperature of the used oil significantly influences the measured parameters of the first order synchronous vibration. The manufacturer defines 9.6 m/s^2 maximum acceleration level for the first order vibration till 90.000 RPM. The maximum measured value during the experiments was 8.1 m/s^2 on the mentioned rotational speed range. By varying the oil inlet parameters and the type of lubricant, the maximum amplitude till 90.000 RPM can be reduced to 4.9 m/s^2. The extent of the amplitude change can easily influence the parameters during the balancing

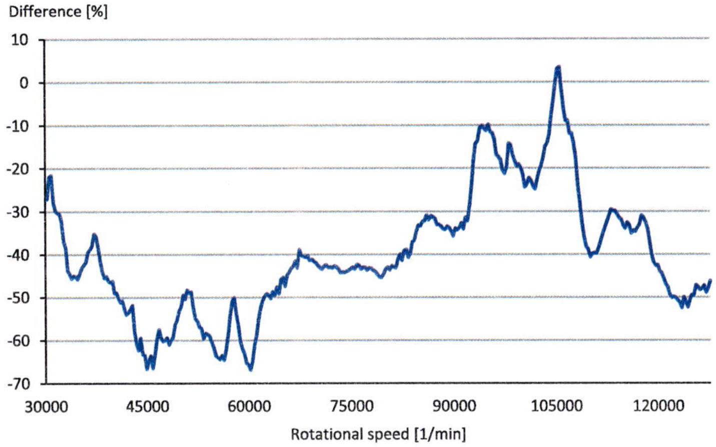

Fig. 8. Change between the first order vibration amplitude values in percentage with 20 °C and 140 °C lubricant temperature (0W20) depending on rotor speed

process and the properties of the used oil can even decide whether the turbocharger is classified as a wastrel product.

Increased oil inlet temperature influences the subsynchronous vibrations of the turbocharger as well. Figure 9 shows the amplitudes of the oil whirling vibrations with 100 °C lubricant temperature. By increasing the oil temperature, an amplitude peak forms at 86.000 RPM. The diagram shows, that in case of the 0W20 and 0W30 viscosity grade oils, the mentioned amplitude peak already formed, while the magnitude of the oil whirling with 5W40 is still lower.

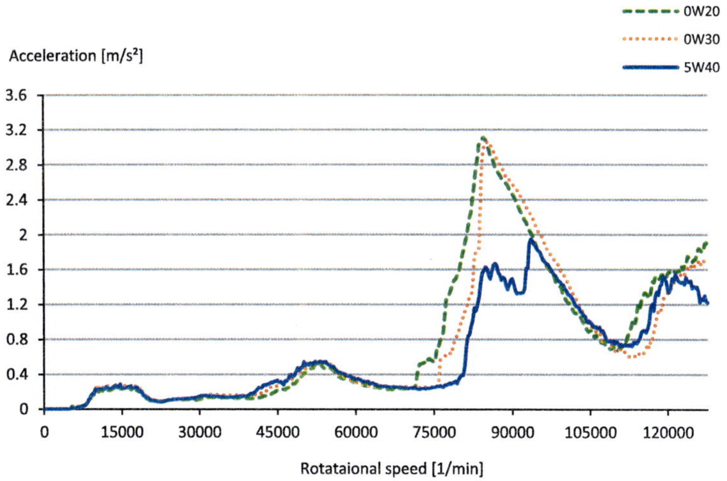

Fig. 9. Amplitudes of the subsynchronous vibrations in function of the rotational speed with 100 °C oil inlet temperature

However, by increasing the oil temperature further to 120 °C, the mentioned amplitude peak forms in the case of 5W40 lubricant as well. As Fig. 9 shows, the amplitude of the oil whirling vibrations can reach even 3.1 m/s^2, which is significant since the vibrations occurring due to imbalance has an amplitude of 2.2 to 3.2 m/s^2 on the same rotational speed range. Thus, the oil whirling vibration, which is affected by the oil properties and input parameters, can influence the turbocharger's overall vibroacoustic behaviour. Even with 100 °C lubricant temperature, the start of the amplitude increase (which happens after the frequency jump of the oil whirling vibration) is affected by the viscosity of the used oil. In case of lower viscosity lubricant (0W20), the amplitude increase starts at 72.000 RPM, while with 5W40 oil, the same phenomenon occurs at 81.000 RPM, which proves, that by changing the properties of the used oil, the magnitude of the turbocharger's vibrations can be influenced.

5 Summary and Further Work

Based on the experiments implemented on the cold-gas operating turbocharger component test-bench, it can be stated, that both the magnitude of the turbocharger's synchronous and subsynchronous vibrations can be influenced by the properties and input parameters of the used lubricant. Using lower viscosity oils with the same temperature can result in lower magnitude of the first order vibrations and amplitude increase in the oil whirling vibrations. In case of lubricants with the same viscosity grade, increased oil inlet temperature to the bearing housing caused a higher magnitude of oil whirling vibrations and lower measured accelerations with the synchronous frequency. In both cases, the volumetric flow of the used oil changed to a great extent. Increasing the lubricant inlet temperature from 20 °C to 140 °C, decreased the amplitude of the first order synchronous vibration by 60% at a rotor speed range from 40.000 to 60.000 RPM. Based on the measured values, the vibrations of the turbocharger significantly change during real environment operation in the vehicle and the used oil parameters (pressure, temperature and the viscosity grade) affect the measured imbalance values (first order vibrations) during the balancing process.

Analyzing the effect of the decreasing lubricant viscosity to the engine components are getting more important in the future since ultra-low viscosity oils (0W12 and even 0W8) are starting to appear on the market. The future aim of the research is to investigate the effect of the ultra-low viscosity lubricants to turbocharger lifetime and NVH. Moreover, ball bearing equipped turbochargers are under development in the automotive sphere, due to the achievable better throttle response, since the ball bearing turbochargers have lower friction at low and mid-range rotor speeds than the journal type bearings [14]. The future experiments will be supported with shaft motion measurements, with the use of high-precision distance sensors, which can detect the exact orbit of the turbocharger shaft during operation, providing advantageous information about the eigenfrequencies, the load and the damping of the bearing system.

References

1. Hiereth, H., Prenninger, P.: Charging the Internal Combustion Engine. Springer-Verlag, Wien (2003)

2. Franzke, B., Voßhall, T., Adomeit, P., Müller, A.: Water injection for meeting future RDE requirements for turbocharged gasoline engines. MTZ Worldwide **80**(3), 30–39 (2019)
3. Romagnoli, A., Manivannan, A., Rajoo, S., Chiong, M.S., Feneley, A., Pesiridis, A.: A review of heat transfer in turbochargers. Renew. Sustain. Energ. Rev. **79**, 1442–1460 (2017)
4. Champagne, N., Obrecht, N., Gangopadhyay, A., Zdrodowski, R., Liu, Z.: Enhanced anti-wear performance induced by innovative base oil in low viscosity engine oil. SAE Int. J. Fuels Lubricants **10**(3), 822–830 (2017)
5. Yang, K., Fletcher, K.A., Styer, J.P., Lam, W.Y., Guinther, G.H.: Engine oil components effects on turbocharger protection and the relevance of the TEOST 33C test for gasoline turbocharger deposit protection. SAE Int. J. Fuels Lubricants **10**(3), 815–821 (2017)
6. Nguyen-Schäfer, H.: Rotordynamics of Automotive Turbochargers. Springer, Heidelberg (2012)
7. IMC Homepage. https://www.imcdataworks.com/. Accessed 02 March 2020
8. Sass, P., Rácz, B.: Investigation of turbocharger tribological system with vibration diagnostics. Tribologie-fachtagung **60**, 662–671 (2019)
9. Kuma, H., Inoue, T., Isogai, T., Shimizu, K., Iida, T., Inagaki, M., Ohara, K.: Development of Reduction Method for Whirl Noise on Turbocharger, Powertrain & Fluid Systems Conference & Exhibition, SAE Technical Paper Series, Rosemont (2007)
10. Sass, P.: Literature research on experimental investigations of automotive turbocharger rotordynamics. Acta Tech. Jaurinensis **12**(4), 268–293 (2019)
11. Sass, P.: Grundlagen für die bewertung des verschleißverhaltens von abgasturboladern. Tribol. Effi. Conf. Győr **5**, 233–252 (2018)
12. Nguyen-Schäfer, H.: Aero and Vibroacoustics of Automotive Turbochargers. Springer, Heidelberg (2013)
13. Leonetti, M., Bargende, M., Kreschel, M., Meier, C., Schulze, H.: An Investigation of Sub-Synchronous Oscillations in Exhaust Gas Turbochargers, 12th International Conference on Engines & Vehicles, vol. 12, SAE International, Capri (2015)
14. Biet, C., Baar, R.: Experimental study on ball bearings for exhaust gas turbocharger. Tribol. Effi. Conf. Győr **4**, 253–272 (2016)

Passive Damping Techniques for Vibration Suppression in Boring Operation with Long Overhangs

Wallyson Thomas(✉), Zsombor Fulop, and Attila Szilágyi

University of Miskolc – Egyetemváros, Miskolc 3515, Hungary
szmwally@uni-miskolc.hu

Abstract. Tool vibration is more critical in the machining of deep holes, where the tool overhang is greater, and, consequently, is the tendency to chatter. This work presents recent designs and dynamic models of boring bars with different passive damping configurations explored in the internal turning in deep holes using hardened materials, where tight tolerances and low surface roughness were always reached. Furthermore, its passive auxiliary systems increased the damping capacity of the tool with a varied physical mechanisms that dissipate mechanical energy.

Keywords: Chatter · Boring bars · Passive damping · Deep holes

1 Introduction

The internal turning operation is one of the most important machining operations to produce either cylindrical or conical internal surfaces of rotating parts. In this operation, workpiece superficial quality is a very important parameter to be controlled, especially in internal turning of hardened steels, where the goal is to reach very good dimensional and superficial quality, similar to that obtained in grinding operations. However, before controlling the surface quality, it is important to control the vibration of the tool holder (or tool bar), which is dependent on cutting parameters, tool geometry and bar rigidity. In this operation, the tool bar must be out of the tool turret in a length (here called tool overhang) some millimeters longer than the machined hole length. Therefore, the longer the machined hole, the less stable is the tool bar, which hinders the machining of long holes.

So, to turn deep holes it is important to find a method to keep the cutting stable in order to obtain good workpiece surface roughness, even using a bar with high length to diameter (L/D) ratio [1].

One way to decrease tool vibration is the use of dampers in the tool. They dissipate part of the energy generated by the vibration movement, increasing the range of the process stability [2].

This study focuses on the vibration damping techniques used in boring operation such as passive damping techniques.

K. Jármai and K. Voith (Eds.): VAE 2020, LNME, pp. 256–264, 2021.
https://doi.org/10.1007/978-981-15-9529-5_22

Therefore, it is very important to use internal turning tool bar with some kind of damping effect, mainly when the turning operation is made in deep holes. There are different vibration dampers used in machining operation. They are divided into three categories: active, passive and hybrid dampers [3].

Active dampers are those in which the attenuator mechanism is controlled based on the system information, measured continuously throughout the process. Passive dampers do not have controls that allow real-time adjustments, but their construction is simpler when compared to active dampers and, therefore, cheaper [3]. When used to reduce vibration in machining, they must provide a high level of damping with a relatively low loss of static stiffness [4]. Several works proved their effectiveness via modelling [5–7] and damping rate tests [8–10].

Several times, however, the passive dampers demand the construction of a mass-spring system in the tool bar, which must work in the frequency the bar, is vibrating. Moreover, it is difficult to exactly determine the damping coefficient for several situations and also to keep the dampers tuned in the desirable frequency, once the space is limited, mainly for internal turning tool bars [3].

Various passive damping strategies can be used to suppress chatter, some that possibilities are shown in the Fig. 1 [11].

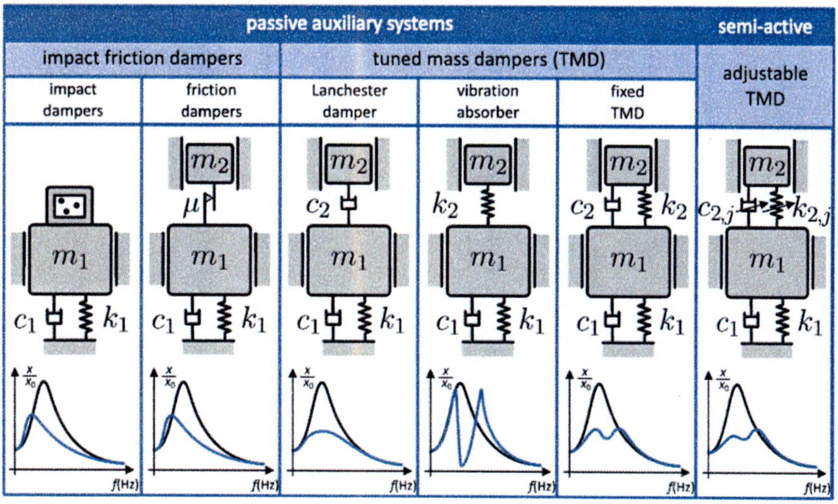

Fig. 1. Passive auxiliary systems [11].

Some passive dampers are:

- Damper Vibration Absorber (DVA) - consists of an additional mass-spring system connected to the bar, which needs to be set up in order to be tuned with the natural frequency of the structure [12]. According to the literature, this kind of damper works efficiently up to a tool bar L/D ratio of 15 [13];
- Tuned Mass Damper (TMD) with viscous elastic material - they are easy to apply in any kind of structure [14]. An example of this tool bar is the one called Silent Tool by

the manufacturer. It consists of an interchangeable head and a body made of heavy metal, supported by elements of rubber and oil [15].

- Friction Damper - it consists of several disks placed inside a cavity of the tool bar. Each disk rubs to other disk and also to the cavity wall in such a way to dissipate vibration energy [16, 17].
- Impact damper - the most usual of these dampers is the Particle Impact Damper (PID), which consists of hundreds of small particles (metallic, ceramic of little sizes) placed inside either the tool bar cavity or the workpiece wall using a reservoir stuck to it. These particles impact against the tool bar cavity while it vibrates and dissipates vibration energy [18]. The behaviour of this kind of damper is non-linear, what implies in some difficulties to control damping parameters like static stiffness (the bar loses some rigidity due to the cavity made in it), the restitution coefficient between the bar and the particles and the gap between the particles and the cavity wall [19]. On the other hand, it can provide high levels of damping along with a large range of frequencies [20].

While the viscous elastic materials dissipate elastic energy to provide damping to the structure, the particle impact dampers dissipate kinetic energy of the structure in a combination of collision, friction and deformation to reach the necessary damping. Metal materials with high density like copper, lead and cemented carbide are suitable materials to be used as particles [21].

One way to attenuate internal turning vibration is to reduce the cutting forces by changing the machining parameters (decreasing depth of cut, cutting speed, and feed rate). These changes reduce productivity dramatically. Another way is to use active or passive dampers attached to the boring bar to avoid or minimize vibrations [22].

2 Tuned Mass Damper (TMD)

A Tuned Mass Damper (TMD) is an inertial mass added to the system to damp via a linear spring of stiffness and damping (tuned to damp the critical mode of the original system that may produce chatter). By matching the natural frequency of the highly damped TMD with the critical frequency of the system both modes can be coupled to increase the damping. Consequently, the original mode is split into two modes with high dynamic stiffness. The TMDs have to be tuned accurately to the targeted frequency, and their positive effect is limited to a certain frequency range [23, 24]. The great advantages of the TMDs are their simplicity and reliability. The TMD should be located in places where the critical mode has large modal displacements. This way, the equivalent mass of the original critical mode is lower [25], and the required mass of the damper can be reduced for the same mass ratio m (Table 1). In general terms, the critical modes have large displacement close to the cutting point where the available space is small. The application of high-density material like lead or carbide can help in this search for a compromise between weight and space [11].

In some cases, the additional mass in that system should be precisely dimensioned and positioned to suppress the vibration of the boring tools. However, in the vertical direction (principal force direction) is well recommended, but hardly suppresses it in

Table 1. Dimensionless parameter of TMD.

Mass ratio	Damping ratio	Frequency ratio	Natural frequency
$\mu = \frac{m_2}{m_1}$	$\gamma = \frac{c_2}{2m_2\omega_1}$	$f = \frac{\omega_2}{\omega_1}$	$\omega_2 = \sqrt{\frac{k_2}{m_2}}$

the horizontal direction (thrust force direction) where the amplitude is extremely small [26, 27].

What tools are really efficient to get, are high overhangs with L/D = 10 and when reinforced with special materials like carbide it can reach L/D = 14 or when added a new degree of freedom with a especial dynamic system it can reach L/D = 15 [28].

It is well known that a TMD system is not so flexible because need to be adjusted to each boring bar condition or are able to operate only in a certain range of diameter or temperature considering that viscous material limitations. Also, that system is expensive because has sometimes complex construction and needs special fixation system. On the other hand, that some of that systems can be seen on the marked like silent tools (Branch Kenametal and Sandvik], and are less complex to install in the machine when compare to active or semiactive boring bar system.

Inside of the dampened tool is a pre-tuned dampening system that consists of a heavy mass, supported by rubber spring elements. Oil is added to increase the dampening. It is important to respect the limits marked on the product (load, temperature, rotation, min/max overhang and pressure) [29]:

- Temperature is highlighted to save the rubber elements in the dampening system.
- Maximum temperature limit depends on the type of product and is marked on the tool, e.g. 75–120 °C (167–248 °F).

There is a big potential for increased productivity using Silent Tools in all types of industrial segments. For components requiring long tools (~6–14 x Body Diameter (BD)) Silent Tools is the only choice for vibration-free machining [29].

Generally, you can use a steel or carbide boring bar for overhangs up to 4 x BD, but even in this range, a Silent Tools bar will give you very productive advantages. Overhangs up to 10 x BD are usually solved by applying a steel dampened boring bar to accomplish a sufficient process, while overhangs over 10 x BD require a carbide reinforced dampened boring bar to deal with radial deflection and vibration [29].

Among the Dynamic damper groups, there are bars, which use viscoelastic materials to aid in the absorption mechanism. With the cutting-edge technology of viscoelastic materials, vibration neutralizers have become easy to produce and apply to almost any structure, no matter how complex it is [30]. The effectiveness of a Silent Tool, outlined in Fig. 2, for a given mass, depends on the vibration amplitude at its point of attachment. Thus, the absorbers are generally installed along the bar. Other factors that determine its effectiveness are the value of the mass (inertial weight) of the bar and the type of viscoelastic material (oil) used inside the bar that make up the damping properties [31, 32].

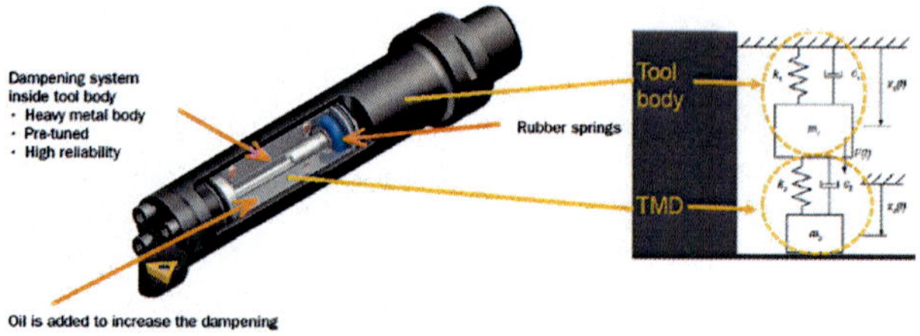

Fig. 2. Tuned mass damper (Silent Tool – Sandvik) and dynamic model [22].

3 Particle Impact Damper (PID)

They are also used to suppress the vibration of the internal turning operation. They are made up of one or more bodies in free movement. They can be composed of a free mass assembled on a cavity with clearance. There is also the particle impact damper, which consists of a container containing thousands of small metallic or ceramic particles or powders. These particles dissipate energy by friction and impact when the container vibrates. The resulting behavior is highly nonlinear, resulting in difficulties in controlling the absorption parameters - such as the stiffness and coefficient of restitution of the bar, and the gap between balls and cavity - each time the amount of mass in the system changes [19]. On the other hand, it can provide high levels of damping over a wide frequency range [20].

While viscoelastic materials dissipate elastic energy to treat damping, particle absorbers focus on dissipating kinetic energy from the balls in a combination of collision, friction and deformation to achieve damping. High-density metal particles such as copper, lead and carbide are the most satisfactory damping improvement materials [33].

Knowing that the sphere size limit is set by the cavity diameter, as the diameter of the balls has grown to the frequency range studied, the utilization limit overhang has also grown, showing that for damping it is better to have greater ball mass and shorter distance between the ball and the cavity wall, with a consequent lower velocity of the impact of the ball against the wall. With the largest ball diameter used it was possible to machine using boring bars with L/D equal to 8, obtaining even holes with surface quality;

In this work, using the kind of ID shown in Fig. 3, the increase of the possible tool overhang was 1.828 times (128 mm/70 mm) from the solid bar to the ID bar.

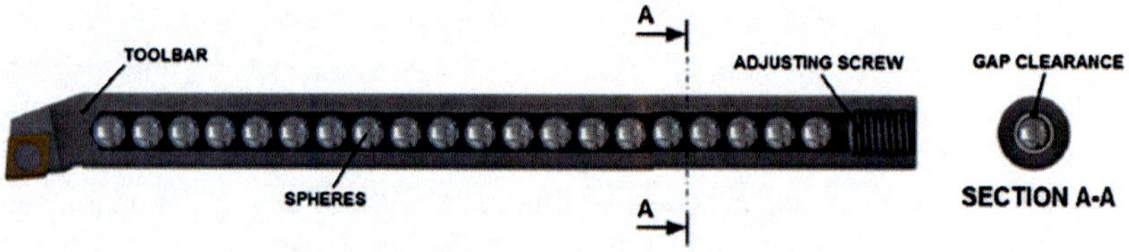

Fig. 3. Particle impact damper [19].

A two degrees of freedom (DoF) system, Fig. 4, was idealized to simulate an impact damper submitted to forced vibration via harmonic base excitation. The system is constituted by a principal mass m_I with stiffness k_I and viscous damping c_I, where a single metallic sphere moves freely without friction between the barriers. When contact occurs the nonlinear model Eq. (1) is applied, then spring and damper forces act providing a momentum transfer between the masses. When no contact occurs, it was considered that the particle movements with constant velocity, i.e. without friction.

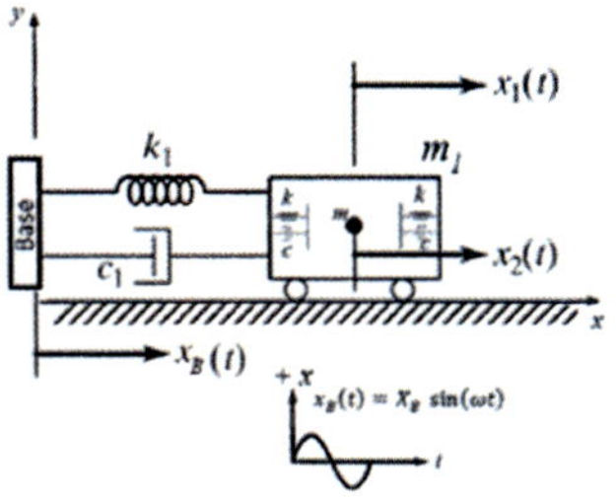

Fig. 4. Two DoF system [34].

The sphere-barrier contact force is modeled by a set of nonlinear spring-damper, based on Hertz theory and proposed by Hunt and Crossley,1975 conforming Eq. (1), where k is the contact stiffness, c is called hysteresis damping factor, δ is relative position and $\dot{\delta}$ is relative velocity [35].

$$F_N = k\delta^{1.5} + c\dot{\delta}\delta^{1.5} \tag{1}$$

The equations of motion contain the contact force and the harmonic base movement, as follows:

$$[M]\{\ddot{x}\} + [C]\{\dot{x}\} + [K]\{x\} + [\Delta]\{\delta^{1.5}\} + [\gamma]\{\dot{\delta}\delta^{1.5}\} = \{F(t)\} \tag{2}$$

$$\begin{bmatrix} m_1 & 0 \\ 0 & m_2 \end{bmatrix}\begin{Bmatrix} \ddot{x}_1 \\ \ddot{x}_2 \end{Bmatrix} + \begin{bmatrix} c_1 & 0 \\ 0 & 0 \end{bmatrix}\begin{Bmatrix} \dot{x}_1 \\ \dot{x}_2 \end{Bmatrix} + \begin{bmatrix} k_1 & 0 \\ 0 & 0 \end{bmatrix}\begin{Bmatrix} x_1 \\ x_2 \end{Bmatrix} + \begin{bmatrix} k & k \\ -k & -k \end{bmatrix}\begin{Bmatrix} \delta_l^{1.5} \\ \delta_r^{1.5} \end{Bmatrix} + \begin{bmatrix} c & c \\ -c & -c \end{bmatrix}\begin{Bmatrix} \dot{\delta}_l\delta_l^{1.5} \\ \dot{\delta}_l\delta_r^{1.5} \end{Bmatrix} = \begin{Bmatrix} k_1(X_B\sin(\omega t)) + c_1(\omega X_B\cos(\omega t)) \\ 0 \end{Bmatrix} \tag{3}$$

where: subscripts l and r indicate left and right wall, respectively. In this case, k and c will only assume value if the contact exists in respective wall.

The contact stiffness, based on Hertz theory, is related to material properties of contact bodies (E_i and v_i) and the surface curvature radius (R_i). According to Goldsmith (1960) apud Flores et al. (2005) the contact stiffness of a sphere plane contact, Eq. (4), is written as follows [36]:

$$k = \frac{4(R_1)^{0.5}}{3\left(\frac{1-v_1^2}{E_1} + \frac{1-v_2^2}{E_2}\right)} \tag{4}$$

where R_i is the radius of sphere, v_i is the Poisson ratio and E_i is the Young modulus.

The hysteresis damping factor c is related to contact stiffness k, initial impact velocity $\dot{\delta}^{(-)}$ and a constant α that varies from each model. Flores et al. (2011) model was chosen to be applied in the simulations of this work and is given by

$$c = \alpha \frac{k}{\dot{\delta}^{(-)}} = \frac{8(1-e)}{5e} \frac{k}{\dot{\delta}^{(-)}} \tag{5}$$

where e is the coefficient of restitution [37].

4 Conclusions

The techniques presented are useful for achieving high accuracy in internal boring operations with long and slender boring bars. The methods will compensate for deflections in the boring bar, and it will eliminate the effect of minor errors in the tool offset.

This review paper reveals that passive damping techniques by using damping particles like steel, plastic granules, etc. can suppress the vibrations during the desired amount. Particle damping is a better passive damping technique that can retain its stiffness within permissible limits. This proves to be a simpler and effective method of vibration reduction. This can be used over a wide range of temperatures and frequencies and shows a wide scope of applications. The review shows a greater damping capability in that particle damping.

References

1. Badadhe, A.M., Bhave, S.Y., Navale, L.G.: Optimization of cutting parameters in boring operation. In: Journal of Mechanical and Civil Engineering, pp. 10–15 (2005)
2. Albuquerque, M.V.: Modelagem e Análise Dinâmica de um Absorvedor de Vibrações por Efeito de Impacto (Dynamic modeling and analysis of a vibration damper by impact effect). Master's Thesis, State University of Campinas, Campinas Brazil (2016)
3. Bankar, V.K., Aradhye, A.S.: A review on active, semi-active and passive vibration damping. Int. J. Curr. Eng. Technol. **6**, 2187–2191 (2016)
4. Waydande, S., Mahajan, D.A., Gajjal, S.Y.: A review on vibration attenuation of boring bar using passive dampers. Int. J. Emerg. Technol. Adv. Eng. **4**(4), 117–122 (2014)
5. Rubio, L., Loya, J.A., Miguélez, M.H., Fernández-Sáez, J.: Optimization of passive vibration absorbers to reduce chatter in boring. Mech. Syst. Sign. Process **41**, 691–704 (2013). https://doi.org/10.1016/j.ymssp.2013.07.019
6. Vinayaravi, R., Kumaresan, D., Jayaraj, K., Asraff, A.K., Muthukumar, R.: Experimental investigation and theoretical modelling of an impact damper. J. Sound Vib. **332**, 1324–1334 (2013). https://doi.org/10.1016/j.jsv.2012.10.032
7. Zhang, C., Chen, T., Wang, X., Li, Y.: Discrete element method model and damping performance of bean bag dampers. J. Sound Vib. **333**, 6024–6037 (2014). https://doi.org/10.1016/j.jsv.2014.07.011
8. Ema, S., Marui, E.: A fundamental study on impact dampers. Int. J. Mach. Tools Manuf. **34**(407), 407–421 (1994). https://doi.org/10.1016/0890-6955(94)90009-4
9. Ema, S., Marui, E.: Damping characteristics of an impact damper and its application. Int. J. Mach. Tools Manuf. **36**(3), 293–306 (1996). https://doi.org/10.1016/0890-6955(95)00073-9

10. Lu, Z., Lu, X., Masri, S.F.: Studies of the performance of particle dampers under dynamic loads. J. Sound Vib. **329**, 5415–5433 (2010). https://doi.org/10.1016/j.jsv.2010.06.027
11. Munoa, J., et al.: Chatter suppression techniques in metal cutting. CIRP Ann. Manuf. Technol. **65**(2), 785–808 (2016). https://doi.org/10.1016/j.cirp.2016.06.004
12. Dimarogonas, A.: Vibration for Engineers, 2nd edn. Prentice Hall Upper Saddle River, New Jersey (1996)
13. Liu, X., Liu, Q., Wu, S., Liu, L., Gao, H.: Research on the performance of damping boring bar with a variable stiffness dynamic vibration absorber. Int. J. Adv. Manuf. Tech. **89**, 2893–2906 (2017)
14. Bavastri, C.A.: Redução de Vibrações de Banda Larga em Estruturas Complexas por Neutralizadores Viscoelásticos (Wide band vibration reduction in complex structures by viscouselastic neutralizers). PhD Thesis, UFSC, Florianópolis Santa Catarina, Brazil (1997)
15. Sandvik Silent tool for turning: overcome vibrations in internal turning (2016). http://www.sandvik.coromant.com/en-us/products/silent_tools_turning. Accessed 15 May 2017
16. Hahn, R.S.: Design of lanchester damper for elimination of metal-cutting chatter. Trans. ASME **73**, 3 (1951)
17. Kyocera Catálogo boring - boring bars with interchangeable heads and anti-vibration dampener system (2017). http://www.kyocera.com.sg/products/cuttingtools/wp-content/uploads/2015/02/F-Boring.pdf. Accessed 12 February 2017
18. Kanase, S.S., Patil, J.S., Jadhav, S.M.: Improvement of Ra value of boring operation using passive damper. Int. J. Eng. Sci. **2**, 103–108 (2013)
19. Booty, C., Bowyer, E.P., Krylov, V.V.: Experimental investigation of damping flexural vibrations using granular materials. In: International Congress on Sound and Vibration Department of Aeronautical and Automotive Engineering. Loughborough University, Loughborough, Leicestershire, England (2014)
20. Sims, N.D.: Vibration absorbers for chatter suppression: a new analytical tuning methodology. J. Sound Vib., pp. 1–16 (2006)
21. Halliday, D., Robert, R., Merrill, J.: Fundamentals of Physics, vol. 9. Wiley, New York (1981)
22. Vasanth, X.A., Paul, P.S., Lawrance, G., et al.: Vibration control techniques during turning process: a review. Aust. J. Mech. Eng. **00**, 1–21 (2019). https://doi.org/10.1080/14484846.2019.1585224
23. Den Hartog, J.P.: Mechanical Vibrations. McGraw-Hill Book Company, New York (1934)
24. Vanherck, P.: Optimisationde l'amortisseur dynamique pour machineoutil. CIRP Ann. **12**, 120–126 (1963)
25. Ewins, D.J.: Modal Testing, Theory, Practice, and Application, 2nd edn. Wiley, Hoboken (2000)
26. Ema, S., Marui, E.: Suppression of chatter vibration of boring tools using impact dampers. Int. J. Mach. Tools Manuf. **40**(8), 1141–1156 (2000)
27. Lawrance, G., et al.: Attenuation of vibration in boring tool using spring controlled impact damper. Int. J. Interact. Des. Manuf. (IJIDeM) **11**(4), 903–915 (2017)
28. Liu, X., et al.: Research on the performance of damping boring bar with a variable stiffness dynamic vibration absorber. Int. J. Adv. Manuf. Technol. **89**(9-12), 2893–2906 (2017)
29. Application guide – silent too acess (2020). https://www.sandvik.coromant.com/sitecollectiondocuments/downloads/global/technical%20guides/en-us/c-1020–17.pdf
30. Bavastri, C.A.: Redução de Vibrações de Banda Larga em Estruturas Complexas por Neutralizadores Viscoelásticos. Tese de Doutorado, UFSC, Florianópolis Santa Catarina, Brasil (1997)
31. de Souza, J.C.S., de Lima Jr., J.J.: Determinação dos módulos de armazenagem e de perda de materiais viscoelásticos-padrão ASTM. CEP **37500**, 903 (2010)
32. Rubio, L., Loya, J.A., Miguelez, M.H., Fernandez-Saez, J.: Optimization of passive vibration absorbers to reduce chatter in boring. Mech. Syst. Sign. Process. **41**(1-2), 691–704 (2013)

33. Khatake, P., Nitnaware, P.T.: Vibration mitigation using passive damper in machining. Int. J. Mod. Eng. Res. **3**(6), 3649–3652 (2013)
34. Marcos, A., Robson, P.: Modeling of an impact damper submitted to forced vibration from a nonlinear contact force model. COBEM 2017 24th ABCM International Congress of Mechanical Engineering Curitiba, Brazil (2017)
35. Hunt, K.H., Crossley, F.R.E.: Coefficient of restitution interpreted as damping in vibroimpact. ASME J. Appl. Mech. **42**, 440–445 (1975)
36. Flores, P., Ambrósio, J., Claro, J.C., Lankarani, H.M.: Influence of the contact-impact force model on the dynamic response of multi-body systems. Proc. Inst. Mech. Eng. Part K J. Multi-body Dyn. **220**(1), 21–34 (2005)
37. Flores, P., Machado, M., Silva, M.T., Martins, J.M.: On the continuous contact force models for soft materials in multibody dynamics. Multibody Syst. Dyn. **25**, 357–375 (2011)

Test Method for Investigation of Reactive Loads on Gear Drives with Supporting Function

Ferenc Sarka(✉), János Bihari, Ágnes Takács, and Zsolt Tóbis

University of Miskolc, Miskolc-Egyetemváros, Miskolc 3515, Hungary
{machsf,machbj,takacs.agnes,machtzs}@uni-miskolc.hu

Abstract. Today's vehicles typically have a number of gear drives in which small plastic gears provide power transmission. During the design of small dimensional plastic gear drives, the effect of reactive loads must be considered in all gear-pairs with supporting function. However, this effect must be tested according to the real loads. Typically affected units are the drives of the air conditioning and the rear-view mirrors. Reactive loads can come from the wind force and in the case of rear-view mirrors from the force of manual adjustment through direct pressing on the mirror, or from pressing the mirror during cleaning. In cases when the gears in the final step of the drive cannot be dimensioned to carry a load such that they are not adversely affected by such forces, the gear unit shall operate in the reverse energy direction. In this case, the inertia of all steps as well as the losses due to friction must also be taken into account in determining the design parameters. Bearings of small plastic gearboxes are typically very simple, so friction losses could be extremely high. In this article, we present experiments where, in the case of large gears with ratios above 1:300, the torque required for reverse rotation is investigated so that the gears are made with the same gears but with different bearings. The torque required for reverse rotation determines the maximum load exerted on the gear pair providing support function.

Keywords: Small plastic gears · Gear drive · Reverse energy direction · Rear-view mirrors

1 Introduction

When designing a gearbox, it is seldom necessary to handle the reaction torques inside the engine, mainly because they can only occur in extreme cases. In such cases, we mainly use external solutions such as safety clutches, brakes, or freewheel clutches. In the applications of small plastic gears, these solutions are rarely applicable. These gears are generally used in structures where a decisive parameter is the cost of production is the simple reason for that. At the same time, reliability is also expected for these structures, while their reasonably predictable faulty use is more typical than the average. Just think that it is useless to prescribe that the mirror of a vehicle should not be pressed hard by

K. Jármai and K. Voith (Eds.): VAE 2020, LNME, pp. 265–272, 2021.
https://doi.org/10.1007/978-981-15-9529-5_23

the user when cleaning. An insect stuck on the mirror is enough to forget this, so the mirror plate will move and the drive that moves the mirror will be loaded with torque in the opposite direction to the torque of the original energy chain. A similar case arises when repairing a vehicle's ventilation system, the mechanic ignores the requirements for pre-electronic resetting of the deflector vanes or does not have the necessary device to do so. The deflectors, which are in a position that makes installation difficult, will of course be adjusted to the correct position by hand, while also turning the actuator drive backwards. Leaving the area of vehicles, such loads can be considered not only extraordinary but also operational in the case of children's toys. A doll's arm or toy crane boom will almost certainly be loaded in the opposite direction of its original movement. A common feature of the above cases is that the gear unit moves a component or unit which, due to its size, allows a high feedback force to be exerted in relation to the forces taken into account when sizing the given system. In such cases, it must always be ensured that the gear unit can rotate backwards without damage due to external forces. But the question is how much load the last step(s) must actually withstand.

2 A Few Words About Designing Small Plastic Gears

When a general flow diagram of the sizing of gearboxes **Error! Reference source not found.** for small plastic gears is prepared [1], it can be seen that these gears are typically designed with a mixture of a conventional and a parameter-based design. In addition to loads, a dimension that is typically limited in space, or often only in one or two planes, is typically an important parameter here.

This means that when selecting or creating any geometric solution, the designer will consider which solution requires the most space, and this will be true not only for the gears but also for the bearings and the design of the housings. If we also take into account that these gears almost always operate in discontinuous operation, i.e. we do not have to reckon with a high degree of heating., the efficiency of the gear is almost never a significant factor, and the fact that cost is almost always an important factor leads a straight path to the application of the simplest possible bearings. These are typically plain bearings, in which the bearing bushes are formed from the housing material by the housing manufacturing technology, usually without post-processing, leaving only the lubricant to influence the slip properties. These are typically made according to similar principles of plastic, or simple, e.g. paired with metal shafts made by cold drawing (Fig. 1).

3 Theoretical Handling of Reaction Torque

If we think about the design of the gears in the usual way and want to take into account the feedback loads, then for each step of the gear we count with the load required to rotate the given step on the gear teeth. In the case of a gearbox with state-of-the-art gear units made of machined metal or plastic or sintered metal and fitted with roller or plain bearings, this load can be easily calculated. In this case, the moment of inertia of

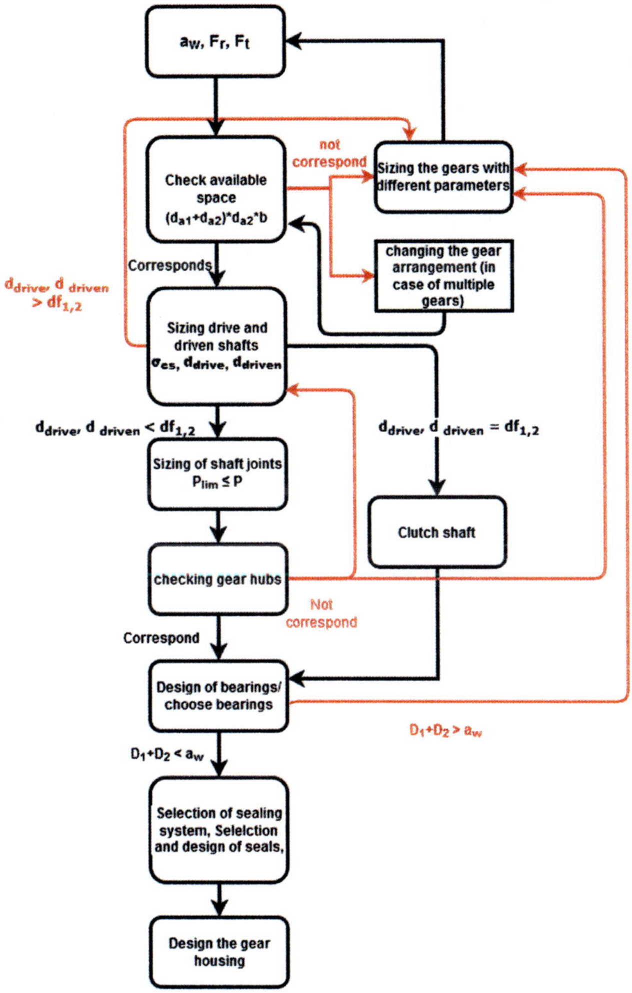

Fig. 1. Combined design process for small plastic gears

the system connected to the gear unit can be multiplied by the angular velocity of the shaft, and the losses of the gear unit can be considered if the gear unit is designed with a small safety factor. However, this would be the wrong approach for many small plastic gears. Here, due to the design viewpoints mentioned above, the losses of the gear unit can be extremely large compared to the usual ones. If this is coupled with a large gear ratio above 1:100, significant torque may already be required to reverse the final stages. Figure 2 shows a general method for handling recirculation torques.

In practice, however, it can be a problem that not all parameters for sizing are known, because here the effect of losses occurs multiplied by a large ratio.

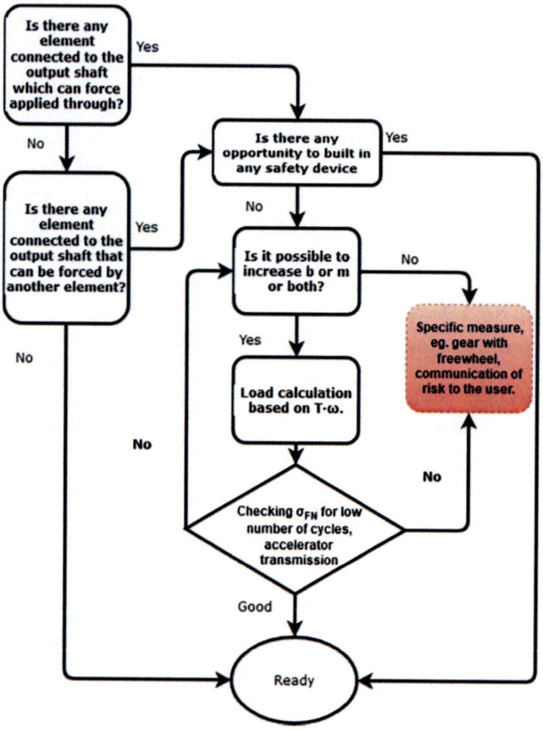

Fig. 2. Theoretical handling of recirculation torques

4 Special Losses of Small Plastic Gears

In the case of small plastic gears, the following losses may be greater than usual:

- Friction losses of bearings.
- Losses from stresses due to large bearing gaps.
- Losses from position error.
- When using stepped gear wheels with a large number of teeth difference, the front surfaces can rub against each other, causing additional losses.

These losses are inaccurate or impossible to calculate, often only experimentally determined. From our point of view, this means that it is difficult to determine the maximum torque on the last steps during reversing. Therefore, we started a series of experiments aimed at determining the different losses. Considering the four main sources of losses listed here, the impact of bearings is presumably the largest for the entire system. After all, in addition to the two losses that can be clearly linked to bearings, inaccurate bearings and bearings that are only accurate enough for normal operation are responsible for a significant part of the position errors. In general, position errors are also responsible for the friction between the front surfaces of the gears. Therefore, in the first part of the series of experiments, we compare bearings.

5 The Structure of the Experiment, the Used Gearboxes

In this experiment, our goal is to determine the difference in recirculation torque when using different bearings. This way for experiments such gearbox had to be built that exclude other characteristic defects of small plastic gears, e.g. inaccuracies in the housings, or skewness of the shafts, circularity error, etc. Because of this in the gearbox only the gears are made of plastic, the housings and shafts are made of metal, with precision machining. 0.5 mm module, 50- and 10-tooth stepped gears were used in arranged one after the other in the gearbox. These are shown in Fig. 3.

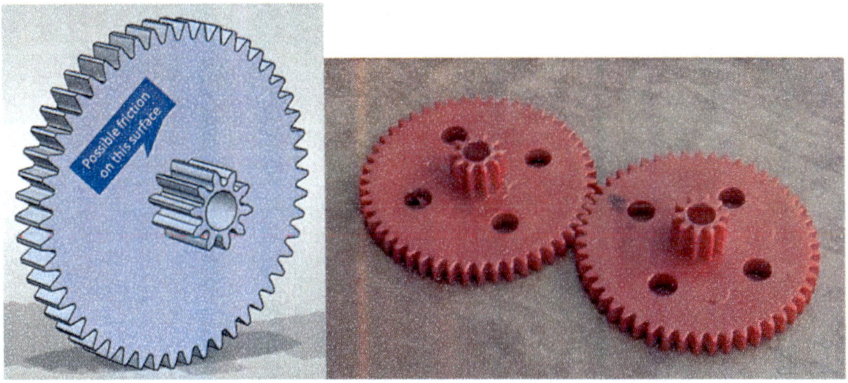

Fig. 3. Gears used for the experiment (CAD and real)

These gears are injection molded and correspond to an accuracy class of approximately 10 according to DIN 3967.[1] The gears were mounted in the housings with ground steel shafts, the torque transmission between the shaft and the gear was ensured in each case only by force locking. The houses are made of aluminum profiles. The installation location of the bearings and shafts was made in one operation on both sides of the housing on a milling machine equipped with a digital odometer to ensure uniaxiality.

We basically made two types of gearboxes. In one type, the shafts rotate in rolling bearings, in the other type in bores that model a plain bearing (Fig. 4).

In the housings the center distances are the same, in the plain bearing housing, the bearing joints correspond to the values used for plastics gears and housings [2–4]. From a comparison of the two solutions, we expected that while plain bearing housings would occur minor stresses (it is frequent in such gearboxes), they would not occur in ball bearing housings, so their impact could at least be estimated. We used the same gears in the houses, installed in the same order. This is important because the manufacturing accuracy of the gears we use is low, and when installed elsewhere, they can give fundamentally different results.

1 Tolerance systems for steel gears cannot be used unconditionally for plastics with a modulus of approx. 0.5 mm. It is advisable to prescribe the tolerances on the basis of the operating parameters [1].

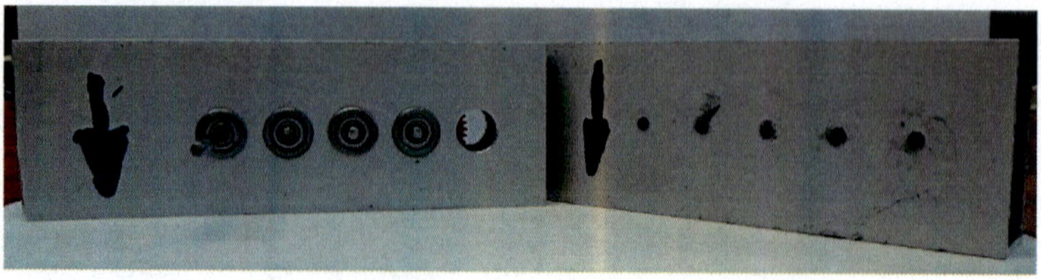

Fig. 4. The two houses of different designs

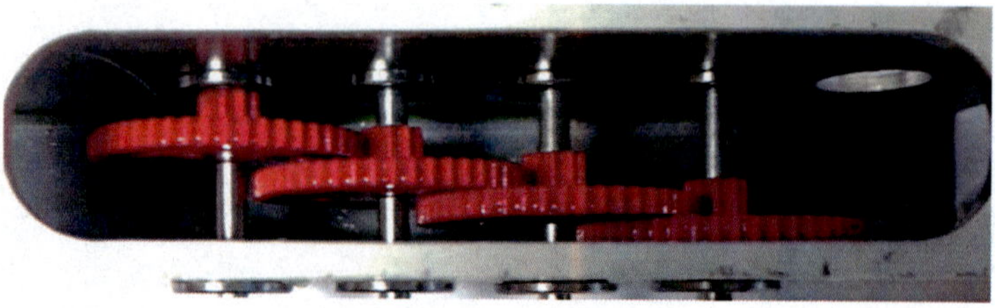

Fig. 5. Arrangement of gears in houses

Of course, another interesting experiment could be the effect of changing the order of the gears on the recirculation torque, since if a less well-connected gear moves forward in the chain (i.e., backward in terms of recirculation torque), the loss due to poor connection means more torque when the output shaft is rotated backwards. The gears were adjusted so that there was a gap of at least 0.2 mm between their plates. The arrangement is shown in Fig. 5.

5.1 Carrying Out the Measurements

Materials and instruments used for measurements:

- TQM-80 static torque measuring shaft with 0.5 s sampling frequency software with continuous recording.
- Gears mounted in two different housings.
- Nicro 607 synthetic creeping oil.
- CRC Brakleen 20 brake cleaning fluid for accurate removal of creeping oil.

The output shaft of the gears was clamped in the chuck of the torque measuring shaft and the gears were rotated at a constant speed. 5 different series of measurements were performed, with 5 measurements each. In the first series, gears mounted in a plain bearing housing were tested without lubrication. This provides a reference value for the next measurement, which already examines an operationally relevant scenario. In the second series, lubricant was applied to the gears of the gearbox and to the bearings in the case of the plain bearing design too, but only from one direction. It is because such gears are typically lubricated at one point with a batch of lubricant during manufacture, and the gear distributes the lubricant to itself during operation.

Table 1. The results of the measurements

Type of the housing (slave or ball)	Slave bearing	Slave bearing	Slave bearing	Ball bearing	Ball bearing
Lubrication	No	Yes, before running in	Yes after running in	No	Yes
Average recirculation torque [Ncm]	7.5	4.8	2.7	2.5	1.3
Maximum recirculation torque [Ncm]	9.2	7.6	3.1	2.7	1.8

The good self-lubricating properties of plastics and the fact that these gears are typically non-power transmissions allow this method not to cause damage to the gearing. However, when installing the gear unit, the installer technician may adjust the element connected to the gear unit to create a recirculation torque. If this is greater than the recirculation torque of the run-in gear, it must be taken into account in the design. In the third series, we examined gear box with slave bearing and after lubrication run-in gears. In the fourth series, ball bearing gear boxes with non-lubricated gears were tested. In the fifth series, we tested ball bearing gearbox with lubricated gears. The results of the measurements are summarized in Table 1. The table shows average and maximum values. The averaging was necessary because the rotation speed was not completely uniform and at such low values the weight of the housing also influences the measurement result. This can be corrected with a more stable grip and better rotation; we plan to develop the experiments in the future. Therefore, the values in the table should not be considered as generally valid values, but their proportions should be monitored.

6 Conclusion

High-quality bearings allow significantly lower return torque even at 125:1 gear ratio, i.e.: more precise bearings should be considered for gear units exposed to such effects. This is especially true in cases where better materials or larger sizes cannot address this problem. Running in after lubrication also significantly reduces recirculation torque, therefore, in the case of point-lubricated gear units, it must be taken care of, e.g. when designing the in-line installation, the recirculation torque for the no-run condition must be taken into account or the gear units must be run-in previously.

Acknowledgement. The described article/presentation/study was carried out as part of the EFOP-3.6.1-16-2016-00011 "Younger and Renewing University – Innovative Knowledge City – institutional development of the University of Miskolc aiming at intelligent specialisation" project implemented in the framework of the Szechenyi 2020 program. The realization of this project is supported by the European Union, co-financed by the European Social Fund.

The authors would like to thank their colleague Attila Potyka for his role in the preparation of the measurement.

References

1. Bihari, J.: Kisméretű műanyag fogaskerekek tervezési és fejlesztési kérdései, doktori disszertáció, Miskolc (2017)
2. VDI 2731 Mikrogetriebe, Grundlagen. BeuthVerlag, Berlin (2009)
3. Zahnräder aus thermoplastischen Kunststoffe. VDI Verlag, Düsseldorf (1981)
4. Kies, T.: Herstellung von Zahnräder in Spritzgussverfahren – Möglichkeiten und Grenzen. Martin-Luther Universität Halle (2016)

Analysis of Wear Curves as Sigmoid Functions

Ferenc János Szabó(✉)

University of Miskolc, 3515 Miskolc, Hungary
machszf@uni-mskolc.hu

Abstract. Sigmoid functions (growth function, logistic function, evolution function, etc.) are used in several fields of science to describe, study and forecast several phenomena of life. Since the sigmoid curves are nonlinear curves, the application of the Fisher- Pry transformation is used for calculating the regression coefficients of the approximated curves. In this paper, the nature of the investigated wear curve makes it necessary to compare the logistic curves and growth function curves. The process of the approximation is based on the principle of least squares: the minimum of the squared sum of differences is searched by the Nelder- Mead unconstrained minimization algorithm. The variables of the optimization are the parameters in the equation of the approximating function. The sigmoid curves can describe mainly the beginning phase and the normal wearing phase of the wear curve, the ending phase of the wear curve is a very quickly increasing function. Therefore on the basis of the results of this study, it could be possible to build a wear- monitoring system, in order to see and follow the differences between the sigmoid curve and the original wear curve, and if these differences are higher than a given limit, this could be the basis of some alert or warning, signaling the possible end of the lifetime.

Keywords: Sigmoid curves · Wear curves · Curve approximation and analysis

1 Introduction

In several fields of our life, curves with S-like shape, describing development, growth or saturation of several phenomena are called "sigmoid" functions, sigmoid curves. On the basis of studies and observations of these curves, it could be possible the modeling or description of growth or saturation phenomena, sometimes it is possible to use them even for extrapolation and forecast, too. These curves are multidisciplinary curves because one can find them during studies or investigations of different disciplines (in biology: population dynamics; in economy: product lifecycle, medicine: growth of tumours; environmental protection: pollution of the air or pollution by plastic in world seas; agriculture: growth of fish populations; forestry: analysis of tree populations; optimization: iteration history of optimum searching algorithms). A large number of examples can be found for the applications of sigmoid curves in the literature:

The discovery and the beginning of the studies of sigmoid curves started at the end of the 1700 years, by Malthus (1798) [1]. According to his statements, the instantaneous

K. Jármai and K. Voith (Eds.): VAE 2020, LNME, pp. 273–281, 2021.
https://doi.org/10.1007/978-981-15-9529-5_24

growth of any population (including human population) is the function of the actual number of the population:

$$\dot{N} = f(N) \tag{1}$$

Malthus introduced a parameter of the growth rate (r) as:

$$\dot{N} = rN; \quad r = \frac{\dot{N}}{N} \tag{2}$$

If the value of the r parameter is constant, then it is possible to write the function describing the growth of the population in function of the time, in the following form:

$$N(t) = N_0 e^{rt} \tag{3}$$

This function can be used for the description of populations growing in large areas, amongst non-disturbed circumstances. An example of this type of growth was given by Gordon E. Moore (1965), who stated the rule that the number of components per integrated circuit is doubled in each 1.5 year time [2]. Another example of the non disturbed growth could be the increase of the number of internet users in function of time. This function gives an exponential increase of the number of the population.

This exponential increase cannot be supported during longtime. Pierre- Francois Verhulst, Belgian mathematician solved and described the case when the growth of the populations has a saturation level K. He proved [3], that in this case the equation describing the growth of the population in function of the time is as follows:

$$\dot{N} = rN\left(1 - \frac{N}{K}\right) \tag{4}$$

and he introduced the nomination of "logistic growth". In Eq. (4), if N is small, then the value of the multiplier (1 – N/K) is close to 1, in this case, the growth of the population is close to the exponential growth. If N is increasing, this multiplier will decrease and around 0 the growth of the population will stop, so the number of the members in the population will stop at a constant number, this is the phenomenon of the saturation. This is the mathematical model of the growth of populations in a constrained space. Pearl and Reed (1920) applied the nomination "logistic growth" or logistic curve for the prediction of the number of habitants of the USA [4]. Equation (4) is a differential equation in function of time for N. The logistic curve fulfilling Eq. (4) is as follows:

$$N(t) = \frac{K}{1 + e^{-rt-b}} \tag{5}$$

Since the shape of this curve is like an S, this was the basis of the nomination of sigmoid (S-shape) curves. The parameter b in the Eq. (5) will change only the position of the curve, not the shape, therefore it can be called as position parameter. If the value of r is positive, this is a logistic increase, if r is negative, this will be a logistic decrease.

During the approximation of the curves, it is important to determine the regression coefficient, in order to check the quality of the approximation. The calculation of the linear regression coefficient is easy to do, therefore it seems to be useful the linear transformation of the sigmoid curves into linear shape, using logarithm. Fisher and Pry (1971) defined this kind of transformation [5]. The Fisher–Pry transformation applies the transformation as F = N/K, in this case the logarithm will show the shape as ln (F/(1–F)) = rt + b, which means that the logistic function can be replaced by a line, having the rise of r. For this linear function, it will be easy to calculate the coefficient of linear regression.

Bertalanffy (1938) made investigations for some cases when during the growing process, the increase is not exponential even in the beginning part of the growth [6]. He created the Bertalanffy- growth function, for the study of the body length of the sharks. Later this function was successfully applied for the description of some fish populations growth and in the forestry, too. Kozuko et al. [7] in 2003 applied this function in medical studies for the investigations of the growth and reproduction of tumour cells. The equation of the Bertalanffy- growth function:

$$N(t) = K - (K - N_0)e^{-rt} = K\left(1 - \frac{K - N_0}{K}e^{-rt}\right) \tag{6}$$

Richards (1959) modified the Bertalanffy- curve and he made it applicable for the studies of the growth of several plants [8].

The results of Mansfield (1961) [9] and Rogers (1962) [10] make possible the application of sigmoid curves during the investigations of the lifecycle of products, and the introduction of new products, in the economy, industry and product design. On the basis of these results, it is possible to study the spreading speed of innovations, or the propagation speed of technology innovations. Jang, Show- Ling et al. (2005) created a mixed model [11] and they investigated the propagation of mobile phones in 29 OECD countries and in Taiwan, also by sigmoid functions.

Meyer (1994) shown that it is possible to find pulsating cases, having several newer and newer increasing parts of the growth [12], these are the bi- logistic, tri- logistic or multi-logistic growth processes. The works of Silverberg and Lehnert (2003) are based on this phenomenon [13], concerning the evolutionary growth of the economy. Nikosz (2009) gives interesting examples of the applications of sigmoid curves for social phenomena [14].

Szabó (2011–2019) applied sigmoid curves and their derivatives and integrals in several fields: He shown (2011) that the curve of world records investigating more than one hundred years for several branches of sport is sigmoid curve [15]. By investigating the equations, derivatives and integrals of sigmoid curves, Szabó built the EBSYQ (Evolutionary Based System for Qualification of Group Achievements) [16] system which is applicable for comparing and qualifying the results of student groups, applicants to grants or jobs or different phenomena showing growth or saturation behaviour. The results obtained by the optimization using these results were used by Zhang, Huang and Yu (2020) during the optimization process of a light van- type electric truck [21]. Other research can be found for human- electric hybrid drives [22] in the work of Sarka and Bihari (2018). The widespread applicability of this system is shown by sigmoid studies of optimum searching algorithms (2018) [17]. Szabó (2019) gives an approximation of the future quantity of plastic contamination of seas [18] by the application of the

EBSYQ system, which shows also the multidisciplinarity and the wide range applicability of the system. For the approximation of the sigmoid curves Szabó used the Fisher-Pry transformation and the principle of least squares, which he transformed into an optimization process, solved by the Nelder–Mead [19] optimization algorithm. Comparing the regression coefficients for different types of sigmoid curves, it is possible to decide which curve describes the best way the given phenomenon. It is possible to find and compare the shape of different sigmoid curves in the paper of Kehl and Sipos (2009) [20] in Hungarian.

New future application field could be to apply sigmoid curves for the approximation of wear curves because by the EBSYQ system it could be possible to compare several elements, being in a different part of their lifetime, or analysing the parameters of the approximating sigmoid curves, an alert could be sent that the given element is near to the end of its lifetime, because of the wear.

2 Approximation of Wear Curves

Figure 1 shows a typical shape of a wear curve in function of time (grinding wheel). The curve has three parts: Part I is the stage of growing, Part II is the saturation stage and Part III is the final stage of the lifetime of the element. Part I and II can form a sigmoid shape function, like a saturation type growth function.

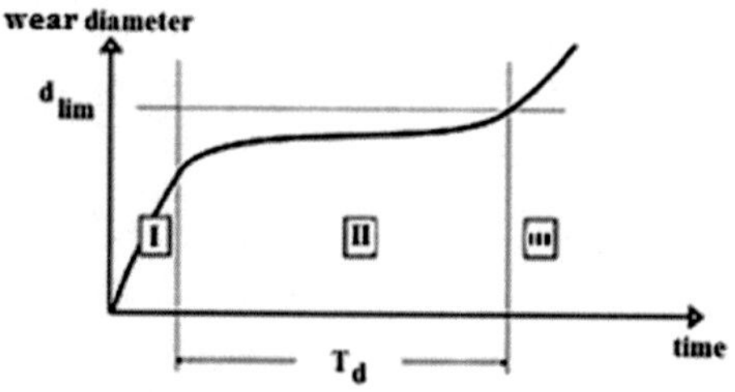

Fig. 1. A typical shape of the wear curve, it has three parts [23]

The third part is an increase comparing to the second part, so monitoring the differences between the real curve and the approximating sigmoid curve, it could be possible to define an alerting situation when we are too close to the ending phase. This could be very useful in some cases if we do not know exactly the limit value of the diameter but we want to avoid to use a damaged tool. Table 1 shows some typical shapes of the sigmoid curves used for the approximation, together with the shape of the integral and of the derivative of the curves.

Figure 2 shows the approximation of the wear curve of Fig. 1, by a logistic function (Pearl–Reed) and by a growth function (Bertalanffy), the data of the curve were measured from Fig. 1. It is very salient from the figure the starting place of phase III, where the end stage of the lifetime starts. Here will start to increase the difference between the growth curve and the original wear curve. If this difference will be higher than a previously defined limit (for example 10–15% of the function value), it is possible to send the alert signal to the operator or to the user of the tool.

Table 1. Some typical shapes of the used sigmoid curves.

curve	derivative	integral
Pearl–Reed		
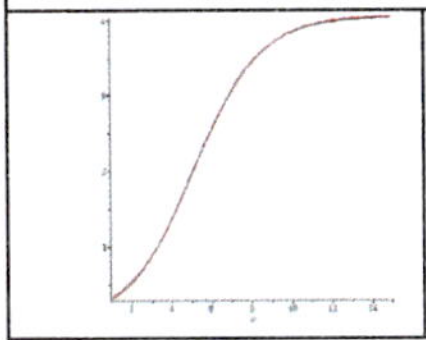	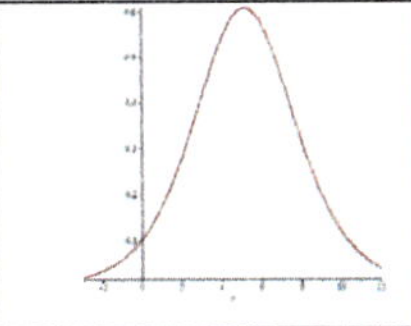	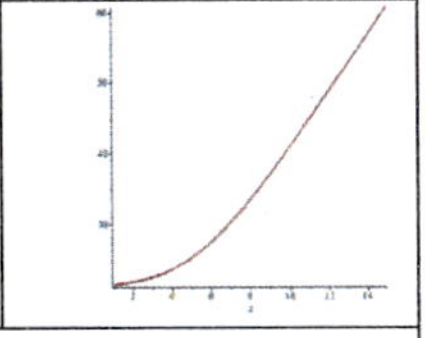
Bertalanffy		
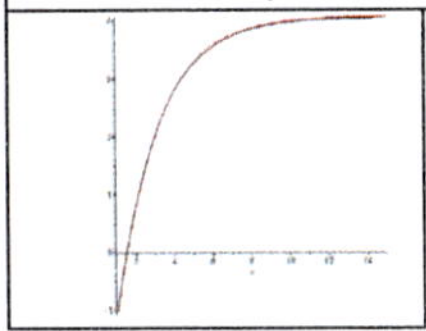	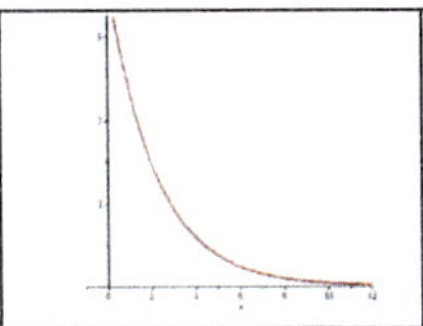	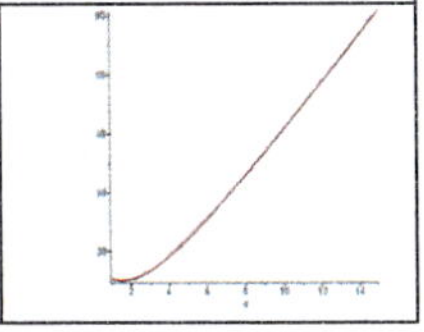

During the approximation process, two types of sigmoid curves will be compared: the Pearl–Reed (logistic) curve and the Bertalanffy (growth) curve. From Fig. 2 one can easily predict that the Bertalanffy curve will be closer to the original curve because in the original curve the exponential part of the logistic growth is missing.

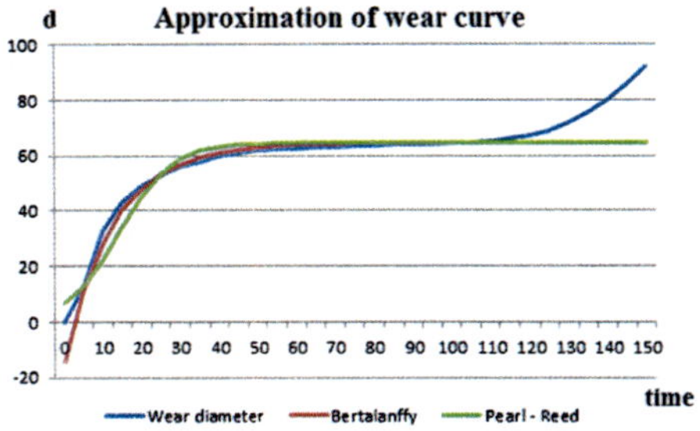

Fig. 2. Approximation of the wear curve of Fig. 1, by two different sigmoid curves

Let us compare the two types of sigmoid curves, regarding their equations, derivatives and integrals. Equation of Pearl–Reed and Bertalanffy curves can be seen in Eq. 7, 8 and 9. More types of sigmoid curves can be found in the paper of Szabó [16].

$$\text{The Pearl} - \text{Reed curve}: y = \frac{K}{1 + ce^{-rx}}, \text{ and Bertalanffy curve}: y = K\left(1 - ce^{-rx}\right) \tag{7}$$

The first derivative of the Pearl–Reed curve: $\frac{dy(x)}{dx} = \frac{Kcre^{-rx}}{(1+ce^{-rx})^2}$, its integral can be calculated as:

$$\int y(x)dx = -\frac{K}{r}\ln\left(e^{-rx}\right) + \frac{K}{r}\ln(1 + ce^{-rx}) \tag{8}$$

For the Bertalanffy - curve, the derivative: $\frac{dy(x)}{dx} = Krce^{-rx}$, the integral can be calculated as:

$$\int y(x)dx = Kx + \frac{Kc}{r}e^{-rx} \tag{9}$$

The shapes of the curves are shown in Fig. 3, the derivatives can be seen in Fig. 4 and the integrals are in Fig. 5. In Fig. 3, 4 and 5, where the d and d_W are denoting to the diameter of wear. The best way to select which curve describes better the wear phenomenon of the element, is to compare the regression coefficient of the curves. Pearl-Reed (left side) and Bertalanffy (right side) curves are nonlinear curves, but the linear regression is very easy to calculate.

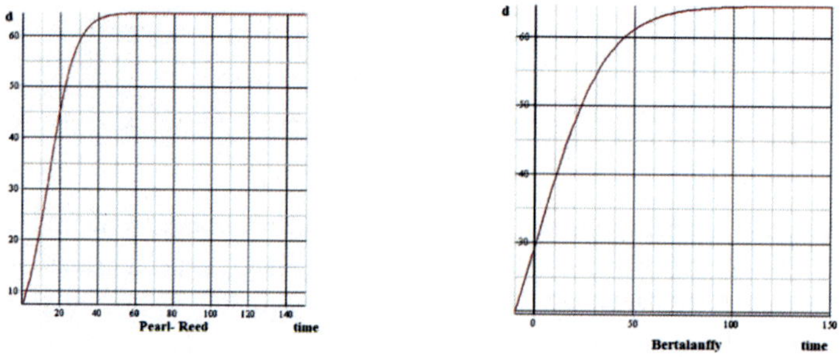

Fig. 3. Comparison of Pearl–Reed (left) curve and Bertalanffy (right) curve

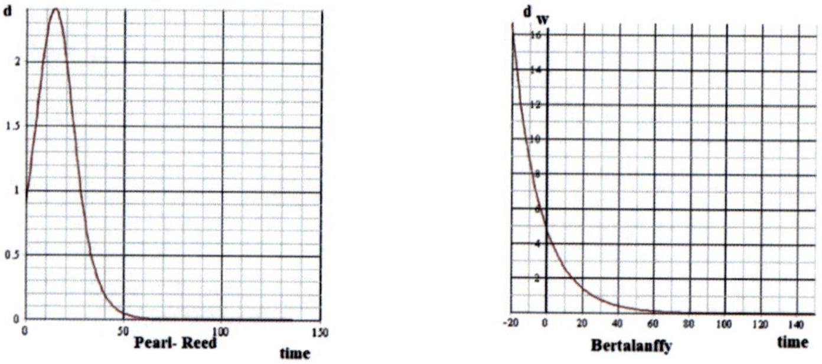

Fig. 4. Derivative of the Pearl–Reed (left) and Bertalanffy (right) curve

Therefore it seems to be very useful to apply the Fisher- Pry transformation which will transform the curves into linear form and the regression coefficient will be calculated on the basis of the transformed shapes of the curves. The details of the application of the Fisher–Pry transformation and the calculation of the regression coefficients can be found in the paper of Szabó [16]. The calculated regression coefficient for the Pearl–Reed curve is 0.96452, and for the Bertalanffy curve is 0.98636. Therefore it is easy to decide, that in the investigated case the Bertalanffy curve is a better approximation for the wear curve than the Pearl- Reed curve.

Table 2 shows the *K, r, c* parameters of the curve equations for both of the curve types used for the approximation.

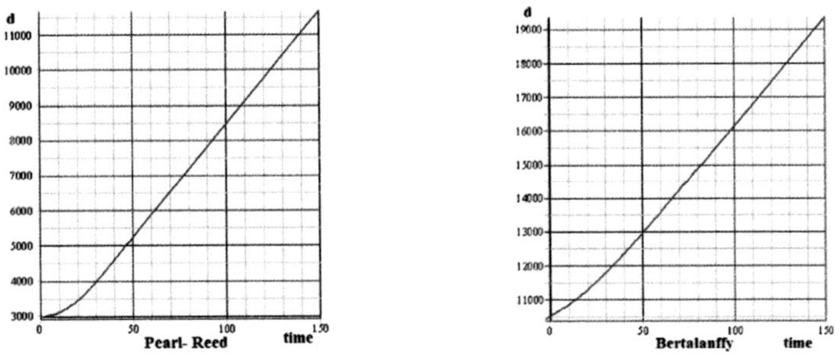

Fig. 5. The integral of the two curves left: logistic curve, right: growth curve

Table 2. The *K, r, c* parameters in the equations.

Curve name	*K*	*r*	*c*
Pearl–Reed	64.56	0.1493	8.61
Betalanffy	64.626	0.0616	1.2213

3 Conclusions

In this paper, a wear curve is approximated by using sigmoid curves. For the approximation the Pearl–Reed (logistic) curve and the Bertalanffy (growth) curve are applied. Since these curves are nonlinear curves but the linear regression coefficient is easy to calculate, the Fisher–Pry transformation is used and the curves are transformed into linear shape. This way the regression coefficient can be easily calculated. The regression coefficient of the Bertalanffy curve is closer to 1, therefore this curve will describe better the phenomenon.

In the 3rd phase of the wear process, the real curve starts to increase and one can observe an important difference between the approximating sigmoid curve and the original curve. If the difference is higher than a previously defined limit value, in this case an alert signal could be sent to the user or operator, that the lifetime of the element is too close to the end. This way it is possible to avoid the usage of wear- damaged element.

Having the shapes and equations of the approximating sigmoid curves, it is possible to calculate their derivatives and integrals, too. The derivative of the curve will show the propagation speed of the wear, while the integral of the curve will give useful information for the calculations of the volume loss because of wear.

In the future, a possible continuation of this research could be to use the facilities of the EBSYQ (Evolutionary Based System for Qualification of Group Achievements) curve analysis system for a comparison of several elements or several wear curves. This comparison and qualification system could give more interesting results showing more information about the characteristics of these curves.

Acknowledgements. The described study was carried out as part of the EFOP-3.6.1-16-2016-00011 "Younger and Renewing University – Innovative Knowledge City – institutional development of the University of Miskolc aiming at intelligent specialisation" project implemented in the framework of the Szechenyi 2020 program. The realization of this project is supported by the European Union, co-financed by the European Social Fund.

The realisation of this project is co-supported by the European Union and co-financed by the European Social Fund and by the EU Lifelong Learning Program.

References

1. Malthus, T.: An Essay on the Principle of Population. J. Johnson in St Paul's Church-Yard, London (1798)
2. Moore, E.G.: Cramming more components onto integrated circuits. Electron. Mag. **38**(8), 114–117 (1965)
3. Verhulst, P.-F.: Deuxieme mémoire sur la loi d'accroissement de la population. Mémoires de l'Académie Royale des Sciences, des Lettres et des Beaux-Arts de Belgique **20**, 1–32 (1847). Accessed 18 Feb 2013
4. Pearl, R., Reed, L.J.: On the rate of growth of the population of the United States since 1790 and its mathematical representation. Proc. Natl. Acad. Sci. **6**(6), 275–288 (1920)
5. Fisher, J.C., Pry, R.H.: A simple substitution model of technological change. Technol. Forecast. Soc. Chang. **3**, 75–88 (1971)
6. Bertalanffy, L.: Principles of theory of growth. In: Fundamental Aspects of Normal and Malignaent Growth, Amsterdam, pp. 137–259 (1960)
7. Kozuko, F., Bajzer, Z.: Combining gompertzian growth and cell population dynamics. Math. Biosci. **185**, 153–167 (2003)
8. Richards, F.J.: A flexible growth function for empirical use. J. Exp. Bot. **10**, 290–300 (1959). https://doi.org/10.1093/jxb/10.2.290
9. Mansfield, E.: Technical change and the rate of imitation. Econometrica **29**(4), 741–766 (1961)
10. Rogers, M.E.: Diffusion of Innovations, 3rd edn., p. 236. The Free Press, Macmillan Publishing Co. Inc., New York (1962)
11. Jang, S.L., Dai, S.C., Sung, S.: The pattern and externality effect of diffusion of mobile telecommunications: the case of OECD and Taiwan. Inf. Econ. Policy **17**, 133–148 (2005)
12. Meyer, W.B., Turner, B.L. (eds.): Changes in Land Use and Land Cover: A Global Perspective, pp. 537 + Xi. Cambridge University Press, Cambridge (1994). ISBN 0521 47085 4
13. Silverberg, G., Lehnert, D.: Evolúciós káosz: növekedés és fluktuációk az "alkotó rombolás" Schumpeter-féle modelljében. In: Nikosz, F. (szerk.) Káosz és nemlineáris dinamika a társadalomtudományokban. Typotex kiadó (2003). (in Hungarian)
14. Nikosz, F.: Növekedési görbék, társadalmi diffúzió, társadalmi változás. (Growth functions, social diffusion and social changing), in Hungarian. https://www.socialnetwork.hu/cikkek/FokaszDiffuzio.pdf. Last visiting of this homepage by the author of this paper: 22 Nov 2019
15. Szabó, F.J.: Analógia a sport-világcsúcsok története és az evolúciós optimáló algoritmusok iteráció-története között. (Analogy between the history of sports world records and the iteration history of optimization algorithms), GÉP, LXII; 9-10, pp. 28–31 (2011). 4 p. (in Hungarian)
16. Szabó, F.J.: Evolutionary based system for qualification and evaluation of group achievements (EBSYQ). Int. J. Curr. Res. **9**(08), 55507–55516 (2017). www.journalcra.com/sites/default/files/21246.pdf. ISSN 0975-833X

17. Szabó, F.J.: Optimumkereső algoritmusok iterációtörténetének vizsgálata. Investigation of iteration history of optimization algorithms. GÉP **69**(4), 82–85 (2018). ISSN 0016-8572. (in Hungarian).
18. Szabó, F.J.: Application of sigmoid curves in environmental protection. In: Szita Tóthné, K., Károly, J., Katalin, V. (szerk.) Solutions for Sustainable Development: Proceedings of the 1st International Conference on Engineering Solutions for Sustainable Development, (ICESSD 2019), pp. 1–7. CRC Press, London (2019). 7 p.
19. Nelder, J.A., Mead, R.: A simple method for function minimisation. Comput. J. **7**, 308–313 (1965). https://doi.org/10.1093/comjnl/7.4.308
20. Kehl, D., Sipos, B.: A telítődési, a logisztikus és az életgörbe becslése Excel parancsfájl segítségével. (Approximating the saturation curve, the logistic curve and the life curve by Excel command file). Statisztikai Szemle **87**(4) (2009). (in Hungarian)
21. Zhang, H., Huang, G., Yu, D.: Numerical modeling for the frame structure of light van - type electric truck. Sci. Prog. **103**(2), 1–27 (2020). https://doi.org/10.1177/0036850420927204
22. Bihari, J., Sarka, F.: Human-electric hybrid drives in medium-sized cities by daily traffic. In: Jármai, K., Bolló, B. (eds.) Vehicle and Automotive Engineering 2. VAE 2018. Lecture Notes in Mechanical Engineering. Springer (2018). https://doi.org/10.1007/978-3-319-75677-6_5. ISBN 978-3-319-75677-6, Print ISBN 978-3-319-75676-9
23. Binder, M., Klocke, F., Doebbeler, B.: An advanced numerical approach on tool wear simulation for tool and process design in metal cutting. Simul. Model. Pract. Theory **70**, 65–82 (2017). https://doi.org/10.1016/j.simpat.2016.09.001

Utilization of Lessons Learned in Product Development

László Soltész[1](✉) and László Berényi[2](✉)

[1] István Sályi Doctoral School, Institute of Machine and Product Design, University of Miskolc, Miskolc, Hungary
[2] Institute of Management Science, University of Miskolc, Miskolc, Hungary
solteszlasz1977@gmail.com

Abstract. New product development is the engine of competitiveness in the automotive industry. Beyond the technical content, the success of a project is significantly influenced by management performance, procedures, and teamwork. Lessons learned within a project and between projects can reduce efforts, development time, and costs. This knowledge sharing contributes to company-level sustainability of success by preventing the reoccurrence of mistakes and errors. 105 experts filled a voluntary online survey about their opinion about lessons learned. Two-third of the respondents use any form of lessons learned. Shared files are evaluated as the most useful way of managing the information flow, followed by targeted databases. Cluster analysis explored three utilization patterns of lessons learned. A minority of the experts found that managing lessons learned are problematic to feasible, but they are the source of information. Next to them, an active and inactive user group is identified.

Keywords: Product development · Lessons learned · Database · Cluster analysis

1 Reason of Lessons Learned

1.1 Introduction

Know how or lessons learned information related database is an increasingly inquisitive tool for delivering successful new product development projects nowadays. Every small, medium or multinational company must evaluate the experience of previous projects, mistakes made, or good practices. Although during product development, project team members must focus on operative tasks to deliver the planned working packages, the lessons of previous projects can help to improve the performance by avoiding repeated mistakes. The standard for organizational project management (OPM) also provides a focus for lessons learned data and sharing of that as an organizational interest. In the knowledge management chapter of the standard, one main organizational chapter is the sharing of lessons learned [1]. Indeed, these lessons learned are interpreted and utilized by persons who perform their tasks in connection with the projects. Exploring

K. Jármai and K. Voith (Eds.): VAE 2020, LNME, pp. 282–292, 2021.
https://doi.org/10.1007/978-981-15-9529-5_25

the personal approach to managing lessons learned can give valuable support for the systematic development of tools and methods in the field.

This chapter gives an overview of the interpretation of lessons learned (Sect. 1.2), including the typology (Sect. 1.3) and the role in the product development process (Sect. 1.4). The empirical results presented in the chapter are based on a survey. Research design (Sect. 2) aims to recognize utilization patterns from user aspects (Sect. 3). Section 4 presents the evaluation and conclusions.

1.2 Requirements

Creating and using lessons learned and historical information is emphasized in the PMI's project management [2] and portfolio management standards [3], defined as the knowledge gained during a project which shows how project events were addressed or should be addressed in the future for the purpose of improving future performance [2]. In a broader approach, knowledge management is a concept involving an organization's deliberate and focused attention to generate, collect, organize, analyze, and disseminate its knowledge in terms of documentation, storage, resources, and people skills [1]. Sharing lessons learned via knowledge management is an inevitable part of effective organizational project management regardless of the tools or frameworks supporting the efforts. During the evaluation of the effectiveness and the value of the project, a critical success factor is how organizations can adopt changes, analyzing the impact, capturing, improve, and use lessons. Listening to customer feedback, the quality improvement steps are needed not only from company management and people from quality departments but also it is the clear requirement of the IATF 16949:2016 standard (Quality management system requirements for automotive and relevant service parts organizations, International Automotive Task Force). The organization shall identify, document, and review manufacturing process design input requirements, including but not limited to the experience from previous development IATF 8.3.3.2.e [4]. Another example of requirement and usage of lessons learned is the U.S. Army Center for Army Lessons Learned (CALL). This organization is responsible and accountable for the entire Army Lessons Learned Program and responsible for the Army's formal lessons learned mission. This program contains information, lessons about the Army's SOPs, tactics, techniques, and procedures [5].

1.3 Types of Lessons Learned

Even though every product development project is different due to, e.g., different project scopes, higher or lower pressure from a timing point of view, project cost should have complied, but the amount of these also different, depending on product type, company, or economic situation, they can learn from each other. Each project will face difficulties or changes, including some new types of requirements from customers or departments who represent customers. The challenges usually require quick reactions and changes that may lead to new issues, mistakes. Otherwise, from an operation perspective of the projects, other mistakes are to find or at least missing the best available practices. Understanding that any problems may occur at any project and at any time that makes product development project team members impossible to deliver in time or quality, a system-oriented approach to managing the issues becomes necessary. Besides, understanding the cross-project benefits boost up the acceptance of the related efforts. The content of the lessons learned can be multifarious. These can be classified into three categories: (1) process improvement opportunities, (2) teaching opportunities, and (3) experience [6] (Fig. 1).

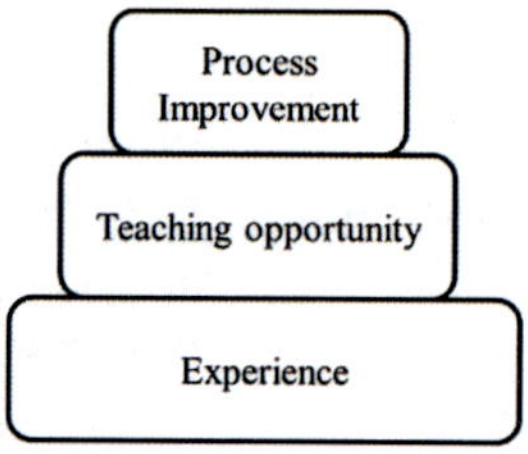

Fig. 1. Chart of lessons types [5]

1.4 Lessons Learned as a Success Factor

Successful and unsuccessful experiments of product development projects may boost the learning of organizations about prototypes, engineering solutions, or specifications. Using these types of knowledge (drawings, manufacturing processes, assembly instructions, and experiences from the decision made) reduce the planning and preparation of projects and provide prevention against previous mistakes, errors [5]. Implicit knowledge is understood as knowledge in the minds of team members. Otherwise, explicit knowledge is know-how that can be stored, e.g., in an external database, or a cloud system [6]. To reach high-level reflexivity of the team, team members must work out detailed plans, foresee problems, and drop obsolete processes in order to use these to meet the demands of a dynamic market better. NPD (new product development) teams that have higher levels of existing knowledge tend to have higher levels of team reflexivity [7]. The conceptual model of new product development success (Fig. 2), summarizes the key success factors and their relations. Existing knowledge as lessons learned definitely influence NPD success via team reflexivity.

Based on the cognitive fit theory, project team members typically have a wide range of available knowledge base what they must discuss for further reflections, ideas [9]. In conclusion, team members can convert the enormous knowledge bases. Effective development teams are holding knowledge database into more useful, productive knowledge by using cognitive activities to speed up team reflexivity [10]. The cognitive fit theory emphasizes that team members may need a broad range of existing knowledge in order to develop task familiarity in a complex context like new product development [11]. To deliver a complex technical solution in new product development normally much more complex than many other areas. New product development team members must be aware of more existing knowledge in their minds and improve continuously for higher levels to be more productive [12]. New product development teams that have higher existing knowledge tend to have higher levels of task familiarity [8].

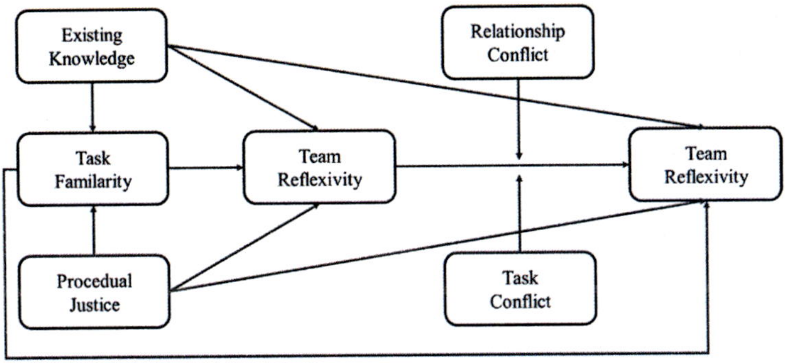

Fig. 2. Conceptual model of NPD success [8]

Lessons learned databases have a fundamental contribution to ensure the success of the product development, but it can be usually considered as the bottleneck of the process. Companies understood the necessity of product knowledge database, especially in line with the requirements of the ISO 9001 and IATF standards, but due to various reasons, there are uncertainties of usage of lessons from previous design works [13].

According to Pacagnella's research [14], the registers of lessons learned can increase the chances of success of product development projects. That means that it is possible to invest an effort to employ the knowledge generated empirically during the project; the team identifies opportunities via lessons learned studies that can be related to technology, management, marketing might bring positive results for output of product development project.

2 Research Design

2.1 Research Goals and Methods

Considering that the contribution of the lessons learned to the new product development depends on the activity of the users (primarily team members), exploring their attitudes to creating and using lessons learned information is a key issue in developing project management in a competitive business environment. The goal of the present research is to describe the experts' approach to lessons learned and to reveal specific profiles.

The research uses a voluntary online survey filled by experts related to product development projects in the automotive industry. The survey asks the respondents to evaluate the corporate practice according to the following statements on a 5-point scale:

- I often add information;
- I often seek information in it;
- Colleagues often add information;
- The information found in it can be used well in my work;
- The database is easy to use;
- Using the database is well regulated and documented.

Groping criteria for the analysis are the type of lessons learned database and the work experience of the respondents. The statistical analysis includes descriptive and variance analysis (ANOVA) and cross-tabulation. User profiles are established by hierarchical cluster analysis. Due to the small extent of the 5-point scale evaluation, the selected clustering method is Ward's method that minimizes the internal standard deviation of the created cluster [15]. Since the evaluation results show significant correlations between the survey items, a principal component analysis with VARIMAX rotation is conducted for dimension reduction [15]. The data collection period is June of 2020, and the data analysis is supported by IBM SPSS 24.

2.2 Research Sample and Limitations

The research sample includes engineering experts in product development related to the automotive industry or electronics industry from Hungary and abroad. Data collection is anonymous, and corporate affiliation is omitted to increase the reliability of the responses. The sampling method is convenient, and the representativeness is not ensured. Due to this characteristic of the research, the generalization of the results is limited. Another limitation is that the corporate practice can be more nuanced than the general evaluation of a survey. However, the respondents are well-trained experts in their field, and open-ended questions allowed to explain further details, unique applications require further consulting.

The sample consists of 105 responses. 71 experts marked that their company manages lessons learned in any format. The work experience of the respondents is summarized in Fig. 3.

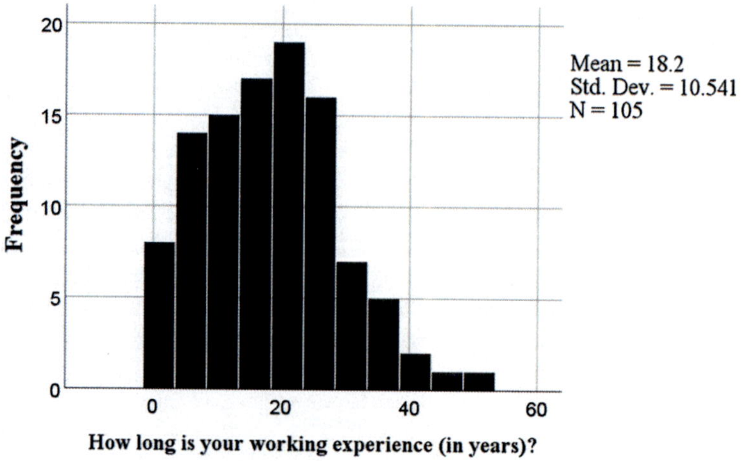

Fig. 3. Distribution of the experts by work experience

3 Results

3.1 Utilization of Lessons Learned

67.6% of all respondents marked that their company manages lessons learned of the product development projects. The results are limited to these cases. The most typical forms are shared files (36.6%) and using a database (29.6%). Meeting minutes include lessons learned in 22.5% of the cases, and sending an e-mail about the lesson is applied by 8.5%. 2 respondents (2.8%) marked other solutions. The utilization of the lessons learned data is moderate on average based on the mean values (Table 1). Although the

Table 1. Utilization of lessons learned (Mean values on a 5-point scale, 1: not typical at all, 5: typical)

	I often add information	I often seek information in it	Colleagues often add information	The information found in it can be used well in my work	The database is easy to use	Using the database is well regulated and documented
Shared file						
Mean	3.12	3.12	3.19	3.35	2.85	2.81
N	26	26	26	26	26	26
Std. Dev.	1.033	1.071	1.167	1.164	1.287	1.327
E-mail						
Mean	3.17	2.33	2.67	2.67	2.00	2.00
N	6	6	6	6	6	6
Std. Dev.	1.722	1.506	1.506	1.211	1.265	1.265
Database						
Mean	2.71	2.57	3.10	3.10	3.00	3.14
N	21	21	21	21	21	21
Std. Dev.	1.146	1.287	.944	1.179	1.049	1.195
Meeting minutes						
MM Mean	2.81	2.44	2.63	2.75	2.25	2.19
N	*16*	*16*	*16*	*16*	*16*	*16*
Std. Dev.	.981	.727	.806	.775	.931	.834
Total						
Mean	2.92	2.72	2.96	3.04	2.66	2.69
N	71	71	71	71	71	71
Std. Dev.	1.105	1.124	1.075	1.114	1.170	1.214

results show differences in the way of managing lessons learned, the significance of the differences is not confirmed by the ANOVA test.

A cross-tabulation analysis is conducted for checking the differences by the used solutions. There is a significant result in the case of the usability of the information found in lessons learned (Chi-square = 27.974, df = 16, sig. = 0.032). Based on the distribution of the evaluations, databases provide useful information for 47.6% of the users, but 38.1% rated it rather insufficient (Table 2). Shared files are found very good by 23.1% of the among its users. However, 38.5% of them found the quality of information moderate and 23.1% rather insufficient.

Table 2. Evaluation results on the usability of the information found in lessons learned (number of markings on a 5-point scale, 1: not typical at all, 5: typical)

Evaluation:	1	2	3	4	5
Shared file	1	5	10	4	6
e-mail	1	2	1	2	0
Database	2	6	2	10	1
Meeting minutes	1	4	9	2	0
Other	1	0	1	0	0
Total	**6**	**17**	**23**	**18**	**7**

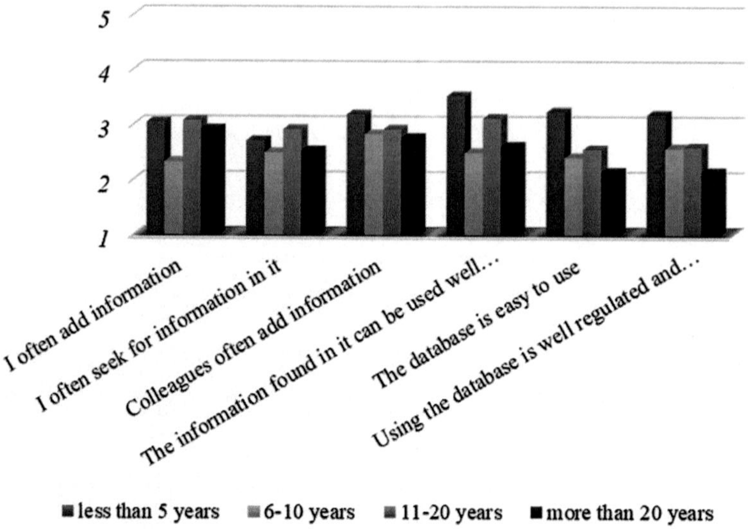

Fig. 4. Utilization of lessons learned solutions by the level of work experience (Mean values on a 5-point scale)

The work experience in product development of the respondents is used as a grouping factor for comparing the mean values. The analysis did not find any significant relationship between the work experience and the evaluation items (Fig. 4).

3.2 Profiling of the Experts

The nonparametric analysis between the evaluation of the utilization shows remarkable and significant correlations between the survey items (Table 3). This result suggests that those with a favorable and unfavorable opinion are well separated in the sample. Hierarchical clustering is selected for defining the profiles of utilization after the dimension reduction of the factors. The dendrogram (Fig. 5) suggests that three clusters are worth creating.

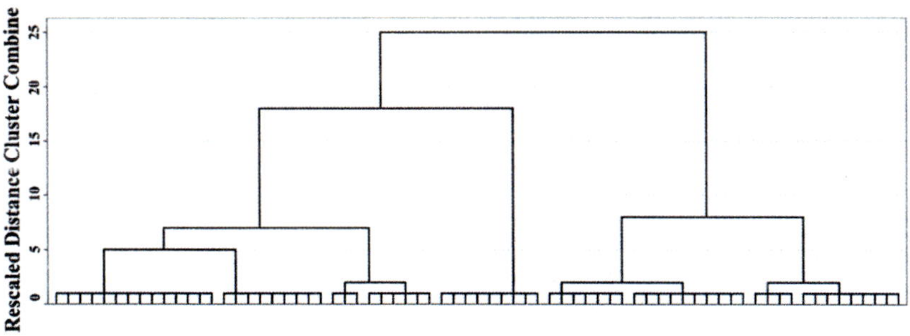

Fig. 5. Dendrogram of hierarchical clustering using Ward Linkage method

Table 3. Spearman's correlation coefficients between evaluation items (n = 71)

		1.	2.	3.	4.	5.	6.
1. I often add information	Coef		.660**	.417**	.353**	.211	.101
	Sig.		.000	.000	.003	.077	.401
2. I often seek information in it	Coef	.660**		.560**	.657**	.466**	.298*
	Sig.	.000		.000	.000	.000	.011
3. Colleagues often add information	Coef	.417**	.560**		.618**	.481**	.387**
	Sig.	.000	.000		.000	.000	.001
4. The information found in it can be used well in my work	Coef	.353**	.657**	.618**		.700**	.501**
	Sig.	*.003*	*.000*	*.000*		*.000*	*.000*
5. The database is easy to use	Coef	.211	.466**	.481**	.700**		.799**
	Sig.	.077	.000	.000	.000		.000
6. Using the database is well regulated and documented	Coef	.101	.298*	.387**	.501**	.799**	
	Sig.	.401	.011	.001	.000	.000	

Correlation (2-talied) is significant at the 0.01 level (**) or at the 0.05 level (*).

Clusters characteristics are as follows (Figs. 6–7):

- Cluster 1: includes 9 experts (12.7% of the sample), 6 of them have more than 20 years of work experience, and they spent more than 10 years with product development tasks. 6 of them has worked as a project manager, 5 of them in testing, and 4 as a product manager. Meeting minutes and shared files are used for knowledge sharing (Fig. 6). The members of this cluster consider themselves as information providers. They find the lessons learned challenging to use and the documentation of the system poor.
- Cluster 2: includes 30 experts (42.3% of the sample), 40% of them have more than 20 years, and 76.7% have more than 10 years of work experience. 53.3% spent more than 10 years with product development tasks. A database is the most commonly used solution for managing lessons learned. 63.3% of project managers belong to this cluster. The members believe the least in lessons learned compared to the others.
- Cluster 3: includes 32 experts (45% of the sample). 50% of them have more than 20 years, and 78,1% have more than 10 years of work experience. 62.3% of the respondents with 1–5 years of work experience in the member of this cluster. 46.9% of them spent more than 10 years with product development tasks. Shared files represent lessons learned knowledge. According to the professional tasks, test engineers (56% of them belong to this cluster) and quality engineers (50% of them belong to this cluster) can be highlighted among the cluster members. The utilization of the lessons learned shows the highest activity (except for adding new information).

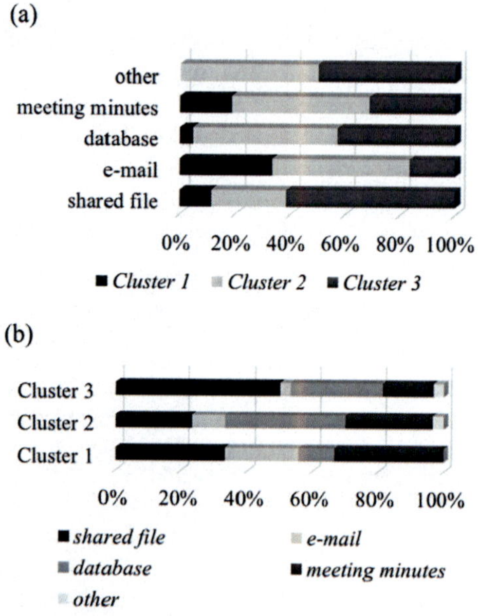

Fig. 6. Lessons learned database use by clusters (a) used forms by clusters (b) distribution of forms within the clusters

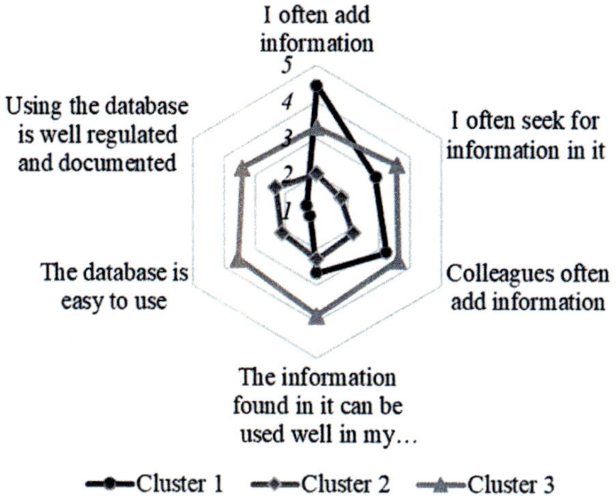

Fig. 7. Evaluation of managing lessons learned (5-point scale, 1: not typical at all, 5: typical)

4 Discussion and Conclusion

Although project management literature emphasizes the project-and company level benefits of managing lessons learned [3], 32.3% of the respondents do not use any solutions in this field. The most popular solutions for managing lessons learned of product development projects are shared files and targeted databases. Besides, meeting minutes are favored. Databases are considered better regulated and the easiest to use, while experts who use file-sharing add or use the information most often. Moreover, they find information the most useful for supporting their job.

The research used different groping factors for exploring the differences in utilization. Although the differences are statistically not significant, the results show that an increase in the work experience the average utilization level of lessons learned information decreases. Experts with less than 5 years of experience in work or in product development projects are outstandingly active.

The significant values of the correlation analysis confirm that active users find managing lessons learned more useful, easier to use, and better regulated. The cluster analysis allows drawing up three patterns of utilization based on the evaluation of lessons learned solutions. A smaller group (Cluster 1, including 9 experts) marks itself as the most active information providers, but there is no common format of the knowledge. They use meeting minutes to the greatest extent compared to the other clusters. The other experts could be classified into two larger groups. Lessons learned databases are the most common form in Cluster 2, but the utilization level of the information is the lowest in the sample. Cluster 3 members member mainly use shared files for lessons learned. The members find their framework the best regulated and the easiest to use. Comparing this result to Table 1 that details the results according to the form of managing lessons learned, this cluster includes experts with higher ratings in these questions than other clusters. Together with the usefulness of the information for supporting the work, these seem to be key drivers of successful managing of lessons learned.

Notwithstanding, knowledge sharing of lessons learned goes beyond the project boundaries. Company-level benefits may appear in the form of faster problem solving and cost reduction. Since the beneficiaries are the future users, and efforts must have been made at the expense of now, generating lessons may be out of the scope of a given project. Consequently, managing lessons learned requires company level coordination and an effective motivation system, which makes the project managers teams interested in making a sacrifice for sustainability. The shared files seem to be the most effective way of knowledge sharing based on the experience of the experts. However, the cluster analysis pointed out that the form of information is not decisive in usability. Considering the limitations of the research, the continuation of the investigations is necessary for exploring the best practices. Since the applied "hard" grouping factors did not show significant differences in the patterns, the "soft" aspects of technical management are worth to involve, like the organization of work or company culture.

References

1. The standard for organizational project management (OPM): Newton Square, PA: Project Management Institute (2018)
2. A Guide to the Project Management Body of Knowledge (PMBOK® Guide). 6th Edition. Newton Square, PA: Project Management Institute (2017)
3. The Standard for Portfolio Management. 3rd Edition. Newton Square, PA: Project Management Institute (2013)
4. Quality management system requirements for automotive and relevant service parts organizations, International Automotive Task Force, IATF 169494:2016, IATF 8.3.3.2.e (2016)
5. Tress, L.: Lessons learned systems people really use - U.S. Army Center for Army Lessons Learned (2010). https://www.apqc.org/resource-library/resource-listing/lessons-learned-systems-people-really-use-us-army-center-army
6. Symon, S., Jansen, M.: Integrating lessons learned throughout the product development process. PMI Global Congress–2011, North America, Dallas, TX. Newton Square, PA: Project Management Institute (2011)
7. Li, B.M., Xie, S.Q., Xu, X.: Recent development of knowledge based systems, methods and tools for one-of-a-kind production. Knowl.-Based Syst. **24**(7), 1108–1119 (2011)
8. Wu, W.-Y., Rivas, A.A.A., Liao, Y.-K.: Influential factors for team reflexivity and new product development. Proj. Manage. J. **48**(3), 20–40 (2017)
9. Lee, L.: The effects of team reflexivity and innovativeness on new product development performance. Ind. Manage. Data **108**(4), 548–569 (2008)
10. Marks, M.A., Mathieu, J.E., Zacvaro, S.J.: A temporary based framework and taxonomy of team processes. Acad. Manag. Rev. **26**(3), 356–376 (2001)
11. Speier, C.: The influence of information presentation formats on complex task decision-making performance. Int. J. Hum. Comput. Stud. **64**(11), 1115–1131 (2006)
12. Goodman, P.S., Leyden, D.P.: Familiarity and group productivity. J. Appl. Psychol. **76**(4), 578–586 (1991)
13. Soltész, L., Berényi, L., Kamondi, L: Analysis and assessment of the product development process. GÉP J. **71**(3–4), 67–71 (2020)
14. Pacagnella Jr., A.C., da Silva, L.S., Pacifico, O., Sergio, de Arruda Ignacio, P.S., da Silva, L.A.: Critical success factors for project manufacturing environments. Proj. Manage. J. **50**(2), 1–16 (2019). Project Management Institute
15. Verma, J.P.: Data Analysis in Management with SPSS Software. Springer, Heidelberg (2013)

Comparison of Center of Gravity Height Estimation Methods

Attila Widner(✉) and Gergely Bári

John von Neumann University, Kecskemét 6000, Hungary
Widner.attila@gamf.uni-neumann.hu

Abstract. Well established vehicle models and simulation methods are more and more important in nowadays technical evolution. With the rise of learning-based techniques in self driving car research, simulated environments have rising importance.

With the advances in vehicle dynamic softwares in the recent years, building models, and defining test cases getting easier, but finding the proper parameters for these vehicle models is usually very labor intensive. One of the most basic parameters of a vehicle model is its center of gravity height.

This paper investigates different center of gravity height estimation methods. The goal is to get a picture about their accuracy, field of suitable application, required time, necessary technical equipment, financial and human resources. We also investigate the possible sources of inaccuracy, and developed procedures to avoid, or at least minimize those.

Three types of estimation methods were examined. First, a static case, when the car is not moving, and the center of gravity height is calculated from the changes in the tire normal force during lifting one axle. These measurements can be carried out in a properly equipped workshop.

Then two dynamic methods are described, where the car is moving and the center of gravity height is calculated from the logged data of an advanced measurement system, that makes possible to log tire normal forces, lateral or longitudinal accelerations, damper potentiometer displacements. These measurements require a lot of sensors in the chassis and suspension and a data logging system. Both calculation methods are based on the dynamic load transfer during accelerations. First, the changes in the measured tire normal forces during longitudinal or lateral accelerations were used to determine center of gravity height, and in the second dynamic case, the tire normal forces were not measured but estimated from damper potentiometers.

The results confirm the widespread use of a well performed "lifted axle method", as turns out to be an efficient choice, without the need for costly sensors and tools. A good comparison is also established about these estimation methods, and detailed procedures for each are developed to avoid mistakes during the different measurements.

Keywords: Center of gravity height estimation

K. Jármai and K. Voith (Eds.): VAE 2020, LNME, pp. 293–301, 2021.
https://doi.org/10.1007/978-981-15-9529-5_26

1 Introduction

The purpose of our project is to create a well-established vehicle model for self-driving algorithm development focusing on the grip-limit performance of the autonomous vehicle. Building models and defining test cases getting easier with the advances in vehicle dynamics softwares but finding the proper parameters for these vehicle models is usually complex and time-consuming activity. One of the most basic parameters of a vehicle model is its center of gravity (CG) height. CG position is a very important parameter because it has a large effect on tire performance by influencing the dynamic load transfer during acceleration. Load transfer effects overall grip due to the degressive relation between normal force and lateral, longitudinal forces.

There are several different methods for estimating the above-mentioned parameter. Each provides different accuracy and the required equipment and resources vary in a wide range. Our goal here is to gather information for finding the most sufficient method for estimating the center of gravity height.

The following information help to get a good picture of the methods:

- accuracy
- field of suitable application
- required time
- necessary technical equipment
- financial and human resources.

We also investigate the possible sources of inaccuracy, and developed procedures to avoid, or at least minimize those.

Three methods were investigated, a static case - when the car is not moving - and two dynamic - when the car is moving - in all three cases the center of gravity height is calculated from the weight transfer.

2 Estimation Methods

Round-Robin Center of Gravity Height Measurement Study introduces three static estimation methods, and a good procedure for the comparison [1]. Three car manufacturer companies were involved in this study, each used different method: “Lifted axle method”, where the CG height is calculated from the static load transfer during lifting one axle; “Tilt-test” where CG height is estimated from the roll angle of the tilt platform; the third measurement was done with the swing configured for pitch inclination (also discussed and improved in [2]).

The results of the study showed significant differences between the results of different laboratories, but the repeatability of results within the individual laboratories was generally good. They examined the individual procedures and expect that variability in the results could be reduced. These recommendations were considered during our work [1].

The “lifted axle method” was chosen as a static method because all the equipment was available in the university, and we also have some experience with this method.

There is an advanced vehicle dynamics measurement system at the department, which is used for vehicle dynamics measurements and vehicle model validation. The core of the system is a Kistler RoaDyn 625 wheel force transducer which is capable of measuring all wheel forces F_x, F_z in the range of ±20kN, F_y in the range of ±15kN and all wheel moments (M_x, M_y, M_z) in the range of ±4kNm [3]. The direction of F_z and F_x is shown in Fig. 1. The idea is to determine the CG height by using the data from the different measurements. Here, two kinds of estimation methods were developed. One calculates directly from the F_z provided by the measurement system and another one where the F_z is also estimated from damper potentiometers.

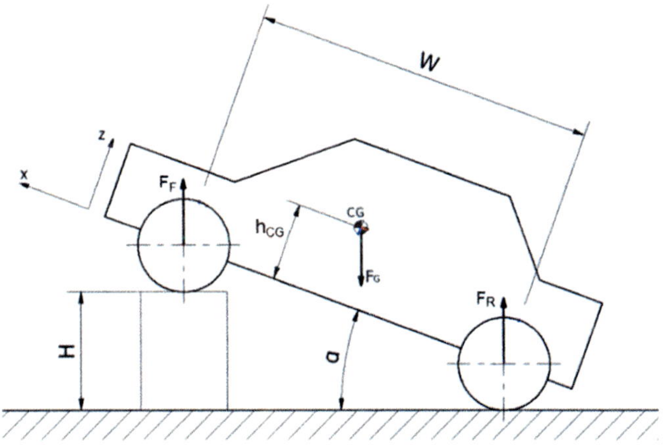

Fig. 1. Lifted axle method - measurement layout

For the measurements, we used a Mercedes-Benz CLA 250 7G-DCT (2014) because it was available in the department. The vehicle has a 2.0 L turbocharged inline 4-cylinder engine and a 7-speed automatic transmission. All measurements were carried out with the following parameters:

- Fuel tank was full to avoid false results caused by the displacement of the liquid.
- All other liquids were on top level (engine oil, steering oil, brake fluid).
- Windscreen cleaner was empty.
- Kistler On-Board Electronics (8 kg) and the battery (13 kg) for the measurement system was placed in the trunk.
- The data acquisition system (10 kg) was mounted on the rear right passenger seat.
- Passengers

  • driver: 75 kg (180 cm)
  • 1 passenger on the front right seat: 70 kg (175 cm)

During all measurements, the vehicle was equipped with the above-mentioned advanced vehicle dynamics measurement system.

Prior to the center of gravity height measurement some preparatory measurements needed to be taken to determine the vehicle's suspension geometry and kinematics (wheelbase, wheel track, damper and damper potentiometer motion ratios, etc.), overall weight and the height and weight of the non-suspended mass as well (Table 1).

Table 1. Vehicle parameters

Parameter	Value
Wheelbase	2,7 m
Static wheel radius	0,312 m
Damper potentiometers motion ratio (front)	0,42
Spring rate (front)	34 N/mm
Damper motion ratio (front)	0,84
Total non-suspended mass	192,5 kg
hCG of non-suspended mass	0,312 m

We assume that the center of gravity of the non-suspended mass is in the axle of the wheel: 312 mm.

3 "Lifted Axle Method"

This method can be carried out in a properly equipped workshop. Prior to the measurement, the vehicle was prepared. The ride springs were changed to an adjustable tube, to avoid spring deflection caused by the load transfer and to provide the possibility to adjust ride height. The measurement process has two steps. First, the tire normal forces are measured on a horizontal flat floor. Then the first axle is lifted, and the tire normal forces and the height of lift values are measured.

From this data the center of gravity height can be calculated using the following equation from [4]:

$$h_{CG} = \frac{W \cdot \Delta m}{m_{total} \cdot H} \cdot \sqrt{W^2 - H^2} + r_{stat} \tag{1}$$

where the

- h_{CG} - center of gravity height [m]
- W - wheelbase [m]
- m_{total} - vehicle mass [kg]
- Δm - weight difference on the front axle between horizontal and lifted state [kg]
- H - front axle height compared to horizontal state [m]
- r_{stat} - static wheel radius [m]

A custom-made Breda Racing weight scale set was used for determining the corner weights. Each weight scale has 2 NBC 16.11053 load cells, and capable of measuring 0–500 kg with an accuracy of 0,1 kg. The measurement requires some basic tools for the spring replacement, weight scale adjustment. The type and size of these depend on the vehicle, and the equipment, but generally the following tools are sufficient: an Allen key set, a wrench set, a ¼" and a ½" socket set with ratchets, ½" torque wrench (40–200 Nm), coil spring compressor and two hydraulic car jack. Measuring devices for the preparatory measurements: tape measure for wheelbase, ruler and Cross Line Laser for the ride height and digital caliper for the spring replacement tube adjustment. In our case, the two-post vehicle lift was equipped with a weight scale holder, which fits properly with the weight scale.

During this measurement process, the car was equipped with the vehicle dynamics measurement system it gives us other sources of F_Z. It was a good opportunity to compare the force values from Kistler with the weight scales. The estimation method was evaluated with both values. CG height was calculated from the values provided by the weight scale and from tire normal forces by Kistler.

The measurement was repeated 4 times with different ride heights. This way we can compare the different results and get some information about the accuracy of the measurement. Adjusting the ride height influences the suspended mass center of gravity height, the height of non-suspended CG stays the same. The effect of ride height change was calculated, 10 mm rise modifies the h_{CG} by 9 mm. The results of the lifted axle center of gravity height estimation method can be seen in Table 2 both estimated values (Kistler, and Breda) are compared to the calculated value, the relative measurement errors are also shown in the table. Possible causes of the errors may the accuracy of the measurement system. Also, as the determination of the non-suspended mass and its center of gravity height are complicated, we used an estimation, it also can cause an inaccuracy that probably can be seen in the relative errors.

Table 2. Estimated and calculated center of gravity height

Ride height [mm]	Calculated CG height [mm]	Estimated CG height (Breda) [mm]	Relative error [%]	Estimated CG height (Kistler) [mm]	Relative error [%]
+20	551.3	553.3	0,37	554.0	0,88
+10	542.3	543.3	0,18	537.0	0,59
base		533.3	–	531.2	–
−10	524.3	525.5	0,22	523.3	0,21
−20	515.3	516.5	0,22	514.3	0,22

Stock ride height was used as a baseline. As it can be seen on the results the estimation method provides good accuracy with both measurement system.

The results confirm the widespread use of a well performed "lifted axle method", as turns out to be an efficient choice, without the need for costly sensors and tools. The measurement can be carried out in one working day with two people.

4 Estimating from Directly Measured Tire Normal Forces

Our goal with the vehicle dynamics measurement system is to create a framework for a proper vehicle model validation. As far as the center of gravity height is considered the idea is that during vehicle tests, we create situations that are suitable for determining the CG height using the F_z values from Kistler wheel force transducer, if this works properly, then there is no need for other time-consuming measurement method.

In this method the equation of dynamic load transfer is used from [5]:

$$F_{Z,F} = \frac{m \cdot a_x \cdot h_{CG}}{W} \tag{2}$$

The above-mentioned equation is only true in steady-state vehicle behavior. The test was performed as braking with constant acceleration.

Motec i2 Pro was used for the data acquisition. We use the following channels: First, we calculate the longitudinal load transfer (channel name: LT_long):

$$\Delta F_{Z,F} = F_{Z,FR} + F_{Z,FL} - F_{Z,F,stat} \tag{3}$$

The height of the center of gravity is calculated with the following equation (channel name: cog_height_straight):

$$h_{CG} = \frac{\Delta F_{Z,F} \cdot W}{m \cdot a_x} \tag{4}$$

- F_Z- Longitudinal load transfer [N]
- W - wheelbase [m]
- m - vehicle total mass [kg]
- a_x - longitudinal acceleration [m/s^2]

As it was mentioned earlier this equation is only true in steady-state. To make sure that we only estimate when the conditions are right some check channels were created. The previous equation uses longitudinal load transfer; therefore, the lateral acceleration must be 0, the first channel gives 1 value if the lateral acceleration is 0 and 0 value if the lateral acceleration is other than 0. Then an acceleration check, this channel gives 1 value if the longitudinal acceleration is above 0,3 G. Then to ensure that the vehicle is in steady-state two other channels are created, one gives 1 value if the derivative of longitudinal acceleration is below 0,1, another one gives 1 when the pitch velocity is below 0,1 deg/sec. These channels are added to the h_{CG} channel as multipliers. Consequently, the h_{CG} channel gives 0 value, if the conditions are not right, and gives the calculated CG height when they are. To sum all results from the suitable data, a CG height mean channel was created, which gives the mean value over the log file, an "If" condition was specified, therefore the calculation is only performed while the CG height channel is not zero.

The results of one braking are shown as an example in Fig. 2. As it can be seen the channel gives zero value when the conditions are not right, when the acceleration is too little, or the vehicle is not in steady-state. On the top left corner the current value is shown which belongs to the blue vertical line (cursor).

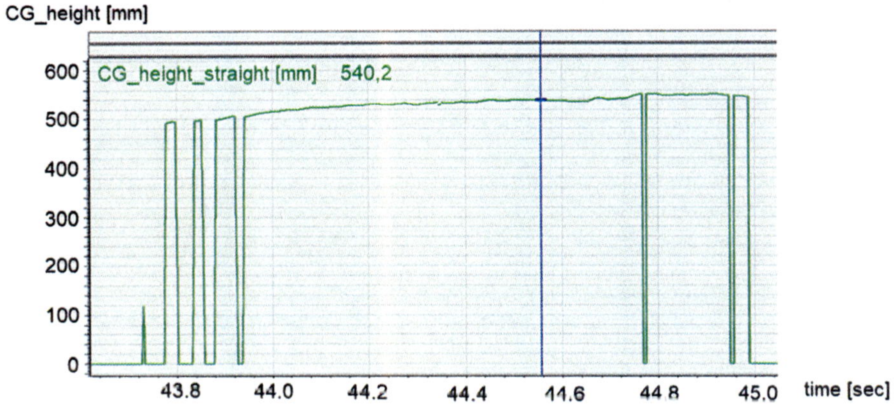

Fig. 2. Results of CG height estimation channel

In the following table, CG height mean values are shown from the different runs (Table 3).

Table 3. Results of tire normal force based h_{CG} estimation

Run ID	Run1	Run2	Run3	Run4	Run5
h_{CG} mean [mm]	545,7	543,2	540,2	544,2	539,3

The average value is 542.5 and the difference between the max and min value is 6.4 mm. During braking there is a jacking force due to suspension kinematics, this has a jacking effect on the suspended mass, therefore the CG height is changing during vehicle motion. There is 10 mm difference between the lifted axle and the dynamic method. Besides the above mentioned phenomenon, there is another thing that could influence the results, during the lifted axle method to simulate the weights of the passengers we put water balloons in the seats - due to lack of human resources.

The advantage of this method is that we can use the logged data from other vehicle dynamics measurements, the channel selects the suitable data then calculates the CG height. Then finally the mean value of the calculations is determined. Of course, this method requires way more resources than the previously presented, therefore it is not efficient by itself. But it can be a good complement of the vehicle dynamics measurement evaluation system.

5 Damper Potentiometer-Based CG Height Estimation

Finally, a damper potentiometer-based estimation method was developed. The goal was to create a similar but more cost-effective system as the previous one, by leaving out the tire force transducer.

The core of the estimation method is the same, the only difference is the dynamic normal force on the front axle which is estimated as the following:

$$F_{Z,\,FR} = s_{damper,\,FR} \cdot K_{s,\,F} \cdot MR_F^2 + F_{Z,\,F,\,stat} \quad (5)$$

MR_F - ride spring motion ratio
$s_{damper,FR}$ - damper potentiometer value
$K_{S,F}$ - stiffness of the ride springs

Figure 3 shows the estimated F_Z and the measured F_Z from Kistler.

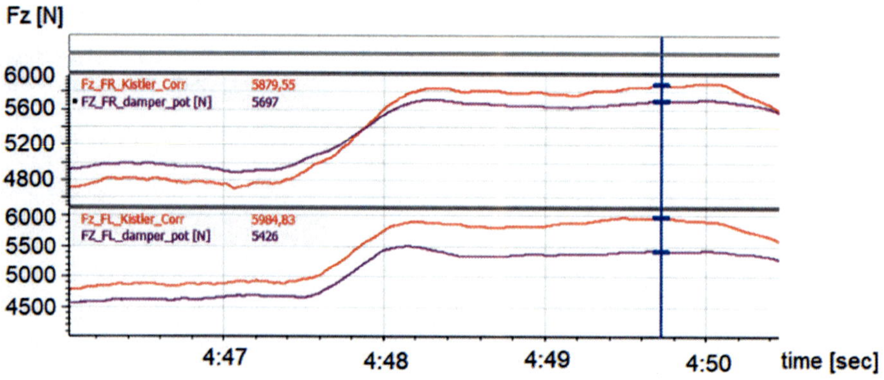

Fig. 3. Estimated and measured F_Z

As it can be seen there are about 300–400N difference between the measured and estimated tire normal forces. The reason for that difference can be found in the suspension kinematics. Pitch center influences the load path of F_X. Anti-dive parameter gives the percentage of the load which "go through the suspension parts" not through the spring. This phenomenon can be the explanation of why the F_Z estimated from the damper potentiometers are smaller than the measured.

The CG estimation is based on the dynamic load transfer, therefore the accurately measured normal force is crucial for the measurement. The previously mentioned 300–400N difference has a large effect on the results, therefore this method does not seem to be usable for estimating the CG height.

6 Conclusion

The widely used lifted axle method seems to be an efficient and accurate solution. Our goal was to create a dynamic method, which does not require a separate measurement, but can be carried out during other vehicle dynamics tests. The dynamic estimation from directly measured normal forces seems to be a good solution. During our work only the longitudinal behavior was analyzed, the next step is using lateral load transfer to calculate CG height. Apparently, the normal force estimation method does not seem to be reliable, because the anti-dive parameter has a large influence on the results.

Acknowledgement. The research presented in this paper was carried out as part of the EFOP-3.6.2-16-2017-00016 project in the framework of the New Széchenyi Plan. The completion of this project is funded by the European Union and co-financed by the European Social Fund.

References

1. Winkler, C.B., Campbell, K.L., Mink, C.E.: Center of Gravity Height: A Round-Robin Measurement Program (1991)
2. Hejtmánek, P., Blaťák, O., Vančura, J.: New Approach to Measure the Vehicle Centre of Gravity Height (2015)
3. Kistler RoaDyn® S625 System 2000 – Wheel Force Sensor for Passenger Cars, Type 9266A datasheet
4. Jazar, R.N.: Vehicle dynamics: Theory and application. Springer, New York (2008)
5. Milliken, W.F., Milliken, D.L.: Race Car Vehicle Dynamics. SAE International, Pittsburgh (1995)

Numerical Solutions of the Kardar-Parisi-Zhang Interface Growing Equation with Different Noise Terms

Okhunjon Sayfidinov(✉) and Gabriella Vadászné Bognár

University of Miskolc, Miskolc-Egyetemváros, Miskolc 3515, Hungary
sayfidinov.ok@gmail.com, v.bognar.gabriella@uni-miskolc.hu

Abstract. The one-dimensional Kardar-Parisi-Zhang dynamic interface growth equation with Gaussian noise and without noise term is analyzed in various initial conditions and its amplitudes. As an analytical point, there some initial conditions with noise term amplitude changes added. From the mathematical point of view, these can be considered as various amplitude distribution functions. Three different conditions and various amplitudes were investigated between Gaussian noise and without noise term. Numerical solutions are evaluated and analyzed for both cases.

Keywords: Gaussian noise · Kardar-Parisi-Zhang equation · Molecular Beam Epitaxy · Initial condition · Amplitude

1 Introduction

Currently surface morphology of the interfaces and understanding it's growth is a highly challenging topic of physics and materials science in different applications [1]. The physics of crystal growth largely depends on the microscopic nature of surfaces. For solids with rough surfaces which growth is often limited by slow mass transport to the crystal (diffusion-limited growth) and solids with smooth atomically flat surfaces which attachment sites are rare, the addition of particles to the surface is not an easy process. Growth may occur either via two-dimensional 2D nucleation or via the attachment of atoms to the preexisting steps [2].

The task of producing a surface with an atomic control is often hampered by the presence of inherent instabilities and/or by the kinetic roughness associated with, for example, shot noise due to the deposition flux. In principle, on a flat surface exposed to a flux in Molecular Beam Epitaxy (MBE) usual fluxes in MBE range from a fraction of a monolayer to a few monolayers per second, each deposited atom has ample time to diffuse and attach to a favorable site e.g., a step before a new atom is deposited [3].

The surface may develop stochastic roughness. Prominent examples of descriptions of this type of roughness are the Edwards-Wilkinson equation [4] and the Kardar-Parisi-Zhang (KPZ) equation [5]. The latter is one of the earliest nonlinear evolution equations

K. Jármai and K. Voith (Eds.): VAE 2020, LNME, pp. 302–311, 2021.
https://doi.org/10.1007/978-981-15-9529-5_27

put forward in MBE growth literature and was derived from symmetries. The KPZ equation reads

$$\frac{dw}{dt} = \nu \nabla^2 w + \frac{\lambda}{2}(\nabla w)^2 + \eta(x,t), \tag{1}$$

where $w(x, y, t)$ stands for the profile height of the local growth [6]. The first term on the right hand side describes relaxation of the interface by a surface tension, which prefers a smooth surface. The second term is the lowest-order nonlinear term that can appear in the surface growth equation justified with the Eden model and originates from the tendency of the surface to locally grow normal to itself and has a non-equilibrium in origin. The last term is a Langevin noise to mimic the stochastic nature of any growth process and has usually a Gaussian distribution [7].

There are numerous studies available about the KPZ equation in the literature in the last two decades. Without completeness, we mention some of them. The foundation of the physics of surface growth can be found in the book of Barabasi and Stanley [1]. Hwa and Frey [8, 9] investigated the KPZ model with the help of the self-mode-coupling method and with renormalization group-theory, which is an exhaustive and sophisticated method using Green's functions. They considered various dynamical scaling forms in one spatial dimension as $w(x, t) = x - 2\varphi C(bx; b^z t)$ for the correlation function (where φ and b and z are real constants). Lässig showed how the KPZ model can be derived and investigated with a field theoretical approach [10]. In a topical review paper, Kriecherbauer and Krug [11] derived the KPZ model from hydro dynamical conservation equations with a general current density relation. Later, Einax et al. [12] published a review study on cluster growth on surfaces.

The aim of this paper is to investigate the numerical solution to KPZ Eq. (1) with and without noise term and to point out the importance of the morphology of the initial surface of the substrate.

2 Theory

Non-linear PDEs has no general mathematical theory, which could help us to derive physically relevant solutions. There are different methods available, beyond the celebrated Lie algebra formalism [13], the most commonly used method is the reduction technique. This means that the original variables of the PDEs like the time t and the spatial coordinate x are used to define a new variable (for example f). Via a variable transformation, the original PDE can be reduced to an ordinary differential equation (ODE). The choice of the form of $f(x, t)$ is basically quite large. Usually, the continuity of first and second derivatives of f in respect of x and t is required. Beyond these continuous models based on partial differential equations (PDEs), there are numerous purely numerical methods available to study diverse surface growth phenomena. Without completeness, we mention the kinetic Monte Carlo [14], Lattice-Boltzmann simulations [15] and the etching model [16].

3 Results Without Noise Term

Simulations have been carried out by MATLAB R2019a. Numerical solution for height profiles are calculated with the following data:

$$x \in [-200,\ 200],\ t \in [0,\ 10000],\ N = 100,\ \Delta t = 1/100,$$

where N denotes the number of division points on the x-axis and Δt is the time step.

In the resulted figures below the complete solutions of the original PDE (1) have been presented showing in different initial condition and various amplitudes. However, for simplicity, the parameters are chosen as $\nu = \lambda = 0.1$ and the initial conditions used are

$$w(x, 0) = (1 \cdot \cos(x/8))(1 + \sin(x/8)) \tag{2}$$

and

$$w(x, 0) = (0.1 \cdot \cos(x/8)) \cdot (0.1 + \sin(x/8)). \tag{3}$$

Figure 1 presents the solutions in the time range [1, 600] for the different amplitudes applied in the initial conditions (2) and (3). It seems that the results are vibrating depending on the parameter value in the initial condition. In Fig. 1(a) it can be seen that $w(x, t)$ is between ± 1.3 and in Fig. 1(b) that it is between ± 0.6 which begins to smooth out in both cases. This examination represents that initial condition amplitudes are affecting only at the initial phase and but later the surface is left in the same flatness.

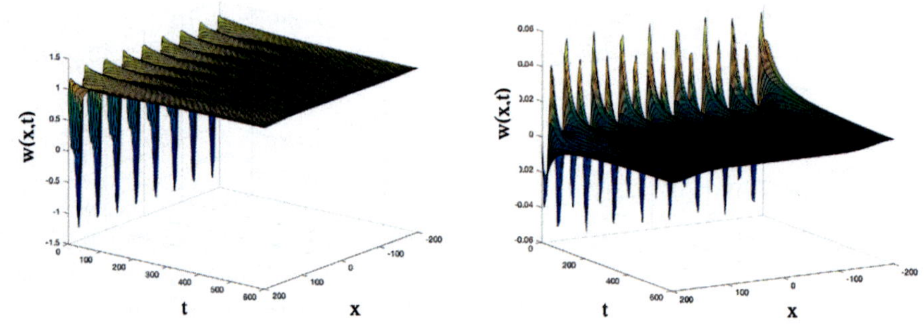

(a) Solution to (1)-(2) (b) Solution to (1)-(3)

Fig. 1. The solutions without noise term with initial conditions (2) and (3)

Figure 2 presents the results in the same parameters for $\nu = \lambda = 0.1$ applying different initial conditions.

Initial conditions are

$$w(x, 0) = 1 \cdot \sin(x/16) + 1 \cdot \cos(x/16) \tag{4}$$

and

$$w(x, 0) = 0.1 \cdot \sin(x/16) + 0.1 \cdot \cos(x/16). \tag{5}$$

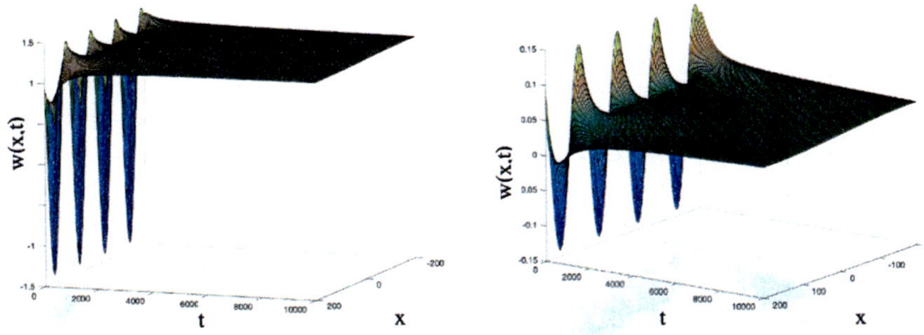

(a) Solution to (1)-(4)

(b) Solution to (1)-(5)

Fig. 2. The solutions without noise term with initial conditions (4) and (5)

The figures presented in Figs. 2(a) and 2(b) are different from each other by the difference in their amplitudes. Here, it can be concluded that when amplitudes of initial condition are higher than 1 then Fig. 2(a) starts from a negative point - with some waves to the flatness but increasing amplitude decrease the wave sides and reaches flatness with an angle. However, small amplitude values increase waviness in $w(x, t)$ line proportionally.

Figures 3(a) and 3(b) exhibit the solution to (1) with initial conditions

$$w(x, 0) = 1 \cdot sin(x/16) \tag{6}$$

and

$$w(x, 0) = 0.1 \cdot sin\left(\frac{x}{16}\right), \tag{7}$$

and also gives the same results as discussed above.

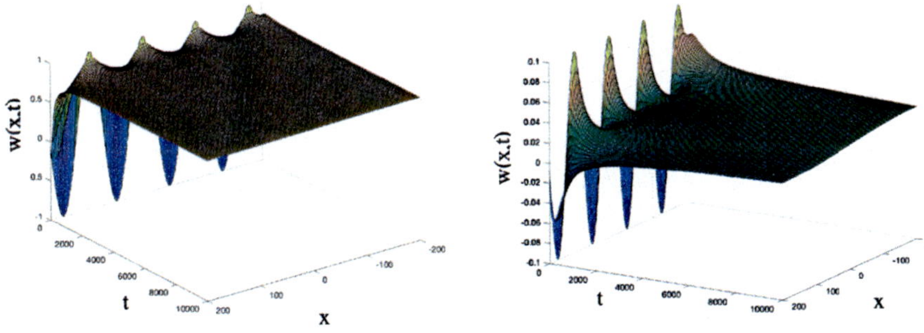

(a) Solution to (1)-(6)

(b) Solution to (1)-(7)

Fig. 3. The solutions without noise term with initial conditions (6) and (7)

The functions illustrate similar structures like Fig. 2 and Fig. 3 depending on amplitude values in the same time interval. The structure of Fig. 1 showed that waviness is changed taking into account of the amplitude inputs. We may say that different numerical values of the initial condition do not drastically change the whole surface as it is.

4 Results with Gaussian Noise

As the main universality classes relevant for kinetic roughening, we focus on the case in which the surface or interface is subject to time dependent noise. In typical applications, these fluctuations arise in those of a driving flux (of, say, aggregating units, atoms or molecules) acting on the system. This is a convenient way to represent the nature of the noise, but it does not by any means imply that its amplitude is directly the square root of the average external flux. For instance, in studies of grows BME for electrochemical or chemical vapor deposition (ECD, CVD, respectively) the noise term appearing in the Langevin equation for the interface is seen to be rather more involved than that. However, and this can never be overemphasized, the universal behavior applies to asymptotic properties, well beyond all existing transients (induced by, e.g. physical instabilities acting on the system) and crossovers (due to competition among various physical mechanisms, each of which is dominant for a different range in time and space). For the type of systems, we are considering, the asymptotic properties are adequately described by equations featuring additive noise, which is Gaussian and uncorrelated in time and space [17].

Applying similarity transformation $w(x,t) = f(\zeta)$ and $\zeta = \frac{x}{\sqrt{t}}$ with Gaussian noise gives us the ODE of

$$\nu f''(\zeta) + 0.5 f'(\zeta)\big[\zeta + \lambda f'(\zeta)\big] + a e^{\frac{-\zeta^2}{n}} = 0, \quad t > 0, \tag{8}$$

where a is in connection with the standard deviation of the Gaussian distribution. There is no general formula available for arbitrary parameters λ, μ, a. Fortunately, if two parameters are fixed e.g. $\nu = \lambda = 0.1$ and $n = 1$, then there is a closed expression (analytical solution) available for the solution.

$$f(\zeta) = -\frac{1}{2\lambda} \ln\left[1 + \tan\left\{\sqrt{\lambda a \pi}\cdot \mathrm{erf}\left(\sqrt{\frac{\zeta}{2}}\right) + c_1\right\}^2\right] + c_2, \tag{9}$$

where erf means the error function and c_1 and c_2 are integration constants, see [15].

Figure 4 presents the numerical results for the given parameters and $a = 1$ and for applying different initial condition amplitudes to Eq. (1) as follows.

$$w(x,0) = (1 \cdot cos(x/8)) \cdot (1 + sin(x/8)) \tag{10}$$

and

$$w(x,0) = (0.1 \cdot cos(x/8)) \cdot (0.1 + sin(x/8)). \tag{11}$$

Figures 4(a) and 4(b) show almost the same structure. The only visible change is that if the amplitude is 1, the ripple appears, which can also be characterized by the maximum value of the amplitude.

Figure 5 shows the solutions of Eq. (1) with the same parameters. The figure shows the change in amplitude and the different representations of the graphs. We examine the

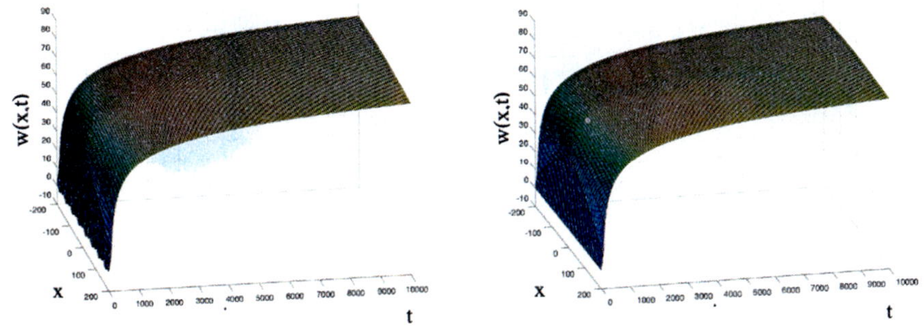

(a) Solution to (1)-(10) (b) Solution to (1)-(11)

Fig. 4. The solutions with Gaussian noise term with initial conditions (10) and (11)

effect of the strength of the Gaussian noise, denoted by a in Eq. (8) providing $a = 0.1$ and $a = 0.01$ in the Gaussian noise term $\eta(x, t)$ together with the initial state.

$$w(x,0) = \left(1 \cdot cos\left(\frac{x}{8}\right)\right) \cdot \left(1 + sin\left(\frac{x}{8}\right)\right). \tag{12}$$

It results the big wavy shape in the surface escaping while keeping an increasing amplitude of noise term.

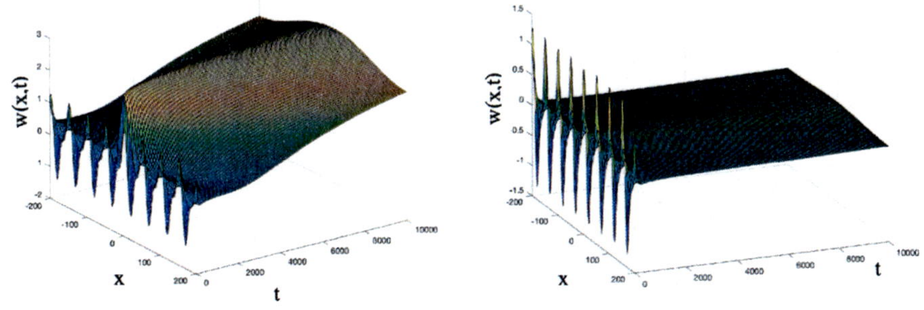

(a) Solution to (1)-(12) $a = 0.1$ (b) Solution to (1)-(12) $a = 0.01$

Fig. 5. The solutions with Gaussian noise term with initial condition (12) and $a = 0.1$ or $a = 0.01$

Figure 6 presents the results in the same output figures as presented in the Fig. 4 in spite of changing initial condition and its amplitudes.

$$w(x,0) = 1 \cdot sin(x/16) + 1 \cdot cos(x/16) \tag{13}$$

and

$$w(x,0) = 0.1 \cdot sin(x/16) + 0.1 \cdot cos(x/16) \tag{14}$$

Figure 7 presents the results of the numerical simulations for the Eq. (1) applying different initial conditions with $a = 0.1$ or $a = 0.01$ in the Gaussian noise term η of Eq. (1).

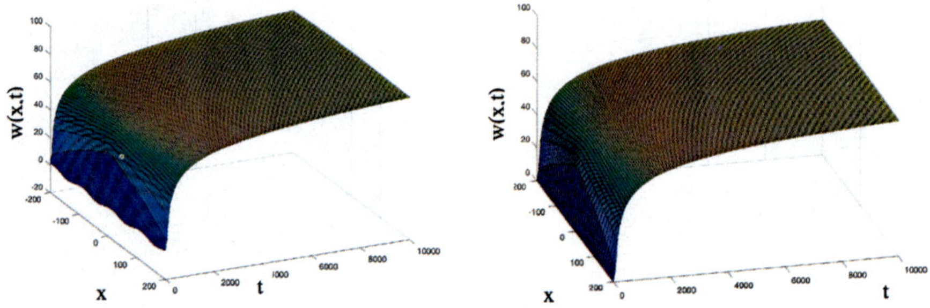

(a) Solution to (1)-(13) (b) Solution to (1)-(14)

Fig. 6. The solutions with Gaussian noise term with initial condition (13) and (14) for $a = 1$

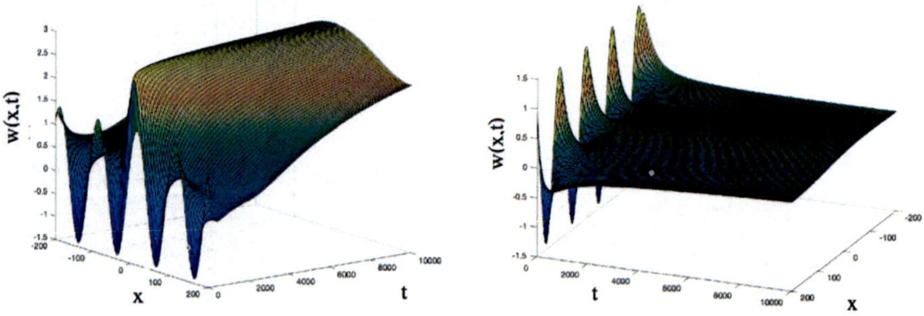

(a) Solution to (1)-(15) $a = 0.1$ (b) Solution to (1)-(15) $a = 0.01$

Fig. 7. The solutions with Gaussian noise term with initial condition (15) and $a = 0.1$ or $a = 0.01$

In this figure different amplitude changes can be seen and how the graphs are variously presented while the initial condition is kept

$$w(x, 0) = 1 \cdot sin(x/16) + 1 \cdot cos\left(\frac{x}{16}\right) \tag{15}$$

Difference between Fig. 5 and Fig. 7 is their wavy steps in the same range x$\in$[-200 200].

Figure 8 introduces the same output figures as presented in Fig. 4 and Fig. 6 in spite of changing the amplitudes in the initial condition

$$w(x, 0) = 1 \cdot sin\left(\frac{x}{16}\right) \tag{16}$$

$$w(x, 0) = 0.1 \cdot sin\left(\frac{x}{16}\right) \tag{17}$$

However, as it is shown that the difference between all these figures is their wavy starting points which are increasing waviness in the same range$x \in [-200\,200]$.

Figure 9 presents the results of changed amplitudes with $a = 0.1$ or $a = 0.01$ in the Gaussian noise term η while the initial condition is kept

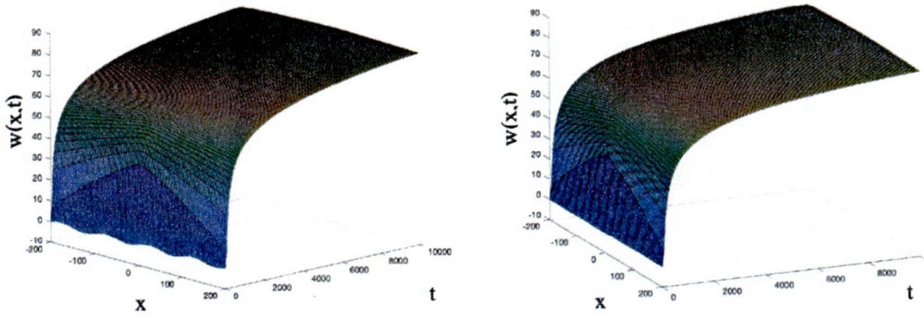

(a) Solution to (1)-(16) (b) Solution to (1)-(17)

Fig. 8. The solutions with Gaussian noise term with initial conditions (16) and (17)

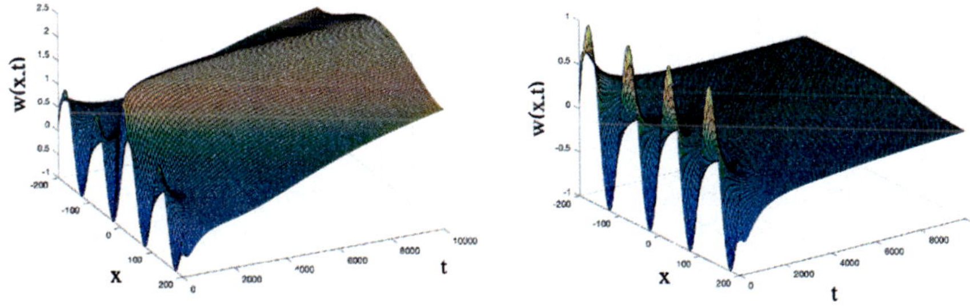

(a) Solution to (1)-(18) $a = 0.1$ (b) Solution to (1)-(18) $a = 0.01$

Fig. 9. The solutions with Gaussian noise term with initial condition (18) and $a = 0.1$ or $a = 0.01$

$$w(x, 0) = 1 \cdot sin(x/16) \tag{18}$$

As the above results represents, the impact of the initial condition and noise amplitudes and show wavy increase in the graphs.

5 Conclusion

In this paper, the numerical solutions to (1) are examined with MATLAB simulations. These results help us to explain the phenomena observed by experiments and validate the mathematical model [8]. So, we follow the stress effects in understanding the physical phenomena. Our different choices for the initial condition are applied to PDE (1).

To conclude, we can say that with an appropriate change of amplitudes in the initial condition and/or strength of noise in the noise term; or examining the problem without noise term, numerical solutions for the KPZ equation can be obtained for one spatial dimension. Numerous shaped figures show that every initial condition effects wavy starting point of surfaces from a smaller range to higher on both Gaussian noise and without noise case in the same way. One of the differences between these two cases is that if we do not apply the noise term then after some time the solution surface is flat, but applying the Gaussian noise the surface has a wavy structure even after $t = 10000$. Reducing the amplitude in the initial condition provides a smoother shape in the surface structure.

Acknowledgement. This work was supported by Project No. 129257 implemented with the support provided from the National Research, Development and Innovation Fund of Hungary, financed under the K_18 funding scheme.

References

1. Barabasi, A.-L., Stanley, H.E.: Fractal Concepts in Surface Growth. Press Syndicate of the University of Cambridge, Cambridge (1995)
2. Bognár, G.: Roughening in nonlinear surface growth model. Appl. Sci. **10**(4), 1422 (2020). https://doi.org/10.3390/app10041422
3. Misbah, C., Pierre-Louis, O., Saito, Y.: Crystal surfaces in and out of equilibrium: a modern view. Rev. Mod. Phys. **82**, 981 (2010). https://doi.org/10.1103/RevModPhys.82.981
4. Einax, M., Dieterich, W., Maass, P.: Colloquium: cluster growth on surfaces: densities, size distributions, and morphologies. Rev. Mod. Phys. **85**(3), 921 (2013). https://doi.org/10.1103/RevModPhys.85.921
5. Frey, E., Täuber, U.C., Hwa, T.: Mode-coupling and renormalization group results for the noisy burgers equation. Phys. Rev. E **53**(5), 4424 (1996). https://doi.org/10.1103/PhysRevE.53.4424
6. Hwa, T., Frey, E.: Exact scaling function of interface growth dynamics. Phys. Rev. A **44**(12), R7873 (1991). https://doi.org/10.1103/PhysRevA.44.R7873
7. Kriecherbauer, T., Krug, J.: A pedestrian's view on interacting particle systems, KPZ universality and random matrices. J. Phys. A: Math. Theor. **43**(40), 403001 (2010). https://doi.org/10.1088/1751-8113/43/40/403001
8. Lässig, M.: On growth, disorder, and field theory. J. Phys.: Condens. Matter **10**(44), 9905 (1998). https://doi.org/10.1088/0953-8984/10/44/003
9. Parisi, G., Kardar, M., Zhang, Y.-C.: Dynamic scaling of growing interfaces. Phys. Rev. Lett. **56**, 889 (1986). https://doi.org/10.1103/PhysRevLett.56.889
10. Martynec, T., Klapp, S.H.L.: Impact of anisotropic interactions on nonequilibrium cluster growth at surfaces. Phys. Rev. E **98**, 042801 (2018). https://doi.org/10.1103/PhysRevE.98.042801
11. Mello, B.A.: A random rule model of surface growth. Physica A **419**, 762–767 (2015). https://doi.org/10.1016/j.physa.2014.10.064
12. Olver, P.J.: Applications of Lie Groups to Differential Equations. vol. 107. Springer Science and Business Media, Berlin (2012)
13. Olver, W.J.F., Lozier, D.W., Boisvert, R.F., Clark, C.W.: NIST Handbook of Mathematical Functions. Cambridge University Press, Cambridge (2010)
14. Sergi, D., Camarano, A., Molina, J.M., Ortona, A., Narciso, J.: Surface growth for molten silicon infiltration into carbon millimeter-sized channels: Lattice–Boltzmann simulations, experiments and models. Int. J. Mod. Phys. C **27**(06), 1650062 (2016). https://doi.org/10.1142/S0129183116500625
15. Barna, I.F., Bognár, G., Guedda, M., Mátyás, L., Hriczó, K.: (2020). Analytic self-similar solutions of the Kardar-Parisi-Zhang interface growing equation with various noise terms. Math. Model. Anal. **25**(2), 241–256 (2020). https://doi.org/10.3846/mma.2020.10459

16. Barna, I.F., Bognár, G., Guedda, M., Mátyás, L., Hriczó, K.: Analytic traveling-wave solutions of the Kardar-Parisi-Zhang interface growing equation with different kind of noise terms. arXiv:1908.09615 (2019)
17. Korutcheva, E., Cuerno, R. (eds.): Advances in Condensed Matter and Statistical Physics. Nova Scientific Publishing. Inc., Hauppauge, pp. 237–259 (2004)

Effect of Cooling Channels to the Press Hardening Tools Temperature

Viktor Gál(✉) and Zsolt Lukács

University of Miskolc, Miskolc 3515, Hungary
metgv@uni-miskolc.hu

Abstract. Environmental and safety regulations strongly influence the developments in the automotive industry. To achieve the fuel-efficient vehicles with the safety standards it needed, higher strength material required. Because of this, the new high strength aluminium, and also the Ultra High Strength Steels (UHSS) like 22MnB5 become commonly used, particularly in the Body in White sheet metal part production.

To achieve a complex shaped structural product from the Ultra High Strength boron steel, a hot stamping forming process needed. The manufacturing of the 22MnB5 sheet metal part has two phases. At first, the blank is heat treated to above 930 °C (to the austenitization temperature), after then a forming operation and a quenching operation (between the closed and cold dies) occur simultaneously. It can be seen the temperature in the viewpoint of the preheated blank, and the change of the temperature in the viewpoint of quenching plays an important role in the stamping process.

The forming process for the new material is given, but because nowadays the finite element analysis becomes as a necessary step before manufacturing the tryout dies, the dedicated FE codes must also keep up with the developments. However, the reliability of the simulations strongly depends on the used constitutive models. An accurate description of the plastic behavior of the material can be complicated for a hot stamping process because of its strain rate and temperature dependence.

At the end of the stamping process the preheated blank contact with the surface of the dies under pressure and due to the heat transfer, the temperature of the die elevated. In case of repetitive manufacturing cycles, the temperature elevation of die can lead to inadequate cooling speed and also inappropriate microstructure of the material. To avoid this phenomenon, cooling channels are applied. The present paper described the effect of the cooling channels to the die temperatures. For this purpose, simulations of hot stamping an A-pillar was carried out with different arranging cooling channels. The simulation was performed with AutoForm R8 FE code, by using thin shell elements.

Keywords: 22MnB5 · Press hardening · FEM

1 Introduction

The development of automotive industry of recent years has basically determined by the reduction of carbon dioxide emissions. With the increased strength of the applied

K. Jármai and K. Voith (Eds.): VAE 2020, LNME, pp. 312–320, 2021.
https://doi.org/10.1007/978-981-15-9529-5_28

materials the mass reduction can be achieved, and the harmful emission also can be reduced. However, the increased strength of steels causes decreases formability. It is well known that higher temperature of the forming process makes higher formability of the material. Based on this, new technologies for sheet metal parts are usually hot forming technologies.

Nowadays the comprehensive process planning is usually aided by a computer which improves the cost - and time-efficiency. The new materials and also the required forming operations on elevated temperatures is a challenge for the software developers. Nowadays there is a possibility to FE modeling the temperature change of the part, the dies and the environment in a hot stamping process.

In this paper, an FE simulation of manufacturing an A pillar made by the ultra high strength 22MnB5 steel sheet was carried out. Based on the FE simulations, the effect of the applied cooling channels to the press hardening tools temperature, in case of a cyclically repetitive manufacturing process, was examined.

2 The Technology of Press Hardening

Press Hardening (also called hot press forming) is a forming process for high strength steels on elevated temperature. The hot press forming includes the austenitization of the complex forming and the rapid cooling of the blank. The order of the technological steps and the parameters can vary widely depending on the needed result. There are two common versions of this technology, the indirect and the direct hot forming. In the case of indirect forming, the process started with a forming in cold conditions, followed by the heating of the part to the austenitization temperature. The process ended with a hot forming with quenching. In the case of direct hot forming, the process begins with austenitic annealing, followed by forming and quenching between cold dies. The final microstructure and the properties of the part mainly depending on the process parameters, like the holding temperatures and the controlled cooling process [1].

In this paper, the technology of direct hot forming was examined. The first step of common practice in the manufacturing 22MnB5 Ultra High Strength steel is to be heated to the austenitization temperature to achieve a homogeneous austenitic texture. The necessity of this is justified by the fact that in this case the originally ferritic-perlitic microstructure with a yield strength of 457 MPa replaced by the austenitic microstructure (it has no precipitates, and holds all the alloys in solution), which has a significantly better formability due to its face centered cubic crystal structure.

The next step is to remove the blank from the furnace and transfer it between the dies. At this stage, time and efficiently the quick positioning the blank to its place, have an important role thanks to the cooling effect of the environment.

After this, due to the closing dies, the part achieves its final geometry. Between the closed dies, the part reaches the martensitic finish temperature with the critical cooling rate due to the heat transfer of the tools. Thereby a significantly larger martensitic microstructure can be achieved with more than 1000 MPa yield strength. The last step of the process is the cooling to room temperature, which is usually done outside the tool. The thermo-mechanic forming process of 22MnB5 can be seen on Fig. 1.

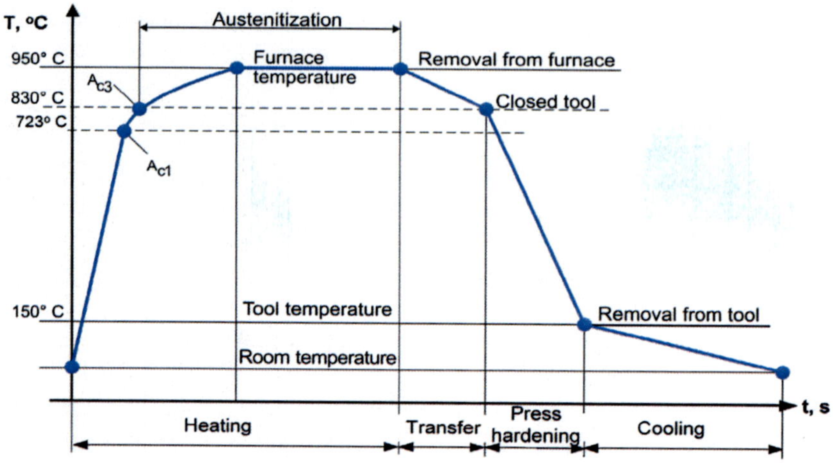

Fig. 1. Thermo-mechanic forming process of 22MnB5 [1]

The most modified process parameter to achieve the proper microstructure and mechanical properties is the cooling rate. As it can be seen in Fig. 2, to reach a purely martensitic structure, a 27 K/s or higher cooling rate needed. When the cooling rate is lower, bainite also appears by the martensite. The result with this is lower strength (and also lower hardness of the final part), but the fracture toughness of the part increases due to the bainite energy absorbing ability [2].

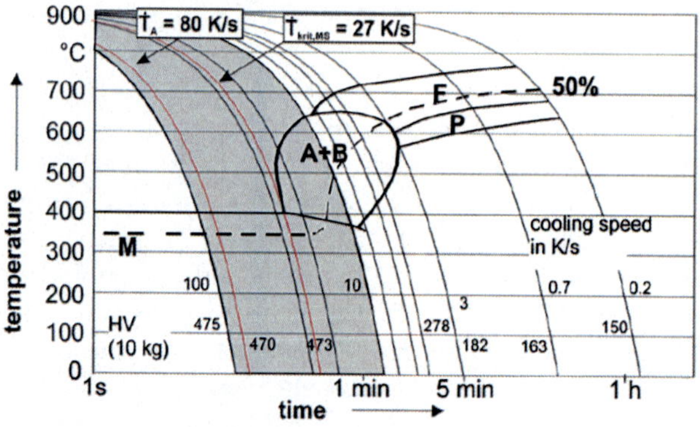

Fig. 2. CCT phase diagram of 22MnB5 with the proper cooling rate [3]

The final step in the process is the laser cutting (where it is necessary), to produce the appropriate contour and the necessary holes of the desired part. In the following, the critical parameters, which are needed to create FE modeling for the above described hot stamping process, will be presented.

3 Numerical Modeling of Press Hardening

To create a healthy simulation which considers all the parameters realistic, we must provide constitutive equations that describe the behavior of a material during plastic deformation. In sheet metal forming, the concept of material card is commonly used, which includes the Flow Curve, the Yield Surface, and the Forming Limit Diagram (FLC).

In case of hot forming, describing the aforementioned parameters is more complicated by the fact that all of them depend on temperature and/or the strain rate. Nowadays there is an opportunity in the dedicated FEM codes to create a press hardening simulation, with the relevant mechanical and thermal properties of the sheet and the dies.

The material card, which is widely used in cold forming simulations, also describes the behavior of the elastic and plastic deformation of the material. The stiffness of the blank can decrease significantly with the increasing temperature, the temperature-dependent Young-modulus is needed to describe the elastic behavior. The plastic anisotropy coefficient (r-value) thus so the Yield Surface also changes with temperature. To describe the plastic deformation the Flow Curve, to define the limit of the formability the FLC is required. The difficulties of modeling a hot stamping process are further increased by the fact that the last two parameters depend not only on temperature but the strain rate too.

With these parameters, a simulation of an isothermal forming process can be described. However, in press hardening, the mode of the temperature change of the examined materials needs to be specified. For this, the volumetric heat capacity and the thermal conductivity need to be defined for the blank and also for the die materials.

For a healthy FEM simulation of a forming process, the interactions between the tool and workpiece surfaces need to be defined as boundary condition. In case of hot forming simulation, the temperature and pressure dependent Interfacial Heat Transfer Coefficient (IHTC) – which defined as the heat transfer coefficient between blank and die surface –, as an additional parameter must be specified [4].

The process becomes more complex if the phase transformation also needs to be modelled. The transformation occurs in the material when the temperature decreases from the γ-zone to the α-zone. The transformation depends on the cooling rate, the chemical composition and the current state of the plastic deformation (which leads to a shift of the phase regions within the CCT diagram to the left [5]).

In order to make a simulation of the phase transformation of a material with a given chemical composition, it is also necessary to describe the dilatation behavior of the material. For this, the thermal expansion of each phase and the phase transformation strain must be considered. For example, the AutoForm calculates [3] phase dependent the temperature dependent volumetric strain, based on the following Eq. (1):

$$\varepsilon_{ii}^{th+tr} = \sum_k V^{(k)}\left(\alpha^{(k)}(T - T_0) + \varepsilon^{(k)tr}\right), i = 1, 2, 3 \tag{1}$$

Where $V^{(k)}$ is the volume fraction of the phases, $\alpha^{(k)}$ is the linear expansion coefficient of phase k, T_0 is the initial blank temperature, $\varepsilon^{(k)tr}$ is the transformation strain of phase k and $\alpha^{(k)}$ is a given parameter.

The provision of the required cooling rate depends on the temperature of the tool. In case of mass production, the die can be heated up due to the cyclically repeated heat loading, so it cannot provide the cooling rate required to create a purely martensitic structure. To avoid this problem, tools must be equipped with a cooling system. For the cooling system Convection Heat Transfer Coefficient (CHTC) also has to be defined, as a heat transfer coefficient between the cooling channel wall and the fluid coolant (water) [4].

4 Modelling of A-Pillar Cooling System

Nowadays it is a real challenge for the automotive developers to provide the technological parameters which are needed for the production of high strength steel sheets. For hot forming processes, it is not enough to simply design the geometry of the tools, the thermo-mechanical parameters for the used materials also need to be known.

In case of manufacturing a single A-pillar (Fig. 3), the temperatures of dies would not change significantly due to the heat transfer associated with its high mass. However, this part is mass-produced and as a result of it, the temperature of the tool starts to rise due to the small but cyclically repetitive heat load. Due to the increase of the temperature, the heat dissipation capacity of the tool will deteriorate, and as a result, it will not be able to provide the expected cooling rate after a certain number of pieces. If the A-Pillar does not have the specified microstructure, it will not be able to provide the load capacity prescribed by the safety regulation. To avoid this phenomenon, the cooling of the tools is carried out by circulating liquid in them.

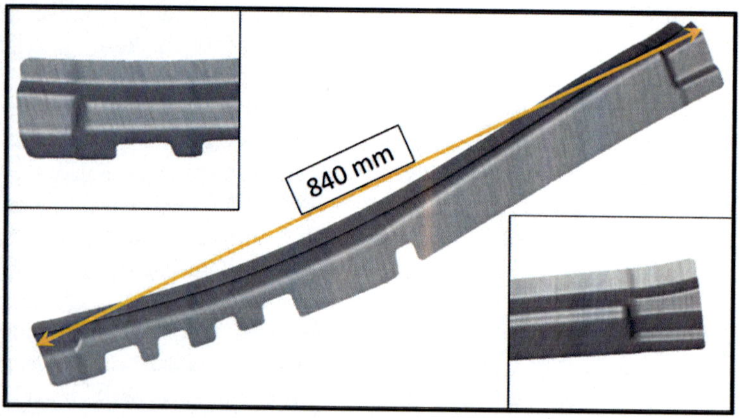

Fig. 3. An A-pillar sheet metal part

In this paper, the effect of the application cooling channels was investigated with FE modelling in the process of manufacturing 22MnB5 A-Pillars for 100 cycles. Auto-Form R8 with thermo solver was used to perform the numerical simulation in hot stamping. The efficiency of each tool concept was based on the comparison of the change in their temperature.

First, the hot stamping technology without cooling was examined. The production of an A-pillar consists of 3 steps. The piece had to be heated to the austenitization temperature, followed by forming between the cold dies. As a final step, the piece is removed from the tools and cooled to room temperature. To prepare the simulation after the tool geometry import, the three technological steps (heating, drawing, cooling) were created. After that the additional technological parameters required to generate the thermal cycle shown in Fig. 1 was provided.

The blank was heated to its austenitization temperature, the time interval the blank spent between the furnace and dies was 5 s. The blank dwelled for an additional 4 s on the punch before the ram motion. After the forming the quenching process was 5 s. As the pressure between the die and part surfaces increases, the IHTC – thus the cooling rate –, increases too. The software describes this phenomenon with a pressure dependent scaling factor for IHTC. Taking advantage of this phenomenon, the tools had a continuous 100 kN quenching force during quenching. The IHTC was 3.5 mW/(mm^2K), the volumetric heat capacity of the die was 4.37 mJ/(mm^3K) and the conductivity was 32 mW/(mmK). Before the removal, the blank had another 10 s on the stamp. The ambient temperature and also the tool temperature at the first cycle were set to 20 °C. The technology ended with cooling the part to 50 °C. In order to determine the temperature of a given tool, it is obvious to apply a mesh on the surface of the tools, which significantly increases the calculation time.

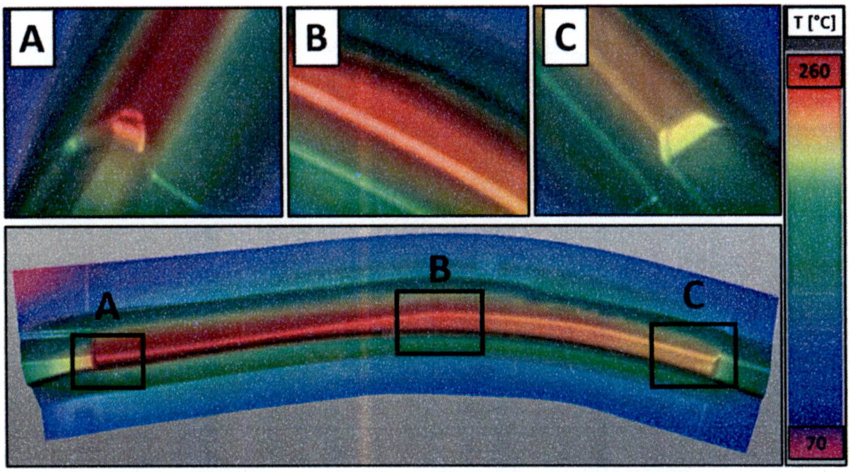

Fig. 4. Punch temperature after 100 cycle without cooling

Figure 4 illustrates the tool without the cooling channel and the temperature distribution observed after 100 pieces of A-pillar was produced. It can also be seen in the figure that the highest temperatures occur at the sharp, small rounded corners of the tool. The heat dissipation capacity of the tool is ensured by the relatively large weight compared to the blank. In addition to their sharpness, the temperature difference between points A and C is also significant because while the height of point A is 74 mm, the high of the C is 65 mm.

To avoid of overheating the dies, application of cooling is needed. The dies are usually cooled with liquid, flowing in the cooling channels. Has to be careful with the location and number of these channels, to make sure the die temperature would not elevate overly.

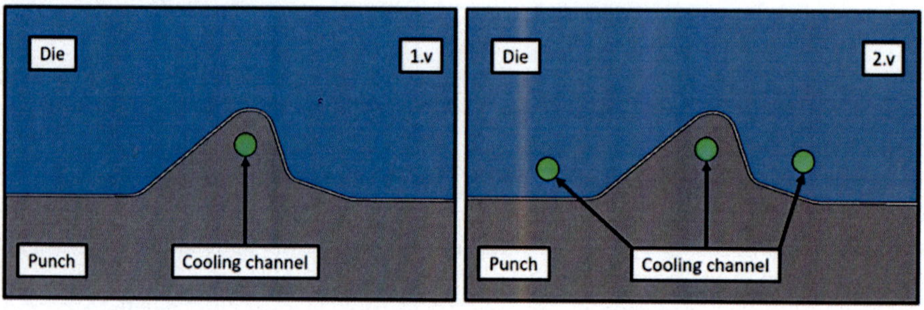

Fig. 5. The cooling channel arrangements on a cross section of the tool

Traditionally, the cooling channels are formed with drilling after the tool segment has been made. Nowadays – thanks to the development of casting technology –, it is also possible to create cooling channels with complex spline shape. Accordingly, the cooling channels were created in both of these simulations by tracking the surface at a specific distance to reach the best cooling effect at the abovementioned critical regions. Two arrangements of cooling channels were examined and can be seen cross sections of them in the Fig. 5. In the first case (1.v) the cooling channels were placed only in the punch, in the other case (2.v) 2 cooling channels were placed in the die as well.

Fig. 6. Punch temperature with cooling channels

The parameters used for these simulations were the same as before at the simulation without cooling channels supplemented with geometric and thermomechanical data for the cooling channel. The coolant was water at 50 °C, the CHTC was 10.0 mW/(mm^{2}K). The simulations were performed for a maximum of 100 cycles (as before) provided by the software, and the temperature change at the critical point of the tool was examined.

Figure 6 shows the results of simulations performed with tools with the different cooling channel arrangements. After simulating the 100 cycles, the temperature in the critical zone (shown in part A of Fig. 4), is approximately 70 °C. The difference between the two cooling channel arrangements is only 6 °C. Based on the small difference, it can be stated, that the cooling of the punch is sufficient.

In all 3 cases, the temperature distribution of the punch, and the die were also examined. The highest temperature point of the tools in all 3 simulations was the part of the punch marked A in Fig. 4. Accordingly, in the following, the conclusions based on the punch was drawn.

5 Conclusions

After running the simulations, the temperature change observed on the critical surface of the punch was examined. Within a given cycle, the temperature of the critical area changes within a given interval (its temperature elevated till the end of every cycle). Examining these intervals, the minimum temperatures for each cycle was used as a basis for comparison, because the higher the minimum temperature is the lower the heat dissipation capacity is, which causes decreasing cooling rate. Figure 7 compares the temperature increase of the three variants (dies *without cooling* system, only punch cooled *01_version*, punch and die cooled *02_version*).

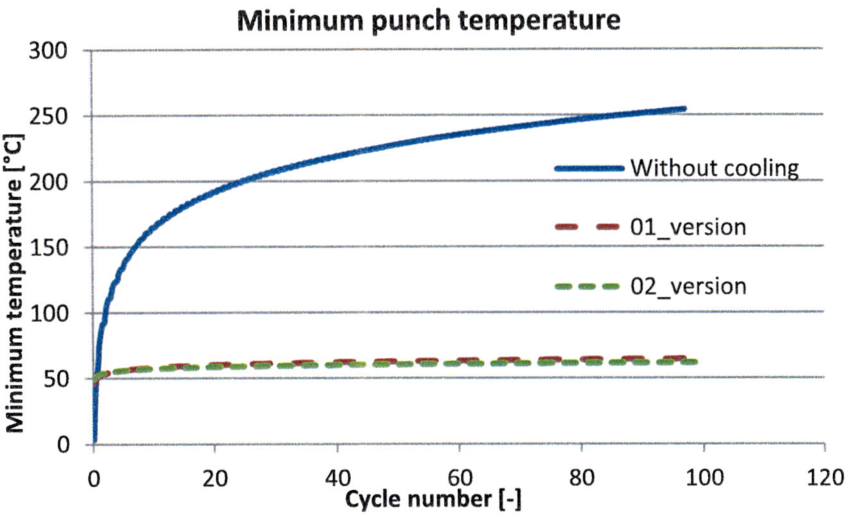

Fig. 7. Comparison of the different cooling methods

As can be seen from Fig. 7, when cooling channels were not used a significant temperature increase occurs after 100 cycles. When cooling channels were used, the temperature increase was negligible by the end of the process, so it is necessary to use them. However, there were no significant difference between the two arrangements, so it can be stated that it is sufficient to use only one cooling channel for high-cycle production.

In summary, modern newly developed materials also require related technologies. However, these new technologies cause complex modeling problems, many more parameters must be defined compared to cold forming, and the determination of these parameters is the result of more complex processes. In total, it is a good idea to be able to model these processes, because if we look at Fig. 6, which shows the maximum temperature of the critical surfaces, it is a pity that it can be modeled and the different versions can be compared at almost no cost.

Acknowledgement. "The described article/presentation/study was carried out as part of the EFOP-3.6.1-16-2016-00011 "Younger and Renewing University – Innovative Knowledge City – institutional development of the University of Miskolc aiming at intelligent specialisation" project implemented in the framework of the Szechenyi 2020 program. The realization of this project is supported by the European Union, co-financed by the European Social Fund."

References

1. Tisza, M., Czinege, I.: Comparative study of the application of steels and aluminium in lightweight production of automotive parts. Int. J. Lightweight Mater. Manuf. **1**, 229–238 (2018)
2. Morbacher, H.: Material concepts and process strategies for the application of heat-treated steels in body engineering. In: Conference on Hot Forming, Dunaújváros, Hungary (2012)
3. AutoForm R8 Software Manual: © 2019 AutoForm Engineering GmbH
4. Li, Y., Ying, L., Hu, P., Shi, D., Zhao, X., Dai, M.: Coupled numerical simulation of hot stamping process and experimental verification. In: AIP Conference Proceedings, vol. 1532, p. 471 (2013). https://doi.org/10.1063/1.4806863
5. Merklein, M., Svec, T.: Transformation kinetics of the hot stamping steel 22MnB5 in dependency of the applied deformation on the austenitic microstructure. In: IDDRG 2010 International Conference, Graz, Austria, pp. 71–80 (2010)

Design and Testing of a Water Injection System of a Turbocharged Spark Ignition Engine in Testbench Environment

Máté Tóth(✉), Attila Gyuris, Balázs Rácz, Péter Sass, and Jan Rohde-Brandenburger

Széchenyi István University, Győr, Hungary
toth.mate@ga.sze.hu

Abstract. In the scope of this article, the design and testing of a water injection system applicable for a spark ignition engine are presented. Increasingly stringent emission standards within the framework of EURO7 require either directly or indirectly the internal combustion engines to be optimized across the entire field of an engine map, therefore they must comply with the emission standards at each operating point. The greatest challenge is expected to be the Lambda = 1 operation on the full field. The conversion efficiency of the exhaust gas after treatment systems is the highest at this point, therefore it is foreseeable that no deviation can be made. As a result, fuel enrichment for performance enhancement and to protect components against thermal load will not be tolerated, so the resulting thermal loads will need to be reduced in other ways. It is possible to reduce the excess thermal loads by using water injection. Evaporation of water in the intake system and combustion chamber takes off heat and the temperature of the contacting components and fluids decreases. The affected components include pistons, combustion chamber, cylinder head, exhaust valves, exhaust manifold, turbine wheel, turbine housing, and as a medium, the temperature of the intake air. Reducing the temperature of some components is important in the aspect of mechanical strength, while for some components the knock limit can be extended. This article presents the detailed design process and testing phase of a water injection system. An important aspect in system design is compatibility with different engine layouts in a cost-effective manner. Injector nozzle testing also includes analysis of mass flow, dispersion and spray pattern. The scope of the work is the implementation of a water injection system, which is capable of performing measurements in testbench environment at the Department of Internal Combustion Engines and Propulsion Technology of Széchenyi István University.

The result of the measurements is the successful cylinder selective application of water injection to the intake system, whereby the addition of water reduces the temperature of the intake air and the exhaust gas, which can be reduced to standard calibration temperature in Lambda 1, without fuel enrichment.

Keywords: Water injection · Design · Water injection system · BSFC · Internal combustion engine

K. Jármai and K. Voith (Eds.): VAE 2020, LNME, pp. 321–333, 2021.
https://doi.org/10.1007/978-981-15-9529-5_29

1 Purpose of the Study

The requirements expected by environmental standards are tightening, which, in addition to the decreasing limits, is also characterized by a change in the measurement methods. The current regulations are aiming to significantly reduce harmful emissions and CO_2 emissions. One option for emission reduction is downsizing, which allows emissions to be effectively reduced with low-displacement turbocharged engines, and by using turbochargers, the engines can meet expectations in terms of performance and efficiency [1–3]. At the same time, the heat load of the engines increase, which increases the thermal load, creating the most important task to be solved, which is the reduction of the knocking tendency and increasing the temperature tolerance of the components [4, 5]. Potential solutions analyzed by the comparison published by FEV Group GmbH (Fig. 1). The purpose of using different methods was to maintain the power density of $\lambda = 1$ and 110 kW/l, in addition to the safety of the components [2].

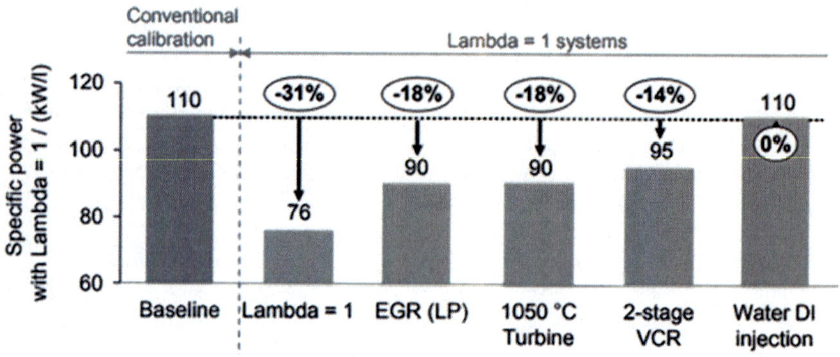

Fig. 1. Results of Lambda 1 concept studies based on literature research [2]

By water injection, the thermal load can be accurately reduced based on the available scientific articles [2, 6]. The evaporation of the injected water into the system removes heat from the environment, thus reducing the temperature of the intake air, the combustion chamber and the exhaust gases well, and the reduced exhaust gas temperature results in a lower thermal load of the components exposed to hot gases such as exhaust valves, turbine wheels and turbine housing moreover, the excessive heat load of the catalytic converter can be reduced [6]. The water injection system entails a number of development tasks, such as solving water supply during vehicle integration without changing driver comfort, thermodynamic application, and testing of the engine's tribological system [7–10].

In this study, the design of a cost-effective water injection system is presented, which demonstrates component analysis as well after a series of implemented testbench measurements. The aim is to reduce the thermal load of the engine in $\lambda = 1$ operating state at a static working point, where the heat absorption capacity of the water is used instead of the fuel.

2 Connection of the Water Injection System to the Engine

The research is based on a turbocharged spark ignition engine. The engine has a power density of 120 kW/l, which typically operates with a rich mixture for high-load work points, which is a major reason for the protection of turbine wheels and turbine housings against thermal overload. The elements of the water injection system have been selected on the basis of literature research, which highlighted that 10–100% of the added water value can be injected in relation to the fuel at the selected working point so that the effect of water injection can be investigated according to the objectives [4, 6–8]. Cost-effectiveness and reliable operation are important considerations during the design process, therefore standard and commercially available parts have been selected. The expectations set for the system are the following:

- Water mass flow control
- Homogeneous water distribution in the intake manifold
- Continuous water supply — deionized water
- Water mass flow and temperature measurement
- Integration possibility into the engine

In addition to the established water injection system, due to the examination of thermodynamic processes, it is necessary to install extra sensors in the engine:

- Cylinder selective intake manifold temperature measurement
- Cylinder selective exhaust gas temperature measurement
- Temperature and pressure measurement before/after: turbine, catalytic converter

2.1 Selection of the Injection Position

Taking into account the selected engine, the budget and the literature research, there were three possible options for the installation of the water injection system [9, 10]:

- Compressor - Throttle section
- Plenum
- Cylinder selective injection in intake tubes

The injection options and the location of the measuring points are shown in Fig. 2.

The selected injection strategy is the cylinder selective intake tube injection, which has been defined by the involvement of literature research and the industrial partner. This method ensures that each cylinder has a known and freely defined amount of injected water. The cylinder selective construction requires minimizing the scattering between the injection nozzles to ensure a homogeneous distribution.

2.2 Selection of Injection System Components

This paragraph illustrates the elements of the water injection system shown in Fig. 3.

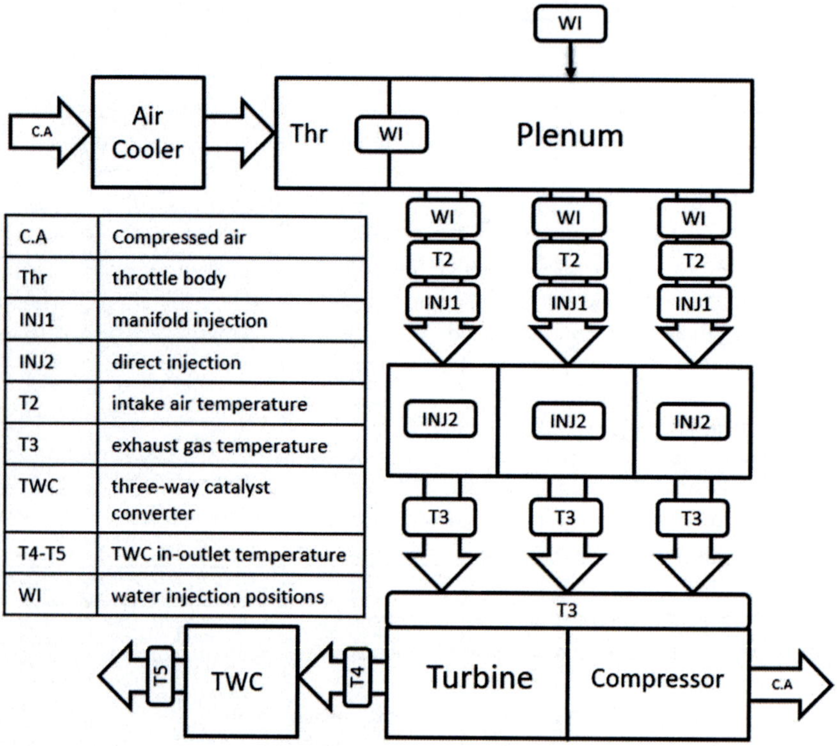

Fig. 2. Structure of the internal combustion engine and schematic presentation of water injection options with retrofitted sensors.

Fig. 3. Components of the water injection system [11–13]

The water tank has 20 L of volumetric capacity, which can be changed to another tank with a capacity of 220 L for long-term experiments. After the tank, the water pump is placed, which must produce the required pressure to have enough volumetric flow rate and pressure for injection. The chosen pump puts out a maximum pressure of 16 bars, but long-term usage above 11 bars of pressure causes overheating due to the increased energy requirement. After the pump, a filter is integrated to save the nozzle from clogging, which will be presented later. The next part is the flow meter, which can measure the volume of water with the required accuracy. Next, a solenoid valve takes place, which serves safety functions. In case of engine shut off, the PUMA system closes this valve, to avoid the engine's filling up with water. The non-return valve prevents the total discharge of the tubes.

The components are connected by polyurethane tubes (also used in pneumatics) that are reliable up to 20 bars. In front of the nozzle, the distributor element can be found, which divides the input into the same number of nozzles as the number of cylinders, and a pressure gauge. Moreover, a PT100 temperature measuring sensor is connected to this distribution system. Regarding the water injection nozzles, the continuous water vapour creating nozzles were chosen. These nozzles are going to be discussed in detail in the following subchapter. The system's final element is a power factor controlling the filling ratio that is able to control the actual power of the water pump with the help of a potentiometer. Water has been added manually during the tests. The peripherals of the designed system were built on a movable unit made of aluminium machine building profiles for easy handling. Their schematic location is presented in the following figure (Fig. 4):

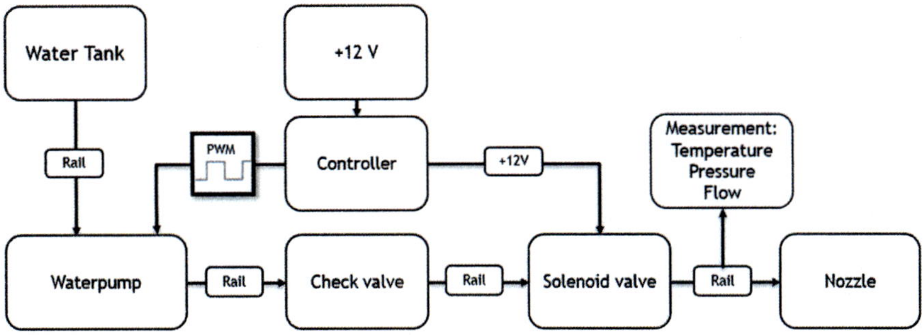

Fig. 4. Systematic construction of the water injection system

After the construction of the system, comparative tests of the nozzles were followed.

3 Testing of the Injection System

The components of the injection system have been tested individually and then built into the system. At first, the diameter of the nozzles was measured, then at constant pressure, a flow rate measurement was taken and finally the spray image was recorded using a high-speed camera.

3.1 Structure of the Injection Nozzles

The nozzles are purchased items from two different manufacturers. They are made up of 4 main components (Fig. 5):

- housing with calibrated diameter hole
- cone and spring for sealing and evaporation
- filter
- nozzle housing and connectors

Fig. 5. Parts of water injection nozzles from two different manufacturers [11, 14]

3.2 Spray Test

The mass flow of the nozzles is determined by the size of the nozzle head hole and the joints of its internal structure. With the help of an electron microscope, the bore diameters of the nozzles purchased in several sizes have been measured. The measured standard deviations are less than the measurement uncertainty, as the contour cannot be clearly determined and therefore a deviation of ± 15 μm can be assumed in the measurement (Fig. 6).

Several nozzles with different flow rates have been purchased, but the goal is to achieve a water-to-fuel ratio (hereinafter WFR) of 10–100%, for which, in case of selective cylinder injection, the nozzle set with the smallest volume flow at predefined operating points is sufficient on the basis of the calculation. The water-fuel ratio was calculated as the quotient of the volume flow of injected water and fuel.

The following diagram shows the purchased nozzles based on the standard deviation of their volumetric flow rate (Fig. 7).

As Fig. 7 shows, in case of a cylinder-selective arrangement, the minimum amount that can be injected is 36 ± 1 ml/min, which corresponds to a volume of ~2.16 l/h broken down into cylinders, these values must be taken into consideration when selecting the operating point. It is also important that fuel consumption has to be based on values without mixture enrichment, around $\lambda = 1$. During the selection of the operating point, T3 temperature and the original application's Lambda value were taken into account.

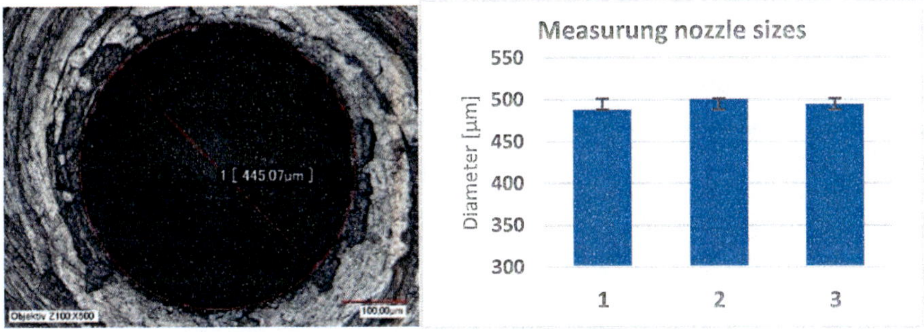

Fig. 6. Microscopic image of the smallest, 500 µm nominal holed nozzle head, and the five nozzles compared to each other.

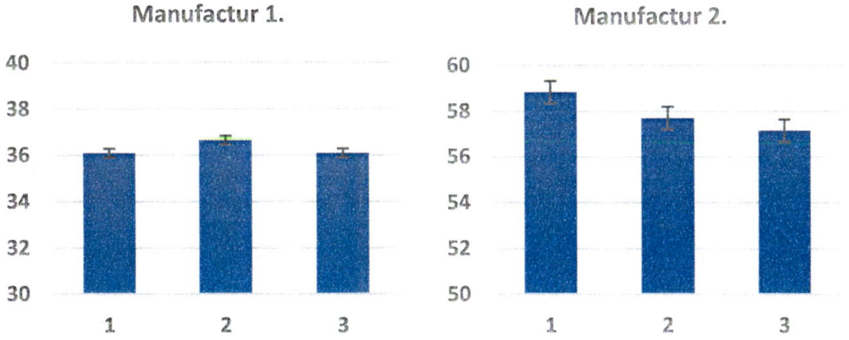

Fig. 7. Mass flow measurement of Manufacture 1 & 2 nozzles analysed at a system pressure of 5 ± 0.3 bar with a water pump filling factor of 30%.

The flow rate can be partly further reduced by decreasing the filling ratio of the water pump, which also affects the injection pressure. Since the measurements were made in laboratory conditions at an absolute pressure of 1 bar, the engine is running in charged mode. Therefore, it is not the absolute pressure but the pressure difference that must be calculated to compare the injected state with the pressure measured during engine operation. Based on the measurement, it can be stated that the maximum deviation in the injection system is less than 1 ml/min for smaller nozzles and 1.7 ml/min for larger nozzles, which is a satisfactory result considering the system elements and costs.

4 Installation of the Injection System into the Intake System

The intake system is equipped with 1.5 mm PT100 thermometer sensors, fixed and sealed with two-component adhesive (Fig. 8).

During the modification of the intake system, the position of the surrounding parts had to be taken into account, therefore the positions of the injectors and sensors were also determined. The vacuum tubes and electronic wiring still fit next to the injectors and sensors, so no other modifications on the engine were required.

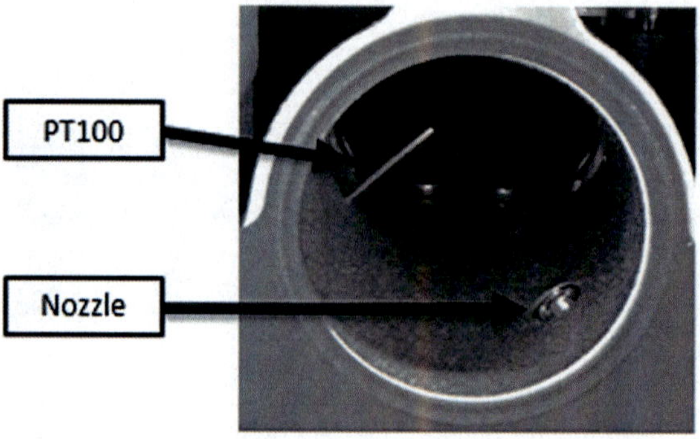

Fig. 8. Temperature sensor and water injection nozzle in intake manifold

4.1 Taking Shots with a High-Speed Camera

The shots were taken by an Olympus i-speed 3 high-speed camera. It is capable of a maximum frame rate of 150.000 fps. The maximum resolution is 1280 × 1024 pixel, 'til 2000 fps, after that, resolution decreases and the light requirement increases. These factors pose the greatest challenge when creating such shots. The proper source of light was provided by an Olympus ILP-2 UHP Light Source. It is a continuous light device for high-speed camera measurements. In order to gain adequate positioning, a fiber optic cable was mounted for proper positioning and a lens was applied at the end. During the first measurements of the water injection nozzles, the nozzles were measured individually mounted on a frame, so a transparent plexiglass cabin had to be used due to the presence of vaporized water. The black background was fixed to the cabin (Fig. 9).

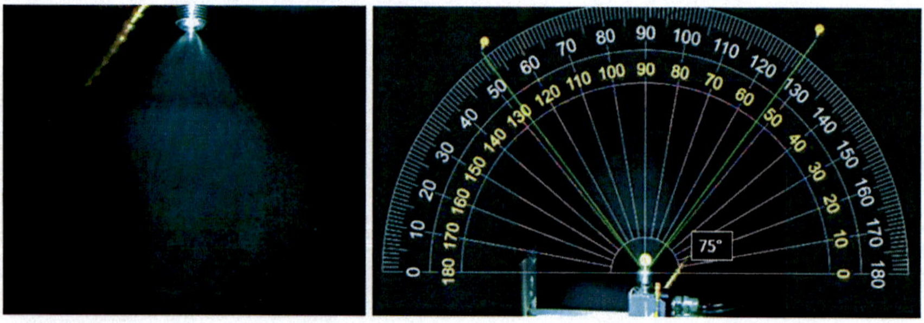

Fig. 9. Capture shot during operation (2000 fps)

In order to achieve applicable results, the measurements had to be performed in the post-machined intake system as well. It was no longer possible to use plexiglass in the intake manifold to record the vaporization image, but the light conditions were much more favorable. The shots could be made from a further distance, with 4000 fps (Fig. 10). In terms of illustration and quick result control, the video records were investigated, whilst in terms of detailed analyzation, the frames were examined separately.

Fig. 10. Capture spray image

During the recordings, the system pressure was 10 bar, the pipe lengths were the same up to the nozzles, only the spray heads and the light conditions were changed. As shown in Fig. 11, the injection angle and the covering of the intake pipe surface with the water vapor can be determined with a suitable approximation on the basis of the images.

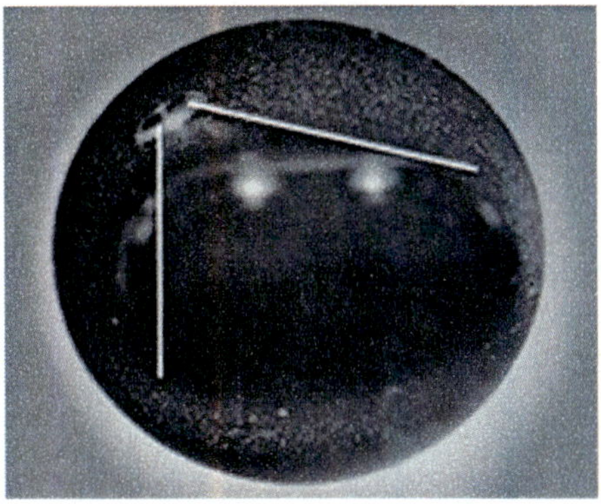

Fig. 11. Shot of a water injector nozzle mounted in an intake manifold (4000 fps)

Based on the image taken in the intake manifold (Fig. 11), it could be seen that the corresponding pressure value of 10 bar was built up under 449 frames, so during the recording of 4000 images per second, it takes 0.11 s to achieve the previously presented evaporation image. After switching the injection off, it takes time for the injection to stop due to the flexibility of the piping system resulting from the stored pressure. Thus, it can be stated that in this form the injection system can only be used for static measurements, however, due to the static dynamometer measurements, this reaction time is not relevant considering the current targets. The position of the injection nozzle is also adequate, the cross-section of the intake pipe is sufficiently covered with water vapor by the nozzle.

5 Integration of the Water Injection System in the Testbench Environment

The used sensors were integrated into the testbench's PUMA software located at the Széchenyi István University. The layout of the user-interface enables that some of the measured parameters can be graphically shown, which are being recorded in the meantime.

The display of the posteriorly set-in sensors shown in a bar chart thus facilitates the visual comparison. As a system improvement, it has been prepared to perform an engine lifecycle test. A further development opportunity is the automatization of the water injection system, controlled by a PID controller, which makes dynamic testing possible.

6 Engine Tests on Dynamometer

The main purpose is the examination of the system's effects on the exhaust gas temperature (T3). The water dosage happened also with the original application and its $\lambda = 1$ version. For consistent and comparable tests, the ignition timing controlled by the Engine Controller Unit (ECU) was recorded during the measurements. The choice of operating point regarding to the factory calibration of the engine, is the operation with a rich mixture, which was modified to $\lambda = 1$ during the measurement.

The selected operating point is 2000 rpm, a load corresponding to an effective mean pressure of 20 bar, controlled by the dynamometer's PUMA system. At this load, the engine with OEM application was operating at a Lambda value of 0.85, which was modified to $\lambda = 1$, which resulted in increased exhaust gas temperature and a decrease in the specific fuel consumption. The water injection was turned on in this state and the results are shown in Fig. 12.

As it can be seen in the results, the T3 with water injection can be reduced below the exhaust temperature of the OEM application, so at this point, the heat dissipation capacity of the excess fuel can be replaced by water (red line). Due to the position of the injection nozzles, when the 25% WFR is reached at the first measured point, the injected water cools the intake charge (T2). This does not change with the increase of volume either, so the lower temperature intake air is likely to help to reduce the knocking tendency even with the addition of 25% WFR. At T3 temperature, the heat dissipation effect of water increases in proportion to the amount at this load. The limit of the injected water mass, in this case, is the cross-section of the nozzles and the maximum pressure of the system, so with this configuration, it was possible to perform measurements up to 73% WFR, which already exceeds the reality of future integration into the vehicle. The factory application temperature limit T3 at low speed is NOx emissions, therefore the factory temperature can be reached with the given system at $\lambda = 1$, if the injected amount is equivalent to 60% WFR. The cooling effect of the water can be clearly seen in the decrease of the exhaust gas temperature and the intake air temperature. The engine was also tested at higher speeds, where the load was 20 bar effective mean pressure and the speed was 4500 rpm. For the reliability of the test, both the ignition time and $\lambda = 1$ were set to a fixed value (Fig. 13).

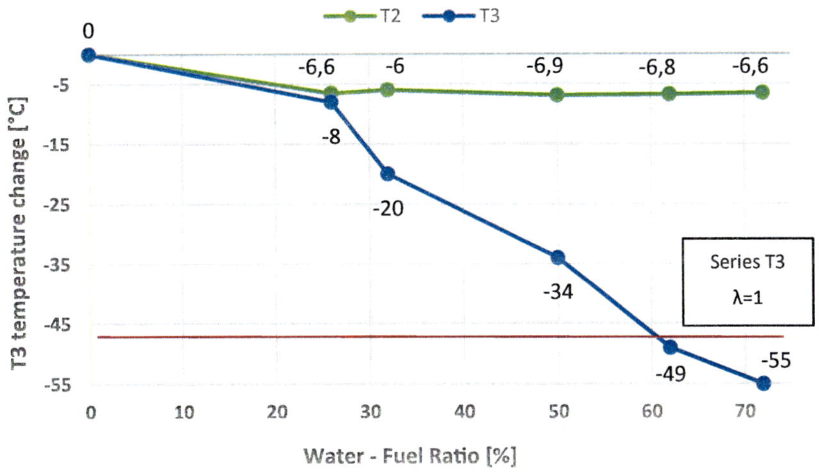

Fig. 12. Intake air and exhaust gas temperature reduction as a function of water injection - fuel ratio (WFR) - 2000 1/min, 20 bar BMEP Lambda at 1 operating point

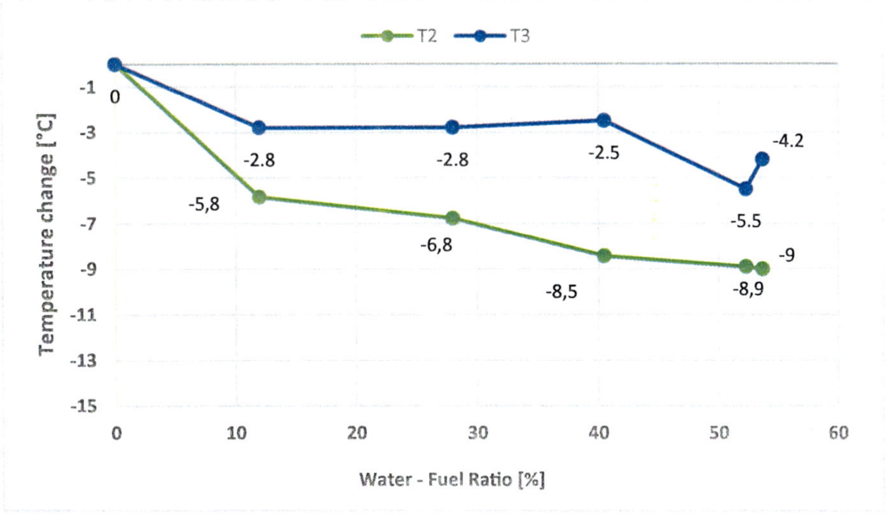

Fig. 13. Intake air and exhaust gas temperature reduction as a function of WFR - 4500 rpm, 20 bar BMEP

Figure 13 shows that at higher speeds and power the temperature-reducing effect of the introduced water is reduced. One of the presumable reasons for this is the excessive fineness of the evaporation, which transmits the heat-dissipating effect in the intake system. With a higher load, the mass flow of the incoming intake air increases significantly, the available time is shorter, which also reduces the time for the combustion chamber to cool down. No misfire occurred during the water injection tests.

6.1 Investigation of Water Usage Efficiency

With the injected water, it is possible to realize that the exhaust gas temperature is the same as the standard temperature achieved with enrichment at certain operating points. According to the literature research, one of the motivations for the application of water injection is the utilization of the high heat of evaporation of water, which has a value almost five times higher than gasoline, so that a fraction of water is sufficient compared to gasoline. However, based on the measurements, much more water was needed compared to the theoretical calculation. The heat of evaporation of water is 2257 kJ/kg, while gasoline average is 440 kJ/kg, so the required energy input is five times more to evaporate one kg of water under laboratory conditions. Engine calibration injects 20% more fuel at the first operating point shown, to maintain the proper exhaust temperature compared to the stoichiometric fuel.

The same result requires the injection of 60% of water compared to the fuel, which does not match the amount of water calculated from the heat of evaporation. The reduction in the cooling effect of the water can be presumably caused by a number of factors, such as the positions of the nozzles in the intake pipe, and the excessive fineness of the droplet size. During the distance between the nozzles and the intake valves, the homogeneous water mist contacts with a significant measure of wall surface, where heat removal takes place. However, approximately a quarter of the cycle is available to the water to enter the combustion chamber, the remaining time is consumed by cooling the wall surface of the intake system and heating the water. In contrast, regarding gasoline injection, the fuel is injected directly into the combustion chamber, so it has a targeted effect on the surfaces of the workspace. In terms of water usage efficiency, it is necessary to examine several injection positions and different evaporation sizes, injection times and pressures in the future, which, in addition to water injection efficiency, also has a serious impact on the tribological system of the engine [2].

7 Summary

Achieving the initial goals, a water injection system was successfully built that met the expectations in terms of both injectable quantity and evaporation quality. The completed system was incorporated into the intake system of a turbocharged spark ignition engine with which, engine dynamometer tests were performed. Based on the measurements, the cooling effect of the water injection is realized, reducing the temperature of the intake charge and the temperature of the exhaust gas, thus reducing the thermal load of the engine. Since the heat load of the components in connection with the exhaust gas can be reduced, it is possible to keep the components at a material-safe point even during $\lambda = 1$ operation. With running at $\lambda = 1$, the system improves efficiency in high-load workplaces with a typically rich mixture and reduces pollutant emissions at the tested operating points.

During the experiments, a higher amount of water was required than the theoretically expected for the corresponding exhaust gas temperature to decrease, thus further development potentials were outlined in the injection method. In terms of water usage efficiency, the optimization is important, which has a significant aftereffect regarding vehicle integration and tribological effects. One possibility for optimization is to replace the water injection nozzles - which produce the current homogeneous water mist - with one that can be controlled with sufficient cycle time, so that water can be injected into the working space through an open valve. During the dynamometer measurements, several observations were made by monitoring the operation of the engine. Due to water injection, depending on oil temperature, oil dilution, corrosion and increase in the amount of blow-by gas has occurred, which affects the lifespan and emission of the engine. There is further research potential in the study of these phenomena, which needs to be investigated before the series production.

References

1. Worldwide Emission Standards and Related Regulations, Passenger Cars/Light and Medium Duty Vehicles, Continental, May 2019
2. Görgen, M., et al.: All lambda 1 gasoline powertrains. In: Liebl, J., Beidl, C., Maus, W. (eds.) Internationaler Motorenkongress 2018, Proceedings. Springer Vieweg, Wiesbaden (2018)
3. Fraidl, G., Kapus, P., Melde, H., Lösch, S., Schöffmann, W., Sorger, H., Weißbäck, M., Wolkerstorfer, J.: AVL List GmbH: Variable Verdichtung – im Technologiewettbewerb? In: 37. Internationales Wiener motorensymposium, 28–29 April 2016
4. Neubauer, M., Kapus, P., Fraidl, G.: Efficiency and emission measures for future high-performance engines. MTZ Worldw. **80**, 54–58 (2019)
5. Hiereth, H., Prenninger, P.: Charging the Internal Combustion Engine. Springer, Wien (2003)
6. Vincenzo, B., Gerd, G., Eric, J., Johannes, W.: Increasing the downsizing level in combination with reduced fuel consumption. In: European GT-SUITE Conference, Frankfurt am Main, 22 October 2012
7. Miroslav, B.: Increasing efficiency of gasoline engines by water injection. Bachelor thesis, Brno, Brno University of Technology (2018)
8. Franzke, B., Voßhall, T., Adomeit, P., Müller, A.: Water injection for meeting future RDE requirements for turbocharged gasoline engines. MTZ Worldw. **80**, 30–39 (2019)
9. Hermann, I., Glahn, C., Kluin, M., Paroll, M., Gumprich, W.: Water injection for gasoline engines - Quo Vadis? In: Günther, M., Sens, M. (eds.) Knocking in Gasoline Engines, KNOCKING 2017. Springer, Cham (2018)
10. Sass, P., Tóth, M.: Vízbefecskendezés alkalmazásának lehetőségei a környezetvédelmi normák teljesítésének elősegítése érdekében. In: Tavaszi Szél Konferencia 2019. Nemzetközi Multidiszciplináris Konferencia Tanulmánykötet, Doktoranduszok Országos Szövetsége, Budapest, Magyarország, pp. 114–125 (2020). 12 p.
11. Snow Performance Homepage. https://www.snowperformance.eu/. Accessed 01 Mar 2020
12. Kobold Homepage. https://www.kobold.com/hu. Accessed 01 Mar 2020
13. IMC Homepage. https://www.imc-tm.hu. Accessed 01 Mar 2020
14. Prometh Homepage. https://prometh.com/. Accessed 01 Mar 2020

Method of Validating the Importance of Aerodynamic (Drag) Parameters for Electric Racing Environments

Gergely Szűcs(✉), Gergely Bári, and Gábor Sipos

John Von Neumann University, Izsáki út 10, Kecskemét 6000, Hungary
szucsi.90.07.09@gmail.com

Abstract. In the 21st century, electric cars are forging ahead on the streets and taking over the place of combustion cars in the traffic. This process can be observed in the world of motorsports too. In the world of open wheelers, Formula E has begun in 2014, and from 2020 this will be an official FIA World Championship while in the world of tin-top racecars, Electric Touring Car Championship (ETCR) will start in 2020 too. One of the key differences between combustion and electric cars are the energy usage. In electric car racing, there is much less energy stored compared to the power demand for a given time, also batteries used nowadays heats up heavily, resulting in the fact that both the stored energy and the temperature of the batteries can be limiting factors on the car performance. Considering these facts, by switching to electric race cars, we need to revise the priorities between the components of performance. This way the effect and importance of drag need to be revised, as it can have more influence on race results than in case of internal combustion engine powered racing. In this paper, a validated simulation model is presented, and based on this model the importance of drag is investigated. According to simulations a very little (1–2%) change in the drag coefficient of the car results countable, because it can be mean 3–4 tenths in laptime, can still result many places on the grid, and can cause big differences during an overtaking maneuver.

Keywords: Electric racecar · Drag · Parameter sensitivity · Efficiency

1 Introduction

One of the most effective limit factors of the performance of an electric racecar is the energy limit [1, 2], therefore efficiency is a very important parameter of the car. The effects of efficiency can be divided into 4 main parts [3]:

1) Powertrain efficiency (Electrical + Mechanical)
2) Rolling Resistance
3) Uphill
4) Drag (Aero)

K. Jármai and K. Voith (Eds.): VAE 2020, LNME, pp. 334–348, 2021.
https://doi.org/10.1007/978-981-15-9529-5_30

In this paper, we want to show the effect of drag to the performance of a touring car. To investigate the effect of the drag, we have created a point-mass model [4] in Optimum Lap software.

First, we had to validate it by setting the main parameters of the car, such as mass, engine-map, rolling resistance, drag- and downforce coefficient…etc. Based on this validated model the importance of drag is investigated using two cases:

- Defining the pure effect of drag to laptime (for internal combustion cars)
- Defining the effect of drag to laptime, while the used amount of energy is kept constant because an accumulator has finite energy (for electric cars)

According to simulations, a very little (1–2%) change in the drag coefficient of the car results countable effect on both. This is important, especially in case of the laptime, because it can be mean 3–4 tenths that can be around only 0.4% in laptime that can still result many places on the grid, and it can help during an overtaking maneuver too. Moreover, with these simulations, we can estimate the required capacity of accumulator.

2 Measuring Main Parameters

We have a manual [5] to TCR car we are simulating, but for more details, and better work we wanted to measure the main parameters of the car, which has many effects to laptime and efficiency.

2.1 Measuring Rolling Resistance

We can measure rolling resistance with coast-down test. During the test, we have to comply with three rules. First, the ground must be quasi flat, without up- and downhills, secondly the speed must be low because we want to exclude aero effects, and thirdly the driver must maintain the steering wheel straight to exclude the resistance of steering [6, 7].

The test consists of two major parts. At first, the driver should accelerate the car and have to maintain 30 km/h until he/she reaches marks, such as cones. From the mark, the driver has to release the throttle, push the clutch and should maintain the steering wheel straight, until the car will stop. Rolling resistance can be calculated from rolled distance assuming constant deceleration.

At first, deceleration can be calculated from rolled distance and starting speed of $v_0 = 30\,\mathrm{km/h} = 8.33\,\mathrm{m/s}$

$$a_x = \frac{v_0^2}{2s} = \frac{8.33^2}{2s} \tag{1}$$

where a_x is the average deceleration in [m/s^2] of the car and s is the rolled distance in [m]. From average deceleration, we can calculate the F_{res} resistance force by multiplying mass (m). The normal forces (F_N) of the car can be calculated with quasi the weight of the car, which is mass multiply the gravity number (9.81 m/s^2). Resistance force divided by normal force, C_{res} rolling resistance coefficient can be calculated [3].

$$C_{res} = \frac{F_{res}}{F_N} = \frac{m \cdot a_x}{m \cdot g} = \frac{a_x}{g} \tag{2}$$

We have measured rolling resistance with different tyre pressure with a results of Table 1:

Table 1. Rolling resistance at each tyre pressures (starting rolling speed is 30 km/h)

Tyre pressure [bar]	Distance of rolling [m]	Average deceleration [m/s^2]	C_{res} [-]
1.6	90	0.386	0.039
1.7	94	0.369	0.038
1.8	96	0.362	0.037
1.9	102	0.340	0.035
2.0	107	0.325	0.033
2.1	114	0.305	0.031
2.2	115	0.302	0.031

In TCR series the optimal pressure is 2 bars, so in our model, we can calculate with 0.033 [-] rolling resistance coefficient.

2.2 Measuring Downforce Coefficient

Downforce can be measured during the car is running if we have damper potential meters on the car and we know the stiffness of springs. Moreover, we have to know the kinematics of the car, especially the motion ratio [4] between spring travel and wheel travel and between wheel travel and damper potential meter travel. With this information downforce can be calculated.

First of all, we measured motion ratios in two ways:

The first measurement was direct, which means we lifted the car with 2 post car lift and put out the springs. After the preparations, we put 5 mm thread steel plates below the wheels while we were measuring the spring and damper potentiometer travels.

The second measurement was indirect, which means we measured the pickup points of the suspension and created a model which can be seen in Fig. 1. With this model, we can investigate our suspension and measure motion ratio.

Both of the two measurements have the same results. The average motion ratio between spring and wheel is 1.1 at the front and rear too. The variation of motion ratio in case of a bump is negligible, so we can calculate with it as a constant number.

$$MR = \frac{wheeltravel}{springtravel} \approx 1.1 \tag{3}$$

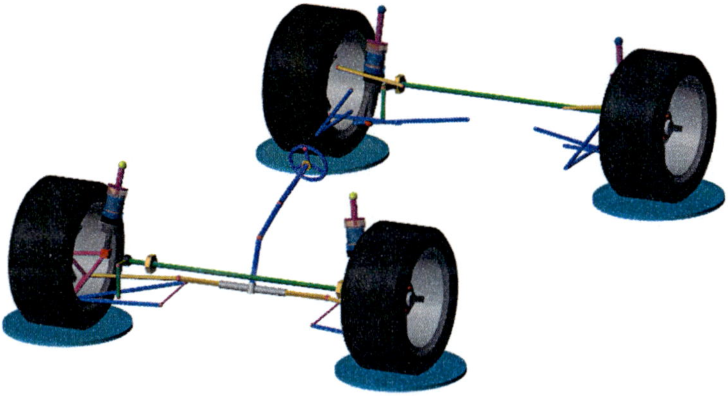

Fig. 1. Kinematic model of the real TCR car.

During the first measurement, we get a result that damper potential meter is set to measure wheel travel - it was calibrated this way. Therefore, if we see damper potential meter channels in logged data, we see wheel travel values.

With this information, we can collect damper potential meter values for calculating downforce. We have to see wheel travel values in two conditions: δ_0 (in [mm]) values at constant low speed in pitlane (not in standing position with zero speed because of the friction of dampers) and δ_{HS} (in [mm]) at high speed straights when there is no longitudinal and lateral acceleration and the ground of the track is quasi flat (no slope, no uphill, no bump). The difference between δ_{HS} and δ_0 is the damper travel because of downforce. Downforce on one wheel can be calculated (in [N]) by wheel stiffness which is K_s spring stiffness (in [N/mm]) divided by square of MR motion ratio at each corner of the car (FL, FR, RL, RR).

$$Downforce = (\delta_{HS} - \delta_0) \cdot \frac{K_s}{MR^2} \tag{4}$$

At the end we can calculate C_L downforce coefficient as it is in Eq. (5):

$$C_L = \frac{\left(\sum_{i=1}^{4} Downforce_i\right)}{0.5\rho A v^2} \tag{5}$$

where i means the corner of the car (FL, FR, RL, RR), ρ is air density in [kg/m^3], A is the frontal area (in [m^2] based on CAD model measurement) and v is the speed of the vehicle in [m/s].

Analyzing a lot of log data at many tracks we got an average downforce coefficient of the car, 0.44 [-], which is the same number in TCR Car manual [5].

3 Validation of the Model

To investigate improving laptime and efficiency we have to create a valid model of our touring car with real parameters of:

- Weight of the car
- Drag coefficient
- Downforce coefficient
- Frontal area
- Air density
- Tyre radius
- Rolling resistance
- Longitudinal friction
- Lateral friction
- Gear and final ratios
- Engine map

The aero and mechanical parameters of our model are based on measurements and TCR car manual, and engine map is based on logged data. The validation results of laptime and energy spent can be seen in Table 2:

Table 2. Simulation vs. log data - laptimes and energy spent track by track

Track	Laptimes		Energy spent [kJ]	
	Simulation	Log data	Simulation	Log data
Monza	1:58.556	1:59.422	18896	20529
Barcelona	1:54.159	1:54.169	15840	15434
Hungaroring	1:52.874	1:53.561	14432	14586
Spa	2:33.425	2:30.739	21513	20874
Assen	1.44:263	1:45.275	15572	14072
Paul Ricard	2:14.124	2:14.118	18808	19048

The precision of the model can be seen on Figs. 2, 3 and 4, presenting the speed and acceleration graphs:

3.1 Energy Calculation – Drag and Rolling Resistance

Energy spent of log data in Table 2 was calculated from vehicle speed channel of a real TCR car. Energy spent consists of 3 major parts:

- Rolling resistance energy
- Drag energy
- Kinetic energy

This method does not calculate with the efficiency of drivetrain, because in Optimum Lap we have a point mass model, and we want to compare the real log data to simulations.

First of all, we have created math channels based on the measured parameter of the car. The power of rolling resistance and drag can be calculated by Eq. (6) and (7).

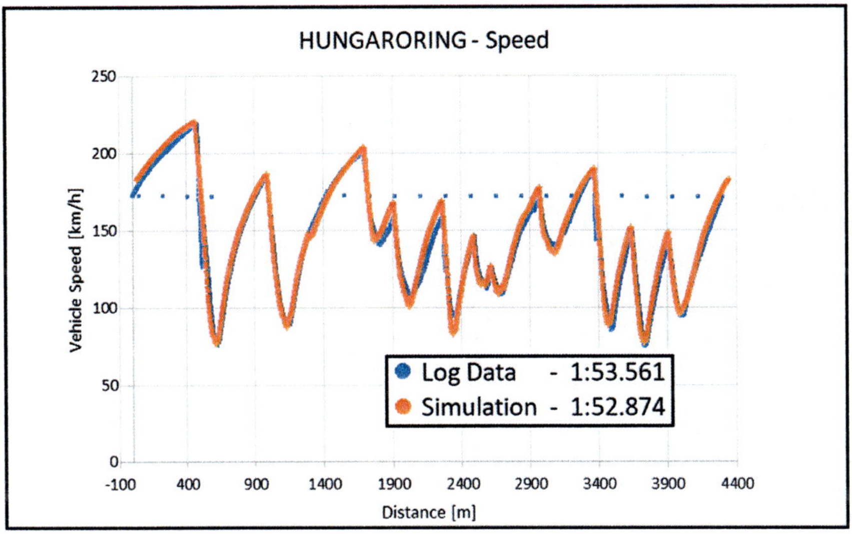

Fig. 2. Simulation vs. log data – Hungaroring (speed)

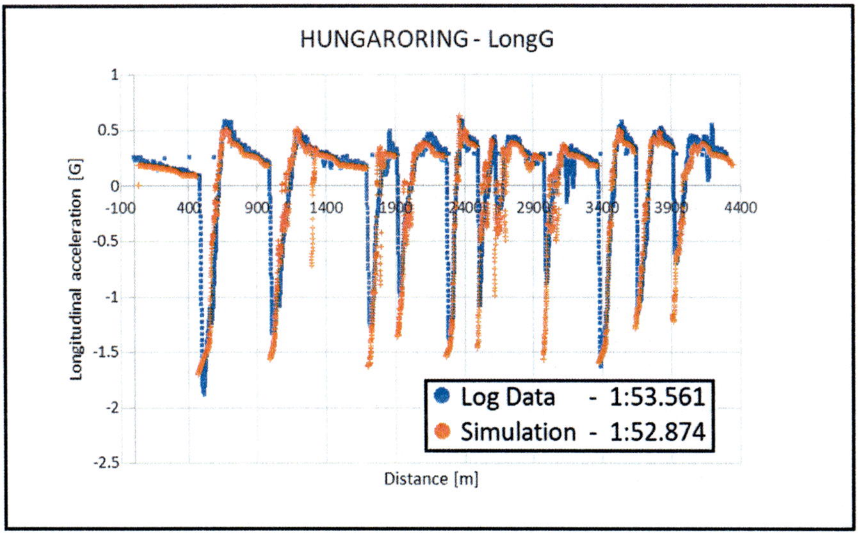

Fig. 3. Simulation vs. log data – Hungaroring (longitudinal acceleration, + traction, – brake)

These equations are correct in case of calculating energy spent, if the car is accelerating, so if the longitudinal acceleration is more than 0.01 G.

$$P_{RollRes} = F_{res} \cdot v = C_{res} \cdot \left(0.5 \cdot \rho A C_L v^2 + mg\right) \cdot v \text{ [4]} \tag{6}$$

$$P_{Drag} = 0.5 \cdot \rho A C_D v^3 \tag{7}$$

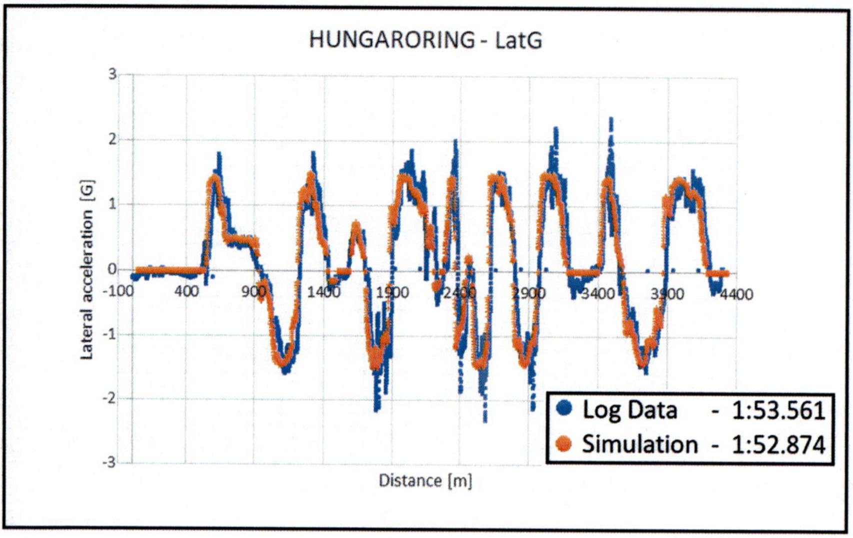

Fig. 4. Simulation vs. log data – Hungaroring (lateral acceleration, + to right, – to left)

In Eq. (6) and (7) parameters were.

- $\rho = 1.2\left[\frac{\text{kg}}{\text{m}^3}\right]$	air density
- $A = 2.3\left[\text{m}^2\right]$	frontal area
- $C_L = 0.44\,[-]$	downforce coefficient (from result in Sect. 2.2)
- $C_D = 0.41\,[-]$	drag coefficient (from [5])
- $C_{res} = 0.033\,[-]$	rolling resistance (from result in Table 1)
- $m = 1300\,[\text{kg}]$	vehicle mass
- $g = 9.81\left[\frac{\text{m}}{\text{s}^2}\right]$	standard gravity
- $v\,[\text{m/s}]$	vehicle speed channel

If we integrate these, we get energy:

$$W_{RollRes} = \int P_{RollRes}\, dt \tag{8}$$

$$W_{Drag} = \int P_{Drag}\, dt \tag{9}$$

3.2 Energy Calculation – Kinetic Energy

The third one of the energies is kinetic energy, which can be calculated by the vehicle speed channel (see Figs. 6 and 7). The process of the calculation of energy is shown for the Hungaroring circuit. Parts of the track where the car is accelerating have to be chosen

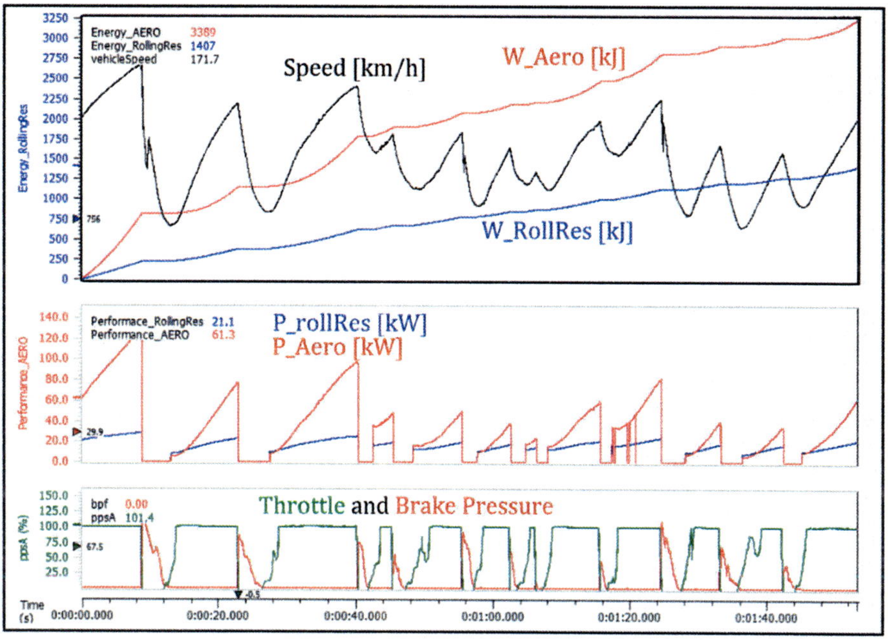

Fig. 5. The energy of rolling resistance and drag during one lap at Hungaroring. As you can see, power of drag and rolling resistance is calculated if the car is accelerating with throttle. The energy which went to overcoming the rolling resistance was 1407 kJ, and the energy which went to overcoming the aerodynamic resistance (drag) was 3389 kJ.

(see Fig. 8), and between these points, we get the maximum and minimum speed. Delta kinetic energy can be calculated from the starting and end speed of the phase.

$$\Delta E_{kin,(i)} = \frac{1}{2} \cdot m \cdot \left(v_{max,(i)}^2 - v_{min,(i)}^2 \right) \tag{10}$$

An example can be seen in Fig. 6, where the first part is from turn1 to turn2, and the second part is from turn2 to turn4 (as it can be seen in Fig. 8 too).

In the end, we can summarize these kinetic energies to the whole lap.

$$E_{kin} = \sum_i \Delta E_{kin,(i)} \tag{11}$$

To summarize the energies of one lap at Hungaroring, as you can see in Fig. 5 and Table 3:

- Rolling resistance 1 407 kJ
- Aero drag 3 389 kJ
- Kinetic Energy 9 790 kJ
- SUM Energy 14 586 kJ

This number can be seen in Table 2 with comparing to the simulation result of 14432 kJ value, the error is 1%.

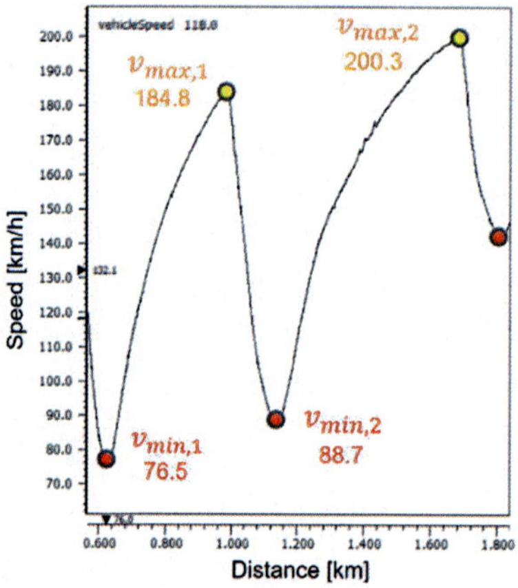

$$\Delta E_{kin,1} = \frac{1}{2} m\left(v_{max,1}^2 - v_{min,1}^2\right) = \frac{1}{2} \cdot 1300 \cdot \left(\left(\frac{184.8}{3.6}\right)^2 - \left(\frac{76.5}{3.6}\right)^2\right)$$

$$\Delta E_{kin,2} = \frac{1}{2} m\left(v_{max,2}^2 - v_{min,2}^2\right) = \frac{1}{2} \cdot 1300 \cdot \left(\left(\frac{200.3}{3.6}\right)^2 - \left(\frac{88.7}{3.6}\right)^2\right)$$

Fig. 6. Calculating kinetic energy from vehicle speed channel as the car is accelerating from 76.5 km/h to 184.8 km/h at the first part, and from 88.7 km/h to 200.3 km/h in the second part.

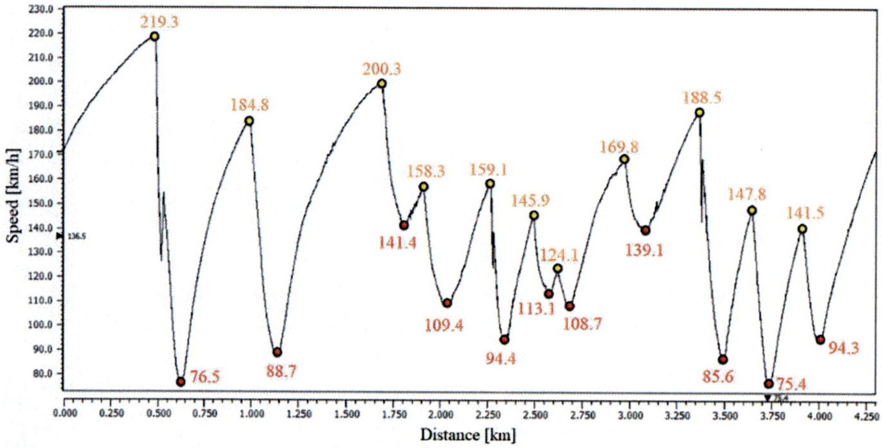

Fig. 7. Hungaroring - speed channel - Calculating kinetic energy from vehicle speed channel for the whole lap. During calculation Eq. (10) is used and the results can be seen in Table 3.

Of course, the calculation above was done to other circuits, as well. Summarize, the energy consumption can be seen in Table 4 regarding to log data of other tracks:

4 Effect of Drag Coefficient

After we have a valid model, we can use laptime simulation for information on parameter sensitivity of drag to laptime and energy spent.

The software we used is Optimum Lap. It is calculating with point mass model, but it can be very useful if we want to know which parameters affect the most on laptime (and efficiency) track by track.

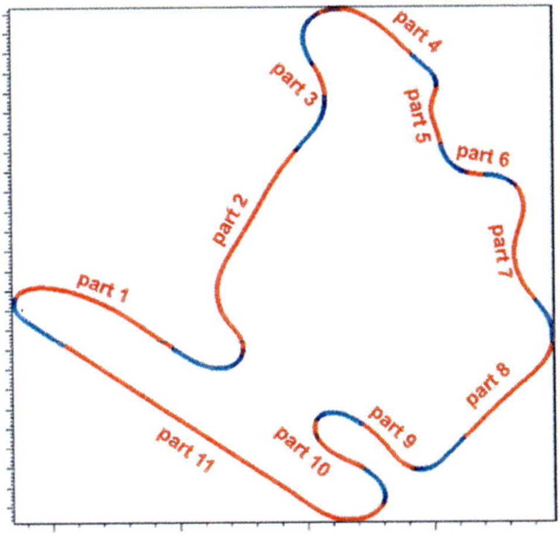

Fig. 8. Hungaroring trackmap. Energy is calculated when the car is accelerating with throttle (red lines on the graph). According to it, this track can be divided into 11 parts.

Table 3. Calculating kinetic energy according to Fig. 7–8 and Eq. 10 ($m = 1300$ [kg])

	v_{min} [km/h]	v_{max} [km/h]	ΔE_{kin} [kJ]
Part1	76.5	184.8	1419
Part2	88.7	200.3	1618
Part3	141.4	158.3	254
Part4	109.4	159.1	669
Part5	94.4	145.9	621
Part6	113.1	124.1	131
Part7	108.7	169.8	853
Part8	139.1	188.5	812
Part9	85.6	147.8	728
Part10	75.4	141.5	719
Part11	94.3	219.3	1966
	SUM Kinetic Energy:		9790

4.1 Only Changing Drag Coefficient

We have done some simulations changing C_D drag coefficient to -4.5%, -3%, -1.5%, Base, $+1.5\%$, $+3\%$ and $+4.5\%$. We wanted to know if we decrease drag how will the laptime improve and how much less energy will be used to complete one lap. The results can be seen in Table 5:

The information in Table 5 can be summarized in a graph, which can be seen in Fig. 9:

Table 4. Energy consumption parts of real log data at TCR Europe events.

Track	$W_{RollRes}$ [kJ]	W_{Drag} [kJ]	E_{kin} [kJ]	SUM energy [kJ]	Fuel cons. [kg]	Laptime [s]
Monza	2231	8027	10271	20529	1.980	1:59.422
Barcelona	1613	4405	9416	15434	1.748	1:54.169
Hungaroring	1407	3389	9790	14586	1.628	1:53.561
Spa	2624	8892	9358	20874	2.510	2:30.739
Assen	1564	4433	8075	14072	1.648	1:45.275
Paul Ricard	2006	6169	10873	19048	2.070	2:14.118

Table 5. Simulation results – Effect of drag coefficient to laptime and used energy

Track	C_D [-]		Laptime [s]	Used energy [kJ]	Track	C_D [-]		Laptime [s]	Used energy [kJ]
Monza	−4.5%	0.392	−0.348	−97	Barcelona	−4.5%	0.392	−0.163	−60
	−3%	0.398	−0.232	−64		−3%	0.398	−0.108	−39
	−1.5%	0.404	−0.116	−34		−1.5%	0.404	−0.054	−17
	BASE	0.41	1:58.556	18 895		BASE	0.41	1:54.159	15 840
	+1.5%	0.416	+0.117	+34		+1.5%	0.416	+0.054	+20
	+3%	0.422	+0.233	+72		+3%	0.422	+0.108	+50
	+4.5%	0.428	+0.350	+99		+4.5%	0.428	+ 0.163	+73
Hungaroring	−4.5%	0.392	−0.111	−55	Spa - Francorchamps	−4.5%	0.392	−0.310	−124
	−3%	0.398	−0.074	−33		−3%	0.398	−0.207	−83
	−1.5%	0.404	−0.037	−16		−1.5%	0.404	−0.104	−37
	BASE	0.41	1:52.874	14 432		BASE	0.41	2:33.425	21 513
	+1.5%	0.416	+0.037	+21		+1.5%	0.416	0.104	+27
	+3%	0.422	+0.074	+41		+3%	0.422	0.209	+66
	+4.5%	0.428	+0.111	+62		+4.5%	0.428	0.313	+100
Assen	−4.5%	0.392	−0.150	−73	Paul Ricard	−4.5%	0.392	−0.240	−117
	−3%	0.398	−0.107	−53		−3%	0.398	−0.160	−68
	−1.5%	0.404	−0.053	−26		−1.5%	0.404	−0.080	−35
	BASE	0.41	1:44.263	15 571		BASE	0.41	2:14.124	18 808
	+1.5%	0.416	+0.053	+23		+1.5%	0.416	+0.081	+27
	+3%	0.422	+0.107	+47		+3%	0.422	+0.161	+66
	+4.5%	0.428	+0.160	+72		+4.5%	0.428	+0.242	+93

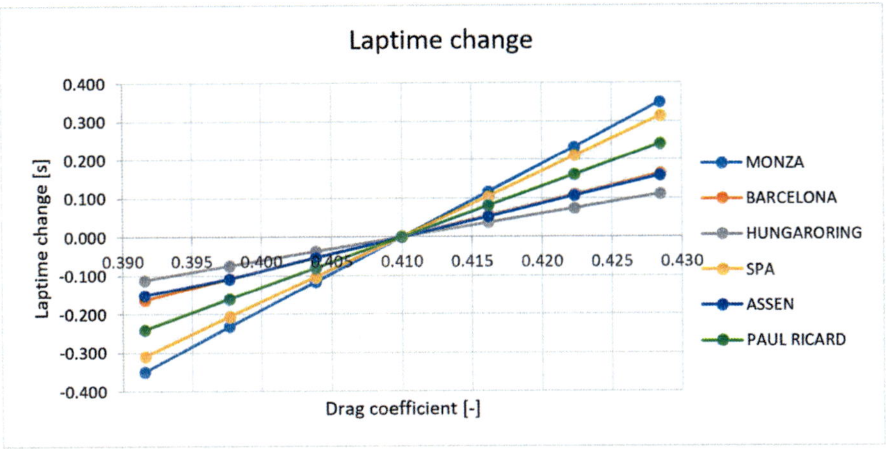

Fig. 9. Laptime change in function of C_D drag coefficient

As we expected, drag has more effect on faster tracks, such as Monza and Spa-Francorchamps. 4.5% reduction of drag (or frontal area) can be result 3 tens in these tracks, so there, drag reduction is the most important object in touring car racing series. It can be reached by flat car with ride heights or as flat wing position as possible.

4.2 Changing Drag Coefficient and Engine Map for Equal Energy Spent

As you can see in Table 5, with reducing drag we could have less energy to complete the race. In case of electric cars, the limiting factor is the amount of total energy we can use. In the next simulations, we have constant energy we can use per lap in case of every drag coefficients. So if we reduce drag and we use less energy, we will add plus power to the engine in order to use the same amount of energy. In the same way, if we increase drag we have to decrease the power of the engine to use equal energy.

According to Table 2, in our simulation we have.

18 895 kJ energy in Monza,
15 840 kJ energy in Barcelona,
14 432 kJ energy in Hungaroring,
21 513 kJ energy in Spa,
15 571 kJ energy in Assen, and
18 808 kJ energy in Paul Ricard.

The results of constant energy simulations can be seen below in Table 6:
The information in Table 6 can be visualized in a graph which can be seen in Fig. 10:

Table 6. Simulation results – Effect of drag coefficient to laptime in case of constant energy

Track	C_D [%]	Engine power scaling factor [%]	Laptime [s]	Used energy [kJ]	Track	C_D [%]	Engine power scaling factor [%]	Laptime [s]	Used energy [kJ]
Monza	−4.5%	100.805	−0.524	18895	Barcelona	−4.5%	100.677	−0.242	15840
	−3.0%	100.513	−0.345			−3.0%	100.441	−0.161	
	−1.5%	100.282	−0.178			−1.5%	100.226	−0.081	
	BASE	100	1:58.556			BASE	100	1:54.159	
	+1.5%	99.733	+0.176			+1.5%	99.713	+0.089	
	+3.0%	99.513	+0.341			+3.0%	99.436	+0.177	
	+4.5%	99.262	+0.514			+4.5%	99.19	+0.262	
Hungaroring	−4.5%	100.726	−0.18	14432	Spa - Francorchamps	−4.5%	100.903	−0.491	21513
	−3.0%	100.495	−0.121			−3.0%	100.63	−0.334	
	−1.5%	100.226	−0.059			−1.5%	100.315	−0.167	
	BASE	100	1:52.874			BASE	100	2:33.425	
	+1.5%	99.723	+0.064			+1.5%	99.783	+0.149	
	+3.0%	99.467	+0.126			+3.0%	99.544	+0.303	
	+4.5%	99.267	+0.184			+4.5%	99.277	+0.463	
Assen	−4.5%	100.79	−0.264	15571	Paul Ricard	−4.5%	101.089	−0.431	18808
	−3.0%	100.574	−0.183			−3.0%	100.726	−0.288	
	−1.5%	100.308	−0.095			−1.5%	100.388	−0.149	
	BASE	100	1:44.263			BASE	100	2:14.124	
	+1.5%	99.723	+0.091			+1.5%	99.723	+0.13	
	+3.0%	99.503	+0.175			+3.0%	99.425	+0.264	
	+4.5%	99.262	+0.262			+4.5%	99.067	+ss0.41	

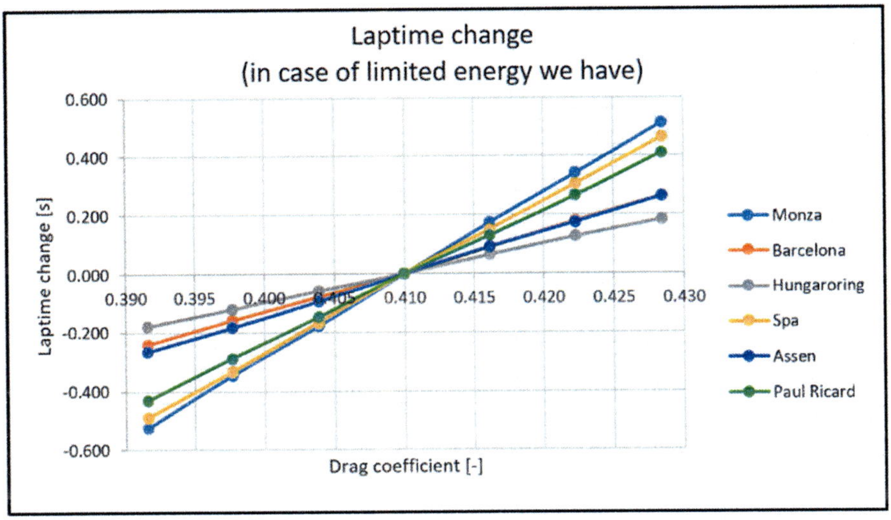

Fig. 10. Laptime change in function of C_D drag coefficient in case of the same energy we have

5 Conclusion

The drag coefficient has more effect in high speed circuits, such as Monza and Spa-Francorchamps. According to Table 5 and Fig. 9, with improving the chassis drag coefficient with 4.5% we can be quicker with 0.350 s which can be mean many places on the grid in touring car racing.

Comparing the results of Fig. 9 and Fig. 10 we can see, drag has more effect in case of limited energy. So in electric touring car racing, it will be much more effective on competitiveness. According to Table 6 and Fig. 10, with improving the chassis drag coefficient with 4.5% we can be quicker with 0.524 s. Thus, electric touring car chassis designers should take attention to reduce the drag of the car as much as possible to reduce laptime.

Acknowledgement. The research presented in this paper was carried out as part of the EFOP-3.6.2-16-2017-00016 project in the framework of the New Széchenyi Plan. The completion of this project is funded by the European Union and co-financed by the European Social Fund.

References

1. Herrmann, T., Christ, F., Betz, J., Lienkamp, M.: Energy management strategy for an autonomous electric racecar using optimal control. In: 2019 IEEE Intelligent Transportation Systems Conference (ITSC), Auckland, New Zealand, pp. 720–725 (2019)
2. Nielson, G., Emadi, A.: Hybrid energy storage systems for high-performance hybrid electric vehicles. In: 2011 IEEE Vehicle Power and Propulsion Conference, Chicago, IL, pp. 1–6 (2011)
3. Zomotor, Á.: Gépjármű menetdinamika (2004)
4. Milliken, W.F., Milliken, D.L.: Race car vehicle dynamics (1995)

5. Hyundai i30 N TCR User Manual
6. Dayman, B.: Tire rolling resistance measurements from coast-down tests. SAE Technical Paper 760153 (1976)
7. Preda, I., Covaciu, D., Ciolan, G.: Coast Down Test – Theoretical and Experimental Approach (2010). https://doi.org/10.13140/RG.2.1.4048.5925

Simulation Environment Developed for Advanced Suspension Design Methods

Gergely Szűcs(✉) and Gergely Bári

John Von Neumann University, Izsáki út 10, Kecskemét 6000, Hungary
szucsi.90.07.09@gmail.com

Abstract. In the 20th century, a lot of methods were based on thumb rules in suspension design, a lot of terminologies were introduced, such as virtual swing arm, roll center, anti-dive, anti-squat, anti-lift. With these concepts, suspension design became easier to understand and visualize, also quantifying the so called anti-geometries and creating graphs about cambergain, roll center height, or toe gain in function of wheel travel became more straightforward.

On the one hand, with the raise of computers, more and more tools became available for suspension design, and the motion of suspension can be calculated correctly and quickly. On the other hand, there is more information from tyres by measurements, so suspension design can focus on to maintain suspension values where tyres produce the highest grip.

In this paper, a novel suspension design method is shown, based on in-plane acceleration direction (G-G direction), instead of the aforementioned characteristic-based methods. Also, a simulation environment is created that assists this method. In this environment, a multi-body suspension model is used that allows suspension kinematics and joint forces to be calculated as a function of the direction of in-plane acceleration, while in the past suspension parameters were calculated as function of wheel-to-chassis relative displacement. Acceleration and steering input from real, measured log data can also be added to this model, and normal forces at the tyre contact patches can be calculated too.

Keywords: Suspension design · Design method · Racecar · Simscape · Grip limit behavior

1 Introduction

Since the early appearance of cars, at the end of the 19th century, suspension design methods were improving both in terms of kinematics and spring-damper calibration. There are a lot of papers about suspension design methods, that target how one should design a "good" suspension [1, 2, 3, 4]. But what does "good" suspension mean?

There are a lot of thumb rules in suspension design.

Some of suspension kinematic investigation is doing parallel wheel travel test, to know how the suspension works in case of bumps. They plot the graphs of camber, toe, roll center height, motion ratio, scrub radius, mechanical trail, kingpin angle, etc.

K. Jármai and K. Voith (Eds.): VAE 2020, LNME, pp. 349–364, 2021.
https://doi.org/10.1007/978-981-15-9529-5_31

The second common method is the opposite wheel travel test. In case of it, one-wheel bumps and the other one rebounds to simulate chassis roll. Chassis roll test is quasi the same. The difference is at the first method chassis is fixed and the ground moves under the wheels. At the second method, the ground is fixed under the wheels and the chassis is rolling during the test. The geometry is investigated in function of chassis roll or wheel travel.

The third common method is steering test. First, to design the Ackermann geometry [3, 5, 6]. Second, camber gain can be sensitive to steering especially in case of higher kingpin and caster angle, so it used to investigate that too [3]. Third, by kingpin and caster angle during steering, we can change the ride height and normal forces too [3, 6].

During these tests a lot of suspension parameters are investigated. A few of them are summarized below.

1.1 Main Parameters

Camber Gain. Almost all papers suggest that camber-gain should be more negative in wheel bump. The idea is that the outer wheel will jounce, and the inner wheel will rebound relative to the chassis in turns. By chassis roll out of a corner, the wheel angle relative to the ground must be compensated, so the change of camber remains in the optimal window.

Until the end of the last century there was no data from tyres, so the optimal camber was unknown, but a rule of thumb was that higher camber means more grip. Nowadays, by tyre measurements, we know optimal camber to each tyres (if we measure it) but it can change as a function of normal force, temperature, pressure, etc. [7, 8]. For example, if the optimal camber value is known to the chosen tyre, suspension can be designed to maintain the camber in optimum in every situation on the so called G-G diagram. But how can we know how much the chassis roll and wheel movements are on the track? Moreover, it depends on roll center position and roll center movement too. The amount of it depends on springs, dampers and anti-roll bars too. Are wheel travel test, chassis roll test, and steering test enough to tune camber gain? Is it enough to know the camber gain in case of wheel travel and steering? For answers, new methods can be developed and used more efficiently.

Roll Center. The roll center is the point in front view which the chassis rolls around [6]. The roll center of a vehicle is the notional point at which the cornering forces in the suspension are acted to the vehicle body. We can draw this point if we know suspension pickup points. In most cases, we want to know the movement of roll center to calculate chassis roll and jacking force [6]. But without knowing roll center position, if we know the chassis roll, each wheel travels, ride heights and jacking force from multibody suspension simulation, do we need to know roll center position?

Roll Center is a theoretical point, it makes the visualization easier for chassis roll, but it is not necessary to know if we know weight transfers, ride height changes, chassis roll, etc.

Anti-geometries. Anti-squat, anti-dive and anti-lift geometries - which will be referred to as "anti-geometry" - are a form of geometry at the front and rear wheels that alters and controls the amount that a car will compress the springs due to acceleration or braking conditions [6]. The load path of weight transfer can go in two ways: Through suspension rods (called geometric weight transfer) and through springs, dampers and anti-roll bars (called elastic weight transfer) [6]. Anti-geometries define - in percent - how much of weight transfer goes through suspension rods and springs. Therefore, these parameters effect the chassis pitch and ride height change as well, which has a high effect in the case of aerodynamic cars. Low anti values mean more elastic weight transfer, so chassis pitch and ride height change are higher [6].

So, during suspension design, we speak about anti-geometries because of chassis pitch during brake and acceleration, but the question is, do we need to know anti numbers if we know ride height changes (and chassis pitch) during longitudinal acceleration or deceleration? Anti-geometries are theoretical values such as roll center. These are used to estimate chassis movements.

Scrub Radius and Mechanical Trail. Scrub radius and mechanical trail are defined as the front- and side view distance between the intersection of the steering axis and the ground measured from the center of the contact patch [6]. Steering torque is proportional to these numbers, for example, higher mechanical trail means more steering torque in case of lateral force at the tyre contact patch [6].

So, knowing scrub radius and the mechanical trail we can estimate steering torque and tie rod forces. From the steering point of view, the important information is not the value of scrub radius or mechanical trail. The question is how much is the steering torque? We can answer it with multibody dynamics suspension analysis too.

Motion Ratio. Motion ratio can be defined between damper or spring travel and wheel displacement, moreover, it can be defined between anti roll bar motion and wheel travel too. With these numbers, we can calculate the spring or anti-roll bar stiffness to wheel rate to know how stiff our suspension or how much is the roll rate distribution between front- and rear axle to estimate elastic weight transfer [6].

To know weight transfer at axles, we do not need to know motion ratios in case of multibody simulations, without these we can get normal forces, roll gradient, ride height and spring deflection too.

Knowing the parameters above is not necessary, but it can be useful during suspension design. It is a higher abstract level, and across tuning these parameters we can get a chassis and wheel movement we want.

1.2 Frequently Asked Questions During Suspension Design

If a car is going on the track, the aim is in motorsport to reduce the lap time as low as possible. To help the driver to achieve the fastest lap we can tune the car with a lot of parameters. These parameters will affect the behavior of the car. With a "good" set of parameters, the handling can be easier and summarized grip can be higher.

The "best" setup is when we can get the maximum possible forces from tyres, so if we want the "best" suspension with "best" setup we must know the tyre we have according to measurements [9]. It is necessary to know which parameters the tyre "likes" to produce maximum forces. What is the optimal slip angle and camber angle at each normal force?

The second thing to get the best setup, if we know how the car is moving on the track. For example, what are the cambers, ride heights and normal forces at each wheel during braking or cornering?

Another question from drivers, can we put a softer spring to the front? With softer springs, wings (in case of formula cars) or splitters (in case of touring cars) can touch the ground in case of braking or cornering. Often, it is hard to answer even in that case if we know motion ratio, wheel rates, anti-dive and a lot of parameters in function of wheel travel.

In case of prototype cars, a frequent question is how much the forces in rods and bearings are. It is necessary to know if we want a lightweight racecar.

For answering these and similar questions we built a simulation environment in Simscape multibody [10].

2 Simulation Environment

For suspension design and investigation, a 3D simulation environment was built in Simscape Multibody. The model input is steering angle and acceleration in lateral and longitudinal direction too.

There is no tyre model, forces are added to the contact patches. As it can be seen in Fig. 1, in case of lateral acceleration, lateral forces are properly distributed according to normal forces of the previous step of the solver. In case of longitudinal acceleration, the longitudinal forces are distributed according to brake balance or acceleration balance to front and rear which can be added by us.

There is only one constrain which connects the whole model to the world. Rotation around vertical axis is zero maintaining yaw movement at zero. The model is held in

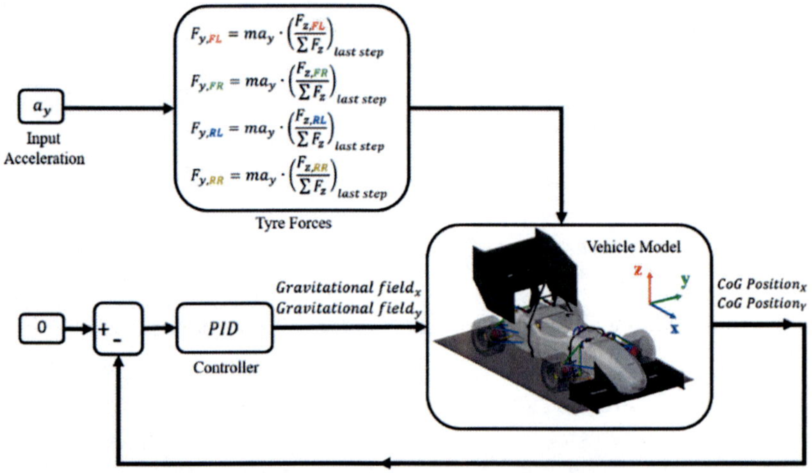

Fig. 1. The model in case of lateral acceleration input. An input acceleration is multiplied with the mass of vehicle to calculate the sum lateral force. It is distributed according to normal forces on each tyres. The model is held in position by PID controller to control the gravitational field representing inertial forces.

longitudinal and lateral direction by the gravitational field, which is constant in vertical direction ($g = -9.81\,\mathrm{m/s^2}$), and it is position controlled by PID controller in longitudinal and lateral direction representing inertial forces.

Besides the 3 main input (lateral and longitudinal accelerations and steering angle) there can be a 4th input. A vehicle input and downforce and drag coefficient can be added to the model to calculate aerodynamic forces. These are external forces that effect the chassis at the front and rear axles. These coefficients can be added as an aero map, in function of ride height at front and rear.

The model consists of 3 main parts, which can be seen in Fig. 2: Rig, Vehicle Model and Test Case boxes. The input accelerations and steering are added in Test Case box and these are transformed to contact patch forces on tyres. These forces are added to the Rig at each contact patch. Contact patches are connected with the vehicle model, so the model will move, but the position of it is going to the Test Case box, where the PID controller calculates the gravitational field to maintain the position of the model in longitudinal and lateral direction.

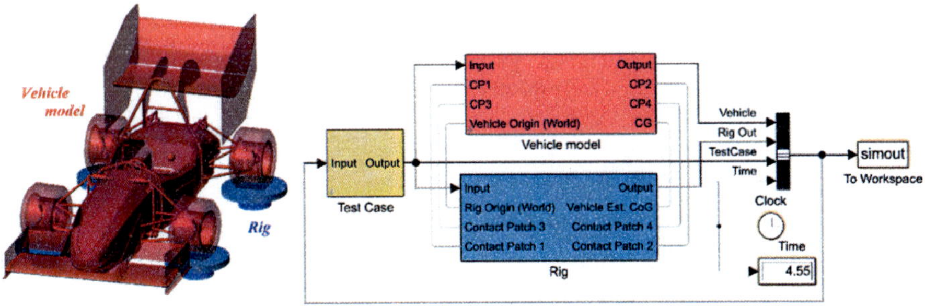

Fig. 2. The simulation environment

2.1 Rig Box

The Rig can be connected to any vehicle model by the center of gravity (CoG) and the tyres contact patches. The rig contains the tyre contact patch model and the chassis constrain to the rig too.

Contact Patch. The logic of tyre contact patch can be seen in Fig. 3. The contact patch of the ground and the wheel is connected with 3 joints. The rectangular joint allows moving the tire contact patch in x and y directions. In case of steering, or track width change in case of bump it is necessary.

The rectangular joint does not allow the movement in z direction. This task is solved by Prismatic joint. It has a stiffness and damping coefficient representing the tyre vertical stiffness and damping coefficient. If the deflection of prismatic joint is multiplied by the stiffness, we get the normal force of the tyre.

At last, a Gimbal joint is needed too because the angle of the wheel to the ground should be changed in case of steering and camber gain. Lateral and longitudinal forces are added after this joint to wheel contact patch as an external force. The contact patch of the vehicle model is connecting to "*Contact Patch Wheel*" in Fig. 3.

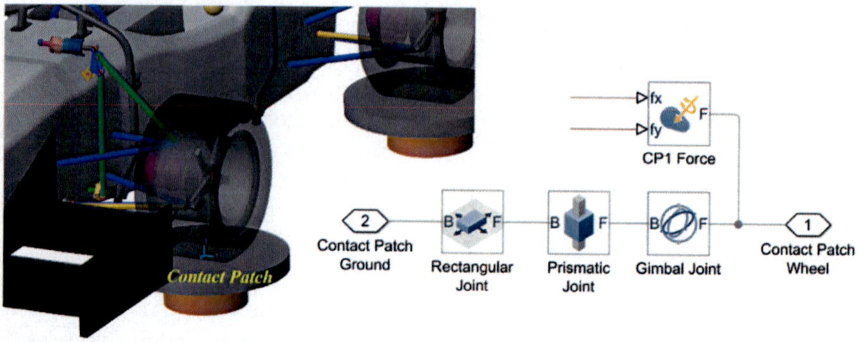

Fig. 3. Modelling of the tyre contact patches. If we do not want any "tyre deflection", prismatic joint can be removed.

Chassis. As it was mentioned, the CoG of the chassis is constrained to the world. Rotation around vertical axis (z) called yaw movement is maintained at zero. This constrain should allow every movement except yaw movement. There is no joint for only keeping the rotation angle at zero around z axis, so for this task, there should be 3 joints. The joints between chassis and the ground can be seen in Fig. 4.

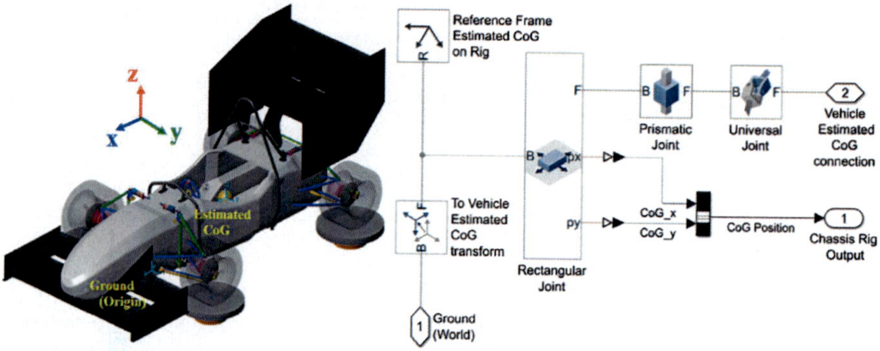

Fig. 4. Modelling of the chassis constrain, maintaining the yaw movement at zero

At first, a Rectangular joint is needed to allow chassis movement in longitudinal (x) and lateral (y) direction, which is controlled by PID controller. For PID, the position of the vehicle CoG should be measured by sensor relative to the estimated CoG on Rig.

At second, a Prismatic joint ensures the CoG height variation in z direction. At last, a Universal joint is needed. It allows the rotations around x and y axis, representing roll and pitch movements of the chassis.

2.2 Vehicle Model Box

The Vehicle model is a double A-arm [2, 3] suspension model, which can be multilink [11] too. In the model, we can define the pickup points, spring and anti-roll bar stiffness, damping characteristic, and geometries of rocker, upright, hub or the position of bearings of anti-roll bar, hub or rocker and so on.

The Vehicle model is connected to the Rig across contact patches and chassis CoG. The model consists of 8 main parts. Four of them are the 4 suspension corner which contains the wheel, hub, upright, brake disc, suspension rods, rocker and spring-damper assembly. Two of eight are the front and rear anti-roll bar. The remaining two parts are the chassis and steering system. The schematic of the model can be seen in Fig. 5.

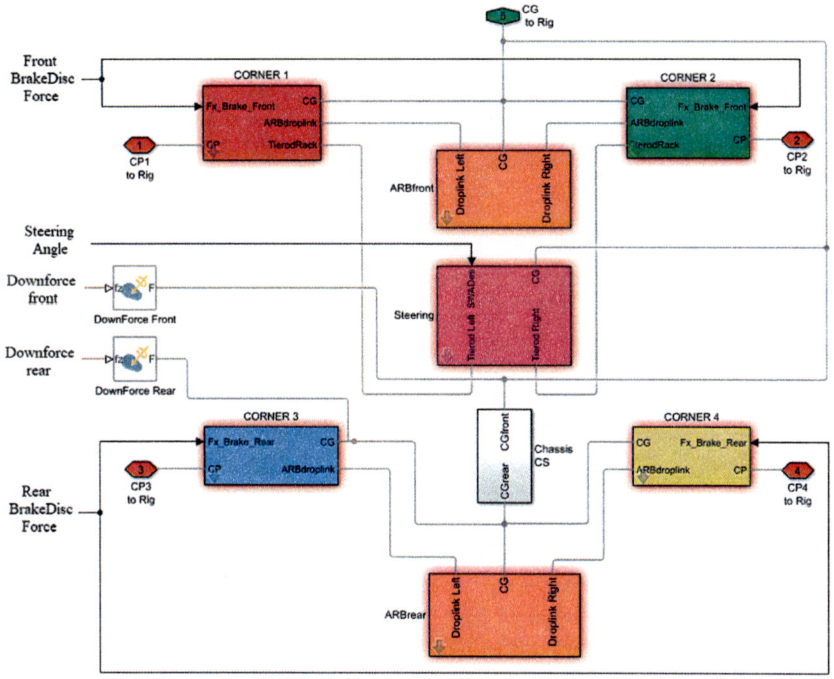

Fig. 5. Vehicle model

Corners. The corner contains the wheel, hub, upright, brake disc, suspension rods, rocker and spring-damper assembly. The corner is connected to the rig across contact patch. Another connection is to the chassis with its CoG. There are transforms from CoG to the pickup points, and these vectors can be defined in each corner, as a "mask" in Simscape [10]. If we want, we can connect an anti-roll bar droplink to the rocker and the upright can be connected to a steering system at the front or to the chassis at the rear.

There is an input force to each corner. In case of braking the upright is loaded by contact patch force and brake force on caliper too. This brake force value is estimated by the moment of the wheel which is equal to the F_x contact patch force multiplied with tyre radius. The caliper modelled as a brick with yellow in Fig. 6.

The wheel, hub, and brake disc are together which are connected to the upright by two bearing joints. One of them is constrained by axial movement avoiding upright to move in lateral relative to the wheel assembly. The upright is in connection with all suspension rods by universal joints. These allow rotation, except the rod axial direction. Corner connects by rods to the chassis with spherical joints, and the rocker with a re-volute joint. The model of the corner with joints can be seen in Fig. 6.

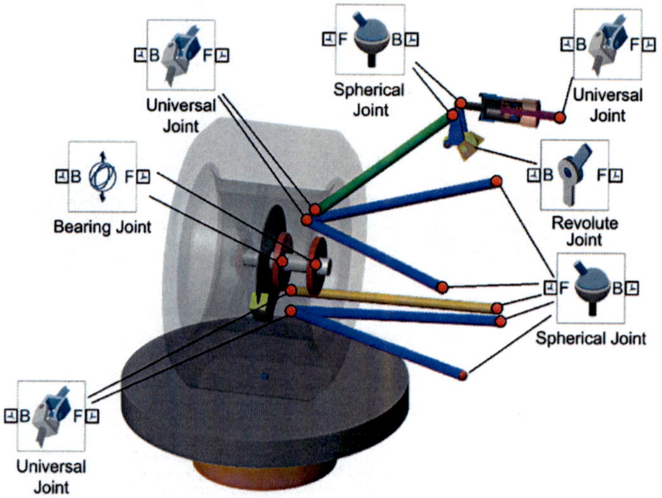

Fig. 6. Model of a corner with its joints

All the joint positions can be added by a coordinate vector, so if we want to modify the geometry, we can do that easily. Rod lengths and any other geometries are automatically modified by writing a new coordinate vector to each joint.

The damper-spring assembly is modelled as a prismatic joint. The position and velocity of the joint go to a look up table, where the spring and damper forces are given in the function of spring deflection and damper velocity. These forces are connected back to the prismatic joint. We can define any spring or damper characteristic we want, even double spring or rebound springs. Of course, spring natural length and preload can be given, moreover, bump- and rebound stop characteristic and damper stroke can be defined too.

Steering System. As it can be seen in Fig. 7, steering system contains the steering wheel, 3 steering shafts, the pinion and the steering rack. The system is connected to the chassis with a cylindrical joint on the lower shaft and with a revolute joint on the upper shaft which has the steering wheel angle input. There is a constrain between rack and pinion which converts the pinion rotation angle to rack travel.

As in a case of corner systems, all the joint positions can be given by a coordinate vector, so by changing the coordinates, rod lengths and rack length are automatically modified.

Anti-roll Bars. An often-used U-type anti-roll bar can be seen in Fig. 8, which is used for increasing the roll stiffness of the car. It contains the torsion bar with blades and

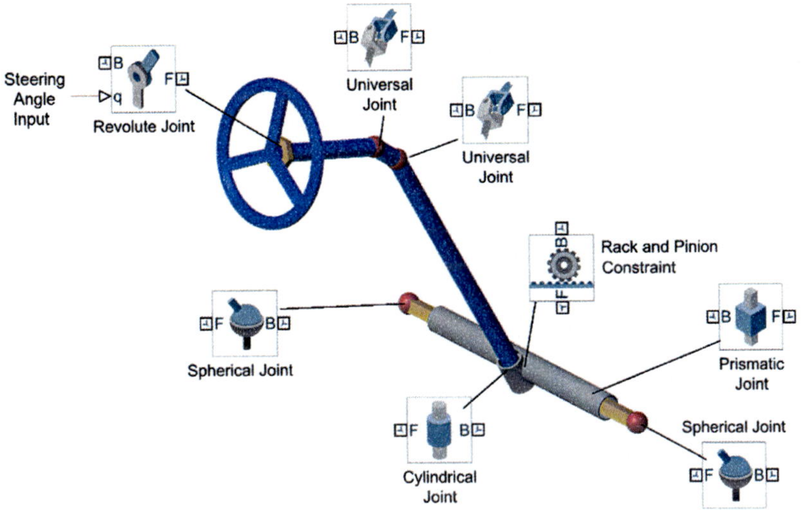

Fig. 7. Model of the steering system with its joints

droplinks. The system is connected to the chassis with a revolute and cylindrical joint that represents the bearings of the torsion bar. The bearing coordinate vector should be given with the length of the torsion bar, blades, and droplinks too. The torsion bar consists of two half parts. These are connected with a revolute joint which has a torsional stiffness, which can be given in the model. Droplink ends can be connected to the rocker of the corner box.

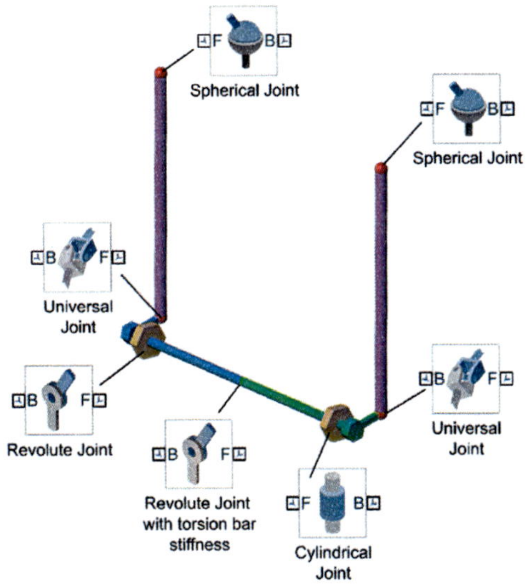

Fig. 8. Model of the anti-roll bar system with its joints

Chassis. The chassis box connects all parts of the vehicle model by its CoG. It is practically a point mass with inertia. The weight and inertia of it can be given to representing all the masses which are suspended mass such as chassis, driver, powertrain and drivetrain systems. Ride height sensor positions can be added to this with its position vectors. There is an option to calculate with chassis torsional stiffness if we define the front and rear parts and connecting them with a torsional stiffness of a revolute joint.

As it can be seen in Fig. 9, for visualization a cad model can be added to the simulation environment which is moving during the simulation.

Fig. 9. The vehicle model without (left) and with (right) a cad model of chassis

2.3 Test Case Box

This box gives the input to the Rig and Vehicle model. The acceleration input is calculated to tyre forces here, and the PID is controlling the gravitational field according to the CoG position data coming from the Rig.

There can be 7 test cases in the model by combining the acceleration, speed and steering wheel angle inputs. These can be switched in the model.

The first test case is the usual bump test. In this case we can give the gravitation field in z direction by a signal builder in function of time. The second possible test case is steering test, where the gravitational field is constant, and the steering input is given by a signal builder. The third can be a vibration test with giving displacement function to the Rig to create a post-rig. With it, we can tune the frequencies and damping ratios. The fourth and fifth opportunity are longitudinal and lateral acceleration test, where we can give the acceleration inputs to a model by a signal builder to test braking or cornering. As the sixth test case we can combine the fourth and fifth case, so we can add the lateral and longitudinal accelerations (or steering input too) and we can simulate the vehicle going through the edge of the GG diagram.

Thinking it further, if we have real log data of accelerations and steering wheel angle, we can read it in Matlab [10] as a csv file, and we can add it to the simulation environment to investigate the kinematics by moving on a real track.

3 Data from Simulation

The vehicle model has a lot of sensors. Forces are measured in all of the joints during simulation, for example, if we put real log data acceleration and steering input to a model, we can see all the suspension rod forces, bearing forces or steering torque too. Moreover, every point coordinate is measured during simulation, so we can look at any suspension kinematic, such as toe angle, camber angle, ride height, CoG position, and roll angle indirectly damper positions and so on.

3.1 Validation

In our example, a double A-arm suspension of a Formula Student [12] racecar is shown. This suspension was designed and built too. During the track speed, accelerations, and steering angle were logged which are the input to our simulation model. Figure 10 shows the PID controlled acceleration field with comparing simulation and real log data.

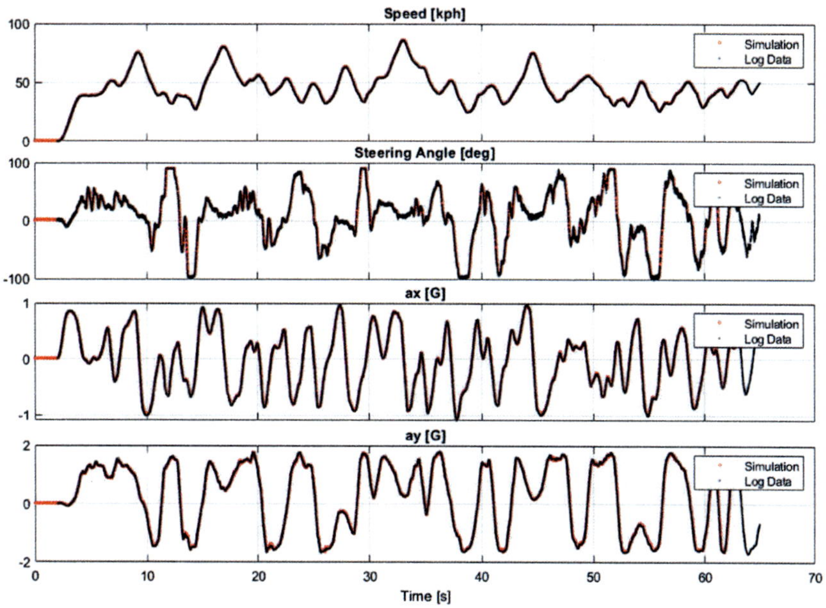

Fig. 10. The real log data as an input for simulation (black) and the simulation input speed, steering angle and accelerations generated by PID control gravity field.

The validation process can be according to damper positions that are measured on the track too. For comparison, the real log data with simulation chassis movement can be seen in Fig. 11. The setup in the simulation was the same in real tests on the track.

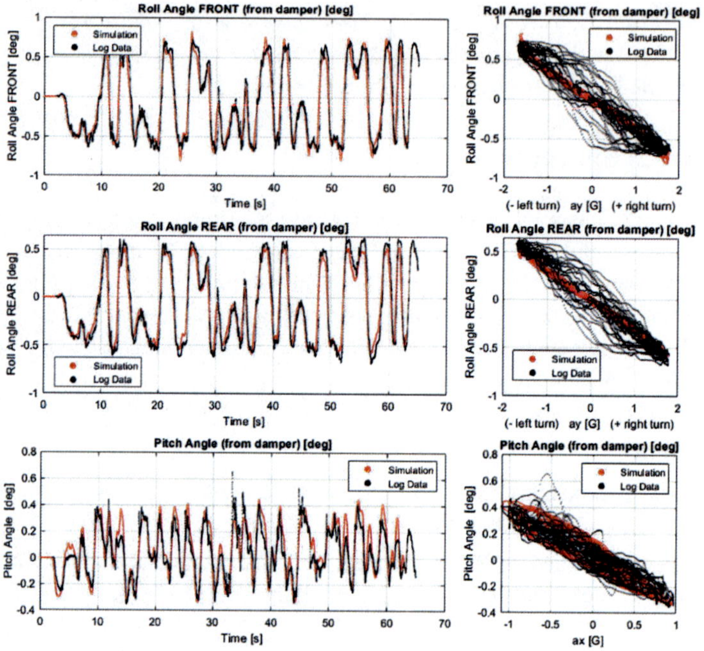

Fig. 11. The chassis roll and pitch angle from damper position in case of real log data (black) and simulation output (red).

Of course, the Simscape model is validated with another suspension kinematic program, such as MSC ADAMS Car. According to validation test results, we can say, suspension kinematics calculated are the same. A rear cambergain comparison can be seen in Fig. 12 in function of wheel travel during a bump test.

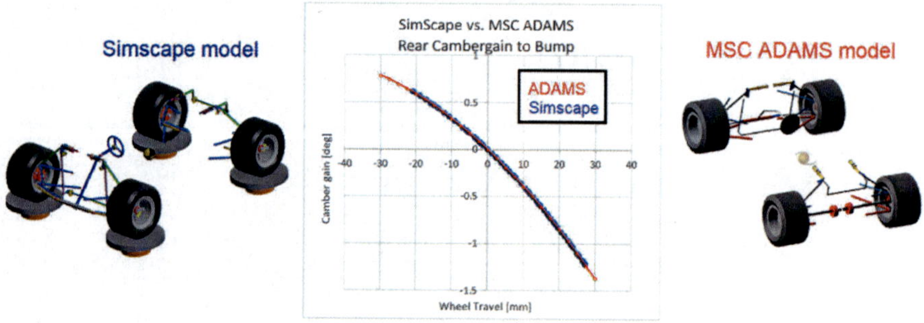

Fig. 12. Comparison test with another suspension kinematic program.

3.2 Setup Comparison

After a validation process, the model seems to work well, so with the help of the model we can design a new suspension, or we can compare suspension setup concepts. For example, we have done four setup concepts to compare on our FS car to see what can we expect in case of weight transfers and ride heights, roll and pitch angle. The difference in four setups can be seen in Table 1. The roll gradient is based on analytical calculations [6].

Table 1. Comparison between setups

	Base	No ARB	No ARB, Stiffer Spring	Base with Chassis Tors. Stf.
Colour on Fig. 13	Red	Blue	Green	Yellow
Spring [N/mm] (front - rear)	39.4 - 35	39.4 - 35	61.3 - 61.3	39.4 - 35
ARB pos. (front - rear)	med - med	med - med	out - out	med - med
Chassis Tors. Stf. [Nm/deg]	Infinity	Infinity	Infinity	3000
Roll gradient [deg/G]	0.41	0.89	0.54	0.41

The simulation result can be seen in Fig. 13. Without anti-roll bar (blue) the chassis roll will be 2 times more than in case of the base (red) setup. Moreover, the rear camber gain is higher too. To compensate the too much chassis roll we put stiffer springs in case of the green setup. The chassis roll is higher than the base, but the pitch of the chassis is quasi half of the base during braking and accelerating. Moreover, the CoG movement because of the jacking force is less in case of stiffer springs too. The fourth setup difference to the base was chassis torsional stiffness. As you can see, there is no considerable difference. If we look at roll gradients in Table 1. Which was calculated analytically, these numbers seem to be the same as we can see in simulation results in Fig. 13 in roll angle graph.

3.3 Suspension Forces

During designing suspension parts, we need information about forces to choose bearings, materials and for topological optimization process and so on. From this simulation environment, the information from load of suspension parts can be known. Some forces of front left suspension can be seen in Fig. 14 during the track test simulation.

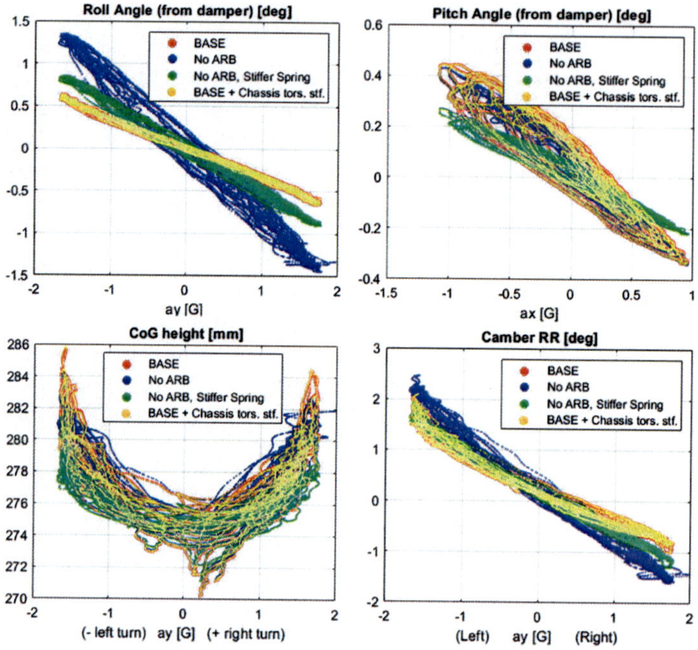

Fig. 13. Comparing four setup concepts.

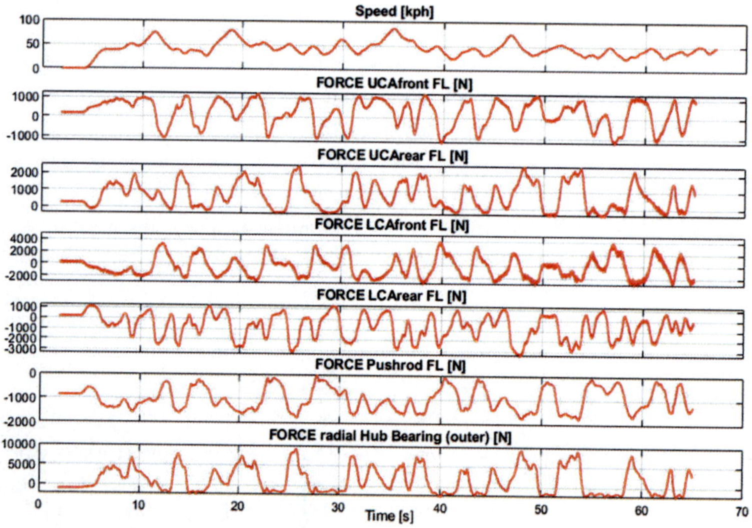

Fig. 14. Front left suspension forces during track test simulation (+ pulled, – pushed) (UCA – Upper Control Arm, LCA – Lower Control Arm).

4 Suspension Types

In this paper, a double wishbone formula student suspension car was introduced, but this model can change easily to other type of suspensions. The model is built of parts, such as anti-roll bar, steering system, damper-spring assembly, so we can assemble these parts as we want. In Fig. 15 another type of suspension can be seen.

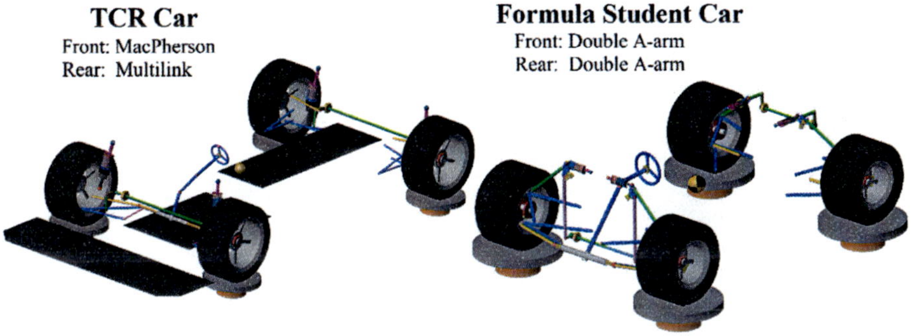

Fig. 15. Model can be changed easily to other types of suspension

5 Summary

With this Simscape simulation environment, we can design a new suspension to any car more easily if we have log data from track. We can investigate suspension kinematics in braking zones, mid-corners and corner exit turn by turn. Moreover, for detailed suspension part design, we can estimate all the forces will load to the part. Therefore, we can choose materials more easily and we can run FEM simulations more precisely.

Besides, if we have bought a racecar in a series, such as TCR or F3 and so on, we just have to measure the suspension coordinates or read it from its manual and we have to write the coordinates into this model for estimation of forces and kinematic movements too.

Acknowledgement. The research presented in this paper was carried out as part of the EFOP-3.6.2-16-2017-00016 project in the framework of the New Széchenyi Plan. The completion of this project is funded by the European Union and co-financed by the European Social Fund.

References

1. Dixon, J.C.: Suspension Geometry and Computation/John C. Dixon. p. cm. Wiley, Hoboken (1948). ISBN 978-0-470-51021-6
2. Kumar, H., Balan, K.N., Raffi, M., Reddy, G.R.V., Yashvanth, U.: Design and simulation of double wishbone (SLA) suspension geometry

3. Theander, A.: Design of a Suspension for a Formula Student Race Car. TRITA-AVE-2004-26. ISSN 1651-7660
4. Jadhao, S., Patil, M.K.: Modelling and simulation of full vehicle to study its dynamic behavior. IJEDR (2017). ISSN 2321-9939. https://www.ijedr.org/papers/IJEDR1704001.pdf
5. Philip, R.C., Varghese, R.G., John, S.G., Cherian, S., Oommen, L.P.: An Approach to Parallel Parking and Zero Turning Radius in Automobiles
6. Milliken, W.F., Milliken, D.R.: Race Car Vehicle Dynamics. SAE, Warrendale (1995)
7. Pacejka, H.B.: Tire and Vehicle Dynamics (Hardcover), p. 627. Society of Automotive Engineers Inc. (2002). ISBN-10: 0768011264. ISBN-13: 978-0768011265
8. French, T., Roland, A., Sluiter, M.: Camber-Car Design & Dynamics. California Polytechnic State University (2013). https://www.zonagravedad.com/images/Carrilanas/Camber_Car_Design_Dynamics.pdf
9. Formula SAE Tire Test Consortium. https://www.fsaettc.org/. Accessed 18 Feb 2020
10. https://uk.mathworks.com/products/simmechanics.html
11. Raafiq, M., Madhu, S.: A review paper on multi link suspension mechanism with compliant joints. Int. J. Appl. Eng. Res. **10** (2015)
12. Schneider, T.: Formula student international ATZextra Worldw. **24**, 68–71 (2019)

Validation of Formula Student Race Car Simulation Environment and Parameter Sensitivity to Race Results

Gergely Szűcs(✉), Zakariás Erõsdi, Gergely Bári, and János Papp

John Von Neumann University, Izsáki út 10, Kecskemét 6000, Hungary
szucsi.90.07.09@gmail.com

Abstract. Formula Student is a competition series for university teams who build their own formula style race cars according to Formula SAE rules. Teams who participate in this event have to design and build a car, and they also have to create the financial background to build the car and get to the competition. Thus, Formula Student is a project management competition too.

From a project management point of view, both financial and human resources are limited. Therefore, every Formula Student team has to decide which parameters should they focus on, and which parameters are less important during the design phase of their race car. To make this decision easier, teams should do a parameter sensitivity analysis to know which parameter has the most effect on the results.

For helping the above, this paper introduces a process about how a team can make a valid model of an electric FS race car and how to calculate the effect of different parameters on the amount of points to be earned. It is also shown, how they can decide which parameters of the car are important and which ones have less effect on the car's performance.

Keywords: Parameter sensitivity · Electric race car · Formula Student · Car concept

1 Introduction

To design, manufacture, build, and operate an electric racecar is a very complex task which requires more people as a team. In this paper we want to show a process, how we could be more effective during the design period to taking attention to those important projects, with you can earn more points in a competition.

First of all, a simulation environment is described. A simple model, based on a single point mass is introduced. Using the model it is shown how to simulate the effect of the main parameters, such as mass, downforce- and drag coefficient, center of gravity height, HV accumulator capacity, gear ratio and friction coefficient on lap time.

Validation of the developed model is also shown. Parameter tuning of the model, based on real car measurements is described. This paper introduces the energy flow measurement of the car too.

K. Jármai and K. Voith (Eds.): VAE 2020, LNME, pp. 365–380, 2021.
https://doi.org/10.1007/978-981-15-9529-5_32

In the end, with a valid model, parameter sensitivity can be determined, and teams can make concept decisions based on these simulations. Therefore, the limited financial and human resources can be assigned reasonably, and the best concept, that has the highest chance to earn the maximum amount of points, can be chosen rationally.

In our example, the competition is Formula Student [1].

1.1 Formula Student

Formula Student is a brand name for races held worldwide, where college and university students compete in designing and constructing a race car. The competition includes creating and presenting a comprehensive business concept and raising funds to run the project.

The series started off in the USA in the 1980's, but as of today, races are held, among others, in the UK, Germany, Italy, Brazil and in Asia-Pacific, too. Participating in our project is not only useful for the students, but for companies working in related industries as well. Formula Student connects employers with fresh graduates, who finish their studies with valuable, up to date knowledge. Symbolizing the reputation of the series, the races are held at such renowned tracks as Silverstone, Hockenheimring or the test track of Ferrari, Fiorano Circuit. In recent years, Formula Student was patronized by household names in the vehicle and racing industry, namely Ross Brawn, technical director in Formula One, and Mario Theissen, former director of BMW Motorsport [1].

Race Events. Several events are held on different days of the competitions, but the first is always the scrutineering, where judges check the compliance of the race car with the safety and technical regulations. Scrutineering is followed by "static" and "dynamic" events. Static events are worth 40% of the total score, while dynamic events account for the remaining 60%.

Dynamic Events. [2] **.** During the dynamic events, the pilots take their car to the track, where both man and machine race against the clock. There are 5 parts in dynamic events:

Acceleration. All vehicles have to travel a distance of 75 m in the shortest possible time, from a standing start Available points equal to 75. The calculation of points can be seen at FS rules [2].

Skidpad. Skid Pad is an eight-figure racetrack, where cornering capabilities of the race car are measured. The driver has to complete 2 laps to left and 2 laps to right, and the 2nd laptime is counted to both direction. Available points equal to 75. The calculation of points can be seen at FS rules [2].

AutoX. This event is held on a short, maximum 1500m, but usually only around 1000m long track, consisting of max 80 m long straights, constant radius turns, hairpins, slalom with cones 7,5–12 m apart, and other elements such as decreasing radius turns, chicanes, etc. The driver has to complete 1 lap as quickly as possible. It is like a qualification lap. Available points equal to 100. The calculation of points can be seen at FS rules [2].

Endurance. This is the hardest event to these prototype cars. During this event, the spotlight is on the durability and energy consumption of the car while it is a race against the clock. Two pilots have to drive the car for 22 km. The track is similar to the one used for AutoX. Available points equal to 325. The calculation of points can be seen at FS rules [2].

Efficiency. During the Endurance event, the cars have to complete 22 km, therefore the most important factors are durability and energy consumption. The latter is scored in the Efficiency event, by measuring which team could complete the distance most efficiently regarding average speed and energy/fuel consumption. Available points equal to 100. The calculation of points can be seen at FS rules [2].

1.2 Goals

A simulation environment has to be created in order to objectively make decisions between different concepts and during the whole design process. We have to know, what are the most effective parameters to race points because we should take more attention to those parameters. Our human and financial resources are limited, so we should divide our resources the best way we can. By doing parameter sensitivity analysis to race points, we can decide how much part of our resources have to be taken to each segment of the car. For example, in view of project management, we have 1000 h to develop our aero kit and reduce the weight of the car. The question is, how much time should we spent on weight reduction from the 1000 h, and how much time should we spent on developing aero kit. After parameter sensitivity analysis, we can answer these main questions.

Moreover, we can answer to basic concept design questions, for example, where shall we place our accumulator? Where should the driver sit? Does a newly designed aero element improve our performance? If so, then how much?

This type of questions can usually be answered by doing field-tests, but it takes a lot of time and human and financial resources. Thus we have created a simulation model of the car in Optimum Lap [3]. This model uses the following parameters:

- Weight
- Lift coefficient
- Drag coefficient
- Height of center of gravity
- Longitudinal friction
- Lateral friction
- HV Battery Capacity
- Final Ratio
- Rolling resistance

Optimum Lap is calculating with point mass model which has no height of CoG, but we have another simulation model [4]. With that model, we can estimate the function between grip (friction) and CoG height, and friction values can be written into Optimum Lap software.

2 Model Validation

Before we are investigating the effect of parameters, we have to validate our simulation environment, because without it we don't know the accuracy of our simulation model [5].

2.1 The Electric Racecar Concept

In our example, it is an electric racecar that complies with the rules of FS [2]. A 600 V DC battery provides the energy to the four motors (21Nm and 35kW per motor) which are connected to each wheel across planetary. So this is an all-wheel driven racecar with a mass of 210 kg without the driver.

2.2 Tire Model

In the Optimum lap, the coefficient of friction of the tire can be specified. These are selected in such a way, that the log data is consistent with the model. The degressivity of the tire to normal force can also be given. This was calculated from Tire Test Consortium (TTC) data [6], where for an increase of 1 kg of normal force, the coefficient of friction decreases by 0.002 in the longitudinal direction and by 0.004 in the lateral direction.

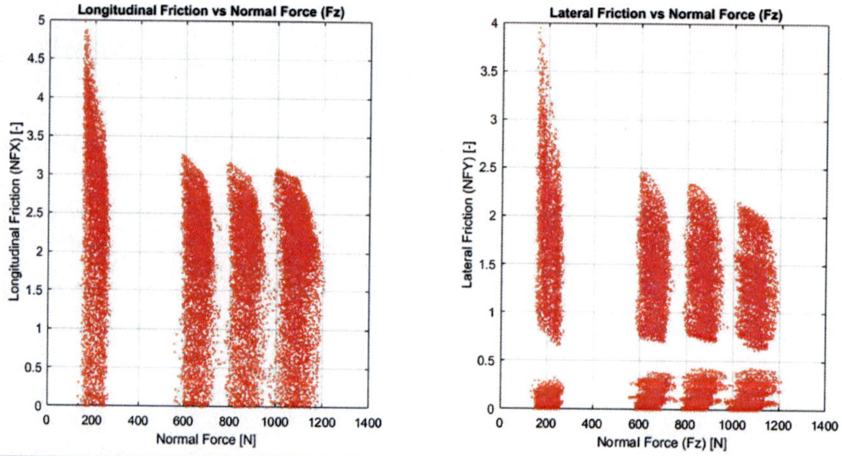

'Normal Force (Fz) [N]'	208	664	886	1109
'mu_x [-]'	4.98	3.21	3.03	2.99

'Normal Force (Fz) [N]'	208	664	886	1109
'mu_y [-]'	3.95	2.45	2.32	2.1

Fig. 1. TTC Tyre measurement data - Friction coefficient in function of normal forces. Longitudinal friction to left, and lateral friction to right.

$$\frac{\Delta\mu_x}{\Delta weight} = \frac{\mu_2 - \mu_4}{\frac{F_{z4}}{g} - \frac{F_{z2}}{g}} = \frac{3.21 - 2.99}{\frac{1109}{9.81} - \frac{664}{9.81}} = 0.0048\left[\frac{1}{kg}\right] \tag{1.1}$$

$$\frac{\Delta\mu_y}{\Delta weight} = \frac{\mu_2 - \mu_4}{\frac{F_{z4}}{g} - \frac{F_{z2}}{g}} = \frac{2.45 - 2.1}{\frac{1109}{9.81} - \frac{664}{9.81}} = 0.0077\left[\frac{1}{kg}\right] \quad (1.2)$$

Based on Fig. 1., the change in friction for 1 kg can be calculated as in Eq. (1.1) and Eq. (1.2). The measurement is conducted on a treadmill, with a special surface with superior friction compared to an ordinary asphalt, so these calculated values have to be normalized with respect to the appropriate values for the asphalt [7].

$$\left(\frac{\Delta\mu_x}{\Delta weight}\right)_{\mu\approx1.1} = \left(\frac{\Delta\mu_x}{\Delta weight}\right)_{\mu\approx3} \bullet \frac{1.1}{3} = 0.0048 \bullet \frac{1.1}{3} = 0.002\left[\frac{1}{kg}\right] \quad (2.1)$$

$$\left(\frac{\Delta\mu_y}{\Delta weight}\right)_{\mu\approx1.3} = \left(\frac{\Delta\mu_y}{\Delta weight}\right)_{\mu\approx2.5} \bullet \frac{1.3}{2.5} = 0.0077 \bullet \frac{1.3}{2.5} = 0.004\left[\frac{1}{kg}\right] \quad (2.2)$$

2.3 Aerodynamics and Rolling Resistance

In the Optimum lap the coefficient of downforce (C_L), drag (C_D) and rolling resistance (C_{res}) can be specified. These are selected in such a way, that the log data is consistent with the model. To specify drag and rolling resistance coefficient we have tested our car during rolling with constant speed.

The advantage of electric cars you can measure the current of motors and you can calculate it to torque according to the datasheet. So in case of constant speed, we can measure how much force (torque divided with tire rolling radius (R_{wheel})) should we have to win the resistance forces such as rolling resistance and drag.

Rolling Resistance. During the rolling resistance test we have to comply three rules. First of all, the ground must be quasi flat, without up- and downhill, secondly the speed must be low (20 km/h) because we want to exclude aero effects, and thirdly the driver must maintain the steering wheel straight to exclude the resistance of steering.

The rolling resistance coefficient (C_{res}) is calculated from sum drive force by motor torques (M_{motor}) divided by the normal forces (F_N) of tires which are estimated by the weight of the car (mg):

$$C_{res} = \frac{F_{drive}}{F_N} = \frac{\frac{\sum_{i=1}^{4} M_{motor,i}}{R_{wheel}}}{mg} \quad (3)$$

With this estimation, C_{res} rolling resistance contains the friction of wheel bearings and gearboxes too. The test was done in case of more tire pressures:

The optimum tire pressure of our race tires is 0.8 bar, so according to Fig. 2. we are calculating with $C_{res} = 0.045$.

Aerodynamic Drag. It was measured with the same method, but the speed was higher. During the high constant speed rolling test, the measurement must be done both ways to exclude the effect of wind. At constant speed (v) we estimated all of the motor torque

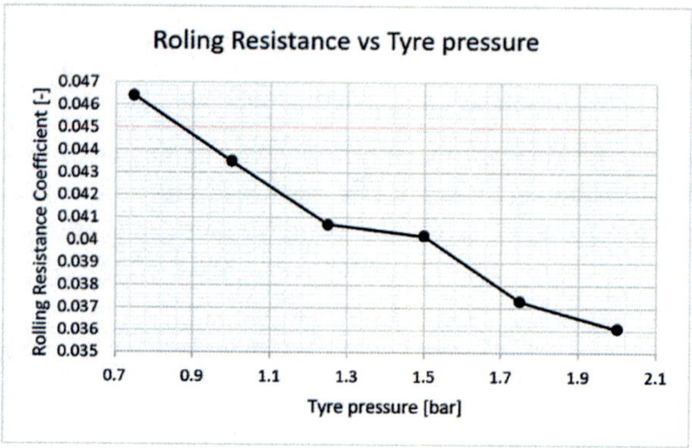

Fig. 2. Rolling resistance of the car.

is to win the resistance forces of drag and rolling resistance (F_{Drive}). Rolling resistance is known from the previous measurement, so drag can be calculated.

$$C_D = \frac{F_{Drive} - F_{RollRes}}{0.5\rho A v^2} = \frac{\frac{\sum_{i=1}^{4} M_{motor,i}}{R_{wheel}} - C_{res}\left(mg + 0.5\rho A v^2 C_L\right)}{0.5\rho A v^2} \tag{4}$$

where ρ is air density, A is frontal area of the car, C_L and the downforce coefficient of the car (this is described below Eq. 5).

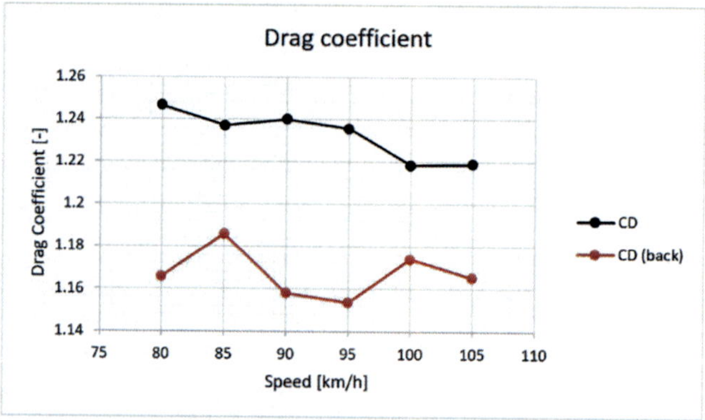

Fig. 3. Drag coefficient of the car in both directions. (wind speed was about 5km/h)

As you can see in Fig. 3. The average value of drag coefficient is 1.2.

Aerodynamic Downforce. It was measured at a constant speed rolling test too. Downforce can be measured during the car is running if we have damper potential meters on the car and we know the stiffness of springs. Moreover, we have to know the kinematics of the car, especially the motion ratio [8]. Between spring travel and wheel travel and between wheel travel and damper potential meter travel. With this information downforce can be calculated.

We have to see damper values in two conditions: At first, δ_0 damper values must be measured at constant low speed, not in standing position with zero speed, because of friction of dampers. At second, δ_{HS} at high constant speed. The difference between δ_{HS} and δ_0 is the damper travel because of downforce. Downforce on one wheel can be calculated by wheel stiffness which is K_s spring stiffness divided by square of motion ratio at each corner of the car (FL, FR, RL, RR).

$$Downforce = (\delta_{HS} - \delta_0) \bullet \frac{K_s}{MR^2} \tag{5}$$

At the end we can calculate C_L downforce coefficient as it is in Eq. (6):

$$C_L = \frac{\left(\sum_{i=1}^{4} Downforce_i\right)}{0.5\rho A v^2} \tag{6}$$

where i means the corner of the car (FL, FR, RL, RR), ρ is air density, A is frontal area (based on CAD model measurement) and v is the speed of the vehicle in [m/s].

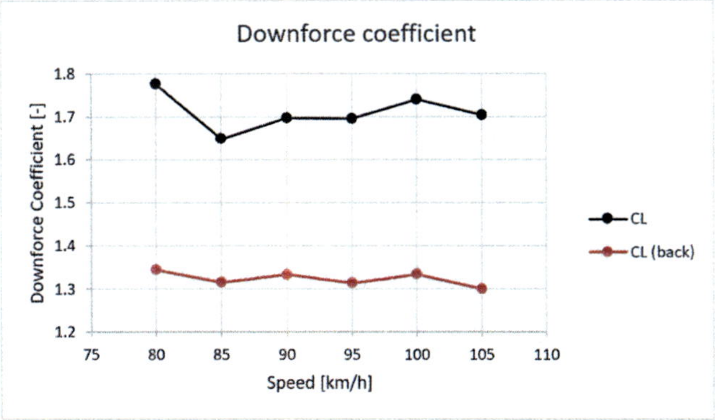

Fig. 4. Downforce coefficient of the car in both directions. (wind speed was about 5km/h)

As you can see in Fig. 4. The average value of downforce coefficient is 1.5.

2.4 Drivetrain

For the engine model, the whole drivetrain has to be considered. Using the typical ideal model of a synchronous motor would lead to a biased model, as the actual drive has significant losses due to frictions in the gear system and other effects. Below the actual torque-rpm diagram can be seen with different motor-load cases (Fig. 5).

This phenomenon can be investigated deeply. Using last year's log data we can determine the energy-balance of the car. Energies were calculating if the car is accelerating with a throttle.

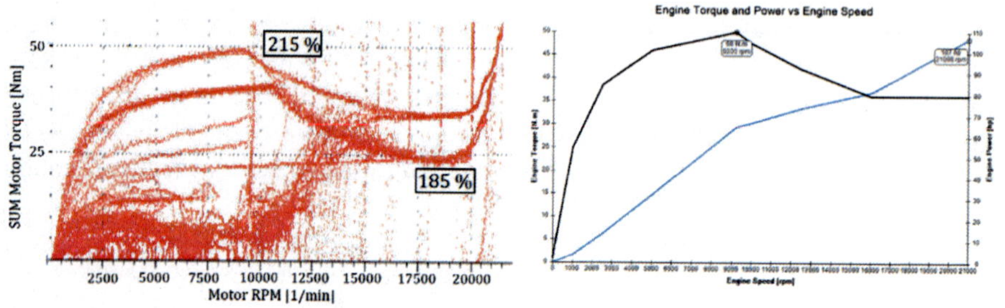

Fig. 5. SUM motor torque - RPM characteristic (Log Data – left, Simulation environment – right)

Total Energy (HV Accumulator). The energy comes from 600V DC HV battery. We can measure the current (I_{HV}) and voltage (U_{HV}) of the accumulator. If we integrate the performance of it to the whole lap, we can see how much energy we use on 1 lap:

$$P_{HV} = U_{HV} \bullet I_{HV} \tag{7.1}$$

$$E_{HV} = \int_0^{T_{lap}} P_{HV} dt \tag{7.2}$$

Energy of the Motors. We can measure the RPM (n_i) of wheels and the current of electric motors that we can calculate to torque ($M_{motor,i}$) according to datasheet [9] on each wheel.

$$P_{Motor} = \sum\nolimits_{i=1}^{4} M_{motor,i} \bullet \omega_i = \sum\nolimits_{i=1}^{4} M_{motor,i} \bullet n_i \bullet \frac{\pi}{30} \tag{8.1}$$

$$E_{Motor} = \int_0^{T_{lap}} P_{Motor} dt \tag{8.2}$$

Losses Due to Rolling Resistance. This estimated energy is based on measurements that you can see above in Eq. (3) and Fig. 2. The loss performance can be calculated with a resistance force ($C_{res} \bullet F_N$) multiplied speed (v).

$$P_{RollRes} = C_{res} \bullet F_N \bullet v = C_{res} \bullet \left(mg + 0.5\rho A v^2 C_L \right) \bullet v \tag{9.1}$$

$$E_{RollRes} = \int_0^{T_{lap}} P_{RollRes} dt \tag{9.2}$$

Losses Due to Drag. This estimated energy is based on measurements that you can see above in Eq. (4) and Fig. 3. The loss performance can be calculated with drag force multiplied speed (v).

$$P_{Drag} = \frac{1}{2} \rho A C_D v^2 \bullet v \tag{10.1}$$

$$E_{Drag} = \int_0^{T_{lap}} P_{Drag} dt \tag{10.2}$$

Kinetic Energy. This energy is used for accelerating the car. This is beneficial energy.

We use the power of accelerating (P_{kin}) when the change of kinetic energy is increasing $\left(\frac{dE_{kin}}{dt} > 0\right)$.

$$E_{kin} = \frac{1}{2}mv^2 \rightarrow P_{kin} = \frac{dE_{kin}}{dt} \text{ if } \frac{dE_{kin}}{dt} > 0 \tag{11.1}$$

$$E_{kin} = \int_0^{T_{lap}} P_{kin} dt \tag{11.2}$$

Other Losses (Heating and Steering). Other, not modeled losses are present as well. One of the most significant is the losses due to steering, as while cornering, the lateral forces on the front tires have a component that "brakes" the car. Another energy is heating of motors and Motor Controller Unit (MCU) which can be estimated as the difference between E_{HV} and E_{Motor} (Fig. 6 and Table 1)

$$E_{Steer} = E_{motor} - \left(E_{kin} + E_{Drag} + E_{RollRes}\right) \tag{12}$$

$$E_{Heat} = E_{HV} - E_{motor} \tag{13}$$

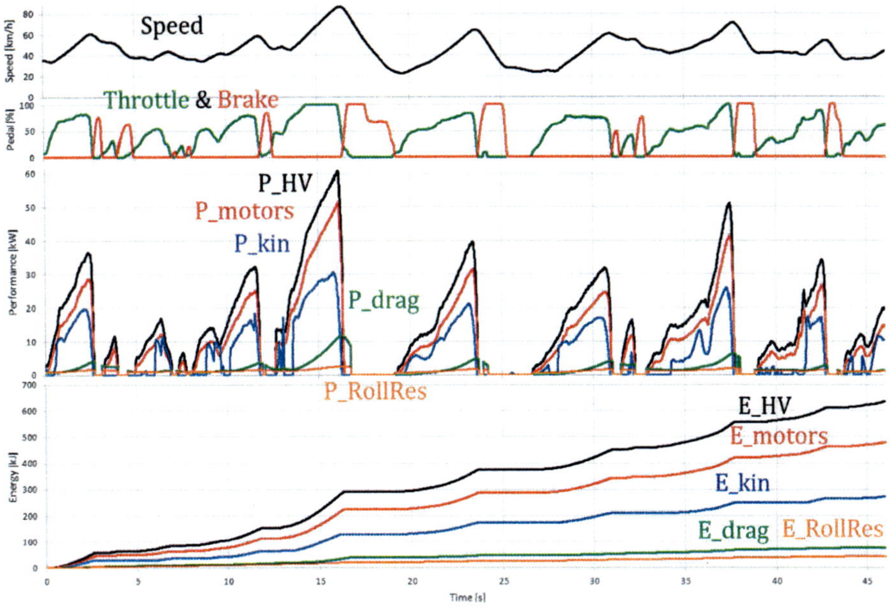

Fig. 6. Speed, Pedal Position, Power (HV, motor, kinetic, drag, rolling res.) and Energies (HV, motor, kinetic, drag, rolling res.) during one lap.

Table 1. Energies on endurance lap by lap

Laps		Lap1	Lap2	Lap3	Lap4	Lap5	Lap6	Lap7	Lap8	Lap9	Lap10
E_{HV}	[kJ]	771	777	770	732	769	749	799	785	749	667
E_{Heat}	[kJ]	205	204	199	188	192	186	193	187	180	164
E_{motor}	[kJ]	566	573	571	544	577	563	606	598	569	503
$E_{RollRes}$	[kJ]	55	55	55	55	55	55	55	55	55	55
E_{Drag}	[kJ]	115	111	114	114	116	115	117	117	116	112
E_{Steer}	[kJ]	81	88	88	73	91	89	121	105	99	63
E_{kin}	[kJ]	315	319	314	302	315	304	313	321	299	273

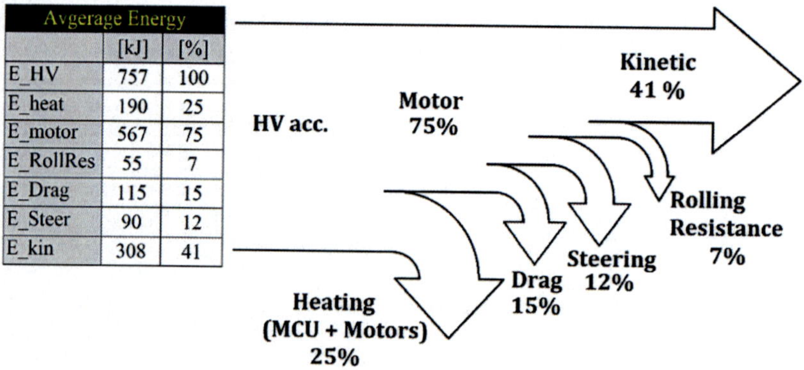

Fig. 7. The average energy flow of the electric Formula Student Car

Optimum Lap software calculates with point mass model. For moving a point mass vehicle model the energy is calculated from kinetic energy, rolling resistance and drag. It is written in Optimum Lap help. *"Energy spent - Amount of energy needed to accelerate during the lap."* [3] So as you can see in Fig. 7. If we want an energy valid model we have to use 63% efficiency. It is detailed at Eq. (14):

$$\eta = \frac{E_{kin} + E_{Drag} + E_{RollRes}}{E_{HV}} = \frac{308 + 115 + 55}{757} = 63\% \tag{14}$$

2.5 Model vs. Simulation

Based on validation measurements above we get the main parameters that can be written into Optimum Lap [3] software and we can run the simulation. We can compare the simulation model and the real log data in Fig. 8:

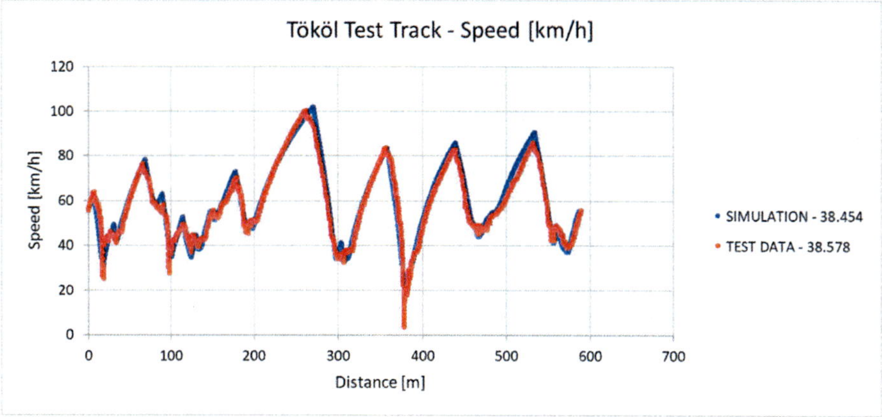

Fig. 8. Speed graph with laptime (Log Data vs. Simulation)

3 Laptime Simulation

With this model, the parameter sensitivity of the car can be evaluated by competing on simulated dynamic events of an FS competition.

The evaluated parameters:

- Weight (m)
- Lift coefficient (C_L)
- Drag coefficient (C_D)
- Rolling resistance (C_{res})
- Final Ratio ($Gear$)
- HV Battery capacity (HV_{Cap})
- CoG height (h_{CG}) (decreasing friction according to MRA Moment Method simulation results) [4]
- Coefficient of friction (μ_x, μ_y) (Table 2).

Table 2. Base parameters of FS car model

m	C_L	C_D	C_{res}	$Gear$	HV_{Cap}	h_{CG}	μ_x	μ_y
300 kg	1.5	1.2	0.045	16	7 kWh	275 mm	1.1	1.35

With these modeling approaches, we have conducted lap-time simulations for FSEast2019 competition. We simulated all the four dynamic events: Acceleration, Skid-Pad, AutoX and Endurance. For endurance simulation, we have to take attention to finite amount of HV energy.

3.1 Endurance Simulation

The HV battery capacity is 7 kWh. Based on logs, 30% of it is recuperated, because of regenerative brake. Optimum Lap does not calculate with regenerative brake. Thus the total capacity that can be used in the model is:

$$HVCapacity_{OptLap} = 7 \bullet 1.3 = 9.1[kWh] = 32760[kJ] \tag{15}$$

The efficiency of our optimum lap model is 63% (see Eq. (14)). For modeling an endurance this has to be considered as well. The energy that can be used during an endurance can be calculated as follows:

$$E_{Endu} = HVCapacity_{OptLap} \bullet \eta = 32760 \bullet 0.63 = 20639[kJ] \tag{16}$$

The endurance is 22 km. It is 24 laps at FSEast2019 track, so the usable energy on one endurance lap is

$$E_{Endu,1lap} = \frac{E_{Endu}}{laps} = \frac{20639}{24} = 860[kJ]. \tag{17}$$

In case of endurance simulation, if the energy spent is more than 860 kJ, we have to decrease the motors scale factor, because, with a 100% scale, we will use more energy than we have.

3.2 Points Calculation

We get laptimes from simulation, but these times have to be calculated to points. This process was according to FS rules [2], the calculation method can be read there. For calculation, we need the best times of FSEast 2019 [10]. These can be seen in Table 3.

Table 3. The best laptimes of FS East. In case of endurance, the total time was 1483.8 s, so it was divided with 24 laps*.

Event	Team	Time [s]
Acceleration	Formula Student Team Delft	3.262
SkidPad	TUfast e-technology	4.6415
AutoX	TUfast e-technology	60.544
Endurance *	TUfast e-technology	61.824

During simulation, we have swept the main parameters between reasonable limits. Example results can be seen in Table 4, where we changed the total weight of the car:

Table 4. Simulation results - The laptimes in function of weight and downforce coefficient

Weight [kg]	Laptimes [s]				E_{Endu}[kJ]
	Accel	SkidPad	AutoX	Endu	
300	3.827	4.915	63.521	64.232	829
290	3.773	4.894	63.209	63.836	827
280	3.720	4.867	62.899	63.494	819
270	3.666	4.841	62.589	63.184	805
260	3.613	4.815	62.279	62.863	788
250	3.574	4.788	61.972	62.552	772
240	3.547	4.762	61.668	62.253	754
230	3.525	4.735	61.366	61.942	737
220	3.508	4.708	61.058	61.632	717
210	3.493	4.681	60.744	61.322	699
C_L[-]	Laptimes [s]				E_{Endu}[kJ]
	Accel	SkidPad	AutoX	Endu	
1.5	3.827	4.915	63.521	64.232	829
1.8	3.827	4.901	63.301	64.072	827
2.1	3.827	4.887	63.084	63.933	824
2.4	3.827	4.873	62.870	63.761	825
2.7	3.827	4.859	62.658	63.590	827
3	3.827	4.845	62.448	63.451	825
3.3	3.827	4.832	62.241	63.291	826
3.6	3.827	4.818	62.038	63.130	828
3.9	3.827	4.804	61.836	62.980	828
4.2	3.827	4.79	61.637	62.841	828

As it was mentioned, these laptimes can be calculated to points according to FS rules [2]. Points can be seen in Table 5.

Of course, we have swept the other parameters and got laptimes which were calculated to points too. The summarized results can be seen in Fig. 9.

If you look at Fig. 9, you can see if we reduce the weight with 10 kg (from 300 to 290) we will get 14 more points at FS competition. When we look at the red line, we can see, if we can design better aero with increasing the downforce coefficient (C_L) with + 1 (from 1.5 to 2.5), we will get 18 points.

To summarize, we can earn + 5 points with this parameter improve according to Fig. 9 and Table 6:

Table 5. Simulation results – Laptimes calculated to FS points

	Points					
Weight [kg]	Accel (MAX 75)	SkidPad (MAX 75)	AutoX (MAX 100)	Endu (MAX 325)	SUM (MAX 575)	Delta points
300	43	54	78	280	454	BASE
290	46	55	80	287	468	+ 14
280	49	57	82	293	481	+ 27
270	51	59	84	299	494	+ 40
260	54	61	87	305	507	+ 53
250	56	63	89	311	519	+ 65
240	58	65	91	317	531	+ 77
230	59	67	94	323	543	+ 89
220	60	69	96	325	550	+ 96
210	61	72	98	325	556	+ 102
CL [-]	Accel (MAX 75)	SkidPad (MAX 75)	AutoX (MAX 100)	Endu (MAX 325)	SUM (MAX 575)	Delta Points
1.5	43	54	78	280	454	BASE
1.8	43	55	79	283	460	+ 6
2.1	43	56	81	285	465	+ 11
2.4	43	57	82	289	471	+ 17
2.7	43	58	84	292	476	+ 22
3	43	59	85	294	482	+ 28
3.3	43	60	87	297	487	+ 33
3.6	43	61	89	300	493	+ 39
3.9	43	62	90	303	498	+ 44
4.2	43	63	92	306	503	+ 49

Table 6. These parameter changes mean +5 points improvement

m	C_L	h_{CG}	μ_y
− 6.8 kg	+ 0.26	− 17 mm	+ 0.082

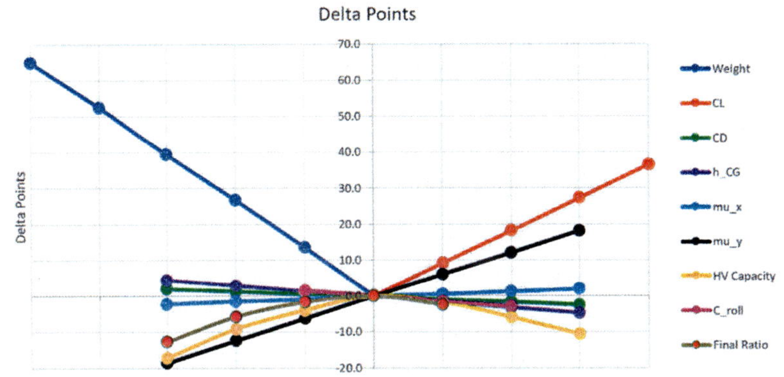

Weight	250kg	260kg	270kg	280kg	290kg	300kg				
CL						1.5	2	2.5	3	4
CD			1.05	1.1	1.15	1.2	1.25	1.3	1.35	
hCG			260mm	265mm	270mm	275mm	280mm	285mm	290mm	
mu_X			1.07	1.08	1.09	1.1	1.11	1.12	1.13	
mu_Y			1.32	1.33	1.34	1.35	1.36	1.37	1.38	
HV_Cap			5.5kWh	6kWh	6.5kWh	7kWh	7.5kWh	8kWh	8.5kWh	
C_res					0.035	0.045	0.055	0.065		
Final Ratio			13	14	15	16	17			
						BASE				

Fig. 9. Parameter sensitivity, delta points in function of parameters

4 Conclusion

Now that we have the tool to examine different aspects of the car we can speak about different concepts. Firstly, weight reduction and aero are one of the most important factors. Thus the overall concept has to be chosen in a way, that it enables the development of an advanced aero package and weight reduction. Besides we have to take attention to suspension setup to increase lateral friction too.

Acknowledgement. The research presented in this paper was carried out as part of the EFOP-3.6.2-16-2017-00016 project in the framework of the New Széchenyi Plan. The completion of this project is funded by the European Union and co-financed by the European Social Fund.

References

1. Schneider, T.: Formula student international. ATZextra Worldw **24**, 68–71 (2019)
2. Formula Student Germany. https://www.formulastudent.de/fileadmin/user_upload/all/2020/rules/FS-Rules_2020_V1.0.pdf. Accessed 10 Feb 2020
3. Optimum G. https://optimumg.com/product/optimumlap/. Accessed 15 Feb 2020
4. Szűcs, G., Bári, G.: Mechanical parameters effect to griplimit behavior of autonomy formula student car. Gradus **6**(1), 204–209 (2019)

5. Setiawan, J.D., Safarudin, M., Singh, A.: Modeling, simulation and validation of 14 DOF full vehicle model. In: International Conference on Instrumentation, Communications, Information Technology and Biomedical Engineering (2009)
6. Formula SAE Tire Test Consortium. https://www.fsaettc.org/. Accessed 18 Feb 2020
7. Villagra, J., d'Andréa-Novel, B., Fliess, M., Mounier,: H.: A diagnosis-based approach for tire–road forces and maximum friction estimation. Control Eng. Pract. **19**(2), 174–184 (2011). https://doi.org/10.1016/j.conengprac.2010.11.005. ISSN 0967-0661
8. Milliken, W.F., Milliken, D.L.: Race car vehicle dynamics (1995)
9. AMK Group. "Technical Data DT" DT5, liquid-cooled motor – DT5-20-10-xxW-3600. https://amk-group.com/en/product/high-torque_motor_series/. Accessed 01 Feb 2020
10. FSEast Competition. https://fseast.eu/wp-content/uploads/2019/07/FS_EAST_2019_Final_Results_EV.pdf. Access 10 Feb 2020

CFD Analyzes of an Axial Fan

Betti Bolló(✉)

Department of Fluid and Heat Engineering, University of Miskolc, Miskolc, Hungary
aramzb@uni-miskolc.hu

Abstract. The main objective of this study is to determine the flow in and around a cooling fan of an automotive engine by means of CFD computations using the commercial software package ANSYS Fluent. Two distinct external disturbances are applied: external flow characterized by the in-flow velocity U and covering plates placed on the suction side of the fan. Two sets of computations are carried out: (1) when U is varied between $U = 0$ and 72 km/h without the covering plate and (2) by covering the full or half of the suction side of the fan at $U =$ 0 km/h. The covering plates represent the structural elements that cover the flow in front of the fan due to its installation. In both external disturbance cases, the aerodynamic noise and force on the axial fan are analyzed. The results show that the aerodynamic noise increases with U and the maximum acoustic sound power is located at the edge of the blade on the pressure side of the fan. By employing covering, the acoustic power level on the suction side increases on the particular blades behind the plate.

Keywords: Axial fan · CFD computation · Free stream velocity · Noise

1 Introduction

Nowadays, the use of fans or blowers is an integral part of a human's life, but the noise generation of rotating machines annoys the users. The noise sources are of two main types: airborne sound sources caused by air fluctuation and structure-born machinery vibration sources. Many researchers have extensively studied an axial fan with the goal of reducing its noise using both experimental and numerical approaches [1–3]. These studies focus primarily on fan noise, uniformity of output flow, impeller force and the analysis of separation zones. Park and Lee [1] presented the source of broadband noise from a typical automotive cooling fan using a hybrid approach with CFD simulations and acoustic analogy. The $k - \omega$ SST turbulence model was used to calculate the turbulent flow, and the generalized acoustic method proposed by Ffowcs Williams and Hawkings [4] was used to determine the noise. The noise source was found to take place at the leading edge of the blade tip and in the vicinity of the fan band. Their computational results were compared to the values obtained from the measurement sequences; good agreement was found.

K. Jármai and K. Voith (Eds.): VAE 2020, LNME, pp. 381–389, 2021.
https://doi.org/10.1007/978-981-15-9529-5_33

The acoustic properties of a cooling fan module were investigated by Lallier-Daniels et al. [2] using the Lattice-Boltzmann Method. The acoustic results showed a very good agreement with their experimental data obtained using the series of microphones arranged in front of the module. Krishna et al. [3] determined the noise source of a motor fan by numerical simulations and measurements. The aerodynamic noise level of the fan was reduced by changing the shape of the blade and the number of blades. They found that there is good agreement between numerical and experimental results.

Aerodynamic noise produced by an axial fan can be reduced by modifying the fan geometry. In addition to noise, it is important to check the efficiency of the device. Lakatos et al. [5] examined that the changes in the geometry of the guide vanes and return guide vanes resulted in lower efficiency in the operating domain of the original design.

The present work focuses on the CFD analyzes of flow around an axial flow fan. The rotation speed of the impeller was fixed at a value of $n = 2500\ \text{min}^{-1}$. In the first part of this study, the axial fan is placed in an external flow and the effect of the incoming flow velocity (hereafter the free stream velocity) ranging between $U = 0$ and 72 km/h is investigated. In the second part, we analyze the effect of the plane cover surface placed in the suction side of the fan. The arrangement under study is the modeling of the cooling fan of a vehicle traveling at a speed of U, when there are structural elements covering the flow in front of the fan due to its installation. In both cases, the aerodynamic forces acting on the blade and the location of the aerodynamic noise source are discussed.

2 Computational Methods

In this study, the three-dimensional, incompressible, Newtonian fluid flow in and around an axial cooling fan placed into a channel is investigated. The equations governing the fluid flow are the three components of the Reynolds averaged Navier-Stokes (RANS) equations and the continuity equation. The realizable $k - \varepsilon$ turbulence model is applied to obtain the Reynolds stresses in the RANS equations.

The axial cooling fan is shown in Fig. 1. The impeller has seven blades rotating clockwise when viewed from the front (see the left in Fig. 1). Behind the rotating blades, there is an asymmetrically arranged support structure. The twelve support vanes located on the back side of the shroud can affect the flow and noise. The axial fan and its parts are placed in the channel with a rectangular cross section which is shown in Fig. 2. In our previous research results [9], where we dealt with the preliminary investigation of the flow conditions around the fan, problems arose. The main problem was that the mesh generates on the computational domain contained almost 58 million cells which made the execution time unreasonably long. On the other hand, the frame mounting the fan was not considered. Therefore, in the present study, a new geometry was created, where the frame mounting of the fan was modelled, and the length of the examined channel section was chosen to be smaller. The length of the channel section was chosen to be 1.5 m and its height and width remained 0.8 m and 1.2 m, respectively.

A schematic diagram of the computational domain is shown in Fig. 2. A flat cover plate is placed on the suction side of the fan which is divided into 13 sub-surfaces. These wall surfaces can be 'activated' or 'inactivated' during computations. In case a sub-surface is activated, no-slip boundary conditions are satisfied on it, otherwise, air can flow through it. If the cover is not activated and the flow passes through it, it is denoted by $\varepsilon = 0$, while the total covering (all surfaces activated) is denoted by $\varepsilon = 1$. (The specific coverage parameter ε shows the ratio between the covering plate and the total surface area of the cover.) At the inlet, uniform velocity is prescribed by giving the free stream velocity U.

In order to solve the governing equations listed before, a mesh has to be generated on the domain shown in Fig. 2. In the part of the computational domain before and after the fan hexagonal elements were used, while a finer tetragonal mesh was created near the fan, the frame, and the cover plate. Two types of mesh were created, i.e. a coarser mesh with approximately 21 million cells and a finer mesh with roughly 28 million cells. Comparing with the measurement values described in our previous study [9]. For the two different mesh designs the difference in absolute pressures on the pressure and suction side of the fan shows barely a 2% relative error, but there is a difference of several hours in running time. We found that the values calculated for the coarser mesh also show a good agreement with the measurement, so further calculations were performed on this mesh.

To solve these highly nonlinear partial equations numerically the commercial software package ANSYS Fluent is applied which uses the finite volume method. The processes in and around the fan are modelled as steady state processes; the rotating motion of the impeller taken into account the "Moving Reference Frame" method. The SIMPLE (Semi-Implicit Method for Pressure-Linked Equation) algorithm was used for pressure-velocity coupling. In spatial discretization of the convective terms, the second order upwind scheme was used in the momentum equation. The Broadband Noise Source (BNS) model is applied to calculate the generated noise.

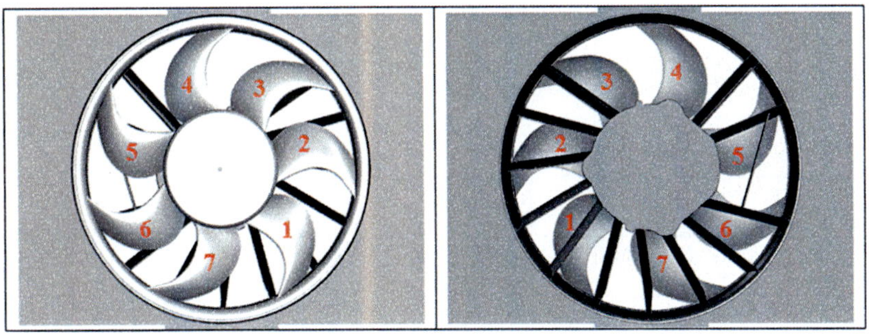

Fig. 1. Geometry of the fan.

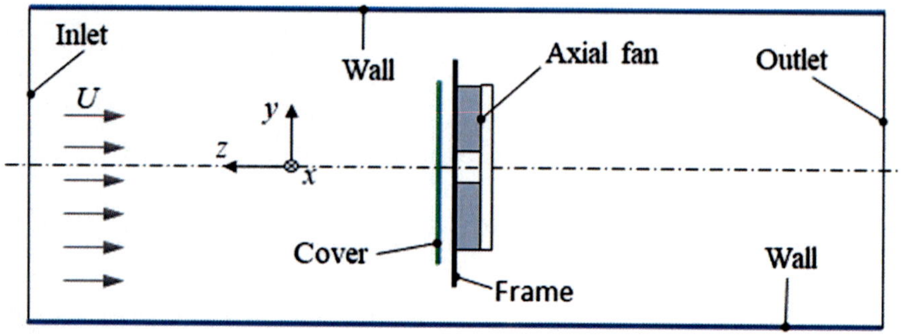

Fig. 2. The schematic diagram of the computational domain.

3 Results and Discussion

The numerical simulations are carried out to investigate the effects of the free stream velocity and the covering plate placed on the suction side of the fan. The rotation speed was kept constant at $n = 2500\ \text{min}^{-1}$, while the free stream velocity was between $U = 0$ and 72 km/h. During the computations, the axial component (in Cartesian coordinate system) of the aerodynamic force (Fz) acting on the fan blades, the pressure distribution, and acoustic sound power on the impeller are investigated.

3.1 The Effects of Free Stream Velocity

In this subsection, the effects of free stream velocity are investigated without the cover plates ($\varepsilon = 0$). During the calculation of the aerodynamic forces, the components F_x and

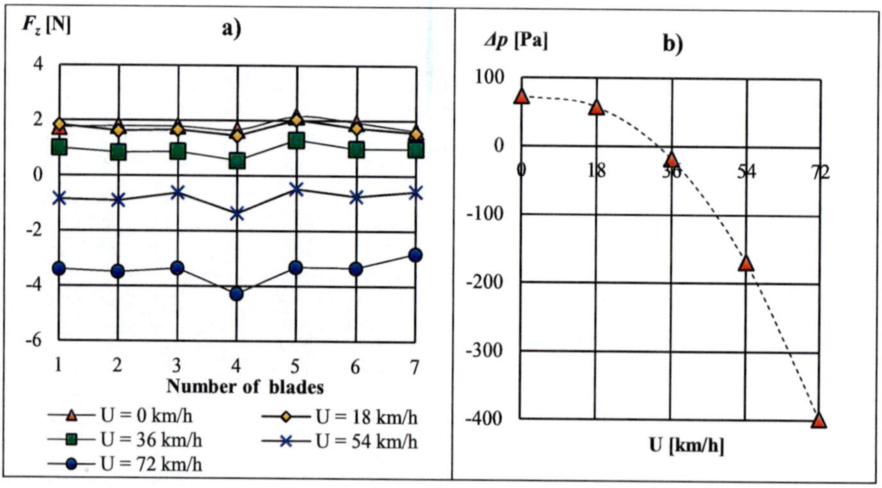

Fig. 3. The axial force F_z acting on the blades of the axial fan (a) and difference between pressure on the pressure and suction side of the fan (b) for free stream velocity values of $U = 0$, 18, 36, 54, and 72 km/h without the covering plate ($\varepsilon = 0$).

F_y can have both positive and negative values [9], i.e. the force acts in the direction of rotation of the impeller or opposite to it. Dorogi et al. [9] showed that the axial force F_z has the most significant effect, so this article details this fluid force component. Figure 3a shows the aerodynamic force component in the axial direction F_z on each blade at U = 0, 18, 36, 54, and 72 km/h. It can be seen that the absolute values of the axial force increase with increasing the free stream velocity. The sign of F_z changes with increasing U: at $U < 36$ km/h the axial force on the blades is positive so that the axial force acts in the same direction as the main flow. In contrast, when U is above $U = 54$ km/h, F_z acts in the opposite direction to the main current.

It can also be seen that the axial force acting on each blade is different at the same velocity, due to the asymmetrically placed support structure behind the rotating blades. There is no support vane behind blade 4 (see Fig. 1 for numbering of the blades), so the value of the axial force acting on the blade is smaller here. This is important because

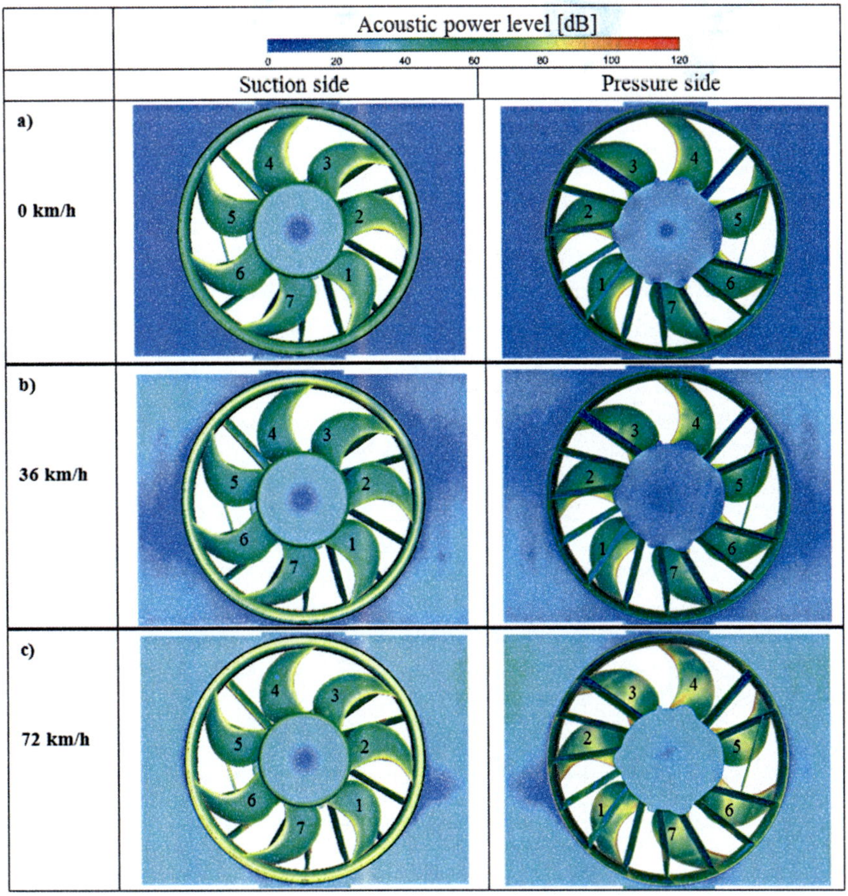

Fig. 4. Acoustic power level on the surface of the fan at $U = 0$ km/h (a), 36 km/h (b), and 72 km/h (c) without the covering plate ($\varepsilon = 0$).

of the unsteady simulation, the force acting on each blade varies between these values, which can lead to mechanical vibration of the fan.

In Fig. 3b the difference between the absolute mean pressures ($\Delta p = p_p - p_s$) on the pressure (p_p) and suction (p_s) side of the fan is shown for different U values between $U = 0$ and 72 km/h. It can be observed that at $U = 0$ and 18 km/h the pressure on the suction side is lower than the absolute pressure value on the pressure side, which is natural for the basic operation of the fan. Increasing the free stream velocity, at $U \geq 36$ km/h p_s is higher than p_p, hence the fan pressure difference is negative. This means that the fan behaves as a resisting body at $U \geq 36$ km/h. The possible reason of this phenomenon is that the air flow due to the high velocity cannot pass through the fan, the air is congested in front of the fan. In this case, the fan is also driven by the wind, a condition which is similar to the operation of a "wind turbine".

The aeroacoustic sound which occurs on the fan surface and the mounting frame is presented in Fig. 4 for $U = 0$, 36, and 72 km/h. The aerodynamic noise generation mechanisms are the turbulence of air flow and laminar and also vortex shedding. The support vanes located on the back side of the shroud affect the flow and noise. It can be seen that the maximum acoustic source power is focused in a small area on the pressure side of the blade at the leading edge. By increasing the free stream velocity, the acoustic sound power level on the blades appear to increase as well.

3.2 The Effects of Covering Plates

CFD simulations are also carried out to analyzed the effects of covering when the free stream velocity in the inlet cross section is set to zero, i.e. $U = 0$km/h. In this computation set, three cases are analyzed where the full cover is "activated" ($\varepsilon = 1$) and the case when half of the covering surface is asymmetrically and symmetrically "activated" ($\varepsilon = 0.5$). The reader is referred to Fig. 6 for the location of the cover surfaces.

In Fig. 5 the axial force acting on the blades of the fan are plotted. F_z shows the highest values when the suction side is fully covered ($\varepsilon = 1$); the value of F_z increases

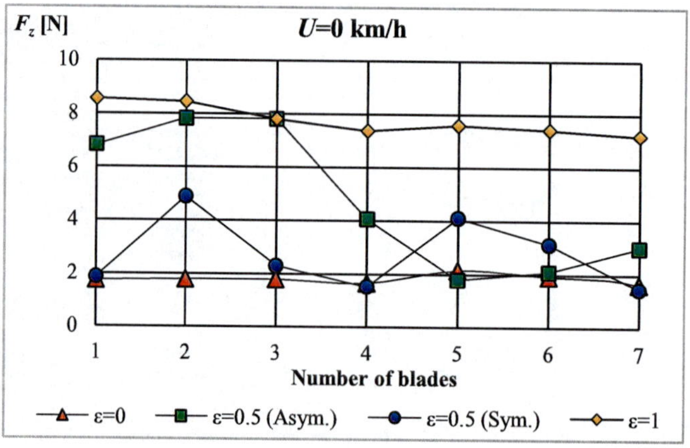

Fig. 5. The axial force on the blades at different covering surface for $U = 0$ km/h.

approximately four times compared to the case without cover ($\varepsilon = 0$). In the asymmetrical case ($\varepsilon = 0.5$) blades 1, 2, and 3 are fully covered and the forces acting on which are higher. In the symmetrical case, only a part of the blade (2, 5, 6) is covered, so the axial force increases to a lesser extent. The uneven force distribution on the blades can cause strong pressure fluctuations, which can lead to mechanical vibration for unsteady simulation, especially in case of the asymmetrical arrangement.

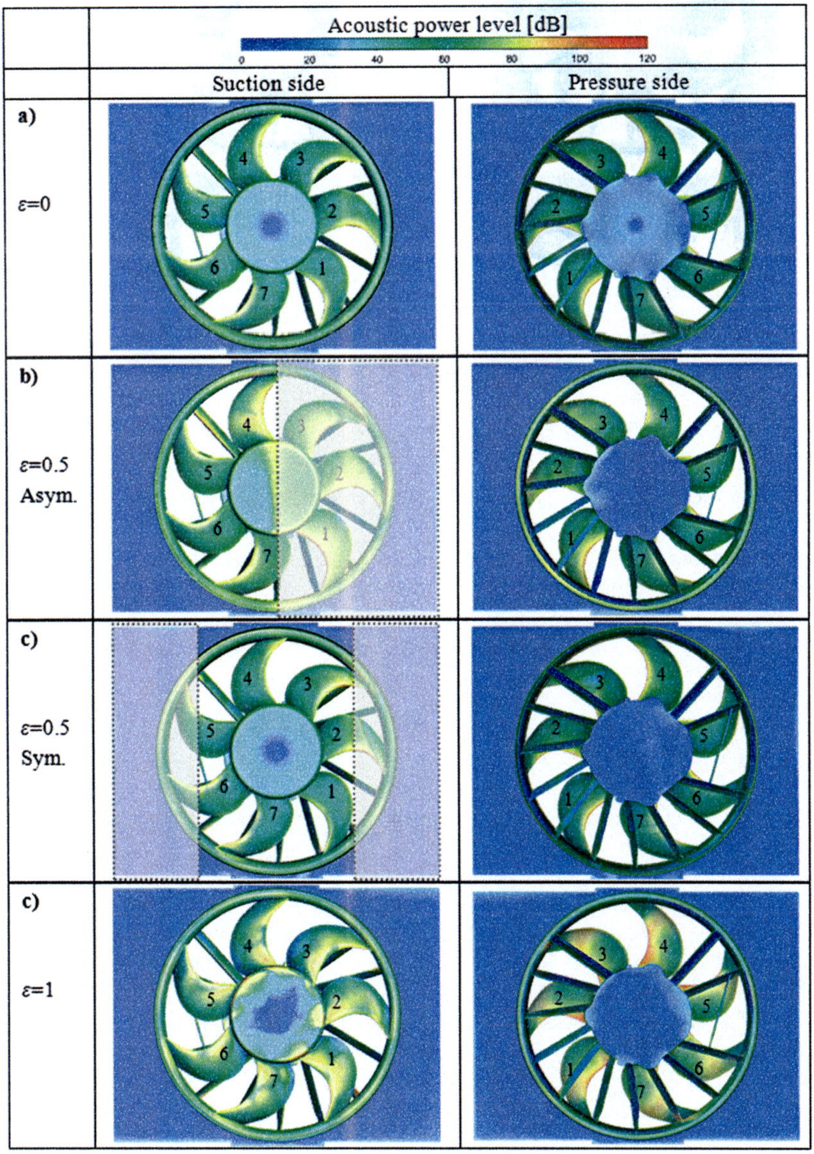

Fig. 6. Acoustic power level in the surface of fan for different covering ($U = 0$ km/h).

The cover also affects the location of the acoustic sound source. The acoustic power level is shown in Fig. 6 on the suction and pressure side surfaces of the fan. It is found that the noise level behind the covered surfaces on the suction side increases. It is interesting to observe in the figure that in the case of full cover, high sound power is generated not only at the edge of the blade but also in its surroundings, especially near the impeller hub. In case of full coverage, the maximum value of the acoustic sound power increases by almost 3 dB compared to the case without cover.

4 Conclusions

In this paper, the flow around an axial cooling fan is numerically studied using CFD simulations. The independent effect of external flow (characterized by the free stream velocity U) and covering plate are investigated in this study. First, the free stream velocity is varied between $U = 0$ and 72 km/h without the covering plate. Second, the influences of the covering plate placed on the suction side of the fan is analyzed at $U = 0$km/h.

The acoustic power level characteristics and aerodynamic force acting on the blade are investigated. The calculations revealed that if an axial fan is placed into a free stream, the magnitude of U has a significant effect on the fan operation, the forces acting of the blades, the pressure difference produced, and acoustic noise. Another important result is that if the free stream path is changed with cover elements, the blades are subjected to a perceptible asymmetrical load, which also affects the vibration characteristics of the fan. The results obtained from the CFD simulation are important starting points for a detailed numerical mechanical analysis, which is the next step of this project.

Acknowledgements. This research was supported by the European Union and the Hungarian State, co-financed by the European Regional Development Fund in the framework of the GINOP-2.2.1-15-2017-00090 project, titled "E-mobility from Miskolc: Improvement of Coolant Pump and Engine Cooling Fan Taking into Account the Higher Quality Requirements in Electric Vehicles.

References

1. Park, M.J., Lee, D.J.: Sources of broadband noise of an automotive cooling fan. Appl. Acoust. **118**, 66–75 (2017)
2. Lallier-Daniels, D., Piellard, M., Coutty, B., Moreau, S.: Aeroacoustic study of an axial engine cooling module using lattice-Boltzmann simulations and the Ffowcs Williams and Hawkings' analogy. Eur. J. Mech. B/Fluids **61**, 244–254 (2017)
3. Krishna, S.R., Krishna, A.R., Ramji, K.: Reduction of motor fan noise using CFD and CAA simulations. Appl. Acoust. **72**(12), 982–992 (2011)
4. Ffowcs Williams, J.E., Hawkings, D.L.: Sound generation by turbulence and surfaces in arbitrary motion. Philos. Trans. R. Soc. Lond. Ser. A Math. Phys. Sci. **264**, 321–342 (1969)
5. Lakatos, K., Szaszák, N., Mátrai, Zs., Soltész, L., Szabó Sz.: Experimental development of guide vanes and return guide vanes of a mini blower. In: XXV. International Scientific Conference, Miskolc, Hungary, pp. 65–72 (2011)
6. Ye, X., Li, P., Li, C., Ding, X.: Numerical investigation of blade tip grooving effect on performance and dynamics of an axial flow fan. Energy **82**, 556–569 (2015)

7. Li, C., Li, X., Li, P., Ye, X.: Numerical investigation of impeller trimming effect on performance of an axial flow fan. Energy **75**(C), 534–548 (2014)
8. Zhang, L., Jin, Y., Jin, Y.: Effect of tip flange on tip leakage flow of small axial flow fans. J. Therm. Sci. **23**, 45–52 (2014)
9. Dorogi, D., Bolló, B., Szabó, Sz.: Effects of external disturbances on the performance of an axial cooling fan. Rev. Fac. Eng. Analecta Technica Szegedinensia **13**, 48–55 (2019)

Robotic Production Oriented Engine Design and Manufacturing

Rabab Benotsmane(✉) and László Dudás

University of Miskolc, Miskolc, Hungary
benrabab1@gmail.com

Abstract. Unmanned, fully robotic manufacturing is one of the trends in Industry 4.0 industrial revolution. To achieve this, the entire product design must prepare for it and the manufacturing and assembly process must be subjected to work without human labor. The article describes the construction of an innovative rotary piston internal combustion engine, which, due to its simplicity and optimized design for standard production equipment, can be easily manufactured and assembled on a production line using only robotic service. The first part of the paper describes the design of the engine structure, while the second part shows the structure of the unmanned automated production line. The innovative engine construction applies standard involute gears as working elements and simple engine body assembly. The advantage of the construction beside the clear rotary motions is the smooth operation owing to the many mating tooth pairs that form working cavities. Due to the simplicity of the parts, the production on automated manufacturing and assembly lines is easy.

Keywords: Rotary engine · Internal combustion · Unmanned production · Robotic service

1 Introduction

1.1 The Object of the Paper

Unmanned, fully robotic manufacturing is one of the trends in Industry 4.0 industrial revolution. The design of new product construction has to aid the unmanned automatic production of the product and the manufacturing plan has to focus on automatic production with robotic service. At first the paper introduces a novel rotary piston internal combustion engine designed for easy automated production while the second part describes a possible arrangement of the unmanned production line.

Though the future in the area of activators is of electric motors, because of the limited capacity of recent batteries justifies the use of internal combustion engines. The innovation of these types of engines focuses on the improvement of rotary engines having rotating working parts only. The introduced construction relates to a rotary piston internal combustion engine having working cavities formed between the teeth of internally toothed ring gear and associated gear, both have involute profile teeth. The field of

K. Jármai and K. Voith (Eds.): VAE 2020, LNME, pp. 390–400, 2021.
https://doi.org/10.1007/978-981-15-9529-5_34

application of such an engine is the conversion of the energy of liquid fuels which form an explosive mixture with air into a rotating motion.

The paper overviews some similar engine inventions demonstrating the active research of such constructions and demarcating the suggested construction from that solutions. Then the novel, automated-production oriented construction will be introduced in detailed form, presenting the build up with figures. In the sequel, the working process will be described shortly.

The second part of the paper presents the arrangement of the automatic manufacturing and assembly line for the unmanned production of the engine. The production is planned for middle series, with flexibility in engine properties and sizes.

Finally, a conclusion closes the paper summarizing the results and delineating the directions of possible improvements in the design and in the planning.

1.2 State of the Art of the Gears-Using Rotary Piston Combustion Engines

Many types of rotary piston internal combustion engines are known in the art. The inventor of the most widely used rotary piston engine, the Wankel engine, originally patented a rotating housing and rotating piston around their own axis construction, however, due to the simpler explosive mixture intake and exhaust gas discharge, the fixed housing, planetary piston design became widespread (Fig. 1).

Fig. 1. The original Wankel engine [1] (left) and the widespread variant modified by Hanns Dieter Paschke [2, 3] (right).

In both embodiments of the motor, both the housing and the rotary piston have regular protrusions leaving the circular shape, a kind of teeth, but it is an essential feature that the designs use a pair of internal and external gears separate from the working surfaces to synchronize the respective rotations. Thus, separate motor elements are used to achieve work and synchronization, which increases the complexity and weight of the motor.

Internal and external mating gears are also disclosed in WO 2012/166079 patent [4], which describes a hybrid internal combustion engine. The purpose of the use of gears here is also to ensure the forced movement of the working elements, thus the aforementioned disadvantages of the separate elements for performing the various functions also appear here.

Similarly, synchronizing gears are used to provide the speed of the working wheels by the Rotary Device patent WO 2011/000050 A1 [5], in which the design, which can be used as a multi-purpose motor, including special rotating gears, uses wheels with special external gears. The production of non-standard teeth is expensive, instead it is preferable to use the proposed standard involute teeth.

U.S. Pat. No. 6,484,687 B1 discloses a rotary machine using a single-toothed internal and external gear and its thermal cycle [6]. Although this solution already features internal-to-external gear design for single-tooth wheels, like the Wankel engine, their speed is provided by separate synchronizing gears on the two rotating elements, increasing the complexity and weight of the engine as Fig. 2 shows.

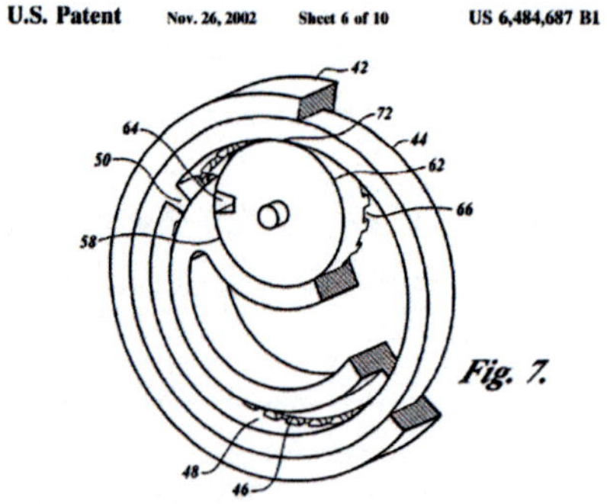

Fig. 2. Separate synchronizing gears on the two rotating elements marked with numbers 46 and 66 in the 7th figure of the patent description [6].

The international patent WO 86/00957, 1986 [7], uses four toothed and associated outer gear wheels. The compression effect between the teeth mentioned by the author cannot be advantageously ensured by the tooth profiles shown with the cavities closing for a very short time. The manufacture of teeth requires non-standard tools. Furthermore, the design also uses separate synchronizing gears due to special tooth profiles that are not

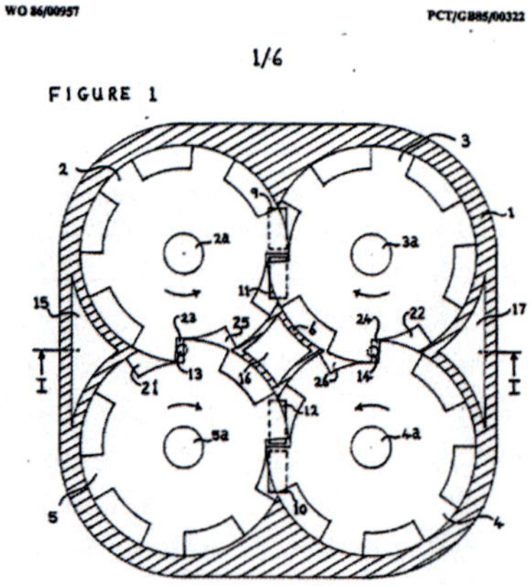

Fig. 3. Construction with gears having non-standard tooth profiles [7].

suitable for driving each other, with their mentioned disadvantages, as can be checked in Fig. 3.

Though the epicyclic tooth profile used would allow the internal combustion rotary engine disclosed in U.S. Patent No. 3,323,499, 1967 [8], to operate without synchronizing gears, the structure, which consists of two external gears forming explosive chambers, also uses separate synchronizing gears, presumably to avoid strong wear of the corners on the tooth sides, as the description emphasizes the advantage that there are virtually no friction parts in the motor. See Fig. 4.

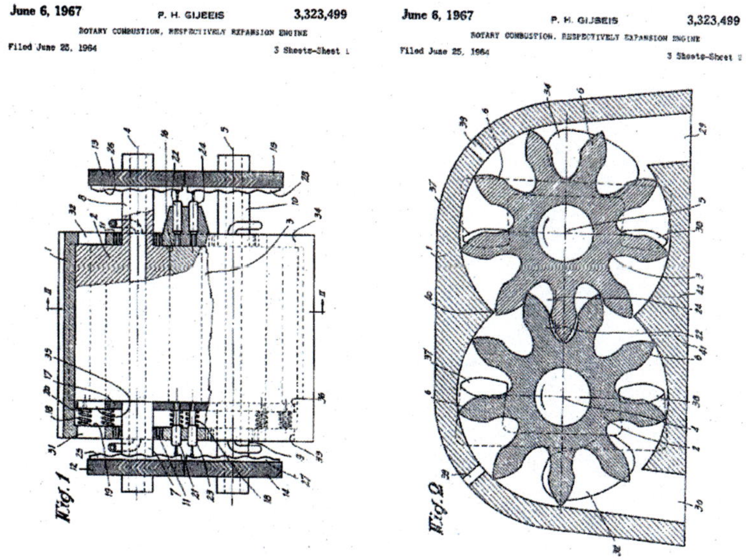

Fig. 4. Synchronizing gears at the two sides of the motor (left) and epicyclic gears forming working chambers (right) [8].

A very similar embodiment to the previous invention is shown in GB Patent Description No. 2313627 A, 1997 from the United Kingdom [9]. The novelty is the use of separate sealing strips built into the teeth, as well as the mention of toothed wheels made of materials other than the usual metal. The solution also uses synchronizing gears in addition to the gears having teeth that delimit the explosive areas.

The above solutions all differ from the design suggested in this paper as the rotational relationship between the toothed rotating elements that partially delimit the explosive spaces is provided by synchronizing gears, thereby increasing the complexity and weight of the structure, with all their consequences, e.g. possibility of failure, need for maintenance. The suggested engine construction uses normal gears as rotary pistons because the manufacturing of such machine elements is easy with the perfect gear cutting or gear profiling machine tools. Sometimes the cutting of a gear is a one-step cutting process with a pull-broach, so can be very efficient.

U.S. Patent No. 1656538 [10], issued in 1928 connects four conventional external gears and uses the interconnected closed cavities and the expanding and shrinking cavities between the tooth sides and the housing wall as the expanding and compressing space. In this invention, it is already recognized that it is not necessary to use separate synchronizing gears in addition to the chamber-forming gears, the two functions can be combined. However, due to the overdetermination of the four-wheel system, it is necessary to take appropriate measures to ensure the simultaneous, correct pressure-resistant sealing of each of the surface pairs. Furthermore, due to the small value of the contact ratio, the soon-to-open closed explosion cavity reduces work efficiency and has a strong dynamic effect. See Fig. 5.

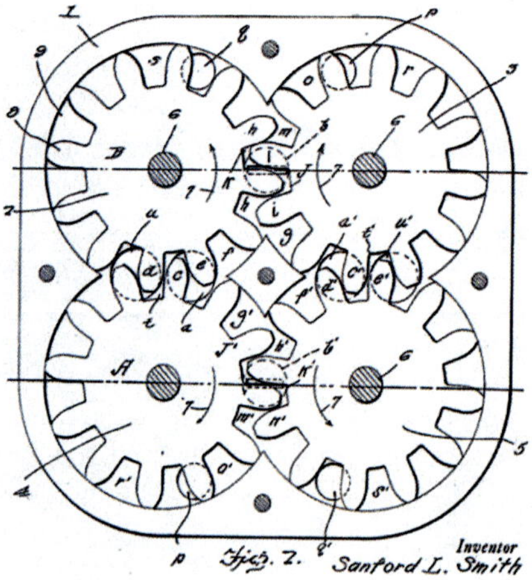

Fig. 5. Engine construction with four conventional gears [10].

The recognition that the teeth of the synchronizing gears can also form the compression and explosion chambers is reintroduced in U.S. 3709199 patent. This construction has the same drawbacks as those mentioned in the invention shown in Fig. 5, the elimination of which is the object and goal of the present suggested design.

2 New Rotary Internal Combustion Engine

From the above patent analysis, it can be concluded that a rotary internal combustion engine operating with variable volume cavities such as compression and expansion chambers enclosed by the teeth of internally and externally connected gears is not known in the art. It is the object of the present paper, on the one hand, to use a previously non-existent internal gear-like design as an internal combustion engine and, on the other hand, to increase the small number of mating teeth achieving longer intertwining arc length.

2.1 The Construction

The suggested design, shown in Fig. 6 has working cavities formed between the teeth of externally toothed and internally toothed mating involute gears (1,2). In an integrated housing consisting of an engine block (3), a bottom cover (30) and a top cover (not shown) with the gear (1) pivotally mounted in the covers and a ring gear (2) pivotally mounted in the engine block (3), has a fuel inlet (29), igniters in the igniter ports (32) and an exhaust port (20). The top cover is identical to the bottom cover (30). In order to ensure the separation of the combustion product from the intake air and the fuel-air explosive mixture, closing gears (16,17) are used to ensure the separation of the combustion product.

A fundamental difference between the solution described in the latter two patents and the internally toothed design described herein is that the mentioned inventions use external toothed gears. In contrast to gears with the same modulus and number of teeth, the advantage of the connection using external and internal teeth is that the teeth of the two gears are intertwined in a larger arc, the contact ratio is higher, which means that there will be more closed cavities for both compression and expansion to realize the rotating

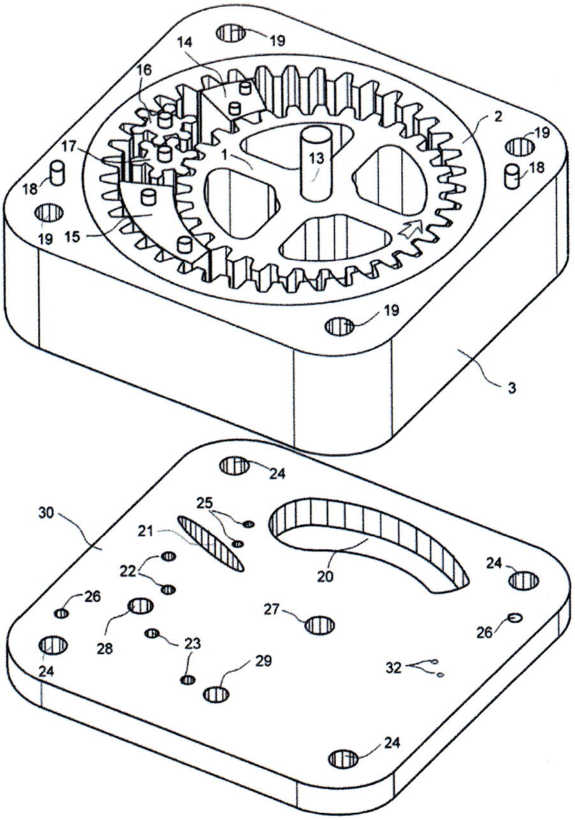

Fig. 6. 3D drawing of the suggested rotary piston engine without the top cover.

effect. The many working cavities result in smooth operation with less dynamic effect at the opening chamber. The object of the present construction was to provide a smooth-running, efficient internal combustion rotary piston engine which can be manufactured in automated manufacturing lines even in large series due to the applied simple parts. Fig. 6 shows the axonometric view of the main parts of the engine.

2.2 The Working

The illustration of the in-use operation of the shown embodiment is made referring to the markings used in Fig. 7.

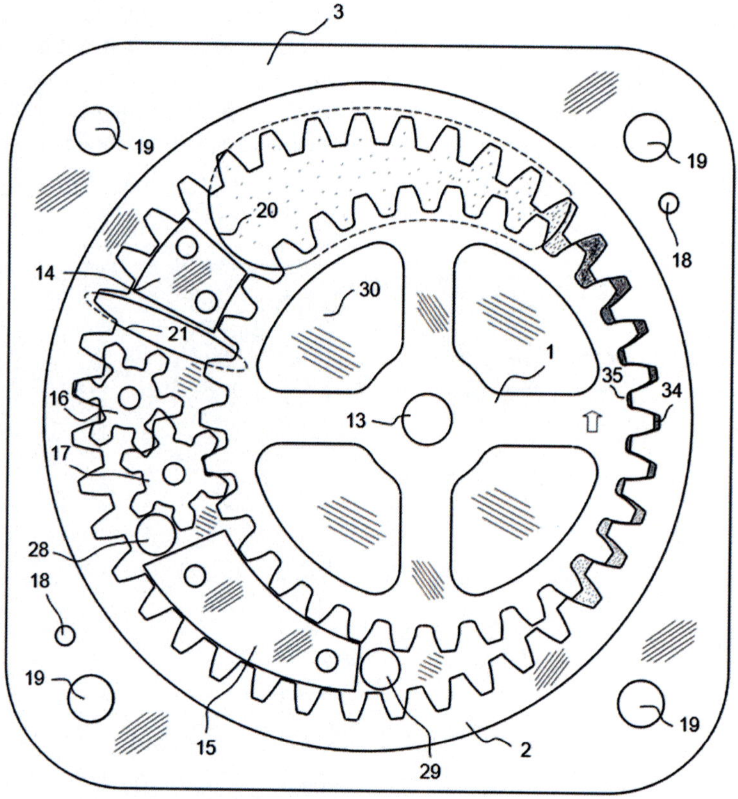

Fig. 7. Top view of the opened engine to introduce the operation.

The engine must be started with external energy by turning the gear (1) in the direction of the arrow by means of the shaft (13). During operation of the motor, as a result of the rotation, the air flowing in through the air intake opening (28) is sucked in by the air-moving effect of the teeth of the rotating gear (1) and the ring gear (2) and by the conveying effect of the tooth cavities between the teeth of gears (1,2) and air deflector (15). The liquid fuel is introduced into the moving air by means of a carburetor, injector or other suitable means placed in the fuel inlet (29), thus forming an explosive mixture with the air. The explosive mixture enclosed in the tooth cavity is compressed by the cavities that narrow during movement until it enters the lower ignition cavity (34) and the upper ignition cavity (35), where it explodes as a result of the ignition used. In the upper mating section of the moving gear (1) and the ring gear (2), after ignition, the cavities expand under the tensioning effect of the high-pressure exploded mixture, which is achieved by rotating the gear (1) and the ring gear (2). The resulting torque can be got from the motor on the shaft (13). As the tooth cavities expand, the teeth open after a time as the teeth move away, and the combustion product, which flows out and is still under significant pressure, leaves the motor through the exhaust opening (20). This is also forced by the smoke blocker (14), which stops its flow. The already heavily sparse combustion product trapped in the teeth of gear (1) and ring gear (2) escapes into the space behind the smoke blocker (14). A residual smoke opening (21) is also used after the smoke blocker (14) and before the closing gears (16,17) to let off the sparse combustion product. The advantage of this embodiment, which results from the use of the upper closing gear (16) and the lower closing gear (17), is manifested in operation. The combustion product passing by the smoke blocker (14) cannot enter the space of the air intake opening (28) because the upper closing gear (16) and the lower closing gear (17) close in front of it. This solution significantly increases the engine efficiency. The retained exhaust gases leave the motor through the residual smoke opening (21).

2.3 Advantages and Disadvantages of the Suggested Engine

The advantages of the introduced rotary piston internal combustion engine are as follows: Compared to designs with connected gears with the same number and modulus of external teeth, the use of internal teeth has a longer mating length so larger contact ratio and correspondingly more seamlessly closed tooth cavities, resulting in higher efficiency and smoother running. The clear rotary motion makes possible high speed. Thanks to the built-in air deflector, the degree of compression is larger, and this increases the efficiency. Due to the closing gears used, there is no mixing of combustion product and explosive mixture as a result of the separated exhaust and intake side, which also increases efficiency. Simple 2D embodiment adumbrate cheap manufacturing and assembly. Because of using standard gears having involute teeth, large-scale production is inexpensive.

The disadvantages of the introduced rotary piston internal combustion engine are as follows: The sealing of the elements needs precise manufacturing, wear-resisting materials with good sliding characteristics.

3 Automated Production of Engine

In this section, the analysis of the buildup of the manufacturing and assembly line will be performed. The engine construction is simple enough to make it possible solving such a task. In the assembly line planning is supposed that the conventional mass-products, like bearings, bolts, assembly materials like grease, etc. are ready and can be get from the market. The parts that have to be manufactured are gear, ring gear, closing gears, engine block, bottom and top covers, smoke blocker, air deflector. These seven types different components can be classified into two homogeneous-parts groups: gears and plates. The gears group includes the gear, ring gear and closing gears, while the plates group contains the engine block, bottom and top covers, smoke blocker, air deflector. For producing the gear-type components the next technology is supposed: cutting, turning, teething, heat treatment, grinding. For producing the elements in the plates group the planned manufacturing steps: cutting, face milling, contour milling and drilling, heat treatment, surface and internal grinding. The last step of production is the assembly.

To accomplish these manufacturing and assembly steps the following machine tools and manufacturing cells are planned: for the gears group – CNC cutting machine, CNC turning machine, CNC gear hobbing and broaching machine, automated heat treatment furnace, CNC external cylindrical machine, CNC gear grinder. For the manufacturing of the members of the plates group – CNC cutting machine, 2.5 axis CNC machining center, automated furnace, CNC surface grinding machine, CNC internal grinding machine. These machines are advantageous for producing middle series of motors. For mass production instead of CNC machines dedicated automated single-purpose machine tools are practical. For performing the logistic tasks programmed robot arms are planned. Fig. 8 shows the buildup of the gear production line with the servicing robots.

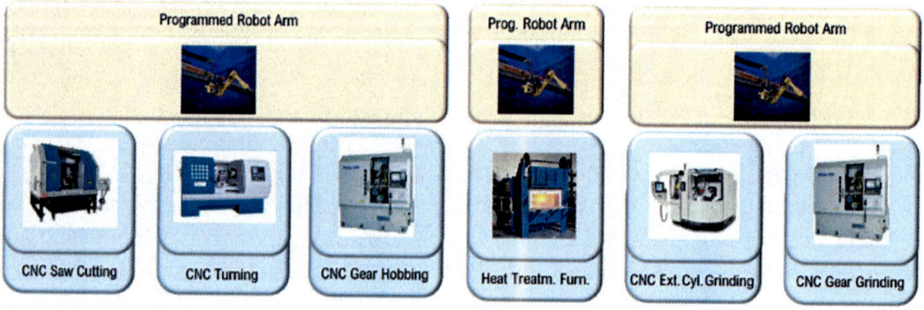

Fig. 8. Manufacturing line components for the gears production.

The buildup of the production line for members of the plates group is shown in Fig. 9.

After manufacturing the engine parts, the fully automatic assembly process follows, also using programmed robot arms. In this process the additional conventional mass-product engine parts, like bearings, etc. also will be building in. The assembly is carried out expediently in an assembly cell where two programmed robot arms are arranged opposite to each other, on the two sides of the assembled engine, corresponding to the mirrored structure of the engine (Fig. 10).

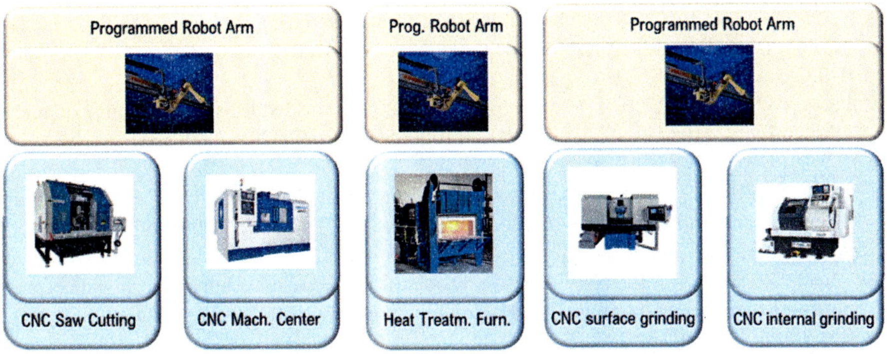

Fig. 9. Manufacturing line components for production of the members of plates group.

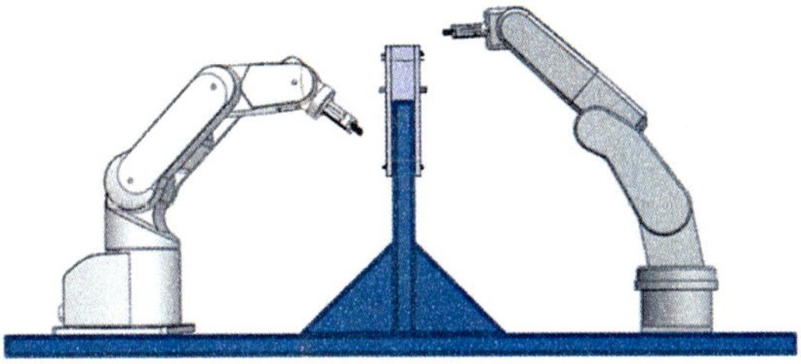

Fig. 10. Automated assembly of the rotary piston internal combustion engine.

4 Summary

The paper presented a novel rotary piston internal combustion engine after discussing the different types of gear-based rotary piston engines. The engine characterized by conventional involute gears as pistons and as a consequence, smooth operation, due to the many teeth acting as small pistons. The engine has only a few elements, as one of the design goals was the simplicity because of the intended fully automatic production. After the detailed description of the design of the engine, the working process was also introduced. Then in the third part of the paper, the analysis of the possibility of the automated production was investigated. The upshot was the giving of the steps of the manufacturing technology parallel to the necessary applied machine tools.

In the future the working simulation of the engine is scheduled, and the refinement of the production process.

References

1. Craig: NSU Wankel rotary engines and cars. https://cp_www.tripod.com/rotary/pg05.htm. Accessed 06 July 2020
2. Der Wankelmotor und sein Erfinder. https://www.nsuro80.ch/das-auto/der-motor/. Accessed 06 July 2020
3. https://totalcar.hu/galeria/totalcar/hirek/2012/06/27/mazda_wankel_galeria/6. Accessed 06 July 2020

4. Patent WO2012166079A1 Hybrid Internal Combustion Engine. https://worldwide.espacenet.com/patent/search/family/056120493/publication/WO2012166079A1?q=WO2012/166079. Accessed 06 July 2020
5. Patent WO2011000050A1 Rotary Device. https://worldwide.espacenet.com/patent/search/family/043410375/publication/WO2011000050A1?q=WO2011/000050a1. Accessed 06 July 2020
6. Patent US6484687B1 Rotary machine and thermal cycle. https://worldwide.espacenet.com/patent/search/family/025309494/publication/US6484687B1?q=US%206484687%20B1. Accessed 06 July 2020
7. Patent WO8600957A1 Rotary Positive Displacement Machine. https://worldwide.espacenet.com/patent/search/family/010564102/publication/WO8600957A1?q=WO%2086%2F00957. Accessed 06 July 2020
8. Patent US3323499A Rotary combustion, respectively expansion. https://worldwide.espacenet.com/patent/search/family/003845501/publication/US3323499A?q=US%203323499. Accessed 06 July 2020
9. Patent GB2313627A Rotary engine. https://worldwide.espacenet.com/patent/search//. Accessed 06 July 2020
10. Patent US1656538A Internal-combustion engine.https://worldwide.espacenet.com/patent/search/family/022418891/publication/US1656538A?q=Us%201656538. Accessed 06 July 2020

Experimental Investigation of the Air-Side Heat Transfer Coefficient on Louver Finned Tube Automotive Radiator

Máté Petrik(✉), Antal Erdős, Károly Jármai, and Gábor Szepesi

University of Miskolc, Miskolc-Egyetemváros, Hungary
petrik@uni-miskolc.hu

Abstract. In this article, we experimentally and analytically studied the heat transfer characteristics of a louver finned automotive radiator. These automotive radiators are one of the most critical devices in the engine cooling systems. These are fluid-gas type heat exchangers, where the fluid phase is cooling fluid, which heated in the engine block, and the gas phase is air from the wind edge. These radiators are commonly used in every vehicle from the simplest scooter to the trucks, but their size is highly depending on the performance of the engine. This study would like to summarize the theoretical background of the louver finned heat exchangers, to show the expressions to calculate the heat performance of the constructions. For the measurements, an experimental setup was made. The airflow provided by fans. In the function of the frequency of the current, the mass flow rate across the fan can be variable. From the results of the measurements, an analytical calculation method can be developed, which fitted to the empirical results.

Keywords: Heat transfer · Finned tube · Automobile

1 Introduction

Finned tube heat exchangers (FTHEs) are widely and commonly used heat transfer equipment in industrial environments, in households in air conditioners and in the vehicles as a part of the engine cooling system. In contrast to other types of heat exchangers, where the aim is to increase the value of the heat transfer coefficient, in case of finned tube heat exchangers the heat transfer surface has been increased by a large value. As a result, the equipment enclosed in a relatively small volume, but a large heat transfer surface is available [1].

Compared to an average heat exchanger, the phenomenon of heat conductivity plays a much more role in case of FTHEs. This is explained by the fact that due to the extended surface the hot fluid from the tube side to the air side in two steps: firstly, by convection, from the fluid to the tube, and secondly by heat conduction from the tube to the fins. This causes a unique temperature distribution on the surface of the fins, and the difference

K. Jármai and K. Voith (Eds.): VAE 2020, LNME, pp. 401–416, 2021.
https://doi.org/10.1007/978-981-15-9529-5_35

of the temperature of fins and air will be the driven force of the heat transfer. Since the temperature of the fins is hard to calculate, a fin efficiency parameter is introduced [2].

This present paper deals with especially with the automotive radiators. The radiator is an essential part of the internal combustion engines. In the combustion chamber, the fuel ignites and some part of generated energy moves the piston. The other part of the energy heats up the engine block, which effect is disadvantageous for the engine since this must be cooled. Efforts are being made to use nanoparticles in the coolant [3, 4]. It follows from their results that all important parameter for heat performance (thermal conductivity, heat transfer rate, Nusselt number and heat transfer coefficient) highly depends on the concentration of the nanofluid in coolant, higher concentration causes higher heat performance, which is advantageous property and higher pumping rate, which is disadvantageous. Instead of nanoparticles, ice slurry also could be a heat performance enhancing additive [5]. In this case, the latent heat required to melt the ice will improve the heat transfer coefficient on the coolant side.

The radiator is also an important component in case of state-of-the-art fuel cell vehicles (FVCs). Gong *et al.* [6] have made comparisons about flat fins, Windward bends, louvered fins and Windward-Louvered fins (WFLs). They proved that WFL is 1.2–1.3 times higher than Windward bends, but the pumping powers are the same, and the heat performances are also enhanced at the tube walls in this case. These radiators can be used not only for FVCs, but also for dense-array concentrator photovoltaic system, as Chong and Tan [7] showed in their paper.

Another important aspect for designers could be the optimal sizing of the devices. This can be done in two ways. First, the pure mathematical method, with analytical relationships and some optimization algorithm [8, 9], or using some computational fluid dynamics software. In case of automotive radiators, louvered fins are the most commonly used, but their geometric modeling and meshing would also exceed the performance of up-to-date computers, so different techniques can be used. The fins can be replaced by porous volume, which reduces the number of elements by orders of magnitude and causes less computational time [10, 11].

2 Car Cooling System

The design and the analysis of the cooling systems in the automotive industry are very advanced. It is expected that for example Formula One racing teams and other researchers carry out extensive research in the field. The available published literature allows to understand the vehicle cooling system operation, the principles of the heat exchangers and their heat load capacity. So, this is essential to analyze the cooling system [12, 13].

Today's high-performance engines extensive amount of heat driven by internal combustion process. So the primary goal of the cooling system is to maintain proper and reliable engine performance between the given temperature range, which is usually around 95 °C, but the thermostat usually starts to operate around 75 °C, this device controls the minimum operating temperature, and the maximum comes from the coolant boiling point and the radiator capacity. But these temperatures depend on the type of the fuel and the type of the car. If the engine operates below, it affects the engine, because the burnout in the engine could not be complete. This leads to unburned fuel in the system, and that could cause serious malfunctions in the engine for example corrosion in the cooling system and the engine. If the engine operates above the recommended temperature range could also cause serious problems, such as oxidation, and varnish. The high temperature also reduces the viscosity of the oil which leads higher friction and it cause performance loss. The problem of the pre-ignition could be also a problem due to the high temperature in the cooling system [12, 13].

The purpose of the heat exchanger is to remove heat from the engine to the ambient air. In theory, the heat generated by the engine is fully utilized. The energy that comes from the combustion of the fuel divides into three parts: one third turns into mechanical work, the remaining tow third goes to heat that goes to the exhaust system and the cooling system.

There are two main types of the cooling system: the difference between them is the method of the heat removal process, so there are air coolers and liquid coolers. The engines using the combination of these two, so this is an air to liquid heat exchanger. The engine cooling is achieved by circulation of water, and to cool back this water there is a heat exchanger is used with fans (usually one or two is used).

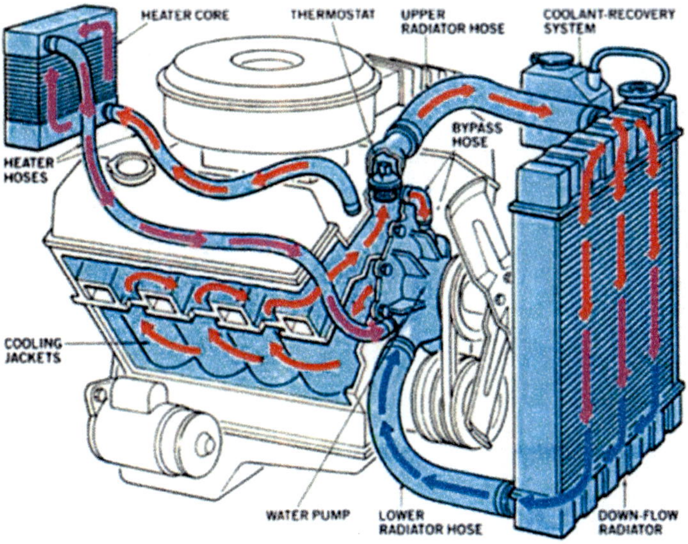

Fig. 1. Vehicle cooling system [13]

This heat exchanger is a louver fin type one. The principal part of it is a combination of harmonica tubes and fins. This has internal and longitudinal openings and being bent zig-zag several times to thereby form operations parallel with each other. For the fabrication of the core of the radiator generally, the core stacking procedure is used, during which make the connection between the fins and the tubes. The number of the layers depends on the desired performance and the dimensions of the radiator. After the core is complete with brazing or welding the next step is to apply the header plates to hold the tubes and the fins in their position. These plates contain the mounting points for the radiator and also the tanks that contain the cooling water. The volume of the coolant is also important for the engine and the design of the radiator.

As it could be seen in Fig. 1 above the radiator consists a tube bank and from which the pipes start, these are brazed or welded to the air-side fins [13]. These fins are mostly made of aluminum which makes the heat transfer more effective, but it made the welding complicated. If the joint is still made by welding, there is a risk of the puncturing of the tubes, which could lead to the overheat of the engine. So, the brazing is safer in this respect, because for the filler material selection a material could be chosen with a lower melting point than the base materials. The brazed joints are strong, aesthetic and corrosion resistant. This technology also enables to make multiple joints at the same time [14].

The welding is also an opportunity to connect the tubes and the fins. For this procedure, there is a special machine, which makes the joint. These machines usually using TIG (Tungsten Inert Gas) welding method, which is one of the low heat input welding technology, with which an aesthetic joint with good mechanical properties can be established and the risk of damage to the base material is reduced due to the flexibility control of the arc. These machines using PLC or other computer control and thanks to which it is produced relatively fast and in the same quality with good reproducibility. The TIG welding is using filler material which is usually rod or wire type, which has to be chosen for the chemical composition of the base material. Filler material manufacturers give recommendations for the selection in their catalogues [15].

Some parts of the radiator, for example pipe connections, fans and other body parts are made of composite. This composite is the most commonly polymer matrix which is reinforced with short line glass fiber. This composite material is also used for door beams, drive shafts and tanks. The glass fiber has good thermo-mechanical properties, which provide a wide range of usability, and it is also cheaper than the carbon fiber. This material could be recycled when it reaches the end of the lifetime of the device. This is usually a mechanical method, where the composite is first shredded and after ground and the pellet could use as a raw material of a future workpiece. Even today, numbers of studies deal with recoverable energy and reusability [16].

3 Experimental Setup

3.1 Materials and Devices

The experimental setup consists of two major parts, which are the hot water tank and the automotive radiator. The schematic drawing of the experimental layout is shown in Fig. 2. The two most common media were used in the measurements, which were water and air. The hot water was produced and stored in a 40 L water tank heated by electricity, which come from the water supply system and it did not receive any treatment. The water was circulated by a pump which first passes through a rotameter so that the volume flow rate can be measured, then passes through the radiator, where can cool down.

The cooling air flows through the radiator by two axial fans, which speed can be controlled. To determine the heat performance of the radiator, the inlet and outlet temperatures of the water were measured with K type thermocouples. The data logger was a

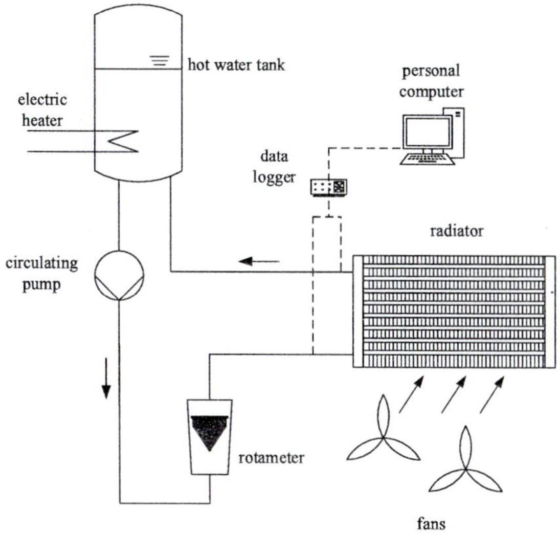

Fig. 2. The experimental layout of the measurement

Quantum X type MX1609 thermocouple amplifier, while the data processing was done with catman®Easy software.

3.2 The Automotive Radiator

The measured device was a radiator of a car, which schematic drawing is shown in Fig. 3. The water flows in 2 × 32 oval tubes with two paths. The radiator had 650 fins with 0,1 mm thickness, the spacing between them were 0,9 mm. The material of tubes and fins were aluminum. The geometric parameters of the radiator shown in Table 1.

Table 1. The geometric parameters of the investigated radiator

Specific parameter	Sign	Value
Length of the radiator	L	650 mm
Height of the radiator	H	415 mm
Width of the radiator	W	35 mm
Thickness of the fins	t_f	0,1 mm
Thickness of tubes	t_t	0,25 mm
Number of fins	n_f	650
Number of tubes	n_t	64
Longitudinal size of a tube	d_{hor}	12 mm
Vertical size of a tube	d_{ver}	2 mm
Horizontal spacing of the tubes	s_{hor}	10,6 mm
Vertical spacing of the tube	s_{ver}	3 mm
Spacing between fins	l	0,9 mm

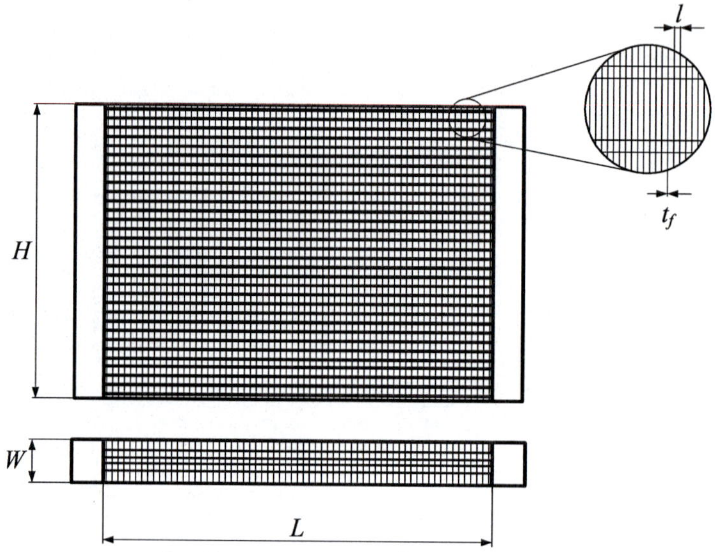

Fig. 3. The investigated radiator with its main parameters

The layout and the spacing of the tubes are described in Fig. 4.

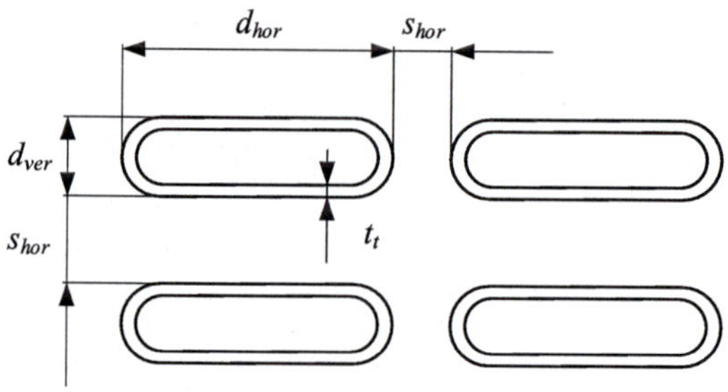

Fig. 4. Tube layout of the investigated radiator

One of the most important specifications of a radiator is the heat transfer surfaces. Because of the fins, more surfaces most be calculated in case of a heat exchanger checking. These surfaces are summarized in Table 2.

3.3 Measurements

During the measurements, three values were detected with the data logger: the inlet and outlet temperature of water (which enters and leaves the heat exchanger) and the temperature of the heated air (which flows through the heat exchanger). Furthermore, the volume flow rate of water was measured with rotameter and these values were recorded manually since in one measurement case the value of the volume flow rate was constant.

Table 2. Heat transfer surfaces required for calculations

Specific surface	Sign	Value
Total heat transfer surface	A	17.94 m^2
Heat transfer surface of fins	A_f	16.96 m^2
Heat transfer surface of 1 bare tube	A_{t0}	0.017 m^2
Heat transfer surface of the inner side	A_i	1.09 m^2
Heat transfer surfaces of the bare tubes	A_0	1.09 m^2

To determine the air velocity a hand-held low-speed vane anemometer was used. As the air flow was provided by two fans mounted on the back of the heat exchanger, a non-homogeneous velocity profile developed.

To determine the average air velocity, an inlet channel was created on the inlet side of the device. This was divided into 4 × 4 equal sized areas, and the velocity of each sub-area was measured at the same fan power. This division is shown in Fig. 5.

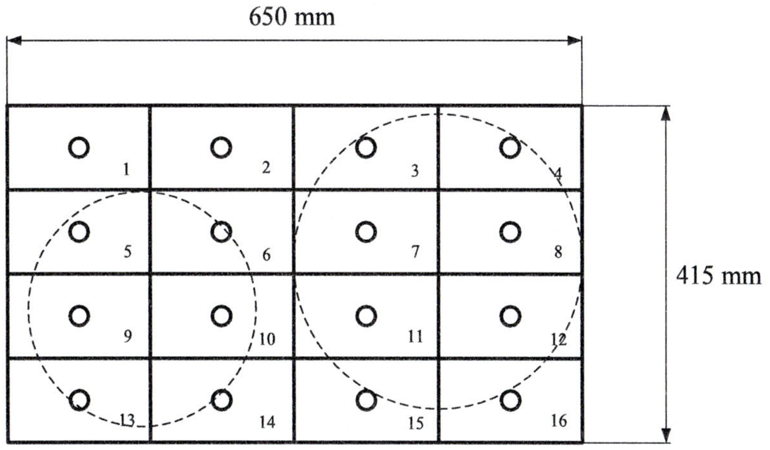

Fig. 5. Division of the inlet channel for air velocity measurement

The measured velocities in two different operating states are given in Table 3 and Table 4 (in the order of the physical positions). These measurement positions have been determined in accordance with engineering practice.

The results show that the velocities near the axis of the fans and the enclosed areas are lower than near the outer diameter of the fan blades. Therefore, the average of the measured velocities must be taken, which is 0.94 m/s for Operation mode 1 (M1–M8), and 1.21 m/s for Operation mode 2 (M11–M18).

Table 3. Measured velocities in Operation mode 1

0.35	0.65	1.07	0.73
1.14	1.04	0.75	0.87
1.46	1.31	0.92	0.91
1.28	1.22	0.85	0.5

Table 4. Measured velocities in Operation mode 2

0.7	0.82	1.34	1.1
1.3	1.3	1.2	1.2
1.7	1.5	1.2	1.1
1.54	1.55	1.03	0.8

3.4 Heat Performances

The aim of the investigations was to determine the performance of the heat exchanger. This performance cannot be measured directly but must correspond to the performance from cooling the water or warming the air. Determining the amount of heat from water cooling is the simplest, which can be done with the well-known relationship in Eq. 1:

$$Q_w = c_{p,w} \cdot m_w \cdot \left(T_{w,in} - T_{w,out}\right), \tag{1}$$

where $T_{w,in}$ is the inlet and $T_{w,out}$ is the outlet temperature of water, c_{pw} is the specific heat of water in the medium temperature (the arithmetic mean of the $T_{w,in}$ and $T_{w,out}$), and m_w is the mass flow rate of the water. The humidity of air cannot be disregarded when calculating the warming of the air. At the time of the measurement, the relative humidity of the room was 60%, and the temperature was 29 °C. From these two data, the values of specific humidity and enthalpy can be determined, which were 14.24 g/kg for humidity and 64.5 kJ/kg for enthalpy. Based on these, the energy required to heat the air can be determined by Eq. 2.

$$Q_a = m_a \cdot \left(H_{a,out} - H_{a,in}\right) \tag{2}$$

The m_a is the mass flow rate calculated from the velocity and the density, $H_{a,out}$ is the enthalpy of outlet state and $H_{a,in}$ is the enthalpy of inlet state. As all the material properties, the enthalpy is also a function of temperature, which can be calculated by Eq. 3 for 14.24 g/kg specific humidity, between 30 °C and 50 °C.

$$H(T) = (1031.299 \cdot T + 35,600.43) \cdot \frac{\mathrm{J}}{\mathrm{kg}} \tag{3}$$

In steady state, the heat performances calculated by Eq. 1 and 2 must be the same, which also means the performance of the heat exchanger. By definition, this can be calculated with the following relationship:

$$Q_{he} = F \cdot U \cdot A \cdot LMTD \tag{4}$$

In Eq. 4, the value A means the total heat transfer surface of the heat exchanger, which is described in Table 2. The U is the overall heat transfer coefficient, the $LMTD$ means the logarithmic mean temperature difference, which can be calculated from the inlet and outlet temperatures:

$$LMTD = \frac{\left(T_{w,in} - T_{a,out}\right) - \left(T_{w,out} - T_{a,in}\right)}{\ln\left(\frac{T_{w,in} - T_{a,out}}{T_{w,out} - T_{a,in}}\right)} \tag{5}$$

The F factor represents the number of passes. Equation 5 refers to counter current, but in this case, it was a cross-current, in addition, the water flows in parallel tubes. Its value depends on the inlet and outlet temperatures and is determined graphically, as it is seen in Fig. 6.

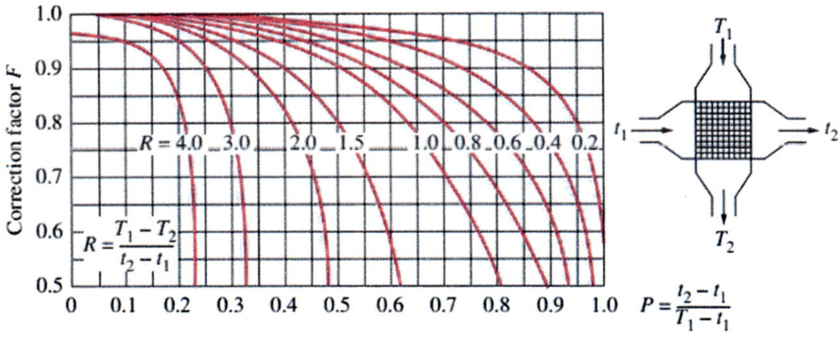

Fig. 6. Graph for determination of F factor [17]

In the measured cases, this modification factor was around 0.95, which was used in the calculations.

4 Results of the Measurements

Several measurements were performed to simulate different operating conditions as well as different atmospheric conditions. During the measurements, the formation of the steady state was expected, and the average of the measured results was used for the calculations. These results are shown in Tables 5, 6 and 7.

In the table, the $\dot{V}$ means the volume flow rate, the T means temperature, $\dot{Q}$ means heat performance, while the w subscripts refer to water and a refers to air. The heat performances were calculated with Eq. 1 and Eq. 2 and $LMTD$ calculated with Eq. 5. The overall heat transfer coefficient in the last column of the table, which belongs to the given flow conditions of the heat exchanger, has been calculated from Eq. 4. The most important issue when designing heat exchangers is to determine this coefficient. In significant part of heat exchanger equipment, this is done by empirical correlations. The aim of the present study is to compare the results of the correlations in the literature with the measurement results. At Case 1, the inlet temperature of air was 29.5 °C, the relative humidity was 50% and its velocity was 0.94 m/s.

Table 5. Results of measurement of Case 1

	$\dot{V}w$ (l/h)	Tw,in (°C)	Tw,out (°C)	$\dot{Q}w$ (W)	$\dot{Q}a$ (W)	$LMTD$ (°C)	U ($Wm^{-2}K^{-1}$)
M1	100	63.88	39.3	2821.30	2822.08	16.78	9.86
M2	200	65.86	49.36	3778.28	3774.17	22.43	9.87
M3	300	66.59	54.22	4243.81	4247.92	24.52	10.15
M4	400	67.28	56.80	4790.50	4788.79	25.18	11.16

Table 6. Results of measurement of Case 2

	$\dot{V}w$ (l/h)	Tw,in (°C)	Tw,out (°C)	$\dot{Q}w$ (W)	$\dot{Q}a$ (W)	$LMTD$ (°C)	U ($Wm^{-2}K^{-1}$)
M5	100	66.74	41.88	2850.24	2847.09	16.18	9.94
M6	200	66.98	50.79	3705.30	3699.93	21.04	10.32
M7	300	67.30	54.80	4287.22	4286.36	22.24	11.31
M8	400	67.69	57.18	4803.39	4799.62	22.64	12.44

Table 6 shows the results of Case 2. At this case, the air velocity was unchanged compared to Case 1, it was 0.94 m/s, but the temperature was 31.6 °C and the relative humidity was 65%. For this condition, a new approximate function must be created, which is described in Eq. 6.

$$H(T) = (1038.091 \cdot T + 47{,}914.48) \cdot \frac{\mathrm{J}}{\mathrm{kg}} \tag{6}$$

Table 7. Results of measurement of Case 3

	$\dot{V}w$ (l/h)	Tw,in (°C)	Tw,out (°C)	$\dot{Q}w$ (W)	$\dot{Q}a$ (W)	$LMTD$ (°C)	U ($Wm^{-2}K^{-1}$)
M11	100	62.82	38.34	2810.97	2805.98	13.59	12.12
M12	200	63.22	45.44	4076.98	4074.92	17.20	13.90
M13	300	65.14	51.62	4642.34	4642.05	20.74	13.13
M14	400	67.68	55.45	5591.58	5592.39	22.65	14.48
M15	100	60.39	35.17	2899.21	2908.06	13.42	12.69
M16	200	60.79	42.67	4159.39	4149.87	17.52	13.91
M17	300	61.88	47.85	4824.58	4816.86	20.08	14.08
M18	400	63.15	50.89	5616.16	5607.71	21.21	15.53

Measurements were made for air velocity of 1,21 m/s, and the results are shown in Table 7.

5 Analytical Model

The thermal design of heat exchangers is iterative process based on empirical correlations. These empirical correlations are the result of several measurements. It is the responsibility of the designer to use the most appropriate correlations for their calculations. The VDI Heat Atlas functions were used for this louvered fin heat exchanger.

5.1 Heat Transfer Coefficient of Water Side

The calculation of the heat transfer coefficient on the water side is similar to other types of tubular heat exchangers. The flow cross section area as well as were the geometric dimensions of pipes were known. Since the tubes were not circular, the hydraulic diameter had to be used as characteristic geometry for the calculation of important dimensionless quantities. The definition of hydraulic diameter is

$$D_h = \frac{4 \cdot A_t}{K_t}, \tag{7}$$

where A_t is the cross section area and K_t is wetted perimeter of the tubes. The measurements were performed between 100 and 400 l/h volume flow rate, and in each case the type of the flow remained within the laminar range, therefore the value of the heat transfer coefficient was considered constant with a value of 1286 $\mathrm{Wm^{-2}K^{-1}}$.

5.2 Heat Transfer Coefficient of the Air Side

The first step in calculating the heat transfer coefficient on the air side is to determine the Reynolds (Re) and Prandtl (Pr) numbers. In the case of the examined fin type, the characteristic geometry should be the fin spacing, since this is the best way to characterize the flow. Another important aspect is the real air velocity. The measured velocities described in Tables 3 and 4 are the velocities in front of the heat exchanger, not the velocities between the fins. This characteristic velocity was higher values, because the free flow cross section was lower due to the fins and tubes. These velocities were 1.26 and 1.61 m/s respectively.

The empirical Nusselt (Nu) number correlation for the investigated inline tube layout is:

$$\mathrm{Nu}_a = 0.22 \cdot \mathrm{Re}_a^{0.6} \cdot \left(\frac{A}{A_{t0}}\right)^{-0.15} \cdot \mathrm{Pr}_a^{1/3}, \tag{8}$$

From this, the heat transfer coefficient α_m can be determined. However, due to the fins this coefficient must be corrected with the fin efficiency. In case of an extended surface, the temperature of the fins will be lower than the tubes, so the driven force will be also lower than without fins. Nonetheless, the heat performance will be higher due to the extended surface. To determine the fin efficiency, two factors must first be calculated:

$$\varphi' = 1.28 \cdot \frac{b_f}{d_{ver}} \cdot \sqrt{\frac{h_f}{b_f} - 0.2}, \tag{9}$$

$$\varphi = (\varphi' - 1) \cdot (1 + 0.35 \ln \varphi') \tag{10}$$

In Eq. 9, the b_f variable is the apparent fin width (17.5 mm), the h_f is the apparent fin height (12.97 mm). In the next step, the ratio of the heat convection and conduction must be formed.

$$X = \varphi \cdot \frac{d_{ver}}{2} \cdot \sqrt{\frac{2 \cdot \alpha_m}{\lambda_f \cdot \delta_f}}, \tag{11}$$

where λ_f is the thermal conductivity and δ_f is the thickness of the fins and the efficiency can be calculated with.

$$\eta_f = \frac{\tanh X}{X}, \tag{12}$$

and the virtual heat transfer coefficient is.

$$\alpha_v = \alpha_m \cdot \left(1 - (1 - \eta_f) \cdot \frac{A_f}{A}\right), \tag{13}$$

Finally, the overall heat transfer coefficient of the heat exchanger is.

$$k = \frac{1}{\frac{1}{\alpha_v} + \frac{A}{A_i} \cdot \left(\frac{1}{\alpha_w} + \frac{t_t}{2 \cdot \lambda_t}\right)}, \tag{14}$$

As the measurements were performed in the operating states on the same equipment, it can be stated that for a given series of measurements the overall heat transfer coefficient can be considered constant, 17.37 $\mathrm{Wm^{-2}K^{-1}}$ for Case 1 and 2 and 19.15 $\mathrm{Wm^{-2}K^{-1}}$ for Case 3.

5.3 Temperature Difference

The calculations show that both on the water and air side, the values of the individual heat transfer coefficients are constant in each series of measurements, therefore the overall heat transfer coefficient is constant. The thermal performance of the heat exchanger only depends on the temperatures of the fluids. In contrast, the values of the heat transfer coefficients counted back from the measurement are much smaller.

Table 8 shows the heat performances and heat transfer coefficients of the measurement and the calculations. It is clearly seen from the results, that the calculated performances give higher values than the measured ones. That means the value of temperature correction factor F chosen based on Fig. 6 is not applicable in this case. It has been shown in the calculations that the heat performance is practically only the function of

Table 8. Comparison of the measurements and calculations

	Measurement		*Analytical model*		
	$\dot{Q}ave$ (W)	U ($Wm^{-2}K^{-1}$)	U ($Wm^{-2}K^{-1}$)	$\dot{Q}calc$ (W)	F
M1	2707.68	9.86	18.75	5928.30	0.46
M2	3776.23	9.87	18.81	7570.99	0.50
M3	4245.87	10.15	18.84	8288.75	0.51
M4	4789.65	11.16	18.85	8516.80	0.56
M5	2848.67	9.94	18.74	5653.99	0.50
M6	3702.62	10.32	18.79	7091.32	0.52
M7	4286.79	11.31	18.81	7506.04	0.57
M8	4801.51	12.44	18.82	7645.84	0.63
M11	2808.47	12.12	20.81	5073.93	0.55
M12	4075.95	13.90	20.86	6438.49	0.63
M13	4642.20	13.13	20.91	7780.74	0.60
M14	5591.99	14.48	20.94	8511.35	0.66
M15	2898.05	12.69	20.80	3923.78	0.74
M16	4157.06	13.91	20.85	5535.98	0.75
M17	4995.20	14.08	20.89	6305.66	0.79
M18	5616.34	15.53	20.92	6891.94	0.81

the temperature difference since the value of the heat transfer coefficient is constant. However, the results show that at higher air velocities the difference between the measured and calculated value is smaller. This means the F factor depends not only the temperatures but on the contact times also. The determination of this set of curves for louvered devices requires further investigations.

In contrast to the data in the literature, the results of the measurements show that the heat transfer coefficient of the water side is not constant despite of the laminar flow. Equation 15 was used to calculate this coefficient.

$$\mathrm{Nu}_w = 1.86 \cdot \left(\mathrm{Re} \cdot \mathrm{Pr} \cdot \frac{D_h}{l}\right)^{1/3}, \tag{15}$$

As a result, the heat transfer coefficients are shown in Table 9.

A new form of Nu number relationship was introduced to create the relationship that best fits the measured data. The ratio of heat transfer surfaces was used only in this Nu correlation, no longer in the calculation of the overall heat transfer coefficient. The classical correlation was used for that,

$$U = \frac{1}{\frac{1}{\alpha_w} + \frac{1}{\alpha_v} + \frac{\delta_f}{\lambda_f}}, \tag{16}$$

Table 9. The water side heat transfer coefficients

$\dot{V}w$ (l/h)	αw $(Wm^{-2}K^{-1})$
100	675
200	850
300	973
400	1071

Applying the least squares regression method to the measured heat transfer coefficients without the preliminary F factors, the C constant, the exponent of Re number and the exponent of the surface ratios were described in Eq. 17, and Table 10 shows the calculated values.

$$Nu_a = 0.817 \cdot Re_a^{0.6} \cdot \left(\frac{A}{A_{t0}}\right)^{-0.39} \cdot Pr_a^{1/3}, \tag{17}$$

The results show that the measurement results are more accessible with this modified method.

Table 10. The result calculated with the modified coefficients

	U_{meas} $(Wm^{-2}K^{-1})$	U_{new} $(Wm^{-2}K^{-1})$	*diff* (%)
M1	9.86	11.32	14.8
M2	9.87	11.33	14.8
M3	10.16	11.34	11.7
M4	11.16	11.34	1.6
M5	9.94	11.28	13.5
M6	10.32	11.29	9.4
M7	11.31	11.30	−0.1
M8	12.44	11.29	−.9.2
M11	12.12	13.10	8.1
M12	13.90	13.13	−5.6
M13	13.13	13.14	0.1
M14	14.48	13.13	−9.3
M15	12.69	13.14	3.6
M16	13.91	13.18	−5.3
M17	14.08	13.19	−6.3
M18	15.53	13.19	−15.1

6 Conclusions

The present paper dealt with the measurement and calculation possibilities of louvered fin heat exchanger mainly used in vehicles. An experimental device was built to measure the heat performances, which can be used to control both the amount of water and air. The measured results confirmed that the overall heat transfer coefficient of heat exchanger with extended surface is relatively low value (in the measurements range from 9.86 to 15.53 $Wm^{-2}K^{-1}$, while in the case of a shell-and-tube heat exchanger they range from 800 to 1200 $Wm^{-2}K^{-1}$). During a series of measurements, the value of heat transfer coefficient was constant despite the change in the volume flow rate of water, however, according to the measured results, this is significant for the exchanged heat. It is also important for the correct calculation of the temperature correction factor, which is not yet available. However, the measurement clearly showed that even a relatively low air speeds (1.21 m/s, which is less than 5 km/h) 5.5 kW heat performance can be achieved, which is relatively high compared to such a small structure.

Acknowledgement. The research was partially supported by the Hungarian National Research, Development and Innovation Office - NKFIH under the project number K 134358.

References

1. Gao, Y., Wang., H., Sasmito, A.P., Mujumdar, A.S.: Measurement and modeling of thermal conductivity of graphene nanoplatelet water and ethylene glycol base nanofluids. Int. J. Heat Mass Transf. **123**, 97–109 (2018)
2. Taler, D.: Experimental determination of correlations for mean heat transfer coefficients in plate fin and tube heat exchangers. Arch. Thermodyn. **33**(3), 3–26 (2012)
3. Nagarajan, F.C., Kannaiyan, S.K., Boobalan, C.: Intensification of heat transfer rate using alumina-silica nanocoolant. Int. J. Heat ass Transf. **149**, 119127 (2020)
4. Abbas, F., Ali, H.M., Shah, T.R., Babar, H., Janjua, M.M., Sajjad, U., Amer, M.: Nanofluid: Potential evaluation in automotive radiator. J. Mol. Liq. **297**, 112014 (2020)
5. Long, Y., Wang, S., Wang, J., Zhang, T.: Mathematical model of heat transfer for a finned tube cross-flow heat exchanger with ice slurry as cooling medium. Procedia Engineering **146**, 513–522 (2016)
6. Gong, C., Shen, J., Yu, Y., Wang, K., Tu, Z.: A novel radiator structure for enhanced heat transfer used in PEM fuel cell vehicle. Int. J. Heat Mass Transf. **157**, 119926 (2020)
7. Chong, K.K., Tan, W.C.: Study of automotive radiator cooling system for dense-array concentration photovoltaic system. Sol. Energy **86**(9), 2632–2643 (2012)
8. Petrik, M., Szepesi, G., Jármai, K., Bolló, B.: Theoretical and parametric investigation of an automobile radiator. In: Vehicle and Automotive Engineering - Proceedings of the JK2016, Miskolc, Hungary, pp. 27–37. Springer International Publishing (2017)
9. Yan, W., Li, M., Pan, X., Wu, G., Liu, L.: Application of support vector regression cooperated with modified artificial fish swarm algorithm for wind tunnel performance prediction of automotive radiators. Appl. Therm. Eng. **164**, 114543 (2020)
10. Kim, T.K., Kang, H.C., Lee, J.S.: A porosity model for flow resistance calculation of heat exchanger with louvered fins. J. Mech. Sci. Technol. **30**(4), 1943–1948 (2016)
11. Petrik, M., Szepesi, G., Jármai, K.: CFD analysis and heat transfer characteristics of finned tube heat exchangers. Pollack Periodica **14**(3), 165–176 (2019)

12. De Silva, C.M., Nor Azmi, M., Christie, T., Abou-Saba, E., Ooi, A.: Computational flow modelling of Formula-SAE sidepods for optimum radiator heat management. J. Eng. Sci. Technol. **6**(1), 94–108 (2011)
13. Mukkamala, Y.: Contemporary trends in thermo-hydraulic testing and modeling of automotive radiators deploying nano-coolants and aerodynamically efficient air-side fins. Renew. Sustain. Energy Rev. **76**, 1208–1229 (2017)
14. Comot, P., Bocher, P., Belanger, P.: Ultrasonic guided wave inspection of Inconel 625 brazed lap joints: Simulation and experimentation. NDT E Int. **91**, 71–78 (2017)
15. Lei, T., Huang, Y., Wang, H., Rong, Y.: Automatic weld seam tracking of tube-to-tubesheet TIG welding robot with multiple sensors. J. Manuf. Process (2020). https://doi.org/10.1016/j.jmapro.2020.03.047
16. Pietroluongo, M., Padovano, E., Frache, A., Badini, C.: Mechanical recycling of an end-of-life automotive composite component. Sustainable Mater. Technol. **23**, e00143 (2020)
17. Cengel, Y.A.: Heat Transfer - A Practical Approach. Second Edition, McGraw-Hill International Editions (2002)

Cutting and Hauling Mining Adapter for Dimension Stone

Zoltán Virág[1(✉)], Géza Fülöp[2], and Viktor Géza Fülöp[3]

[1] University of Miskolc, Miskolc-Egyetemváros, Hungary
gtbvir@uni-miskolc.hu
[2] FÉMGER Kft., Fertőd, Hungary
[3] University of Sopron, Sopron, Hungary

Abstract. The MIOCÉN project aims to develop a small volume mining production and transport adapter. The mining tool to be developed for both production and transport can reduce the number of machines to be used in the mine. It is planned to reduce the expenses of mining businesses, which include mining operations and costs of maintenance. During the extraction of the block stone, the cut rock material can be lifted and moved with the help of the product. Based on technology, the adapter can be divided into two main machine units. One is the rock chain saw unit combined with a feed consisting of hydraulic cylinders for feeding and rotating and a hydraulic rotating unit. The other is used to support and stiffen the first main machine unit during slotting. This paper shows a prototype of a dimension stone mining adapter. The prototype adapter is under design protection.

Keywords: Mining · Development · Adapter · Prototype

1 Introduction

The MIOCÉN project aims to develop a small volume mining production and transport adapter suitable for limestone. The research optimizes the mining production of small volume dimension stone products with the re-thinking of the use of production and transport equipment in one. The determination of the mined rock environment is essential for the design of the mining adapter to be developed. The rock selected for the project is limestone Leitha in the Sopron Mountains, which is the main component of this area. The limestone in Fertőrákos (Hungary) and Margitbánya (Austria) has been a popular building material for millennia. The Celts and the Romans used this stone as well but it became a popular building material from the 18th century. These stones were easy to mine and transport. Massive amounts of blocks were transported to the buildings of nearby big cities. Much of the houses in Sopron, Vienna and Bratislava were made of this material. We find buildings, monuments and statues built from this "Rákosi" stone in almost every village around Sopron.

The prototype tool to be developed for both production and transport can reduce the number of mining machines to be used in the limestone quarries (see Fig. 1). This

K. Jármai and K. Voith (Eds.): VAE 2020, LNME, pp. 417–423, 2021.
https://doi.org/10.1007/978-981-15-9529-5_36

development is planned to reduce the expenses of running mining businesses, which include mining operations and costs of maintenance. The main goal is to be cheaper and to be more competitive with the mining companies in the market [1].

Fig. 1. A building stone mine.

There are several methods of the dimensional stone production such as a) Block production with a large borehole blasting technology, b) Cleavage of stone blocks using chemical swelling energy, c) Use of mechanical, hydraulic and combined rock setting wedges placed in small-diameter holes, d) Creating blocks by drilling, e) Rock sawing. Since the gentlest production method is rock sawing, and to reduce the number of machines, we plan to solve the production with a design that is capable of winning and transport.

2 The Basic Machine

The basic mechanics of the mining prototype is provided by a Construction Loadall (see Fig. 2). Due to its high load capacity and unique design, the machine is used for different works in many areas. It is extremely manoeuvrable making the machine easy to operate in confined spaces. The vehicle has three selectable steer modes. There is two-wheel steer, which is ideal for travelling at high speed e.g. on the road. The others are four-wheel steer for working in tight spaces, and the crab steer for manoeuvring close to walls.

Its multifunctional applicability is ensured by the hydraulic system of the requested machine provides with the main pump capacity of 140 l/min and the system operating pressure of 260 bar to perform the task. Front stabilisers are fitted for increased stability and lift performance.

These new loadalls use up to 7% less fuel than the previous ones. It makes the machine more eco friendly than other machines. Its economic engine can be adapted to operate with lower-grade diesel oils, making the machines useful across different territories [2].

Fig. 2. The basic Construction Loadall machine of the project with the prototype adapter.

3 The Mining Adapter

The MIOCÉN small volume dimensional stone mining adapter (see Fig. 3) is suitable for the extraction of block stone, primarily for the extraction of limestone or other homogeneous rock not exceeding six on the Mohs hardness scale.

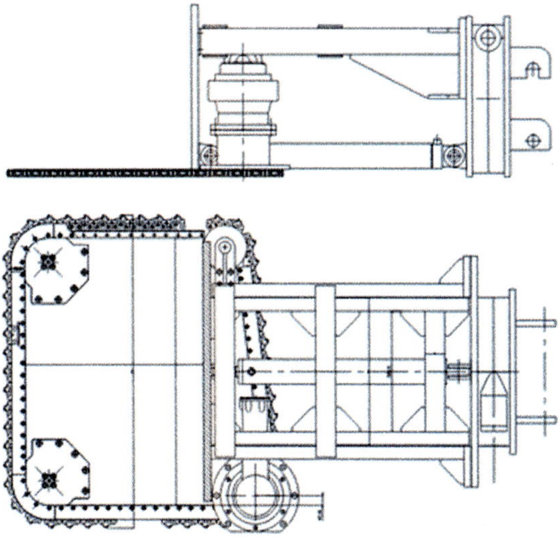

Fig. 3. CAD drawings of the prototype adapter.

Based on the technology of the mining device, the adapter can be divided into two main units. One is the rock chain saw unit combined with a work consisting of hydraulic cylinders for feeding and rotating and a hydraulic rotating unit. This part of the adapter is capable of slitting from three sides perpendicular to the extraction front. The unique structure of the rock chain saw includes custom-made elements. The teeth that make up the chain follow the chain in a specific order and form a repetitive chain of 5 and 6 teeth, specially developed for the limestone rock quality (see Fig. 4). The shape of the cutting tool is very important because of the high cutting forces and loads [3, 4]. Reducing the stresses are significant because high loads can produce unwanted effects on the mechanics [5]. During the design, the guide plate carrying the chain was dimensioned separately, taking into account the design of a lubrication system to help reduce friction during operation.

The other unit is used to support and stiffen the first main machine unit during slotting. This unit provides stability and elevation of the excavated block with an object table which is made of a special alloy in a direction perpendicular to the front. Blocks are slid on wear-resistant carbide sheets, supplemented with several modular elements that are subject to heavy wear during mining applications [6].

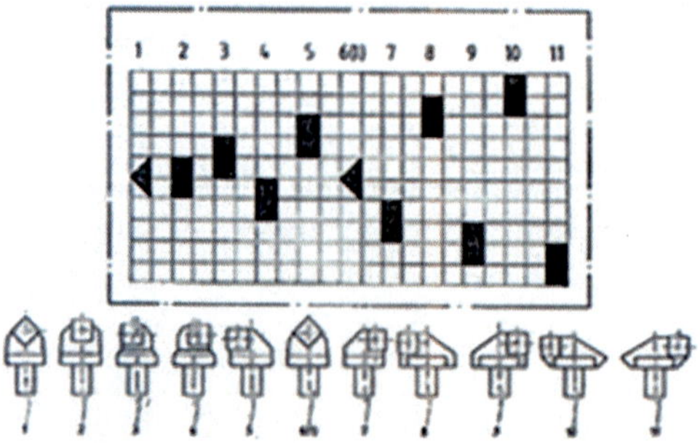

Fig. 4. The repetitive chain saw teeth.

3.1 Cutting Forces and Power Demand

Determining cutting forces and power demand is an important task during the design. The following numerical example for the prototype adapter shows a brief calculation.

Building on the experiences obtained in our earlier works, the specific cutting force is determined [7]. For limestone, we calculate with a value of $F_{cutting} = 24\ \text{N/mm}^2$. The chain speed is $v_{chain} = 1.5$ m/s and teeth pitch is $l_{pitch} = 0.1$ m. The cutting length of the adapter is $l_{cutting} = 1.2$ m.

The number of cutting chain teeth at the same time

$$N_c = \frac{l_{cutting}}{l_{pitch}} \tag{1}$$

The overall cutting force is

$$F_o = N_c A_{chip} F_{cutting} \tag{2}$$

where A_{chip} is the chip area.

The power demand is

$$P = F_o v_{chain} \tag{3}$$

Since the friction of the chain and the efficiency of the machine has not been taken into account, the value of the power must be increased by 25%. Table 1. Shows the cutting forces and power demands for different feed speeds and chip areas.

Table 1. Cutting forces and power demands.

Feed speed [mm/s]	Chip area [mm^2]	Cutting force for one tooth [N]	Overall cutting force [N]	Power demand [W]	Increased power demand [W]
10	4	96	1152	1728	2160
15	6	144	1728	2592	3240
20	8	192	2304	3456	4320

The results are shown in Table 1. Are appropriate for the prototype machine.

3.2 Finite Element Analysis

The design of the structure the adapter is another special task. The greatest load is generated during the transport of the excavated stone. We use FE analysis to determine stresses and displacements of the adapter. The prototype is designed for a 1 m^3 dimensional stone which has a density between 2000 and 2600 kg/m^3. We calculate with the highest density for safety reasons. Distributed load acts on the object table and the ends of the beams are fixed. The results are shown in Figs. 5 and 6.

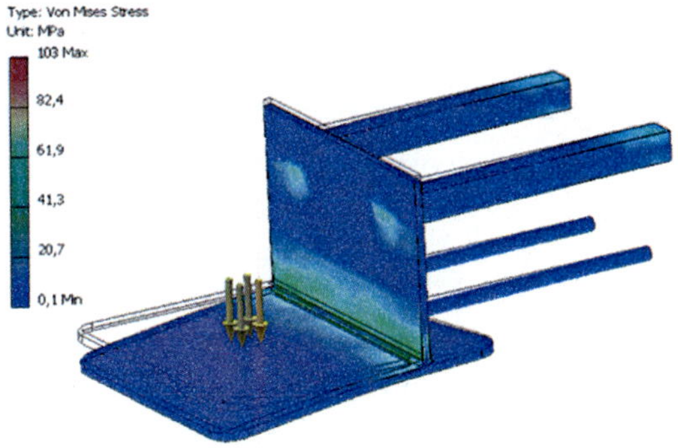

Fig. 5. Effective stress distribution in the adapter.

The Figs. 5 and 6. Show maximum stress of 103 MPa and maximum displacement of 8 mm which are small values.

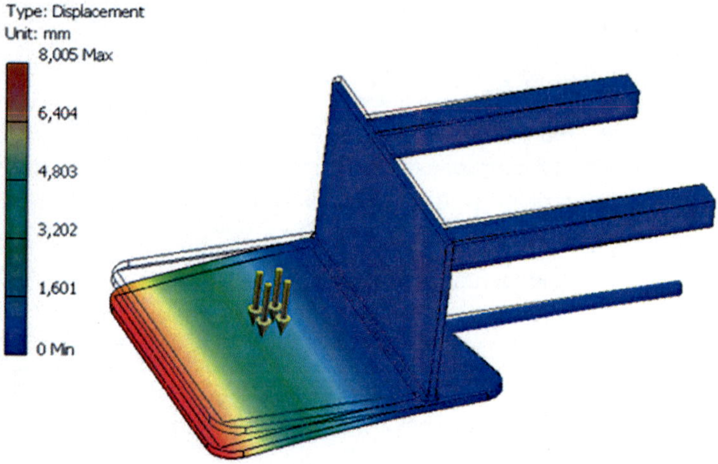

Fig. 6. Displacement distribution in the adapter.

4 Conclusions

The future reconstructions of historic buildings and memorials need good quality limestones. It was the aim of we start this project to design a mining adapter for dimensional stone. To develop a rock winning technology, it is imperative to know the geological knowledge of the site and the use of the extracted material.

For good quality building materials, the appropriate mining procedure is the rock sawing since this is the gentlest production method. To reduce the costs we have to reduce the number of machines. That is why we plan to solve the production with a design that is capable of winning and transport.

The next step is the comprehensive testing of the prototype (see Fig. 7) which is inevitable during the development period. It will be carried out at the production workshop and in the limestone quarry. During the testing, the static and dynamic stresses of the machine units during the use of adapters will be studied, even in extreme cases, to determine more precise technical parameters. Tests will be performed on materials of different qualities and sizes for cutting and transport.

Fig. 7. The prototype adapter.

References

1. Virág, Z., Fülöp, V., Molnár, J.: Initial steps to develop a cutting and hauling adapter for dimension stone mining. Ann. Univ. Petroşani Mech. Eng. **20**, 121–126 (2018)
2. Kovács, G., Gubán, M.: Planning of optimal fuel supply of international transport activity. Periodica Polytechnica Transp. Eng. **45**(4), 186–195 (2017)
3. Tomus, O.B., Rada, A.C.: Study of the cutting forces and specific cutting resistance from the point of view of the cutting direction relative to stratification of the coal seam. Ann. Univ. Petroşani Mech. Eng. **19**, 143–148 (2017)
4. Virág, Z., Szirbik, S.: Examination of an optimized replaceable cutting tooth of excavator. Geosci. Eng. **1**, 337–342 (2012)
5. Andras, I., Radu, S.M., Andras, A.: Study regarding the bucket-wheel excavators used in hard rock excavations. Ann. Univ. Petroşani Mech. Eng. **18**, 11–22 (2016)
6. Kovács, G.: Optimization of structural elements of transport vehicles in order to reduce weight and fuel consumption. Struct. Eng. Mech. **71**(3), 283–290 (2019)
7. Ladányi, G., Sümegi, I., Virág, Z.: Laboratory rock cutting tests on rock samples from Visonta South Mine. Ann. Univ. Petroşani Mech. Eng. **9**, 209–218 (2007)

Optimization

Optimum Design of Solar Sandwich Panels for Satellites Applications

Alaa Al-Fatlawi[1,2](✉), Károly Jármai[1], and György Kovács[1]

[1] Faculty of Mechanical Engineering and Informatics, Miskolc University, Egyetemváros, Miskolc 3515, Hungary
{altkovac,jarmai,vegyalaa}@uni-miskolc.hu
[2] Faculty of Mechanical Engineering, Kufa University, Al-Najaf, Iraq

Abstract. The aim of the article was to verify the optimum design of solar sandwich panels for microsatellites applications. The sandwich panel consists of aluminum honeycomb core and aluminum materials facesheets. In this study, a methodology for a combined weight and/or cost optimization for sandwich panel with aluminum facesheets and honeycomb core is presented. The fmincon Solver Constrained Nonlinear Minimization/Interior Point Algorithm was used to solve the single objective function the weight or the cost of the honeycomb sandwich panel. The Multiobjective optimization technique was applied to minimize the weight and the cost of the sandwich panel included the Weighted Normalized Method with Excel Solver program and Genetic Algorithm Solver with Pareto front in Matlab program. The weight and/or the cost of the sandwich panel are the objective functions subjected to required constraints based on total stiffness (bending and shear stiffness), total deflection (bending and shear deflection), facing skin stress (bending load), core shear stress, facing skin stress (end loading), overall panel buckling (critical bending and shear buckling load), shear crimping load, skin wrinkling (critical stress and load) and intracell buckling (facesheet dimpling). The design variables are thicknesses of core and facesheets. The use of sandwich construction results in light structure.

Keywords: Honeycomb sandwich panel · Satellite sandwich panel · Minimum weight and cost · Multiobjective optimization

1 Introduction

The aim of insert structural parts like honeycomb sandwich panel because the solar panels of satellite requires use several holes for connection, installation, and fixing. The honeycomb cores have a small density and high specific stiffness, which can meet the stiffness, deflection, skin stress, core shear stress, panel buckling, shear crimping, skin wrinkling and intracell buckling requirements and reduce the weight of the sandwich panel [1]. The elaborated structural model can be used for satellite sandwich panels to fulfill the requirements of Aerospace such as satellite. The basic concept of honeycomb sandwich construction is to use thin, dense, strong facing materials bonded to a thick,

K. Jármai and K. Voith (Eds.): VAE 2020, LNME, pp. 427–442, 2021.
https://doi.org/10.1007/978-981-15-9529-5_37

lightweight honeycomb core [2]. The core materials of honeycomb type have been developed and used primarily in aerospace applications. The honeycomb materials can be manufactured in a variety of cell shapes, but the most commonly used shape is the hexagonal shape [3].

The sandwich panel consists of three or more constituents, the faces, the core and the adhesive joints [4]. The sandwich panels, consisting of a core covered by facesheets, are frequently used instead of solid panels because of their high bending stiffness to weight ratio. The high bending stiffness is the result of the distance between the facesheets, which carry the load, and the lightweight is due to the lightweight of the core [5]. Jun & Dai developed a new lightweight sandwich structure by reinforcing the web of insert with high strength carbon composite to increase the loading capability with reduced mass [6]. Craig L. created a model of the ION–F stack to verify the integrity of the structure at the launch loads and proved that the HokieSat structure will survive all environmental loads with no yielding or failures [7].

Daisuke et al. designed as a practical configuration for a hybrid panel of sandwich structures [8]. Byoung & Dai designed a satellite structure from composite sandwich panels, which consist of two carbon fiber/epoxy composite faces and an aluminum honeycomb core to reduce structural mass and to improve static and dynamic structural rigidity [9]. Boudjemai A. et al. is proposed a versatile, robust and enhanced genetic algorithm for structural topology optimization and examined all of the technologies and their benefits for satellite structural designs [10].

Cho & Rhee studied the vibration characteristics of spacecraft structure using the FEA code with experimental data. The satellite structure consists of composite sandwich panels composed of carbon fiber reinforced laminate facesheets and aluminum honeycomb core [11]. Virág & Szirbik investigated the dynamic behaviour of the hybrid stiffened plate under longitudinal compression using the finite element analysis [12]. Alaa et al. developed a new lightweight sandwich structure consists of aluminum foam core and composite facesheets, which can be to manufacture air containers elements. The study included theoretically and numerically using Matlab and Digimat programs [13]. The connection, installation, and fixing of solar panels require the use of several holes. So, it is necessary to insert structural parts such as honeycomb core to support the panel structure. The local stress concentration will influence the structural strength of the panel and will also increase the weight of the plate [14]. Alaa et al. investigated the replacement of the current aluminum base plate of air container with a sandwich plate. The Interior Point Algorithm in Matlab program was used to solve the single objective function (weight or cost). The Genetic Algorithm Solver by Matlab program was applied to minimize multiobjective functions (weight and cost) [15]. Virág minimized the cost function for the stiffened and unstiffened circular cylindrical steel shell. The constraints are shell buckling, panel stiffener buckling and maximum deflection of the bridge. The design variables are shell thickness, dimensions and number of stringers [16]. Rion et al. studied the ultra-light photovoltaic composite sandwich structures. The goal of this work was to investigate the failure mechanisms of such ultra-light sandwich structure and their correlation with microstructure, processing pressure, and strength in order to obtain optimal design and processing [17].

2 Solar Sandwich Panel Construction for Satellite

The sandwich structures are often utilized in solar panel applications. A sandwich structure consists of two thin face sheets bonded to both sides of a lightweight core. The design of sandwich structures allows the outer face sheets to carry the axial loads, bending moments, and in-plane shears, while the honeycomb core carries the normal flexural shears. The sandwich structures are susceptible to failures due to large normal local stress concentrations because of the heterogeneous nature of the core/face sheet assembly. Therefore, component mounting must employ potted inserts to distribute the point loads from connections. The sandwich panel face sheets are usually manufactured using aluminum alloy or composite materials. The core is typically manufactured using a honeycomb or aluminum foam construction [1]. The satellite sandwich panel is simply supported and has aluminum face sheet and honeycomb core. The sandwich panel is subjected to a uniform pressure (p = 50 kPa) and deform (2 mm) at any point of the sandwich panel. The optimum facesheet and core thicknesses for a minimum weight and cost design are calculated. The upper and lower facesheets are assumed to have the same thickness. The technical data of satellite sandwich panel as shown in Table 1 (see Fig. 1 and 2) [3].

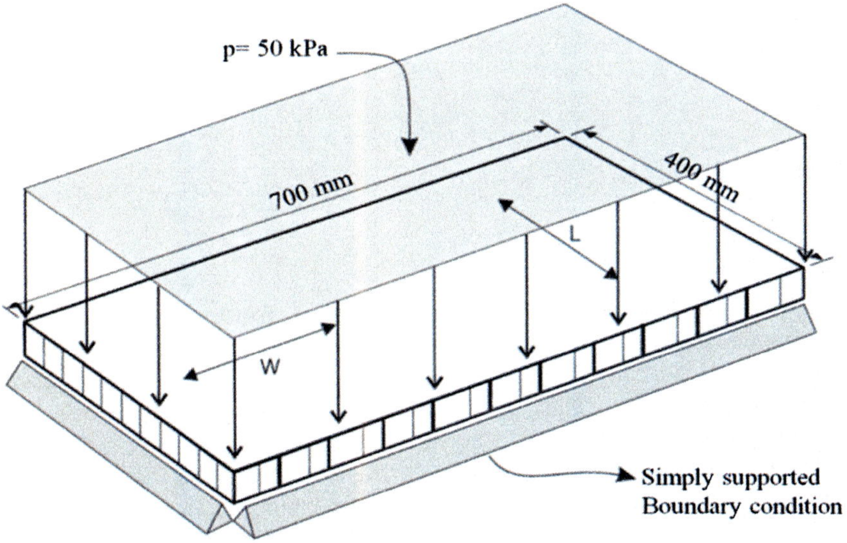

Fig. 1. Honeycomb sandwich panel with simply supported boundary condition on all four sides with a uniformly distributed load of (50 kN/m^2) applied on the upper facesheet.

Table 1. Technical data of the satellite sandwich panel [3].

Length	Width	Deflection	Pressure	Load
l [mm]	b [mm]	δ_{max} [mm]	p [kPa]	P [kN]
700	400	2	50	14

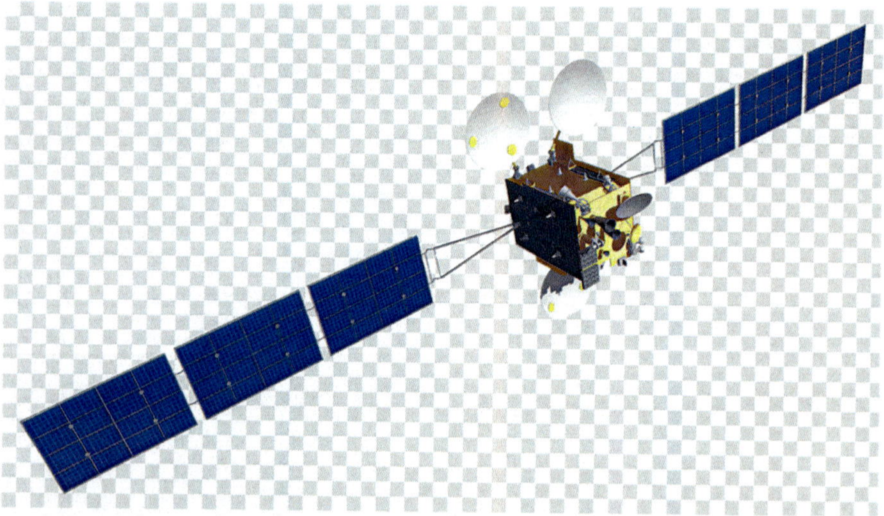

Fig. 2. Satellite panels structure (ultra–high stiffness and strength per unit weight).

2.1 Aluminum Honeycomb Core Structure

The honeycomb sandwich panels are the lightest option for compressive or bending loads in specific applications. The honeycomb sandwich cores are manufactured using thin strips formed into honeycomb cells. The honeycomb geometry is nonisotropic, with greater stiffness in the longitudinal direction. However, the core acts nearly isotropically for in-plane loads when assembled in a sandwich configuration [1].

The aluminum honeycomb core is used for several applications and in different sectors such as public transport industry, nautical sector, building industry, etc… As core material, the aluminum honeycomb core is used in sandwich panels and it is utilized in floors, roofs, doors, partitions, facades, working surfaces for automatic machines and for all products which require an optimal stiffness to weight ratio. The aluminum honeycomb as panels' core has several advantages as lightweight, stiffness, fire resistance, compression and shear and corrosion resistance flatness. The aluminum honeycomb core can be used as a deflector for laminar flow ventilation, and as crash absorber for kinetic energy. The density of the honeycomb core depends on the thickness of the foil and the diameter of the cells. The engineering properties of the honeycomb core as shown in Table 3 (see Fig. 3) make it ideal for many applications like satellite sandwich panels [18].

2.2 Upper and Lower Aluminum Facesheets

Aluminum alloys are the most commonly used metallic materials in spacecraft manufacturing. The advantages are high strength to weight ratios, high ductility, and ease of machining, weldability, and availability at low cost [18].

The upper and a lower face sheet of solar sandwich panel for a satellite consisted of aluminum (alloy 5251 H24) materials. Table 2 shows the properties of typical facing materials of the panel.

Table 2. Mechanical Properties of typical facing materials for sandwich panel [18].

Facing material aluminum alloy (5251 H24)	
Typical strength Tension/Compression [MPa]	150
Modulus of elasticity tension/compression [GPa]	70
Poisson's ratio	0.33
Typical cured ply thickness [mm]	0.5
Typical weight per ply [kg/m^2]	1.35

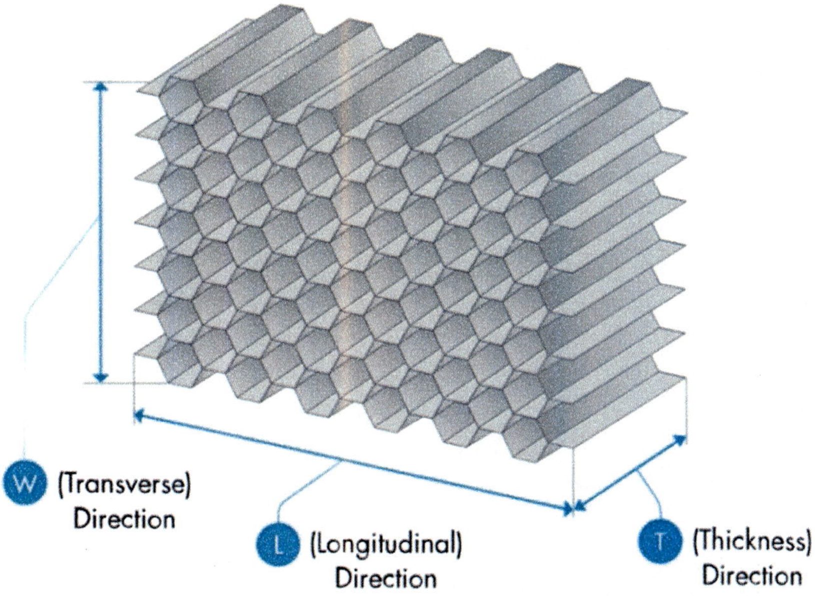

Fig. 3. Aluminum honeycomb structures.

Table 3. Mechanical properties of aluminum honeycomb core materials [18].

Product construction		Compression		Plate shear			
Density	Cell size	Stabilized		L – direction		W – direction	
		Strength	Modulus	Strength	Modulus	Strength	Modulus
kg/m^3	mm	MPa	MPa	MPa	MPa	MPa	MPa
83	6	4.6	1000	2.4	440	1.5	220

3 Optimization Techniques (Mathematical Modeling)

The optimization processes for the aluminum solar sandwich panel of the satellite were presented. The optimum design variables were thicknesses of facesheet t_f and honeycomb core t_c to minimize the weight and/or the cost of the sandwich panel. During the optimization techniques, nine design constraints were taken into consideration. The constraints of the optimization problem are total stiffness (bending and shear stiffness), total deflection (bending and shear deflection), facing skin stress (bending load), core shear stress, facing skin stress (end loading), overall panel buckling (critical bending and shear buckling load), shear crimping load, skin wrinkling (critical stress and load) and intracell buckling. The sandwich panel's total stiffness (bending and shear stiffness), total deflection (bending and shear deflection), facing skin stress (bending load), core shear stress, facing skin stress (end loading), overall panel buckling (critical bending and shear buckling load), shear crimping load, skin wrinkling (critical stress and load) and intracell buckling for every facesheets and core thicknesses were calculated to compare with yield stresses and applied loads. The steps of optimization procedure formulate the objective functions for the weight and/or the cost of the honeycomb sandwich panel, formulate the constraints and defined the boundaries for the design variables, solve the single objective function problem to minimize the weight or the cost separately by using Matlab program (Interior Point Algorithm) and Excel Solver program. Solve the multiobjective optimization problem to minimize the weight and the cost simultaneously by applying Genetic Algorithm Solver with Pareto Front in Matlab program and Excel Solver program (Weighted Normalized Method).

3.1 Single Objective Functions by Using Excel Solver Program and Matlab Program (Fmincon Solver/Interior Point Algorithm)

Weight Objective Function

The total weight of the sandwich panel, which including the weight of the upper and lower aluminum facesheets as well as a honeycomb core with considers neglecting the weight of the adhesive bond were minimized by using the Excel Solver program and Matlab program.

$$W_t = 2\rho_f lbt_f + \rho_c lbt_c \quad (1)$$

Where W_t is the total weight of the sandwich panel, ρ_f is the density of the facesheet, l is the length, b is the width of the sandwich panel, t_f is the thickness of the facesheet, ρ_c is the density of the honeycomb core and t_c is the thickness of the honeycomb core.

Cost Objective Function
The total cost of the sandwich panel, which including the cost of the upper and lower aluminum facesheets materials and the cost of the aluminum honeycomb core was minimized by using the Matlab program and Excel Solver program.

$$C_t = 2\rho_f lbt_f C_f + lbt_c C_c \tag{2}$$

Where C_t is the total cost of the sandwich panel in Euro, C_f is the cost of the facesheet material per unit weight (4.61 €/kg) [19]. C_c is the cost of the honeycomb core material per unit volume (20 €/m^2 in case of 18 mm core height) (€/m^3).

3.2 Weight and Cost Multiobjective Functions

Excel Solver Program (Weighted Normalized Method)
The weight and the cost of multiobjective optimization using the Excel Solver program (Weighted Normalized Method) were presented:

$$f(x) = \sum_{i=1}^{r} w_i f_i(x)/f_i^0 \tag{3}$$

Where $w_i \geq 0$ and $\sum_{i=1}^{r} w_i = 1$. The condition $f_1^0 \neq 0$ is assumed.

Matlab Program (Genetic Algorithm Solver)
The tool of Multiobjective Genetic Algorithm Solver in Matlab program corresponds to the gamultiobj function. The Gamultiobj Solver attempts to create a set of Pareto optima for a multiobjective minimization [20].

3.3 Design Variables

To achieve the acceptable performance, modify the core thickness and the skin facesheets thickness.

$$10\,\text{mm} \leq t_{c,\,opt} \leq 100\,\text{mm} \tag{3}$$

$$0.1\,\text{mm} \leq t_{f,opt} \leq 5\,\text{mm} \tag{4}$$

3.4 Constraints

Total Stiffness

The constraint of the total stiffness of the aluminum sandwich panels includes the bending and shear stiffnesses.

$$D_{f,x} \geq D_{min} \tag{6}$$

$$D_{f,x} = (E_f t_f h^2 \cdot b)/2(1 - \nu_f^2) \tag{7}$$

$$D_{min} = (EI)min = K_b P_{max} l^3/\delta_{max} \tag{8}$$

Where, $K_b = 5/384$, $S = b\ h\ G_c$, $G_c = G_W$ and $h = t_c + t_f$.

The calculated stiffness of the sandwich panel in global coordinate $D_{f,x}$ must be greater than the minimum stiffness of sandwich structure D_{min} was calculated using given data (δ_{max} and P_{max}) as shown in Table 1. Where, S is the shear stiffness, K_b is the bending deflection coefficient, E_f is the modulus of elasticity of facing skin, ν_f is Poisson's ratio of face material and h is the distance between facing skin centers [18].

Total Deflection

The constraint for the total deflection of the aluminum sandwich panel is included the bending and shear deflection:

$$\delta = (K_b P l^3/D_{f,x}) + (K_s P l/S) \leq \delta_{\max} \tag{9}$$

Where, $K_b = 5/384$ and $K_s = 1/8$.

The calculated total deflection δ must be less than the honeycomb sandwich panel deflection δ_{max} of (2 mm) as shown in Table 1. Where, K_s is the shear deflection coefficient. The panel is simply supported on all four edges.

Skin Stress

The constraint for the skin stress of the honeycomb sandwich panel is:

$$\sigma_f = M/ht_f b \leq \sigma_{f,\mathrm{y}} \tag{10}$$

The calculated skin stress σ_f must be less than the facesheet material typical yield strength $\sigma_{f,\mathrm{y}}$ of (150 MPa) as shown in Table 2. Where, $M = Pl/8$ is the maximum bending moment [18].

Core Shear Stress

The constraint for the core shear stress of the honeycomb sandwich panel is:

$$\tau_c = F/hb \leq \tau_{c,y} \tag{5}$$

The calculated core shear stress must be less than core material typical panel shear stress in the transverse (W) direction $\tau_{c,y}$ of (1.5 MPa) as shown in Table 3. Where, $F = P/2$ is the maximum shear force.

Skin Facing Stress, End Loading
The constraint for the skin facing stress of the honeycomb sandwich panel is:

$$\sigma_f = P/(2t_f b) \leq \sigma_{f,y} \tag{12}$$

The calculated skin facing stress must be less than the facesheet material typical yield strength $\sigma_{f,y}$ of (150 MPa) as shown in Table 2.

Panel Buckling
The constraint for the panel buckling of the honeycomb sandwich panel is:

$$P_{b,cr} = \pi^2 D_{f,x}/(\beta l^2 + \pi^2 D_{f,x}/S) \geq P \tag{6}$$

Where $\beta = 1$ and $P_{b,cr}$ is the critical buckling load. The calculated load at which critical buckling would occur is greater than the end load being applied P of (14 kN) as shown in Table 1.

Shear Crimping
The constraint for the shear crimping of the honeycomb sandwich panel is:

$$P_{cr} = t_c G_c b \geq P \tag{7}$$

The calculated load at which shear crimping would occur is greater than the end load being applied P of (14 kN). Where P_{cr} is critical shear crimping load, G_c is core shear modulus and E_c is the compression modulus of core as shown in Table 3.

Skin Wrinkling
The constraint for the skin wrinkling of the honeycomb sandwich panel is:

$$\sigma_{f,cr} = 0.5(E_f E_c G_c)^{1/3} \geq \sigma_{f,y} \tag{8}$$

$$P_{cr,min} = t_f(2t_f E_f E_c/3t_c(1 - \nu_f^2)) \geq P/b \tag{9}$$

The stress level at which skin wrinkling would occur is well beyond the skin material typical yield strength of (150 MPa) as shown in Table 2, so skin stress is more critical than skin wrinkling. The calculated load at which skin wrinkling would occur is greater than the end load per width being applied (P/b) of (35 kN/m).

Intracell Buckling
The constraint for the intracell buckling of the honeycomb sandwich panel is:

$$\sigma_{cr} = 2E_f(t_f/s)^2/(1 - \nu_f^2) \geq \sigma_{f,y} \tag{10}$$

The stress level at which intracell buckling would occur is well beyond the skin material typical yield strength $\sigma_{f,y}$ of (150 MPa), so skin stress is more critical than intracell buckling as shown in Table 2. Where σ_{cr} is critical facing skin stress and s is cell size [18].

4 Evaluation Optimization Results

The final results are optimum core thickness $t_{f,\mathrm{opt}}$, optimum face thickness $t_{c,\mathrm{opt}}$, minimum weight W_{min} and cost C_{min} as shown in Tables 4, 5, 6, 7, 8 and 9 (see Figs. 4, 5, 6 and 7).

4.1 Theoretical Results by Using Excel Solver Program

Optimum Results of Single Objective Function

Depending on the theoretical results by using the Excel Solver program, the minimum weight and cost with optimum facesheet and core thicknesses of the honeycomb sandwich panel for satellite application which consists of aluminum facesheets and aluminum honeycomb core as shown in the Tables 4 and 5.

Table 4. Minimize the weight objective function, disregard the cost objective function by using the Excel Solver program for honeycomb sandwich panel of the satellite.

W_{min}	$t_{f,\mathrm{opt}}$	$t_{c,\mathrm{opt}}$
[kg]	[mm]	[mm]
2.293473661	0.487460114	66.97220177

Table 5. Minimize the cost objective function, disregard the weight objective function by using the Excel Solver program for honeycomb sandwich panel of the satellite.

C_{min}	$t_{f,\mathrm{opt}}$	$t_{c,\mathrm{opt}}$
[€]	[mm]	[mm]
21.65734	1.025124	46.64538

Optimum Results of Multiobjective Function

The weight and the cost of multiobjective function are presented by using the Excel Solver program (Weighted Normalized Method) as shown in Table 6 (see Fig. 4, 5 and 6), where W_1 is the weighted of the objective function 1 (weight objective function) and W_2 is the weighted of the objective function 2 (cost objective function).

4.2 Theoretical Results by Using Matlab Program

Optimum Results of Single Objective Function

Depending on the theoretical results by using Matlab program obtained by utilized fmincon Solver Constrained Nonlinear Minimization/Interior Point Algorithm, the minimum weight or cost of the sandwich panel of the satellite which consists of aluminum facesheets and aluminum honeycomb core as shown in the Tables 7 and 8.

Table 6. Minimize the weight and the cost objective function by using the Excel Solver program (Weighted Normalized Method) for sandwich panel of satellite consists of aluminum honeycomb core and aluminum facesheets.

Type	Aluminum alloy (5251 H24)		W_{min}	C_{min}	$t_{f,opt}$	$t_{c,opt}$
No	W_1 (%)	W_2 (%)	[kg]	[€]	[mm]	[mm]
1	50	50	2.3575184	22.428431	0.688089	56.675031
2	60	40	2.3335075	22.701801	0.643268	58.557914
3	70	30	2.3134871	23.055143	0.596447	60.742672
4	80	20	2.3030815	23.351422	0.563624	62.430378
5	90	10	2.2956145	23.740916	0.526742	64.508660

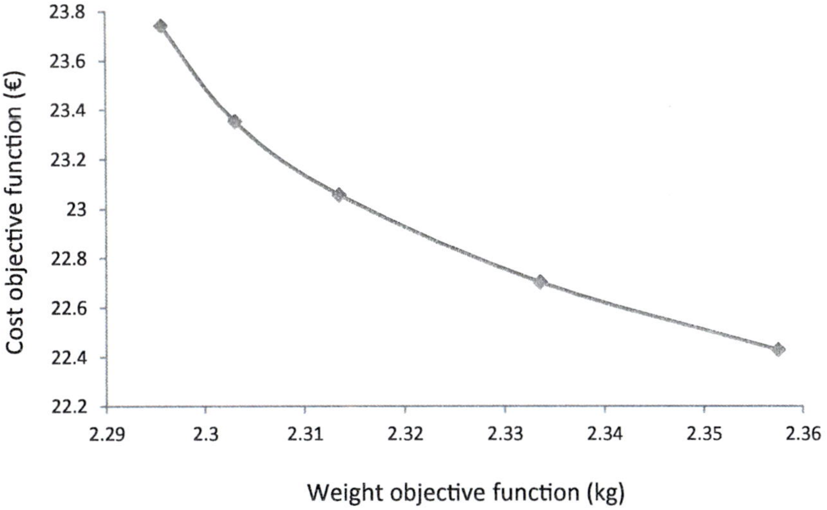

Fig. 4. Minimum weight versus minimum cost objective function by using Excel Solver program (Weighted Normalized Method) for sandwich panel of satellite consists of aluminum honeycomb core and aluminum facesheets.

Results of Multiobjective Optimization

The tool of Multiobjective Genetic Algorithm Solver in Matlab program corresponds to the gamultiobj function. The gamultiobj solver attempts to create a set of Pareto optima for a multiobjective minimization as shown in Table 9 (see Fig. 7).

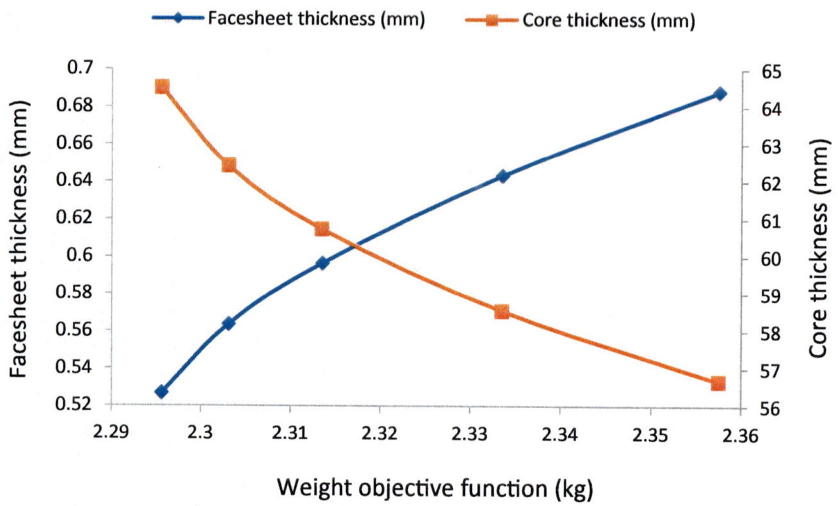

Fig. 5. Minimum weight objective function versus optimum facesheet and core thicknesses by using Excel Solver program (Weighted Normalized Method) for sandwich panel of satellite consist of aluminum honeycomb core and aluminum facesheets.

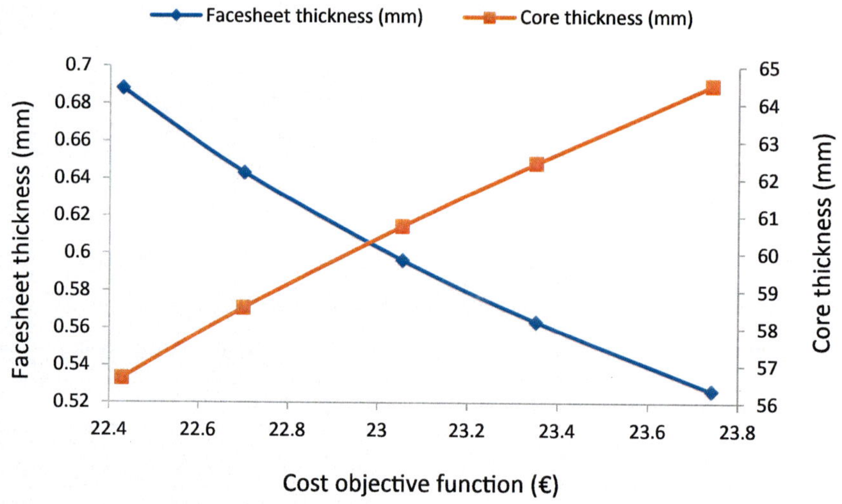

Fig. 6. Minimum cost objective function versus optimum facesheet and core thicknesses by using Excel Solver program (Weighted Normalized Method) for sandwich panel of satellite consist of aluminum honeycomb core and aluminum facesheets.

Table 7. Minimize the weight objective function, disregard the cost objective function by using fmincon Solver/Interior Point Algorithm for honeycomb sandwich panel of the satellite.

W_{min}	$t_{f,opt}$	$t_{c,opt}$
[kg]	[mm]	[mm]
2.23973	0.505007	63.51806

Table 8. Minimize the cost objective function, disregard the weight objective function by using fmincon Solver/Interior Point Algorithm for honeycomb sandwich panel of the satellite.

C_{min}	$t_{f,opt}$	$t_{c,opt}$
[€]	[mm]	[mm]
21.6589	1.025687	46.63779

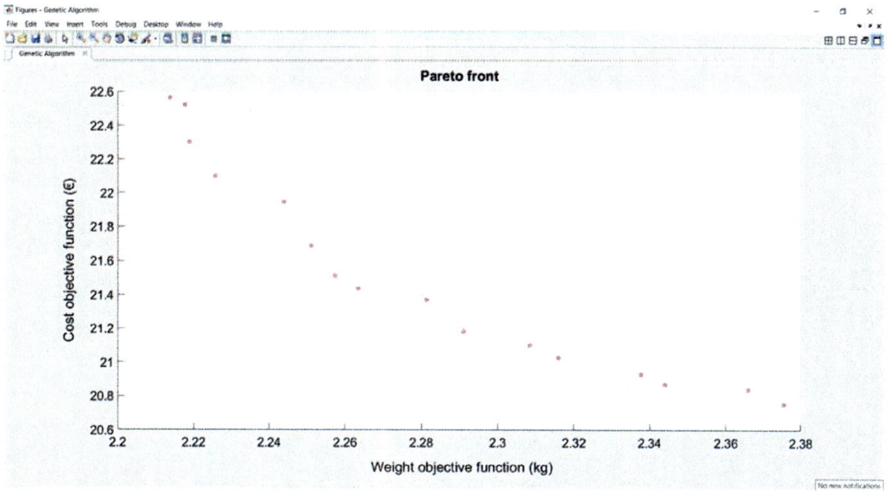

Fig. 7. Pareto front set for multiobjective functions (weight and cost) by using Matlab program (Genetic Algorithm Solver) for sandwich panel of the satellite consists of aluminum honeycomb core and aluminum facesheets.

Table 9. Minimize the weight objective function and the cost objective function by using Multiobjective Genetic Algorithm Solver for honeycomb sandwich panel of the satellite.

Index	W_{min} [kg]	C_{min} [€]	$t_{f,\,opt}$ [mm]	$t_{c,\,opt}$ [mm]
1	**2.213722**	22.561573	**0.5330**	60.5777
2	2.21767	22.5198278	0.5401	60.2838
3	2.225761	22.0990166	0.5800	58.0379
4	2.375451	**20.7545828**	0.8323	**48.0636**
5	2.243922	21.9433244	0.6101	56.8641
6	2.365845	20.8395664	0.8162	48.6973
7	2.281312	21.3681759	0.6911	53.1994
8	2.218824	22.3005114	0.5578	59.1825
9	2.308381	21.0991162	0.7387	51.2685
10	2.263419	21.4345405	0.6681	53.9291
11	2.3159	21.0257705	0.7518	50.7390
12	2.375451	20.7545828	0.8323	48.0636
13	2.251073	21.6858396	0.6367	55.4402
14	2.213722	22.561573	0.5330	60.5777
15	2.34405	20.8691365	0.7920	49.3349
16	2.257304	21.5090363	0.6563	54.4326
17	2.290992	21.178749	0.7151	52.0519
18	2.337654	20.9283943	0.7811	49.7700

5 Conclusions

The purpose of the article was to design a lightweight sandwich panel, which can be applied in the industry of satellite application because, the solar panels of satellite requires uses several holes for connection, installation, and fixing. The sandwich panel consists of an aluminum honeycomb core with upper and lower aluminum facesheets. The optimization techniques were elaborated for a sandwich panel by using the Excel Solver program and the Matlab program (fmincon Solver Constrained Nonlinear Minimization/Interior Point Algorithm for single objective function and Genetic Algorithm Solver for multiobjective functions). The objective functions were the total weight and/or cost of the panel of the satellite. The design constraints were taken into consideration were the following: total stiffness (bending and shear stiffnesses), total deflection (bending and shear deflections), facing skin stress (bending load), core shear stress, facing skin stress (end loading), overall panel buckling (critical bending and shear buckling load), shear crimping load, skin wrinkling (critical stress and load) and intracell buckling as well as the size constraint for design variables. The optimum values for sandwich panel were presented by using two different programs (Excel Solver program and Matlab program).

For single objective function by using Excel Solver program, the optimum aluminum sandwich panel which ensuring the minimum weight is (2.2934 kg), with optimum thicknesses of aluminum facesheet and honeycomb core are (0.4874 mm, 66.9722 mm) respectively, as well as the optimum aluminum sandwich panel which ensuring the minimum cost is (21.6573 €) with optimum thicknesses of aluminum facesheet and honeycomb core are (1.0251 mm, 46.6453 mm) respectively.

Whereas, for single objective function by using Matlab Program (fmincon Solver Constrained Nonlinear Minimization Interior Point Algorithm), the optimum aluminum sandwich panel which ensuring the minimum weight is (2.2397 kg) with optimal thicknesses of aluminum facesheet and honeycomb core are (0.505 mm, 63.518 mm) respectively, as well as the optimum aluminum sandwich panel which ensuring the minimum cost is (21.6589 €) with optimum thicknesses of aluminum facesheet and honeycomb core are (1.02568 mm, 46.6377 mm) respectively.

As for multiobjective function by using the Excel Solver Program (Weighted Normalized Method), the optimum aluminum sandwich panel, which ensures the minimum weight and cost are (2.3575 kg, 22.4284 €) respectively, with optimum thicknesses of aluminum facesheet and honeycomb core are (0.688 mm, 56.675 mm) respectively.

Whereas, for multiobjective function by using Matlab Program (Multiobjective Genetic Algorithm Solver), the optimum aluminum sandwich panel which ensuring the minimum weight is (2.2137 kg) with the cost is (22.5615 €), and the optimal thicknesses of aluminum facesheet and honeycomb core are (0.533 mm, 60.5777 mm) respectively, and the minimum cost is (20.7545 €) with the weight is (2.37545 kg), and the optimal thickness of aluminum facesheet and honeycomb core are (0.8323 mm and 48.0636 mm) respectively.

The results give good agreement between two programs (Excel Solver program and Matlab program) as well as between two methods in Matlab program (Interior Point Algorithm and Genetic Algorithm Solver), about (2.343%) for single objective function and (6.1%) for multiobjective functions.

Acknowledgements. The described article was carried out as part of the EFOP-3.6.1–16-2016-00011 "Younger and Renewing University – Innovative Knowledge City – institutional development of the University of Miskolc aiming at intelligent specialization" project implemented in the framework of the Szechenyi 2020 program. The realization of this project is supported by the European Union, co-financed by the European Social Fund.

References

1. Gasser, F., Nader, A., Ahmed, H.: Finite Element Analysis for Satellite Structures. Springer, London (2013)
2. Bitzer, T.: Honeycomb Technology: Materials, Design, Manufacturing, Applications and Testing. Chapman and Hall, London (1997)
3. Zenkert, D.: An Introduction to Sandwich Construction, Student Chamelon Press, Oxford, London (1995)

4. Zenkert, D.: The Handbook of Sandwich Construction. Engineering Materials Advisory Services Ltd. (EMAS). Emas Publishing, London (1997)
5. Kollár, L., Springer, G.: Mechanics of Composite Structures. Cambridge University Press, London (2003)
6. Jun, W., Dai, G.: Development of the hybrid insert for composite sandwich satellite structures. Compos. Part A **42**(8), 1040–1048 (2011)
7. Craig, L.: Design, Analysis, Fabrication, and Testing of a Nanosatellite Structure. Faculty of the Virginia Polytechnic Institute and State University, Virginia (2002)
8. Daisuke, S., Noboru, Y., Koji, T.: Thermal design of photovoltaic/microwave conversion hybrid panel for space solar power system. IEEE J. Photovoltaics **7**(1), 374–382 (2017)
9. Byoung, J.K., Dai, G.L.: Development of a satellite structure with the sandwich T-joint. Compos. Struct. **92**(2), 460–468 (2010)
10. Boudjemai, A., Bouanane, M.H., Merad, L., Si Mohammed, A.M.: Small satellite structural optimisation using genetic algorithm approach. In: 3rd International Conference on Recent Advances in Space Technologies, Institute of Electrical and Electronics (IEEE), Istanbul, Turkey (2007)
11. Cho, H.K., Rhee, J.: Vibration in a satellite structure with a laminate composite hybrid sandwich panel. Compos. Struct. **93**(10), 2566–2574 (2011)
12. Virág, Z., Szirbik, S.: Finite element modal analysis of a hybrid stiffened plate. Ann. Univ. Petroşani Mech. Eng. **21**, 115–120 (2019)
13. Alaa, A., Jármai, K., Kovács, G.: Optimum design of honeycomb sandwich plates used for manufacturing of air cargo containers. Editura Politehnica Acad. J. Manuf. Eng. **18**(2), 116–123 (2020)
14. Teng, L., Zheng, X., Jin, H.: Performance optimization and verification of a new type of solar panel for microsatellites. Int. J. Aerospace Eng. **2**, 1–14 (2019)
15. Alaa, A., Jármai, K., Kovács, G.: Theoretical and numerical comparison study of aluminum foam sandwich structure. Pollack Periodica Int. J. Eng. Inf. Sci. **15**(2), 1–15 (2020)
16. Virág, Z.: Determination of optimum diameter of a welded stiffened cylindrical shell. Pollack Periodica **4**(1), 41–52 (2009)
17. Rion, J., Månson, J., Leterrier, Y.: Ultra-light photovoltaic composite sandwich structures. Lausanne, EPFL, France (2008)
18. Hexcel Homepage. https://www.hexcel.com/user_area/content_media/raw/Honeycomb_Sandwich_Design_Technology.pdf. Accessed 10 Mar 2020
19. Bode, W.: Evaluation of a lightweight composite bottom panel for air cargo containers. Master Thesis, Faculty of Aerospace Engineering, Department of Aerospace Structures & Materials, Netherlands (2016)
20. Achille, M.: Optimization in Practice with Matlab for Engineering Students and Professionals. Cambridge University Press, United States of America (2015)

Survey on New Trends of Robotic Tools in the Automotive Industry

Rabab Benotsmane(✉), László Dudás, and György Kovács

University of Miskolc, Miskolc-Egyetemváros 3515, Hungary
Benrabab1@gmail.com

Abstract. In the last decade, industrial robotics sector has known an evolution at level of its technological platform, especially with the appearance of Industry 4.0, therefore we are entering a new era of automation where intelligent production and digital factories produce increasingly complex and high-volume quality products with less human effort. The automotive industry which is already revolutionizing the transition to electric cars has to deal with the change in construction and the tools used to produce them. The use of evermore collaborating intelligent robots with machines and humans of smart manufacturing and assembly lines represents a new level of development in automation. This article provides an overview of the latest tools, directions and intelligent methods available today in this area, from virtual reality to wearable machine intelligence devices for human-driven systems, all the while focusing on robotic applications. Describes the latest industry trends in each field, by highlighting detection systems designed to achieve security, response speed, detection completeness, and reliability, introducing security vision systems of robotic systems, ultrasonic and laser sensors and their efficiency issues. In the final stage, the paper will compare the state of the art in automotive robotics by analyzing the development of leading car manufacturers in Europe, the United States and the Far East.

Keywords: Industry 4.0 · Automotive industry · Industrial robotics · Collaborating intelligent robots · Smart manufacturing

1 Introduction

In today's world, Industrial robotics has become one of the most valuable keys for factory progress and industrial innovations. Especially in last years, where the industrial platform has faced a new revolution age named Industry 4.0 that means the Fourth Industrial Revolution which origins from the strategic project of the German government, that was highlighted for the first time at the Hannover Messe in 2011 [1], this concept leads to integrate the production processes [2] with the latest information technologies and techniques based on nine pillars related to computer science and robotics fields [3]. One of them is the Industrial Internet of Things, which realizes the communication between machines and other systems sharing a huge amount of information on the industrial network collected from all the processes, which has to be protected using Big data and

K. Jármai and K. Voith (Eds.): VAE 2020, LNME, pp. 443–457, 2021.
https://doi.org/10.1007/978-981-15-9529-5_38

Cybersecurity technologies [4]. The production chain in Industry 4.0 expects to use autonomous systems regrouping Multi Agent Systems [5] where those agents can represent machines, sensors, controllers, industrial robots and human operators working and collaborating continuously to finalize the product considering the lead time expected. The novelty of those systems in Industry 4.0 appears in the integration of Artificial Intelligence technologies, the manner to make a machine smart and taking decisions by itself. AI can be used not only to control systems but also to handle the problems ensue from Big data [6], it helps to manage and make a profitable use of these data between departments. AI can handle the “3V” of Big data, namely Velocity, Variety, Volume with uncertainties, volatility, complexity in the automotive industry and related market. AI techniques have been used in multiple segments of the networked enterprises. They have taken a prominent role to integrate people, information and products across dynamic networked enterprise boundaries including management of various manufacturing, logistics and retailing operations. 3D Simulation tools and Augmented reality [7–8] are the most important keys to observe the real process how it works in virtual scene where the operator can detect the errors and avoid the accidents that can happen, also those simulation tools are able to generate several scenarios to optimize the manufacturing processes. The main objective of Industry 4.0 is transforming the factories to a smart environment based on a digital platform, focusing on the idea of manufacturers that the product satisfies the individual needs of customers where companies have to achieve the efficient progress regarding the product and its quality taking into account the lead time needed in order to cope up with the customer’s requirements [3]. Nowadays automotive corporations such as Audi, BMW, Mercedes aim to emerge the concept of Industry 4.0 in their companies with the use of one of the previously mentioned pillars of the Autonomous Systems. Last decades’ automotive industries started installing industrial robots in the production chains as parallel robots, manipulator arms, hybrid robots to accomplish a concrete task [9]: pick and place, packaging, welding [10] or painting and with the appearance of AI and MAS concepts the idea was developed to use intelligent industrial robots [11] instead of controlled ones collaborating together and take a decision in smart way to achieve their tasks. Usually, industrial manipulator arms are the most used in the automotive industry because of their advantages to use them in different situations, also their ability to carry a heavy product. There is a system that regroups the structure of arm itself – there is a specific controller to command automatically the arm – and a teach pendant to control the arm manually, the controller presents a black box that receives inputs as information from the sensors installed in the arm and sends outputs generating torques to produce the accelerations, velocities and positions needed to travel a planned trajectory or to carry objects according to the tasks planned. From literature review, the interest of cooperating industrial robots in production cells were discussed recently in many articles regarding the wide range benefits offered by such topic like optimization operations in the industry [12]. Cooperating of multi manipulator arms guarantee the achievement of tasks planned more easily than single manipulator arm, it is known that the control of a single robot arm was always a trivial task comparing to the control of multi robot arms which presents a real deal for scientist, the development of controller design for such structure was proposed in several articles [13].

This article contributes describing an overview of how the transformation of factories can be achieved from the traditional paradigm to the smart one based on Industry 4.0 concept, where the purpose of this survey is to present the latest techniques of our time from virtual reality to the use of intelligent machines and the role of their collaboration with human operators, taking as an example the sector of the automotive industry, all the while focusing on robotic applications, highlighting detection systems designed to achieve security, response speed, detection completeness, and reliability, introducing the security vision systems of robotic systems, ultrasonic and laser sensors, and their efficiency issues. Finally, the paper will close the survey with a comparison of the state of the art in the automotive industry concerning Europe, the United States, and the Far East by analyzing the development of leading car manufacturers in each area.

2 Automotive Industry Background from the Past to Now

Through the historical background, the automotive industry was first invented and perfected in Europe in the 19th century, in the first half of the 20th century the United States completely dominated the automobile industry through the invention of mass production techniques especially after embrace T model invented by Henry Ford based on the idea of that cars would be produced in larger volume at lower prices than in Europe [14], therefore the use of cheap raw materials and a chronic shortage of skilled labor early encouraged the mechanization of industrial processes in the United States. In the second half of the century, the situation altered sharply as western European countries and Japan became major producers and exporters.

The modern automobile presents a complex integration of many subsystems, with specific design functions that deal with different technological fields such as high tech systems involving auto electronic controllers and engine computers used to do everything from regulating fuel to diagnosing problems, where they are consisting of thousands of components, parts. The specifications of these components are defined by Original Equipment Manufacturers (OEMs) or car manufacturers - the use of high-strength plastics and new alloys of steel and nonferrous metals [15]. Due to the lightweight of plastics used in, cars will be more fuel-efficient, while the uses of modern materials like aluminum and carbon fiber are helpful, but the wise allocation of plastics is making an increasing difference. Other subsystems have come about as a result of factors such as safety legislation, and foreign competition. In general way vehicles are designed and manufactured to respond to market needs, which are translated to requirements and are mainly driven by product lifetime cost taking into consideration how to avoid environmental issues like air pollution resulted. Figure 1 describes the essential parts in a car.

2.1 Comparison of Traditional and Future Automotive Industry

In the growth of automobile models and the development of new technologies, several automotive companies started to upgrade their resources in order to achieve the next industrial revolution that guarantees the three keys factors (higher productivity - good quality - short lead time) for a successful economic market and to respond rapidly to changing customer demands [16]. With the appearance of Industry 4.0 concept many

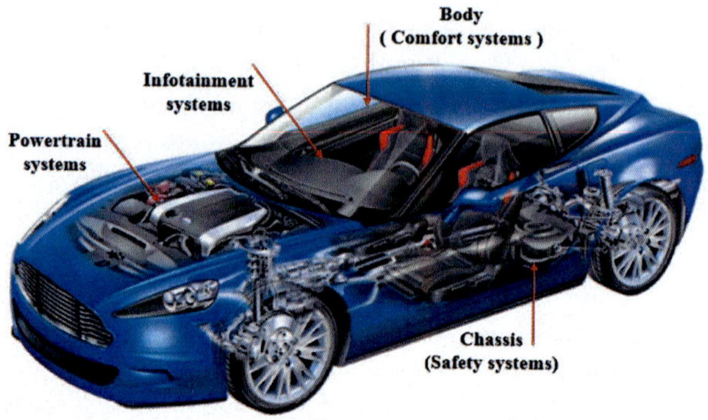

Fig. 1. Essential parts in a modern car

literature reviews started to deal with it to understand what does a factory need to transform its modern paradigm to a flexible and smart one, it is about the future industry, where till now and with all the technologies presented in manufactories, the leaders still want to achieve the perfection in the supply chain. Nowadays most automotive companies installed in their production lines the advanced robotics technology for the majority of the operations where these machines need the maintenance and the prediction of the breakdowns which requires human operator intervention.

Table1. Comparison between traditional and future automotive industry

Traditional automotive industry	Future automotive industry
- The resources are limited and not sustainable for a long time - Fixed production lines and non-reconfigurable machines where the flexibility is limited - Limited products where the car models are always proposed by the companies - The machines have downtimes and the maintenance is harder to predict before breakdowns - Intervention of human operator in case of problem or maintenance - The product based on gasoil causes the pollution for the environment - Lower productivity comparing to the future industry in large lead time which results in less competitiveness - Lack of use of the concept of IoT to collect data analytics for processes and achieve the cybersecurity of the company - A larger workforce has to be applied resulting a higher labor costs for the enterprises - The use of heavy materials at higher costs	- Sustainable production due to the use of efficient resources - Application of flexible and reconfigurable production lines and logistical systems - Diversity of products which can satisfy the customer demands - The use of Augmented reality and Simulation tool to predict errors and malfunctions - The use of cooperating concept machine to machine or machine to human with AI application - The product itself (car) is electrified and autonomous which emit less exhaust fumes and noise into its environment - Higher productivity with a good quality in short lead time which results in more competitive enterprises - The efficient Application of IoT and Cybersecurity to ensure better operation and security of stored data - Installation of intelligent devices instead of human operators, and use a highly skilled workforce for the control of those smart systems - Increasing use of lightweight materials in manufacturing components and vehicle structures

Table 1 presents a deep comparison between the modern and future automotive industry.

2.2 Challenges Faced the Modern Automotive Industry

The modern automotive industry is catching out the basic pillars of Industry 4.0 that aims to form a structured supply chain and make the industry smarter, the development of car design and manufacturing is already enhanced by the high degree of customization, electrification, digitalization, automation, and mobility, where driving will become easier, safer, cheaper, and more comfortable, to accelerate those changes and get such product, the future industry suggest to emerge flexible and decentralized manufacturing processes where Cloud computing system will present a robust storage for complex automotive systems and their design data, process information, simulation, etc., in other hand it is important for cyber physical systems as humans, robots, sensors, machines equipment work in resonance, the role of human will be limited characterized for setting the tasks program in the production line for the whole processes where the machines can take their decisions easily using AI technics therefore a car getting fully assembled without any physical human intervention.

From literature review, the challenges of the modern automotive can be faced are addressed by the OEMs presented as the following [17]:

- Complexity and cost pressure. Diversity of modular systems and cost pressure where prices in markets are likely to be flat.
- Diverging markets where OEMs need to develop their technologies regarding the new revolution.
- Digital demands to satisfy consumers have to be more flexible and require more safety.
- Shifting industry paradigm, different technologies and innovative solutions will take apart which creates the competitiveness between countries of Europe, China, US.

3 Tools and Technologies Used in Modern Automotive Industries

The automotive sector regroups different types of companies alongside auto manufacturers. Some of them are suppliers who aim to produce the component parts installed on a car (OEMs), where other companies are responsible for other operations as vehicle sales, repairs and maintenance.

Original Equipment Manufacturers:
OEMs play an important role in the challenges that face the economical and technological growth of the modern automotive industry as presented earlier, where three kinds of suppliers are responsible to manufacture all parts used in automotive specializing on material and electronics used in a body and chassis of a car. Although there exist several automakers that prefer to produce some of their own equipment and assembled with other parts ordered it from OEMs. Figure 2 presents the different suppliers of OEMs.

The assembly line in the automotive sector is introduced as the most important process that involves two main phases, body and chassis [18]:

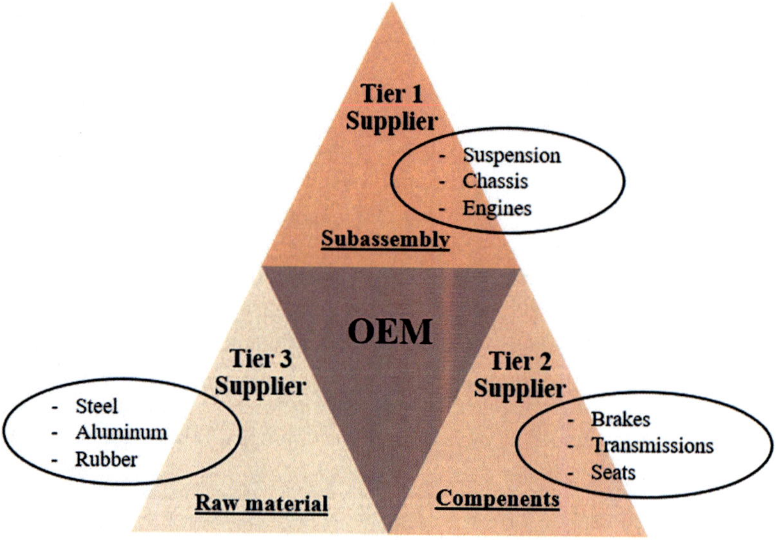

Fig. 2. Different tiers suppliers in OEMs

- The first line manufactures the body panels using press technology, cutting [19–20] and welding technology, then assembles the painted body with the windows and the doors.
- The second line, the frame adds – among others - the landing gear and the power train that should be composed of different components like wheels, springs, steering gear, and power train including the engine and the transmission.

The development of the modern automotive sector is improved regarding its root to incorporate the consumer's demand. It involves transformation of production processes including smart cyber physical systems as Sensors – AGVs – Industrial robots – IoT. Decentralization of controllers to minimize manual work, reduce wasting time and create flexible manufacturing processes.

It is well known that the production of a car presents a large operation that includes many sub-processes, Fig. 3 describes the units responsible to finalize such a product.

In the following, we present the different tools and technologies used in modern automotive industries.

3.1 Industrial Robots and AGVs

Within the concept of a flexible automotive industry where the cyber physical systems have to be agile and respond rapidly in an intelligent way, industrial robots and AGV systems take a large part as presented in Fig. 4 [21–22]:

- **Industrial robots:** are a combination of a mechanical structure, sensors, and intelligent controllers, which replaces a human worker and can complete tasks and resolve

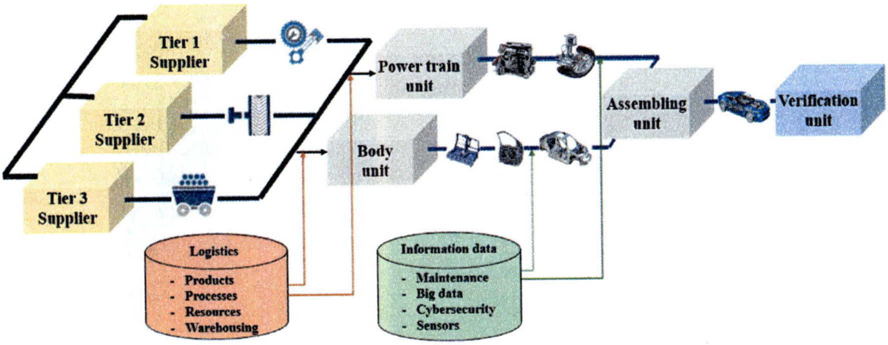

Fig. 3. Automotive industry chain value

the problems in a fast way. They can be a serial type of manipulator's arms, parallel type, or hybrid one.

- **AGVs (Automated Guided Vehicles):** are autonomous robot systems controlled by following along marked long lines or wires on the floor, or use radio waves, vision cameras, magnets, or lasers for navigation where they react intelligently to supply the right material or component at the right location at right time.

Figure 4 presents Industrial robots and AGVs systems in V-rep virtual environment:

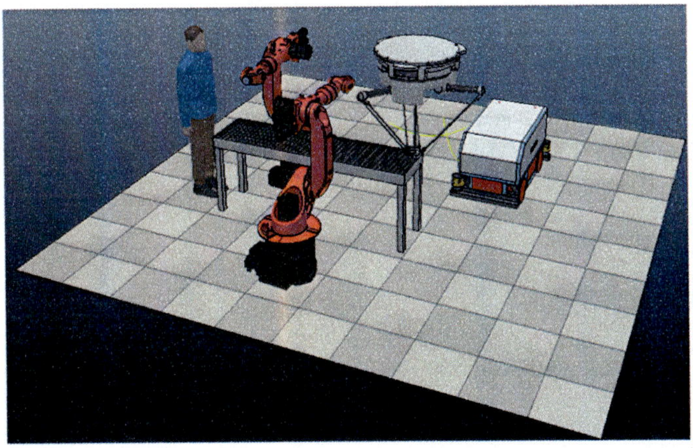

Fig. 4. Robots and AGV in an assembly cell.

3.2 Collaborating Multi Agent System Concept

Multi agent system [23] is a group of several cooperating agents to achieve a determined task. Usually, an agent can present a computer system located in a dynamic environment such as in robots or intelligent machines and capable of executing independent or intelligent actions using AI techniques, or it can be a human operator. The collaboration between agents may mean two well-known concepts:

- **Collaborating Human – Robot (HCR)**: in the modern automotive industry, this concept is already realized where human operators and robots working together to manufacture a car product [24], the concept requires to study several aspects as collision, safety, and respect the robot workspace aiming to enable versatile automation steps and increase productivity. It is an additional element that combines human capabilities with the efficiency and precision of machines. Figure 5 illustrates HCR scenario in V-rep software.

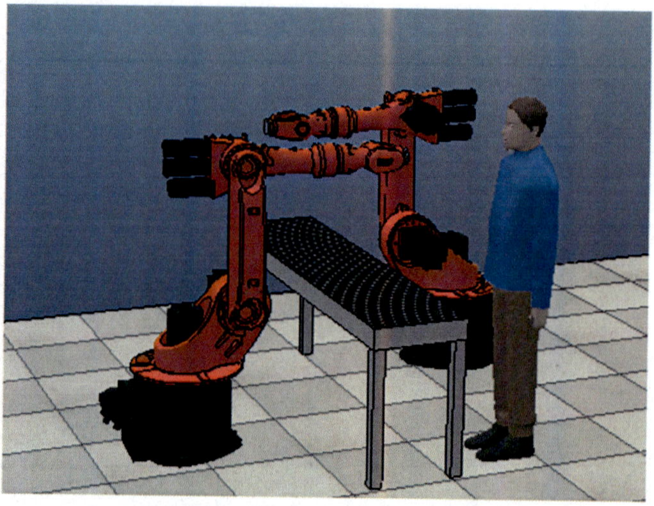

Fig. 5. Human-robot collaboration in V-rep software.

- **Collaborating Robot – Robot (Cobot)**: in future industry is highly recommended to transform the assembly chain to a flexible and structured line without any human interaction where multiple robots working cooperatively in a redundant way that will offer new possibilities in the execution of complex tasks in dynamic workspaces if some malfunction or an error occurs in the production line [25]. The robotic manufacturing or assembly cell can react and make a decision easily to continue the task planned for. Figure 6 presents cooperative industrial robots working together on the part of a production line.

3.3 Simulation Modeling and Augmented Reality Tools

Since many years, engineers and leaders in industries upgrade their capabilities in Simulation tools and modeling processes regarding the benefits obtained by these technologies

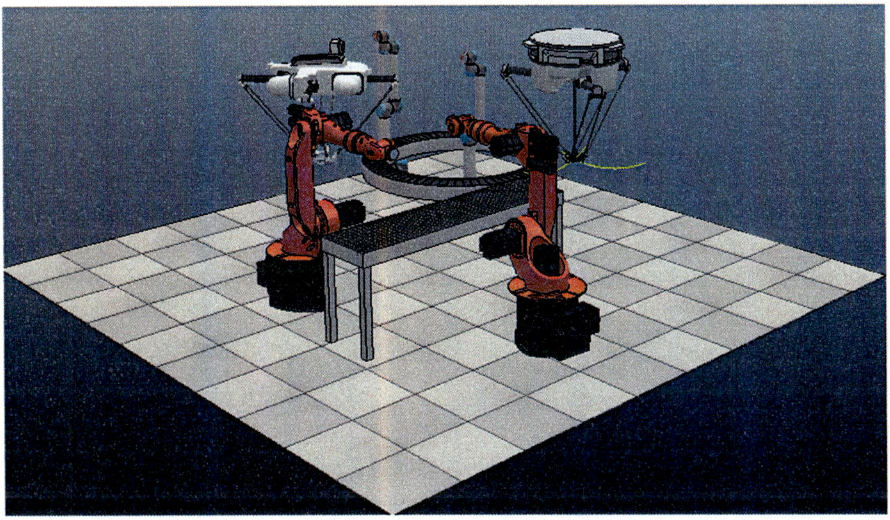

Fig. 6. Cooperative industrial robots working together in a production line.

where *in manufacturing line view*: modeling a process and visualize how it works allows us to study well the cycle time needed, optimize energy consumption from the process and determine the errors that can occur. In the same time simulation modeling accord companies to design precisely different car models and fulfil customer demands. *In logistic view*: simulation analyses all the statistics of the process where it defines the information needed about the warehouse, the resources and defines the cost, cycle time and the quality of the product in the production process.

In last years Augmented reality (AR) was highlighted as a new technology in this field projecting as a data glasses or tablets and smartphones, it provides the possibility of visualization of manufacturing processes by transforming all the details of the real environment to a virtual one. In the modern automotive industry AR represents an important tool to minimize the real industry land shops and put it in the hands of the specialists, this advantage gives the opportunity to visualize and control all resources presented. Eventually, this technology reduces the planning costs for materials where AR devices will count the number of packages, GLTs or pallets and scan them to collect data and information on the level they are full the enlarged reality, not only enriches the production of automobiles, it also enriches their marketing considerably by bringing more emotions to car advertising when a company lunch a new car model.

3.4 Artificial Intelligence in the Framework of Optimization Processes

The concept of artificial intelligence has appeared in the last decade where it was defined as a new methodology to think better and in a smart way, it was addressed at first for the machine with the aim to make it take a decision easily as a human using machine learning topic. Nowadays AI has taken a large part of our life where it is used in most of the mobile phones and internet applications, many research seeks to emerge AI in the human brain in the aim to develop the IQ of human thinking. For modern companies, AI represents the digital brain of Industry 4.0 and is its driving force. Actually, AI is much more powerful in industry networking where it collects and recognizes all data received from sensors installed in the whole process, it analyses and optimizes the calculation operations guaranteeing the protection of this information under cloud computing and cybersecurity. For the future Industry 4.0, leaders aim to emerge AI in the cyber physical system where devices as industrial robots and AGVs can take decision easily in a case of normal process, the production line appliances are collaborating and can optimize more the lead time and increase the production by working in a parallel way and continually. In case of a malfunction these cyber physical systems can react without human intervention, the progress will be appearing on the link between the production line and the product where they both are smart and communicate together using IoT technology, then if an error occurs the assembly chain, the robotic cell and the product will collaborate together to find the optimal solution to accomplish the task and perform the quality control, using vision processing and machine learning in the cloud. This scenario can be realized with the presence of sensor technology where robots can undertake multiple tasks with remote reconfiguration of the Programmable Logic Controllers (PLCs).

3.5 Additive Manufacturing and 3D Printing Concept

Over the past decade, additive manufacturing (AM) known as 3D printing is an innovative technology to build three-dimensional objects by adding layer upon layer of a given material, it puts a significant change in production process theory where the product has to be designed, developed, manufactured, and distributed. This new technology based on designing the prototype of the product using a CAD software and printing the model in 3D using a high-quality plastic material. The potential application of AM in the modern automotive industry results in new prototypes and products characterized by the specification of designs, which are cleaner, lighter and safer with shorter lead times and lower costs. Challenges that can face this technology is sophisticated customization where advanced 3D printing would be a feasible way for realization. That needs a lot of development in the printing technology to make them fast and accurate. Quality at its peak is a must, as printing practices don't run on economies of scale like a stamping press in a plant. There will not be any time to reject the parts or events to inspect the quality. The real advantage of AM for the automotive sector is its ability to break existing performance trade-offs in two fundamental ways: First, AM reduces the capital required to achieve economies of scale. Second, it increases flexibility and reduces the prototype manufacturing time. There are four tactical paths that companies have to take regarding this technology [26]: Path I: Companies focus on AM technologies to improve value delivery for current products within existing supply chains. Path II: Companies take advantage of scale economics offered by AM as a potential enabler of supply chain transformation of the products they offer. Path III: Companies take advantage of the scope of economics offered by AM technologies to achieve new levels of performance or innovation in the products they offer. Path IV: Companies alter both supply chains and products in the pursuit of new business models.

3.6 Achieving the Safety Working Using Sensors

Sensing technology *in the industry* presents the basic connection between the human and machine where receiving information and observing the state of process is the major target for employees. The modern automotive industry provides many types of sensors regarding the task, where we can find kind of sensors used for the supply chain to control the whole process in order to achieve high quality products and realize the safety regarding the resources especially with the use of HRC human robot cooperation.

The future automotive industry requires an improved level sensor technology as the need of smart machines that perform better than the human operators. Since data sharing in clouds and IoT are the essential pillars for this revolution, therefore, the installation of intelligent sensors is the potential target where they support production on a highly flexible manufacturing line where real-time tasks will be executed locally and the wasted time on uploading data and information exchange will be optimized. Sensors in the whole supply chain will validate the quality of the product, aid in reducing the waste and perform an inspection of parts.

Some kinds of sensors are installed *in the product* where the car has to achieve all the requirements of safety for the driver. The rising trend of Autonomous Things is largely driven by the move towards the Autonomous car, which both addresses the main existing safety issues and creates new issues. The autonomous car is expected to be safer than existing vehicles, by eliminating the single most dangerous element - the driver.

4 Automotive Industry Development Statistics in Europe, US and Far East

The global automotive industry is in higher progress than it was five years ago, currently, it presents a powerful key where motor vehicle production and sales are one of the major indexes of the state of the economy in those countries. For such countries as Europe, US, and China, motor vehicle exports are essential to the maintenance of healthy international trade balances. In the following we highlight about the development statistics of automotive industry market in different area from the world. Figure 7 presents Global light vehicle sales in the world.

4.1 Automotive Market Forecast in Europe

Today, the evolutionary path of car design and manufacturing is already disrupted by the high degree of electrification, digitalization and automation. Maritime and air-transport demonstrated electric and hybrid technologies at small scale, preparing the ground for real transport vehicles in the years to come. The automotive industry plays an important role in European Economy. EU was traditionally the largest automotive producer but the last decade brought some changes [27]. The competition in the market of materials for automotive applications is needed since nowadays the environmental concern has focused on the production of lighter vehicles for lower fuel consumption and also for the need for recycling. These recent pressures bring opportunity for product innovation, technological advancements, and maintaining quality standards in which they present

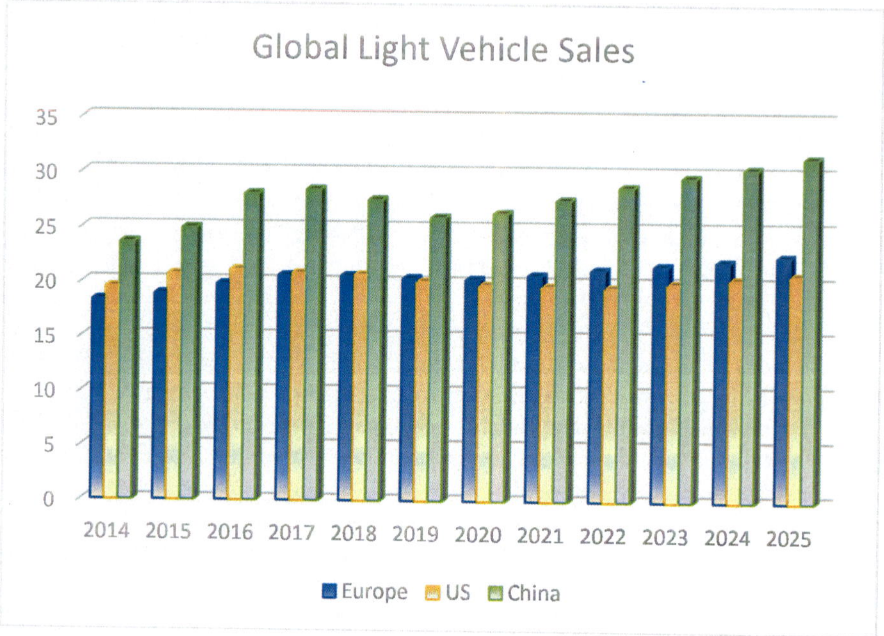

Fig. 7. The global light vehicle sales in Europe, US and China.

the key strategies adopted by leaders to maintain a stronghold in the global automotive market. Most developing countries have tariff policies that make imported cars extremely expensive where generates requirements that a substantial part of the components used in local assembly plants be of domestic origin. Creating an internal market for vehicles is the aim of the EU's policy in the automotive sector by ensuring a high level of green environmental protection and safety and provide a strengthen competitiveness. Challenges faced in the EU are the decreasing demand for new vehicles and the growing competition from Chinese and Indian manufacturers where the pressure on development of new alternative powertrain technologies with lower-emission is a large challenge for the European automotive market. The fastest growing demand for transport services will be in low income countries (mainly Africa and Asia). By 2025 emerging markets will generate over two-thirds of the total automotive profit (China will be the leader).

4.2 Automotive Market Forecast in US

The United States has one of the largest automotive markets in the world [14], where in 2018, light vehicle sales reached 17.2 Mio. Units [28]. Generally, the United States is the world's second-largest market for vehicle sales and production. Since Honda opened its first US plant in 1982, almost every major European, Japanese, and Korean automaker has produced vehicles and invested more in the United States. Additionally, many automakers have US-based engine and transmission plants, and conduct R&D, design, and testing. Total foreign direct investment in the U.S. The automotive industry is also at the forefront of innovation. New R&D initiatives are transforming the industry to better respond to the opportunities of the 21st century. In 2018, the United States exported 1.8 Mio. New light vehicles and 131,200 medium and heavy trucks to more than 200 markets around the world, with additional exports of automotive parts. With an open investment policy, a large consumer market, a highly skilled workforce, available infrastructure, and local and state government incentives, the United States is the premier market for the 21st-century automotive industry.

4.3 Automotive Market Forecast in China

From statistics, the largest assembly plants can be found in the Asia-Pacific region, where China started to attract the attention of the world's major automotive companies since 1990. Somewhat relaxed governmental controls on private ownership and the consequent rise of entrepreneurial enterprises provided a burgeoning market in China for automobile ownership by individuals [29]. This potential, plus local-component requirements, led to the establishment by automakers and component manufacturers of complete manufacturing facilities in China rather than limited local assembly operations.

China's global market share in auto production was close to 30 percent before experiencing a slump in production amid the coronavirus epidemic. By 2020, it is expected that Toyota will reclaim the spot as the world's largest automobile producer, outperforming the Renault-Nissan and Mitsubishi alliance as well as Volkswagen [30].

5 Conclusion

The appearance of Industry 4.0 concept has a huge effect on making the revolution in the automotive industry units were the developments of new applied technologies mounting the current and the future state of production manufacturing, this paper presents an overview on the new changes that occurred in the automotive industry units to achieve different trends in car manufacturing. Different tools were discussed combined with digitalization concept and the latest technologies, in the final part, the paper gives a global view about automotive industry development statistics in Europe, US and China, where it summarized that the future of the automotive industry will beholder by China automotive markets.

References

1. Sung, T.K.: Industry 4.0: a Korea perspective. Technol. Forecast. Soc. Change **132**, 40–45 (2017)
2. Dima, I.C., Kot, S.: Capacity of production. Industrial Production Management in Flexible Manufacturing Systems. Book News Inc, Portland (2013)
3. Benotsmane, R., Dudás, L., Kovács, G.: Collaborating robots in Industry 4.0 conception. In: Proceedings of XXIII International Conference on Manufacturing, vol. 448, 012023. IOP Conference Series: Materials Science and Engineering, Kecskemét (2018)
4. Benotsmane, R., Kovács, G., Dudás, L., Economic, social impacts and operation of smart factories in Industry 4.0 focusing on simulation and artificial intelligence of collaborating robots. Soc. Sci. **8**, 143 (2019)

5. Thomassey, S., Zeng, X.: Artificial Intelligence for Fashion Industry in the Big Data Era. Springer, Singapore (2018)
6. Ding, H., Wang, W., Qiu, M., Dong, J.: Artificial Intelligence Techniques for Networked Manufacturing Enterprises Management. Springer Series in Advanced Manufacturing, London (2010)
7. Xu, J., Huang, E., Hsieh, L., Lee, L.H., Jia, Q.S., Chen, C.H.: Simulation optimization in the era of Industrial 4.0 and the industrial internet. J. Simul. **10**, 310–320 (2016)
8. Regenbrecht, H., Baratoff, G., Wilke, W.: Augmented reality projects in the automotive and aerospace industries. IEEE Comput. Graphics Appl. **25**, 48–56 (2005)
9. Karabegović, I.: The role of industrial robots in the development of automotive industry in China. Int. J. Eng. Works **3**(12), 92–97 (2016)
10. Ranky, P.G.: Collaborative, synchronous robots serving machines and cells. Ind. Robot **30**(3), 213–217 (2003)
11. Parodi, F., Gerio, P.: AURA: an example of collaborative robot for automotive and general industry applications. Procedia Manuf. **11**, 338–345 (2017)
12. Tuli, T., Manns, M.: Hierarchical motion control for real time simulation of industrial robots. Procedia CIRP **81**, 713–718 (2019)
13. Binbin, Z., Jun, W., Liping, W., Zhenyang, Y.: Accurate dynamic modeling and control parameters design of an industrial hybrid spray-painting robot. Robot. Comput. Integrated Manuf. **63**, 101923 (2020)
14. ENCYCLOPAEDIA BRITANNICA Homepage. https://www.britannica.com/technology/automotive-industry. Accessed 25 June 2020
15. Dajsuren, Y., Van den Brand, M. (eds.): Automotive Systems and Software Engineering. Springer International Publishing, Switzerland (2019)
16. Pwc Global Homepage. https://www.pwc.com/gx/en/industries/automotive/publications/eascy.html. Accessed 28 June 2020
17. McKinsey & Company. https://www.mckinsey.com/industries/automotive-and-assembly/our-insights/the-road-to-2020-and-beyond-whats-driving-the-global-automotive-industry. Accessed 28 Jun 2020
18. Williams, E., Ülgen, O.: Use Cases of Discrete Event Simulation. Springer, , Heidelberg (2012)
19. Maros, Z.: Machining of different materials with abrasive waterjet cutting. IOP Conf. Ser. Mater. Sci. Eng. **448**(1), 012009 (2018)
20. Kundrák, J., Pálmai, Z., Varga, G.: Analysis of tool life functions in hard turning. Tehnicki Vjesnik Technical Gazette **27**(1), 166–173 (2020)
21. Jaiganesh, V., Jayashankar, D.K., Girijadevi, J.: Automated guided vehicle with robotic logistics system. Procedia Eng. **97**, 2011–2021 (2014)
22. Garcia, M., Rojas, R., Gualtieri, L., Rauch, E., Matt, D.: A human-in-the-loop cyber-physical system for collaborative assembly in smart manufacturing. Procedia CIRP **81**, 600–605 (2019)
23. Yan, Z., et al.: A survey and analysis of multi-robot coordination. Int. J. Adv. Robot. Syst. **10**(12), 399 (2013)
24. Magrini, E., Ferraguti, F., Jacopo Ronga, A., Pini, F.: Human-robot coexistence and interaction in open industrial cells. Rob. Comput. Integrated Manuf. **61**, 101846 (2020)
25. Makris, S., Eytan, A.: Cooperating robots for reconfigurable assembly operations: review and challenges. Procedia CIRP **3**, 346–351 (2012)
26. https://www2.deloitte.com/us/en/insights/focus/3d-opportunity/additive-manufacturing-3d-opportunity-in-automotive.html. Accessed 28 June 2020
27. European Commission Home page. https://ec.europa.eu/transparency/regexpert/index.cfm?do=groupDetail.groupDetailDoc&id=34594&no=1. Accessed 27 June 2020
28. SELECTUSA Homepage. https://www.selectusa.gov/automotive-industry-united-states. Accessed 30 June 2020

29. Tang, R.: The Rise of China's Auto Industry and Its Impact on the U.S. Motor Vehicle Industry. Federal Publications (2009)
30. Statista Homepage. https://www.statista.com/topics/1050/automobile-manufacturing-in-china. Accessed 30 June 2020

Trial - and - Error Optimization Method of Pick and Place Task for RV-2AJ Robot Arm

Rabab Benotsmane(✉), László Dudás, and György Kovács

University of Miskolc, Miskolc-Egyetemváros 3515, Hungary
iitrabab@uni-miskolc.hu

Abstract. In recent years, industrial robots have played an important role in the revolution of a production line in factories, and especially in the growth of Industry 4.0 concept, due to their flexibility to execute tasks and cooperate with their environment fluently, today manipulator arms takes a large part in the production chain especially in the automotive sector where the robot can be configured due to the control terminal for the different task process as welding, painting, pick and place heavy parts. Manipulator arm used in the industry is usually combined 6 degrees of freedom to have a large workspace and manipulation capability. In this article we present an optimization approach regarding a pick and place application for RV-2AJ robot arm which has five degrees of freedom in order to execute different movements, the approach aims to build a card house using one manipulator arm and a support element. Trial – and – error optimization method proposed in this article highlights a good solution regarding the positioning problem for RV-2AJ arm, which has five degrees of freedom that limits its workspace.

Keywords: Manipulator arm · Industry 4.0 · Pick and place · Workspace

1 Introduction

In recent years, industrial robotics play an important role in the efficiency improvement of the production sector [1], due to the evolution of automation technology and computer science where machines and robots in the manufactories had the ability to collaborate together and control themselves in a flexible way [2, 3]. This new technology boosted traditional manufacturing systems to be modern and smart, to create the concept of Industry 4.0 [4, 5], which present a new revolution on the industrial level, this transformation based on regrouping newest technologies of different fields in the same platform, to achieve a higher level of facilities and quality product, where aims to the technical integration of cyber-physical systems (CPS) into production and logistics, using Internet of Things to connect all the systems and collecting data [6]. Since the 19th century, Industrial robotics field has taken a large part in different sectors due to its role to improve the production line. Nowadays industrial robots can be presented in different structures and types as manipulator arms, parallel robot, Hybrid robots, where the International Organization for Standardization (ISO) describes an industrial robot

K. Jármai and K. Voith (Eds.): VAE 2020, LNME, pp. 458–467, 2021.
https://doi.org/10.1007/978-981-15-9529-5_39

as a system that is automatically controlled, multi-application, reprogrammable, multi-purpose manipulator, programmable in three or more axes. It is well known that serial arm is the most used in industries especially in the automotive factory [7, 8] where it needs to carry heavy products and achieve different task processes: cutting, welding, painting, and assembling. Manipulator arm can guarantee the flexibility and precision in manufacturing environments that depend on its physical configuration such as degree of freedom, which limits the workspace. Usually, they are of six-degree of freedom and the end-effector that can be reconfigurable regarding to the task allowed [9, 10]. It is characterized by high product diversity and rapid evolution of production demand. The control of manipulator robots is based on two essential phases: 1) execution of the task that can be for example: pick and place, welding, painting; 2) execution of the trajectory indicated to the end effector of robot arm, which depends on the joints motion of each link.

The system is equipped by an arm structure that involves engines usually servomotors in each joint that characterized the motion type and the degree of freedom – transmissions to increase the joint torque – sensors to transfer data, the arm receives and send information as inputs and outputs data from the controller is mainly responsible for the kinematics calculation, motion planning and interpolation calculation of the robot, and transmitting the user's motion control instructions to the actuator that is by sending orders to the arm automatically using commands and a specific interface or manually using teaching box, the controller in manipulator arm presents the brain of the whole system, whose task is to drive the mechanical structure of arm to complete specific tasks according to the operator instructions, it is described as a black box composed of electronic circuits. From literature review [11] practically there are three classification methods for industrial robot control systems according to the degree of system openness [12]: the closed control system, the open control system, and the hybrid control system. With the appearance of Cobot that improves the flexibility of robotic cell, in the dynamic control field, the control of a single robot arm is always an easy task comparing to the control of multi-robot arms, where it needs to study more deeply the structure of each robot taking into account different parameters, theoretically, from literature view [13] many researchers were interested in this topic. Braun et al. [14] developed a control scheme using an adaptive fuzzy controller. Tzierakis and Koumboulis [15] developed a control law providing direct control of the position and internal forces separately in each direction of the task space. Caccavale et al. [16] developed a general impedance control scheme that combines centralized and decentralized control strategies for compliance and internal loading force control. For robust and stable controllers, Al-Yahmadi et al. [17] used a sliding mode scheme to enable cooperating arms to handle a flexible beam, García et al. [18] and Jeon et al. [19] have investigated sensor fusion principles for accurate force feedback necessary in cooperative robot systems. The main major task for a manipulator arm is to execute the path from a starting point to a final point precisely and in a specific cycle time [20], this mission needs the study of motion, which is composed of those data and can be studied on two classes in joint space where it is limited by the following constraints (velocity, torque, workspace limits) or in operational space where it depends on (precision, obstacle) [21, 22], by using inverse kinematics algorithm robot can determine the joint angles needed to maintain such position. The nature of motion is plotted by the transition from point to another one where different interpolation can be mentioned: joint interpolation movement – continuous interpolation movement – circular interpolation, etc. In robotics area, the trajectory optimization of a path for robot arm to execute a specific task can reach many advantages regarding: reducing time and energy, increasing the productivity. This article contributes a real experiment of an RV-2AJ Mitsubishi robot arm made up of five rotational joints plus the motion of the pneumatic gripper, where RV-2AJ robot arm has to build up a card house building without any human interaction. First of all, the task application will be described after that the results of experiments will be presented in order to highlight a real problem that can be caused by such a robot arm with a limited degree of freedom, in the final part an innovative solution will be discussed and realized to solve this kind of problem, the approach is based on trial - and - error optimization method with checking the effect of the different robot placement variations (RV-2AJ arm in vertical placement – RV-2AJ arm in horizontal placement) and their effect on precise card elements placement, the results will prove that the horizontal robot placement can eliminate the card placement error where it needs a minor cost without changing the robot to a new one having more degree of freedom which is a cost effective solution for the arisen card element positioning problem.

2 Motion Control of an Industrial Robot Arm

Motion control of an industrial robot arm is executed precisely by adjusting the position and saving its coordinate system in the interface of the controller, the position can be defined - in Joint space where the controller receives joints angles vector containing the angle of each joint and by using forward kinematics, the position vector is calculated in Cartesian space where the position is defined regarding the coordinate system of the gripper on (X, Y, Z) plus the rotations A, B, C around X, Y and Z. Using inverse kinematics, the controller can calculate the joint angles needed for each articulation. The motion from point P1 to another point P2 can take different paths as we mentioned linear motion or circular one.

The experiment presented in this article realized using Mitsubishi industrial robot arm RV-2AJ as presented in Fig. 1, the robot arm is characterized by the following specifications presented in Table 1 and the operating range described in Table 2, length of links highlighted in Table 3.

3 Task Application: Card House Building

The main target of the task application is building a card house. This task to be achieved usually needs two hands, which require a collaborating of two manipulator arms. In this paper, the execution of the task will be presented using only one manipulator arm Mitsubishi RV-2AJ robot by applying an innovative technique using a simple support element. The task is executed in the real environment and modeled in the virtual environment using SolidWorks software [24].

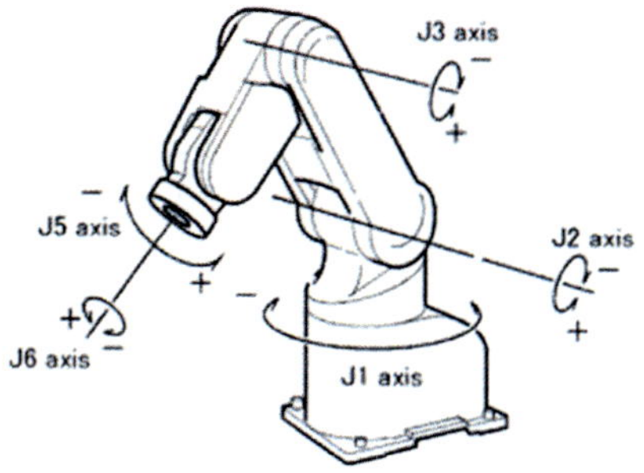

Fig. 1. Axes of RV-2AJ arm [23]

Table 1. Specifications of RV-2AJ arm [23].

Feature	Value
Number of joints	5
Payload	2 kg
Repeatability	± 0.020 mm
Reach length	410 mm
Programming language	MELFA BASIC
Gripper	Pneumatic

Table 2. Operating range for RV-2AJ arm.

Joint	Angle
J1: Waist	−150° to +150°
J2: Shoulder	−60° to +120°
J3: Elbow	−60° to +120°
J4: Wrist pitch	−90° to +90°
J5: Wrist roll	−200° to +200°

Figure 2 describes the task application prototype with the elements used, where (a) Card house building prototype, (b) the slanted card elements, (c) the horizontal card elements and (d) the support card element are presented.

The components of the card house are the following:

- 12 cards (slanted cards) having a special design to create A shape and guarantee the stability of the structure,
- 3 cards in horizontal position having rectangular shape located between the levels of the card house,
- one card which is a support element to help to fix the single slanted cards.

Table 3. Length of links.

Links	Distance
Waist to Shoulder	300 mm
Shoulder to Elbow	250 mm
Elbow to Wrist pitch	160 mm
Wrist pitch to roll	72 mm
Gripper	70 mm

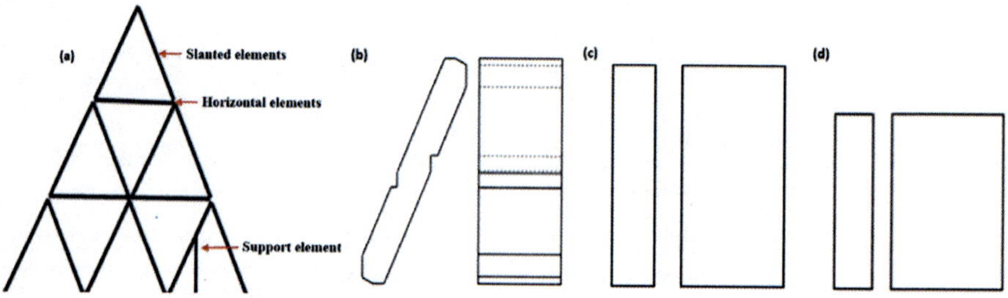

Fig. 2. Card house building prototype and elements needed.

The cards presented in Fig. 3 were manufactured from wood material using CNC milling machine.

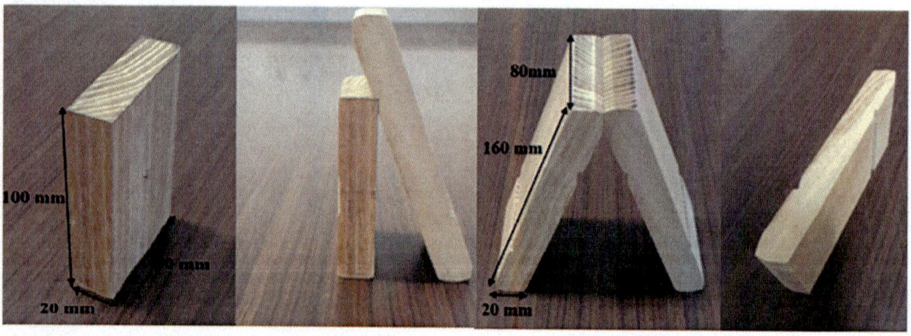

Fig. 3. Card elements manufactured using CNC machine.

3.1 Operation in the Real Environment

The task is divided into subtasks starting by realizing the first small A shape composed of two cards where RV-2AJ arm executes the following scenario to achieve the final process:

- RV-2AJ arm picks the support element and places it in a known position.
- RV-2AJ arm picks again the slanted card and places it according to the support element.
- RV-2AJ arm picks a new slanted card and places it according to the previous card that forms an A shape.

Figure 4 illustrates Card house to be built and Fig. 5 shows Pick and place scenario by RV-2AJ.

Fig. 4. Card house to be built.

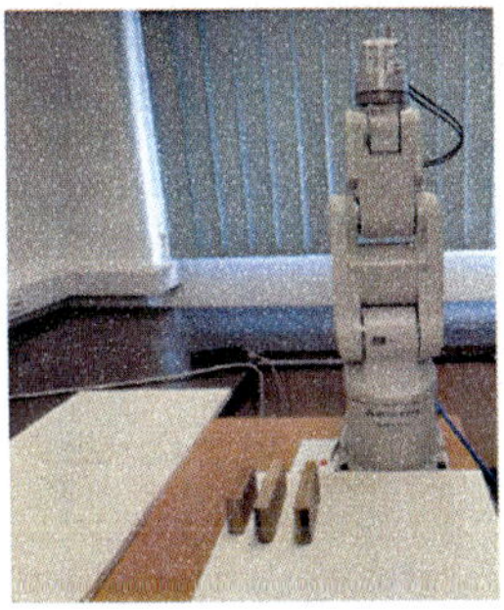

Fig. 5. Pick and place scenario by RV-2AJ arm.

3.2 Result of Card House Building Process

The results of the previous scenario presented in Fig. 6 where we can see that RV-2AJ arm was able to place the three elements in the positions expected but the imprecise contact between two cards appeared, the experience was repeated many times even with the correct calibration of RV-2AJ and setting the precise zero position "reference point", this problem arises from the limited freedom of the robot arm and creates instability where we were unable to build up the card house.

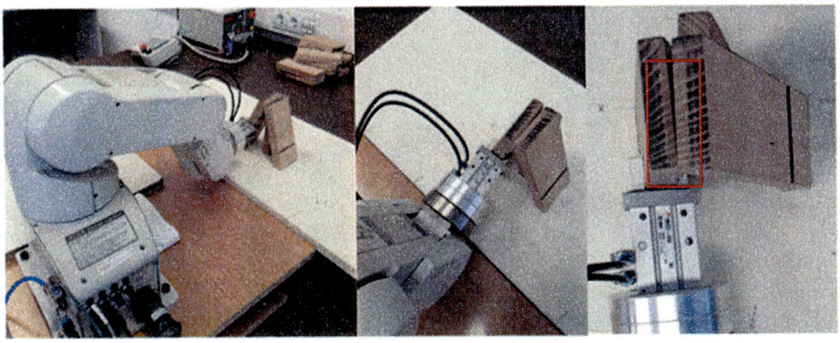

Fig. 6. Imprecise contact between two cards.

4 Positioning Solution for Card House Building Task

The imprecise contact between the two slanted cards is caused due to the angle problem where RV-2AJ arm has a five-degree of freedom, the missing of the 6th joint makes RV-2AJ robot incapable to execute such positioning and coincide the two slanted cards in a good fixation. In order to solve the problem of parallelism precision trials were made, the solution based on creating a new placement for RV-2AJ where the robot will be changed from normal vertical into horizontal placement. Figure 7 illustrates the design of the mechanical holder prototype where RV-2AJ arm will be fixed on it.

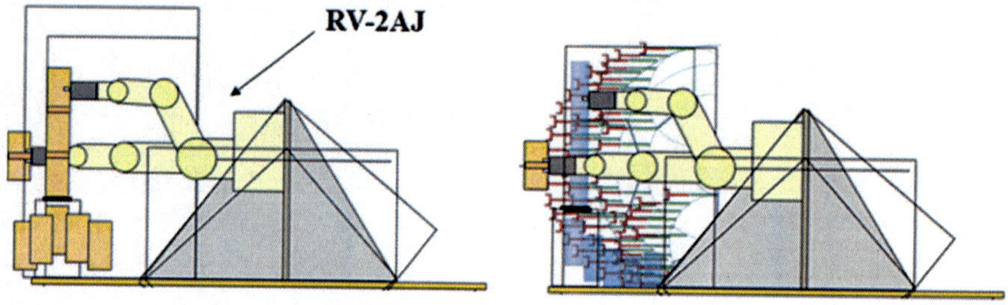

Fig. 7. The mechanical holder prototype geometry development.

From Fig. 7 is clear that RV-2AJ arm in the new horizontal placement could achieve all the position required for card house building, the solution for this problem was based on the optimal formation of the workspace where the arm in the horizontal placement can move and rotate the end effector more precisely to eliminate the positioning errors providing stability of the card house. To prove the success of the new placement a simulation test has been performed using SolidWorks software by modeling the whole process and testing the accessing of RV-2AJ gripper to all the cards. Figure 8 shows the Card house building by RV-2AJ robot arm in the real environment (Fig. 8a) and a virtual environment of SolidWorks software (Fig. 8b) [25].

Figure 8 describes the mechanical holder with the new placement of RV-2AJ robot arm where it is clear that the holder is strong enough to carry on such a body.

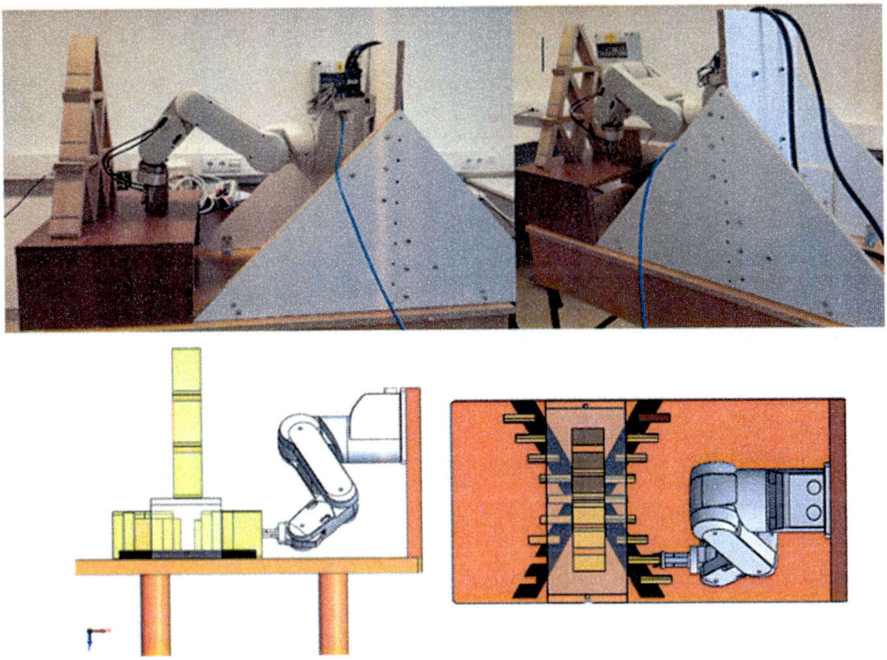

Fig. 8. Card house building by RV-2AJ robot arm in the real environment (a) and in the virtual environment of SolidWorks software (b).

5 Conclusion

This article suggests a new idea regarding the task application of Card house building achieved by only one manipulator arm "Mitsubishi RV-2AJ robot" using an innovative technique based on a support element. In the second part of the paper, a problematic was described regarding the positioning problem of RV-2AJ arm that has five-degree of freedom, the solution for such a problem was highlighted using a mechanical holder as an optimal stand on changing the orientation of RV-2AJ arm from vertical placement to a horizontal one by creating a strong mechanical holder and fix the arm on it. By this solution, the first joint will be able to correct with the fifth joint the error of parallelism between two slanted cards where it needs a minor cost without changing the robot to a new one having more degree of freedom which would be an expensive solution for the arisen card element positioning problem. In the future research, we would like to optimize the motion of the robot arm by the application of the CAD modeling with SolidWorks and developing a new algorithm using Python which defines all the steps of the motion. Using V-rep software as a simulation tool, we can synchronize the CAD model and our optimization program. The result of the optimization will be the reduction of the total cycle time of the building process.

References

1. Dima, I.C., Kot, S.: Industrial Production Management in Flexible Manufacturing Systems. Book News Inc., Portland (2013)
2. El Zoghby, N., Loscri, V., Natalizio, E., Cherfaoui, V.: Robot Cooperation and Swarm Intelligence. Wireless Sensor and Robot Networks: From Topology Control to Communication Aspects. World Scientific Publishing Company, pp. 168–201 (2014)
3. Alessio, C., Maratea, M., Mastrogiovanni, F., Vallati, M.: On the manipulation of articulated objects in human-robot cooperation scenarios. Rob. Auton. Syst. **109**, 139–155 (2018)
4. Gilchrist, A.: Industry 4.0: The Industrial Internet of Things. Apress, Berkeley (2014)
5. Benotsmane, R., Kovács, G., Dudás, L.: Economic, social impacts and operation of smart factories in industry 4.0 focusing on simulation and artificial intelligence of collaborating robots. Soc. Sci. **8**(5), 143 (2019)
6. Holubek, R., Kostal, P.: The intelligent manufacturing systems. Adv. Sci. Lett. **19**(3), 972–975 (2013)
7. Forrai, K.M., Kulcsár, G.: Modeling and solving an extended parallel resource scheduling problem in the automotive industry. Acta Polytechnica Hungarica **14**(4), 27–46 (2017)
8. Simões, L.M.C., Jármai, K., Virág, Z.: ε-optimum reliability-based cost design of longitudinally stiffened welded steel plate. In: Proceedings of the 2nd VAE2018. Lecture Notes in Mechanical Engineering, Vehicle and Automotive Engineering 2, Miskolc, Hungary, pp. 649–665 (2018). https://doi.org/10.1007/978-3-319-75677-6_56
9. Yim, M., Shen, W., Salemi, B., Rus, D., Moll, M., Lipson, H., Klavins, E., Chirikjian, G.S.: Modular self-reconfigurable robot systems. IEEE Robot. Autom. Mag. **14**(1), 43–52 (2007)
10. Koren, Y., Heisel, U., Joyane, F., Moriwaki, T., Pritschow, G., Ulsoy, G., Van Brussel, H.: Reconfigurable manufacturing systems. CIRP Ann. Manuf. Technol. **48**(2), 527–540 (1999)
11. Wang, R., Chen, Z., Zhang, W., Zhu, Q.: Proceedings of the 11th International Conference on Modelling, Identification and Control (ICMIC2019). Springer, Singapore (2020)
12. Wang, N., Qu, D.: The open architecture of industrial robot control system. Robotics **03**, 256–261 (2002)
13. Norman, A., Schönberg, A., Gorlach, A., Schmitt, R.: Cooperation of industrial robots with Indoor-GPS. In: Dimitri, D. (ed.) Conference 2010, International Conference on Competitive Manufacturing, Stellenbosch, South of Africa, pp. 215–224 (2010)
14. Braun, B., Starr, G., Wood, J., Lumia, R.: A framework for implementing cooperative motion on industrial controllers. IEEE Trans. Rob. Autom. **20**(3), 583–589 (2004)
15. Tzierakis, K., Koumboulis, F.: Independent force and position control for cooperating manipulators. J. Franklin Inst. **340**, 435–460 (2003)
16. Caccavale, F., Chiacchio, P., Marino, A., Villani, L.: Six-DOF impedance control of dualarm cooperative manipulators. IEEE/ASME Trans. Mechatron. **13**(5), 576–586 (2008)
17. Al-Yahmadi, A., Abdo, J., Hsia, T.: Modeling and control of two manipulators handling a flexible object. J. Franklin Inst. **344**, 349–361 (2007)
18. García, J., Robertsson, A., Ortega, J., Johansson, R.: Sensor fusion for compliant robot motion control. IEEE Trans. Rob. **24**(2), 430–441 (2008)
19. Jeon, S., Tomizuka, M., Katou, T.: Kinematic Kalman Filter (KKF) for robot end-effector sensing. J. Dyn. Syst. Meas. Contr. **131**(2), 8 (2009)
20. Cservenák, Á., Szabó, T.: Positioning measurements of two industrial robots. Acta Mechanica Slovaca **19**(4), 38–43 (2015)
21. Wissama, K., Etienne, D.: Modélisation identification et commande des robots, 2nd edn. Harmes, New Castle (1999). Lavoisier, France
22. Liu, X., Qiu, C., Zeng, Q., Li, A.: Kinematics analysis and trajectory planning of collaborative welding robot with multiple manipulators. Procedia CIRP **81**, 1034–1039 (2019)

23. Mitsubishi Electric France Factory Automation. https://fr3a.mitsubishielectric.com/. Accessed 15 July 2020
24. Onwubolu, G.: A Comprehensive Introduction to Solidworks 2013. SDC publications, Mission (2013)
25. Rohmer, E., Singh, S., Freese, M.: V-REP: a versatile and scalable robot simulation framework. In: IEEE/RSJ International Conference on Intelligent Robots and Systems, Tokyo, Japan, pp. 1321–1326 (2013)

Optimization Algorithms for Inverse Kinematics of Robots with MATLAB Source Code

Hazim Nasir Ghafil[1(✉)] and Károly Jármai[2]

[1] University of Miskolc, Miskolc, Egyetemváros 3515, Hungary
vegyhnr@uni-miskolc.hu
[2] University of Kufa, Najaf, Iraq

Abstract. This work presents a methodology to solve the inverse kinematic problem for any kind of robot arm using optimization algorithms. Forward kinematic is usually a straightforward analysis for any robot while inverse kinematic is hard to be solved for many cases. Thus, depending on a set of the forward kinematic equation, the objective function can be formulated to be minimized to find the inverse position. This methodology makes the inverse kinematic very simple operation for all types of the robot, even for those who are complicated with a high degree of freedom. A particle example of 5DOF revolute joint arm was used to present this methodology with source code written in MATLAB for the objective function. Dynamic differential optimization algorithm DDAO was used to minimize the objective. DDAO has promising usage for embedded systems when prototyping a controller that estimates the inverse kinematic as per user request.

Keywords: Inverse kinematics · Optimization algorithms · Robotics · Dynamic differential optimization algorithm · Particle swarm optimization

1 Introduction

Inverse kinematics [1] is the cornerstone for articulated robots in all daily life applications because all the rest of the robotic processes depends on its output. Articulated arm robot moves by giving joint input variables to the actuators [2], and accordingly, the tip of the arm moves in Cartesian space. This is called forward kinematic, and it is a straightforward operation that does not need serious computations or optimizations. We deal with the motion in Cartesian space because that what is desired for most of the application while the movement in the joint space still in the shadow. The most difficult process is when the inputs are the Cartesian coordinates, and the desired is to find the corresponding joint variables. This is called inverse kinematics which is what we usually deal with most of the robotic applications. In the automotive industry, robots follow a specific trajectory of Cartesian points to do some achievement like welding, cutting, grinding, painting, etc. [3]. Thus, inverse kinematics maps the motion of the tooltip (or just the extreme tip of the arm) from Cartesian space where the tip swims to the joint space where the actuators perform. It is worth mentioning that for simple topology robots, inverse

K. Jármai and K. Voith (Eds.): VAE 2020, LNME, pp. 468–477, 2021.
https://doi.org/10.1007/978-981-15-9529-5_40

kinematics can be simple and solved by many methods like geometric or analytical solutions. For a complicated high degree of freedom robots, the process will be hard or even impossible to be solved by traditional methods [4]. In this work, we are presenting the formulation of the objective function for inverse kinematics to be solved by any of the optimization algorithms. Dynamic annealed optimization algorithm (DDAO) [5], specifically, proposed to find the inverse solutions for robot arms. What is interesting in this algorithm is that it is independent of the population size, and that makes DDAO perfect for embedded systems applications as we will express in the respective sections.

2 DDAO

Dynamic differential optimization algorithm (DDAO) is a physically inspired optimization algorithm that mimics the process of production dual-phase steel. The mathematical model of the algorithm is expressed as follows:

$$S^k = \left(Sc_i - Sc_j\right) + Sr.f \tag{1}$$

$$f = \begin{cases} 1 & if\ rem(iteration, 2) = 1 \\ random\,[0, 1] & if\ rem(iteration, 2) = 0 \end{cases}, \tag{2}$$

where rem is the remonder after division on 2, we suggest the same procedure depending on the probability formula described by SA algorithm

$$P = e^{\frac{-\Delta E}{T}}, \tag{3}$$

$$\Delta E = \frac{Cost(S^k) - Cost(S_L)}{Cost(S_L)}, \tag{4}$$

where S^k is a new solution proposed for the iteration number (k), $k = 1 \ldots n$ where n is the number of iterations, and Sc_i and Sc_j, are randomly chosen solutions from the population with random (i) and (j) indices. Sr is a randomly generated solution within the search space of the problem out of the population. P is the probability of accepting a new solution, ΔE is the difference between the objective value of the proposed solution from Eq. (1) and the objective value of the solution S_L, which is a solution of index L in the population, $L = 1,\ldots,$ population size. T is the temperature variable, which should start with high value and be updated during iterations constantly to a lower value. The proposed solution can be accepted if $P >$ random number $\in [0, 1]$. At the beginning of the search, T starts with high value; consequently, P will be close to one, according to Eq. (3). This means that a wide range of random numbers can be less that one and the solution will be selected. At the low value of T, the probability P will be close to zero; according to Eq. (3), this means that a very narrow range of random numbers could be less than P and the solution is less likely to be selected. The pseudocode illustrated below.

```
Initialize population Xi (i = 1,2,…,n)
Initialize parameter T, cooling rate
Calculate the cost of each solution
Xb= The best solution
While (t < Max iteration)
Initialize sub-population S
Calculate the cost of the sub-population
Sort sub-population
Sr= Best solution in sub-population
Choose two random solutions Xm and Xn from population
Calculate Sk from Eq. (1)
Sort population X
foreach solution in population X
if there is an improvement
Xi= Sk
otherwise, replace the worst solution in population X using
Eqs. (3) and (4)
endif
endfor
Update Xb
T = T*cooling rate
t = t+1
endwhile
return Xb
```

DDAO has a unique characteristic which is that it is independent of population size, this means that it uses the minimum size of the RAM when considering the population size of three individuals. Of course, other algorithms like particle swarm optimization [6], genetic algorithm [7], grey wolf optimization [8]. These algorithms can be used by setting population size as minimum as possible and what is the best algorithm for the inverse kinematic problem is left to the reader for future works.

3 Practical Example

In this study, LabVolt 5150 robot manipulator [9] is used to apply the proposed methodology of calculating the inverse kinematics using optimization algorithms. It is a 5 DOF manipulator; its rotational axes are base, shoulder, elbow, pitch and roller rotation, the manipulator equipped with a griper and all the revolute joints actuated by fife stepper motors. Figure 1 shows this robot manipulator, Fig. 2 expresses the configuration pace of the robot, and Fig. 3 reveals the frame assignment. According to the model shown in Fig. 2, the spatial parameters are estimated for each link, as shown in Table 1.

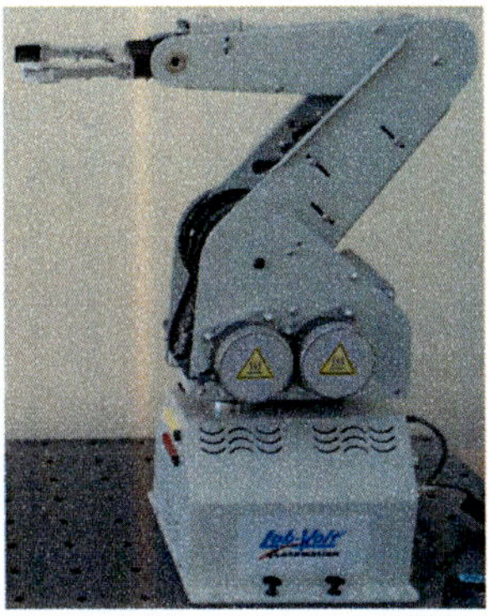

Fig. 1. Lab-volt 5150 manipulator

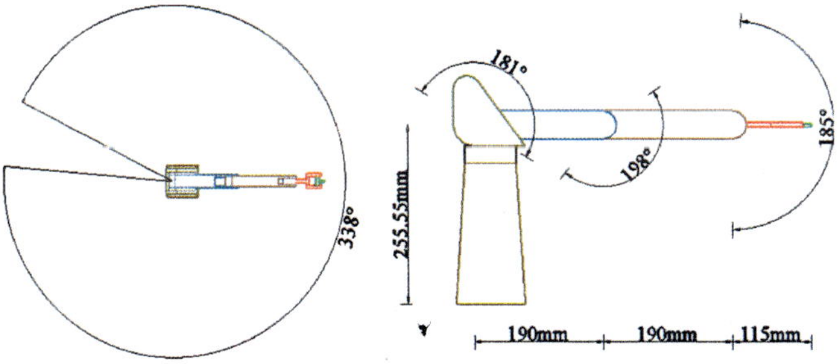

Fig. 2. Operative ranges and specifications of lab-volt 5150

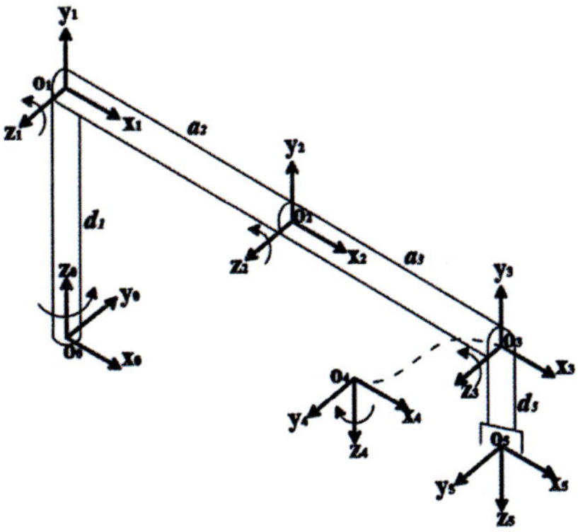

Fig. 3. Operative ranges and specifications of lab-volt 5150

Table 1. Spatial parameters of the lab-volt 5150 manipulator

Link ID	Frame	ϕ	α	a	d	*limits*
1	$o_0x_0y_0z_0$–$o_1x_1y_1z_1$	0	90	0	d_1	−185, 153
2	$o_1x_1y_1z_1$–$o_2x_2y_2z_2$	0	0	a_2	0	−32, 149
3	$o_2x_2y_2z_2$–$o_3x_3y_3z_3$	0	0	a_3	0	−147, 51
4	$o_3x_3y_3z_3$–$o_4x_4y_4z_4$	0	90	0	0	−5, 180
5	$o_4x_4y_4z_4$–$o_5x_5y_5z_5$	0	0	0	d_5	−360, 360

By using Denvit-Hartenberg convention [10], the homogenous transformation matrix HTM for the links are calculated as follows:

$$H_1^0 = \begin{bmatrix} C_1 & 0 & S_1 & 0 \\ S_1 & 0 & -C_1 & 0 \\ 0 & 1 & 0 & d_1 \\ 0 & 0 & 0 & 1 \end{bmatrix} \tag{5}$$

$$H_2^1 = \begin{bmatrix} C_2 & -S_2 & 0 & a_2.C_2 \\ S_2 & C_2 & 0 & a_2.S_2 \\ 0 & 0 & 1 & 0 \\ 0 & 0 & 0 & 1 \end{bmatrix} \tag{6}$$

$$H_3^2 = \begin{bmatrix} C_3 & -S_3 & 0 & a_3.C_3 \\ S_3 & C_3 & 0 & a_3.S_3 \\ 0 & 0 & 1 & 0 \\ 0 & 0 & 0 & 1 \end{bmatrix} \tag{7}$$

$$H_4^3 = \begin{bmatrix} C_4 & 0 & S_4 & 0 \\ S_4 & 0 & -C_4 & 0 \\ 0 & 1 & 0 & 0 \\ 0 & 0 & 0 & 1 \end{bmatrix} \tag{8}$$

$$H_5^4 = \begin{bmatrix} C_5 & -S_5 & 0 & 0 \\ S_5 & C_5 & 0 & 0 \\ 0 & 0 & 1 & d_5 \\ 0 & 0 & 0 & 1 \end{bmatrix} \tag{9}$$

$$H = H_1^0 \times H_2^1 \times H_3^2 \times H_4^3 \times H_5^4 \tag{10}$$

equation (10) is 4×4 matrix holds the orientation and position vector of the end-effector with respect to the base frame, as revealed in Eq. (11) and Eq. (12).

$$R_5^0 = \begin{bmatrix} S_1.S_5 + C_1.C_5.C_{234} & S_1.C_5 - C_1.S_5.C_{234} & C_1.S_{234} \\ -C_1.S_5 + S_1.C_5.C_{234} & C_1.S_5 + S_1.C_5.C_{234} & S_1.S_{234} \\ C_5.S_{234} & S_5.S_{234} & -C_{234} \end{bmatrix} \tag{11}$$

$$P_5^0 = \begin{bmatrix} x = d_5C_1S_{234} + a_2C_1C_2 + a_3C_1C_{23} \\ y = d_5S_1S_{234} + a_2C_2S_1 + a_3C_{23} \\ z = -d_5C_{234} + a_2S_2 + d_1 + a_3S_{23} \end{bmatrix} \tag{12}$$

4 Objective Function

This is the problem of finding the joint variables from the given position and orientation of the end-effector. While forward kinematics is detecting the position and orientation of the end effector from the given set of joint variables, inverse kinematics is the inverse operation, but it is somewhat complicated.

$$\begin{bmatrix} \theta_1 \\ \theta_2 \\ \vdots \\ \theta_n \end{bmatrix} \begin{matrix} \Rightarrow \\ \Leftarrow \end{matrix} \begin{bmatrix} r_{11} & r_{12} & r_{13} & x \\ r_{21} & r_{22} & r_{23} & y \\ r_{31} & r_{32} & r_{33} & z \\ 0 & 0 & 0 & 1 \end{bmatrix} \tag{13}$$

While forward equations are a straightforward process, we will rely on these equations to establish the objective function for the inverse problem.

Here we looking for an optimum set of joint variables that can lead to the minimum of a cost function, the only thing to do is developing cost function for the inverse Kinematic. Consider Fig. 4, for a specific robot configuration, the current position vector of the end-effector can be represented by the distance from the base of the end-effector of the manipulator while the desired position vector represents the task point. Obviously, if the difference between these two vectors is zero, then the tooltip will be in the right position at the task point, and this is the objective function f of the inverse problem

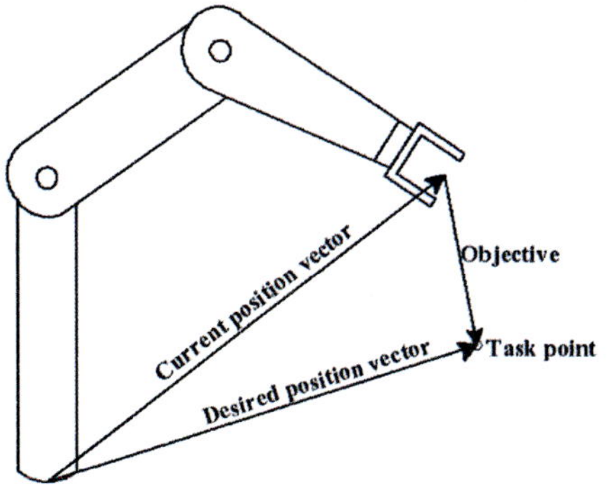

Fig. 4. Representation of the objective function for inverse kinematic problem

$$f = \|Ci - De\| \tag{14}$$

Where Ci denotes the instantaneous position vector, and De is the desired position vector. In other words, Eq. (14) is the function that has to be minimized as much as possible, and it just the distance between the end-effector and task point.

$$f = \sqrt{(x_{Ci} - x_t)^2 + (y_{Ci} - y_t)^2 + (z_{Ci} - z_t)^2} \tag{15}$$

Where t refers to the task point coordinates which is given for the inverse kinematic problem.

If Eq. (15) has been used alone as an objective function, we may get the end-effector in the task point but with many choices of orientations.

5 The Procedure of the Objective Function

In this section, we shall model the objective function for the inverse kinematic of any robot manipulator. Figure 5 shows a schematic diagram for the inverse problem; it is more descriptive to explain the procedure by a set of notes as follows.

1. Optimization algorithm sent the candidate solution, which is a set of possible joint variables, to the cost function to evaluate its fitness.
2. Cost function contains the desire task point coordinates; it sends the possible solution to Forward function to get x, y, and z coordinates of the tooltip.
3. Forward function contains all the forward kinematic equations of the robot arm, by substituting the candidate solution to that equations we can get the overall homogenous transformation matrix by a repeated call for HTM function. The output of the Forward function is the position vector of the total transformation matrix.
4. Cost function will receive the position vector and apply Eq. (15) to the candidate vector and the desired task position vector. The result is the fitness of the solution that will be back to the main optimization algorithm.

This process is constant for all types of manipulators; the only thing to change is the forward kinematic equations and the task position vector. It is worth mentioning that not all optimization methods can return a guaranteed solution; this depends on the efficiency of the algorithm itself.

The implementation of this methodology has a great significance on robot autonation, consequently, facilate and increase the productivity especially in aumotive engineeering. The automotive enginnering has wide applications for robots where they can be used in many cases and many operations [11, 12].

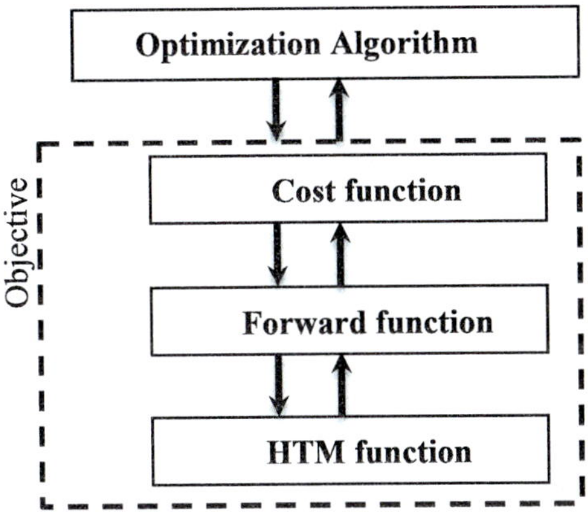

Fig. 5. Objective function scheme for inverse kinematic problem

6 Implementation of the Objective Function

The problem for LabVolt 5150 consists of five variables with lower and upper limits shown in Table 1, for a given point in space $v \in R^3$ the objective is finding the best corresponding joint angles that drive the end-effector to that point. From optimization algorithm, the candidate solution (`sol`) is transferred to the cost function (`cost`):

```
function obj = cost(sol)
t1 = sol(1);
t2 = sol(2);
t3 = sol(3);
t4= sol(4);
t5= sol(5);
[x,y,z] = Forward(t1,t2,t3,t4,t5);
% Cartesian point coordinate [x,y,z]
v = [495,0,255.55];
% Objective function; equation 11
obj = sqrt((x-v(1))^2+(y-v(2))^2+(z-v(3))^2);
end
```

For each candidate solution, we have to calculate the corresponding forward kinematic to find the position vector in Eq. (12) to be used in Eq. (15) to estimate the objective. Thus, the candidate solution v is transferred to the forward kinematic equations function (`Forward`). In this function, the spatial parameters are defined for each link considering data in Table 1:

```
function [x,y,z] = Forward(t1,t2,t3,t4,t5)
alpha_1 = 90;            %twist angle for link 1
a_1 = 0;                 %distance between z1 and z2
d_1 = 255.55;            % distance between x1 and x2

alpha_2 = 0;             %twist angle for link 2
a_2 = 190;               %distance between z2 and z3
d_2 = 0;                 %distance between x2 and x3

alpha_3 = 0;             %twist angle for link 3
a_3 = 190;               %distance between z3 and z4
d_3 = 0;                 % distance between x3 and x4

alpha_4 = 90;            %twist angle for link 4
a_4 = 0;                 %distance between z4 and z5
d_4 = 0;                 % distance between x4 and x5

alpha_5 = 0;         %twist angle for link 5
a_5 = 0;             %distance between z5 and z6
d_5 = 115;           % distance between x5 and x6

H_1= HTM(t1,alpha_1,a_1,d_1); % HTM 1
H_2= HTM(t2,alpha_2,a_2,d_2); % HTM 2
H_3= HTM(t3,alpha_3,a_3,d_3); % HTM 3
H_4= HTM(t4,alpha_4,a_4,d_4); % HTM 4
H_5= HTM(t5,alpha_5,a_5,d_5); % HTM 5
TH = H_1*H_2*H_3*H_4*H_5;     % over all HTM
x=TH(1,4);
y=TH(2,4);
z=TH(3,4);
end
```

For general usage, we have developed separated function (HTM) to return the homogenous transformation matrix described by Denavit matrix.

```
function [T] = HTM(t,alpha,a,d)
T = [cosd(t),-
1*sind(t)*cosd(alpha),sind(t)*sind(alpha),a*cosd(t);...
    sind(t),cosd(t)*cosd(alpha),-
1*cosd(t)*sind(alpha),a*sind(t);...
    0,sind(alpha),cosd(alpha),d;...
    0,0,0,1];
end
```

The source code described above is general, one can solve the inverse kinematic of any robot just by replacing the proper spatial parameters and define them in function (Forward). According to the number of links, less or more Denavit matrices can be estimated H_6,..., H_n. The function (HTM) is still valid for all types of robot manipulators without changes.

7 Conclusion

Optimization algorithms are proposed to find the solution for the inverse kinematic problem for robots of any type by optimizing the minimization objective function. The proposed optimization algorithm is dynamic differential annealed optimization, which is simple, fast, and uses low space of the memory of host machines or target devices. A practical example of 5DOF revolute joints manipulator, LabVolt 5150, was considered for the inverse problem. The described methodology is quite simple and can simplify the hard problem of inverse kinematic greatly and make it simple, straightforward operation. The proposed DDAO does not promise a perfect solution, and many other optimization algorithms should be tested to find a solution for the inverse problem, and that is proposed for future works.

Acknowledgments. The research was supported by the Hungarian National Research, Development and Innovation Office - NKFIH under the project number K 134358.

References

1. Spong, M.W., Hutchinson, S., Vidyasagar, M.: Robot modeling and control (2006)
2. Craig, J.J.: Introduction to robotics: mechanics and control, 3/E. Pearson Education India (2009)
3. Ghafil, H.N., Jármai, K.: Optimization for Robot Modelling with MATLAB. Springer, Cham (2020). https://doi.org/10.1007/978-3-030-40410-9
4. Lee, C.S.G.: A Geometric Approach in Solving the Inverse Kinematics of PUMA Robots–College of Engineering. The University of Michigan, Ann Arbor (1983)
5. Ghafil, H.N., Jármai, K.: Dynamic differential annealed optimization: new metaheuristic optimization algorithm for engineering applications. Appl. Soft Comput. **93**, 106392 (2020)
6. Jiang, Y., Hu, T., Huang, C., Wu, X.: An improved particle swarm optimization algorithm. Appl. Math. Comput. **193**(1), 231–239 (2007)
7. Whitley, D.: A genetic algorithm tutorial. Stat. Comput. **4**(2), 65–85 (1994)
8. Mirjalili, S., Mirjalili, S.M., Lewis, A.: Grey wolf optimizer. Adv. Eng. Softw. **69**, 46–61 (2014)
9. Al-Saedi, F.A.T., Mohammed, A.H.: Design and implementation of chess-playing robotic system. Int. J. Sci. Eng. Comput. Technol. **5**(5), 90 (2015)
10. Bi, Z.M., Gruver, W.A., Zhang, W.-J., Lang, S.Y.T.: Automated modeling of modular robotic configurations. Rob. Auton. Syst. **54**(12), 1015–1025 (2006)
11. Jármai, K., Bolló, B.: Vehicle and Automotive Engineering 2: Proceedings of the 2nd VAE2018, Miskolc, Hungary. Springer (2018). ISBN 978-3-319-75677-6
12. Jármai, K., Bolló, B.: Vehicle and Automotive Engineering: Proceedings of the JK2016, Miskolc, Hungary. Springer (2018). ISBN 978-3-319-51189-4

Newer Manufacturing Technologies and Their Costs in Automotive Structures, A Review

Károly Jármai(✉)

University of Miskolc, Miskolc, Egyetemváros 3515, Hungary
jarmai@uni-miskolc.hu

Abstract. The costs of a construction consisting of material, production, transport, installation and maintenance costs. The manufacturing costs of welded structures include cutting, preparation (assembly and stitching), welding, ancillary (electrode changing, deslagging, cleaning) and painting costs. In this paper, a review is made to show some newer manufacturing technologies and their cost calculation opportunities. To calculate the costs for comparison purposes, we rely on internationally measured times and multiply them by a variable cost factor.

Keywords: Welded structures · Cost calculation · Laser welding · Plasma cutting · Optimisation

1 Introduction

It is difficult to formulate a cost function that is of general application because costs change over time and are highly dependent on the state of the art in a given country, and on the equipment, overheads, etc. of a company within a country. To calculate the costs for comparison purposes, we rely on internationally measured welding times and multiply them by a variable cost factor.

Laser welding is characterised by a concentrated energy input that allows high speed and thus reduces shrinkage. Compared to arc welding, laser welding applies to many different materials and can be performed in a single layer up to a thickness of 20 mm. Tandem GMA welding is also accelerating the speed of welding. Cold metal transfer welding can reduce the used energy and welding shrinkage. Hybrid welding, like plasma-MIG, has better features than the used technologies alone. Water jet cutting can be applied to a wide range of materials with high-pressure water jets or jets of water and abrasives.

At plasma cutting the plasma temperatures can range from 5500 °C to 28, 000 °C. The gas is used with air, water or carbon dioxide. The bending time of the sheet elements in the shell shape depends on the shell shape, radius and thickness of the sheet. Punch riveting and clinching are suitable to fix two different materials.

K. Jármai and K. Voith (Eds.): VAE 2020, LNME, pp. 478–485, 2021.
https://doi.org/10.1007/978-981-15-9529-5_41

In this article, we would like to show the applicability and the cost of these technologies.

The increasing demand put on the automotive industry to reduce car costs and fuel consumption has led to the design of innovative technological solutions. These new concepts aim at higher productivities, better quality, and improved performances of the final product. The use of tailor blanks in the automotive industry is an example of a successful innovation, which is nowadays used in many companies worldwide. The main objective of the present study is to assess laser welding of tailor blanks in steel, comparing the performance of the three types of high-power lasers, CO_2, Nd/YAG, and fiber laser, in terms of productivity, costs, and weld quality. Mechanical properties have also been targeted as these are essential for the formability of the welded blanks.

2 Laser Welding

Tailor blanks are composed of two or more dissimilar sheet metals, with different thicknesses, shapes, strengths, or materials that are butt-welded together before being formed. This allows the use of thicker or stronger materials in the critical regions of a component, so as to increase the local stiffness. In comparison, thinner or lighter materials are used in other areas to reduce the component overall weight. In the past years, the use of Tailor Welded Blanks has increased in the automotive industry due to the need of obtaining lighter automotive vehicles that are more fuel-efficient, produce fewer emissions, and deliver improved handling and occupant protection.

Cost reduction, along with aspects such as vehicle appearance, comfort, convenience, safety, functionality, and reduction of the number of parts, without compromising the strength or stiffness of the final part has also contributed to this increase.

Three different laser welding technologies can be considered: YAG. CO_2 and Fiber laser. Assunção et al. [1] have made a comparison between them (Table 1). They have made the welding of 0.7–1.2 mm steel plates, which are common in car production.

Table 1. Welding speeds obtained for fiber, Nd: YAG, and CO_2 lasers

Laser type	Plate thickness combination (mm)	Welding speed (m/min)
Fiber laser	0.7–1.0 mm plate to 0.9–1.2 mm plate	7.0–7.8
Nd: YAG laser	0.7–1.0 mm plate to 1.0–1.2 mm plate	6.5–7.0
CO_2 laser	0.8–0.9 mm plate to 0.9–1.2 mm plate	5.5–6.5

Measurements show that fiber laser has the highest productivity and its market share is relatively large [2].

They have provided some approximations on laser speed parameters depending on material properties and technological settings (Table 2). These parameters help designers to estimate the best laser welding speed and power.

Table 2. Laser speed parameters

Designation	Equation	Definitions/Unit
Laser speed parameter	$LSP = \frac{ws\ x\ b}{\alpha}$	*LSP* dimensionless *ws*—welding speed [m/s] *b*—weld bead width [m] α—diffusivity $\alpha = \frac{K}{c_p}$ [m²/s] *K*—thermal conductivity [W/m K] ρ—density [kg/m3] c_p—specific heat capacity [J/kg K]
Laser power parameter (LPP)	$LPP = \frac{P}{h\ x\ K\ x\ d}$ $d = T_m - T_0 + \frac{H}{c_p}$	*LPP*=dimensionless *P*—power [W] *h*—weld bead penetration [m] *K*—thermal conductivity [W/m K] T_m—melting temperature [K] T_0—room temperature [K] *H*—melting latent heat [J/kg] c_p—specific heat capacity [J/kg K]

3 Hybrid Laser-Arc Welding

Hybrid laser-arc welding (HLAW) is a combination of laser welding with arc welding that overcomes many of the shortfalls of both processes. This important book gives a comprehensive account of hybrid laser-arc welding technology and applications. Industrial robotic application of laser-GMAW and laser-Tandem hybrid welding. It was found, that using hybrid welding one can reduce the heat input, while the quality of the welded joint remains the same [3].

Laser beam welded structures offer great opportunities for the lightweight design of fuselage structures to reduce structural weight for increased fuel efficiency. Laser beam welding (LBW) technology provides the best opportunities in terms of weight reduction, production time and energy consumption for manufacturing aircraft components. To this end, a comparison in terms of energy, process time, cost and carbon footprint is assessed against the 'conventional' manufacturing process of riveting, to prove that LBW is an environmentally friendly process. They have shown the manufacturing of a four-stringer stiffened flat subscale component and its cost calculation. The LBW process has been broken down into several sub-processes and activities according to the Activity-Based Costing (ABC) methodology and the weight reduction, production time and energy consumption results were compared against the respect of the riveting process. It was proved that for the specific subscale LBW component, it consumes half the energy and can be processed in less than half the time needed (in serial processing of the element) with riveting.

Activity-Based Costing (ABC) has become a popular cost estimation method due to the poor results of the traditional costing systems. The ABC model is composed of both the cost assignment view and the process view with activities as the intersection of these two views [4]. ABC analysis provides an understanding of how costs are driven by the demands for actions within a process and allow the identification of value and non-value-added manufacturing operations as well as how resources are consumed [5, 6]. More specifically, the ABC method was introduced by Kaplan and Cooper [7] of Harvard Business School as an alternative to traditional accounting techniques.

Hybrid laser – metal active gas (MAG) arc welding is an emerging joining technology. This technique combines the synergistic qualities of the laser and MAG arc welding techniques, which permits a high energy density process with fit-up gap tolerance. As the heat input of hybrid laser – arc welding (HLAW) is higher than in laser welding but much smaller than in MAG arc welding, a relatively narrow weld and restricted heat-affected zone (HAZ) is obtained, which can minimise the residual stress and distortion. Furthermore, adding MAG arc can increase the penetration depth for a given laser power, which can translate to faster welding speeds or fewer number of passes necessary for one-sided welding of thick plates [8].

The prospects for the use of lightweight materials (aluminium alloys, magnesium alloys, and titanium alloys) in high volume vehicle manufacturing are discussed. Laser welding of galvanised steel is compared to resistance spot welding of galvanised steel, along with ongoing efforts to improve the quality of laser welding of galvanised steel by altering the weld configuration, changing the elemental composition, utilising a pulse laser, and removing the zinc coating. The feasibility of implementing these techniques in the industrial setup is discussed. Microstructure changes and defects encountered during laser welding of these materials are described, and mechanical properties of welds such as hardness, shear and tensile strength are analysed [9].

In the automotive industry, one of the oldest and most widely used processes is resistance spot welding. The number of spot welds in a single vehicle can range from 2000 to 5000, which signifies the importance of resistance spot welding in an automotive assembly.

Welding times for resistance spot welding are given in Table 3 [10].

Table 3. Welding time for resistance spot welding [10].

Metal sheet thickness (mm)	Welding speed (welding spots/min)
0.7–0.8	40–60
0.9–1.0	30–50
1.0–1.2	30–40
1.25–1.6	20–30
1.6–2.0	15–25

Laser welding of dissimilar metals, such as steel to aluminium alloy or steel to magnesium alloy, is difficult due to the formation of brittle intermetallic phases. Table 4 shows some examples of spot welding of galvanised steel sheets [10].

Table 4. Summary of the laser welding of dissimilar materials investigations (material grades according to [11–13].

Material type	Thickness [mm]	Laser type	Welding conditions
DC04 steel-AA6016, DC04 steel-AA6056	1.2–1 1.2–1.3	Nd: YAG	2.25–3.5 kW, 4–6 m/min
JCS270CC steel-AA6111-T4	0.8–1.2	(1) CW Nd: YAG (2) PW Nd: YAG	(1) 390 W, 0.06 m/min (2) Peak 2.61 kW, 0.06 m/min
H220YD steel-AA 6016 Filler: Al–5%Si	1.2–1.15	Fiber laser	2.3–2.6 kW, 1 m/min
SP781 steel-AZ31B	1.2–3	Disk laser	2 kW, 2–4 m/min

4 Our Cost Calculation System

In our cost calculation system, the cot function looks like the following. The total cost function can be formulated by adding the previous cost functions together (depending on the structure; some can be zero) [14–17].

$$\frac{K}{k_m} = \rho V + \frac{k_f}{k_m}(T_{w1} + T_{w2} + T_{w3} + T_{FP} + T_{SP} + T_P + T_{CG} + \ldots) + \frac{K_{pi}}{k_m} + \ldots + \quad (1)$$

where subscripts mean: CG = cutting and grinding, w = welding, P = painting, FP = flattening plates, SP = surface preparation, pi = Intumescent painting

Taking k_m = 0.5–1.5 \$/kg, k_f = 0–1 \$/min. The k_f/k_m ratio varies between 0–2 kg/min. If $k_f/k_m = 0$, then we get the minimum mass. If $k_f/k_m = 2.0$ it means a very high labour cost (Japan, USA), $k_f/k_m = 1.5$ and 1.0 means a West European labour cost, $k_f/k_m = 0.5$ means the labour cost of developing countries. Even if the production rate is similar to these cases, the difference between costs due to the different labour costs is significant [18, 19].

Figure 1 shows the laser welding time/unit length [min/m] in the function of plate thickness t [mm] for structural steel materials.

A water jet cutter can cut a wide variety of materials using a very high-pressure jet of water, or a mixture of water and an abrasive substance. Figures 2 and 3 show the waterjet cutting time of carbon and stainless-steel sheets [min/m] in the function of plate thickness t [mm].

Plasma cutting uses an extremely high temperature, high-velocity stream of ionised gas to cut the metal. Plasma temperatures range from about 5500°C to 28,000°C. Depending upon the material to be plasma cut, the gases used include standard compressed shop air, oxygen, argon and hydrogen, or nitrogen and hydrogen. Gas shielding is accomplished with air, water, or carbon dioxide.

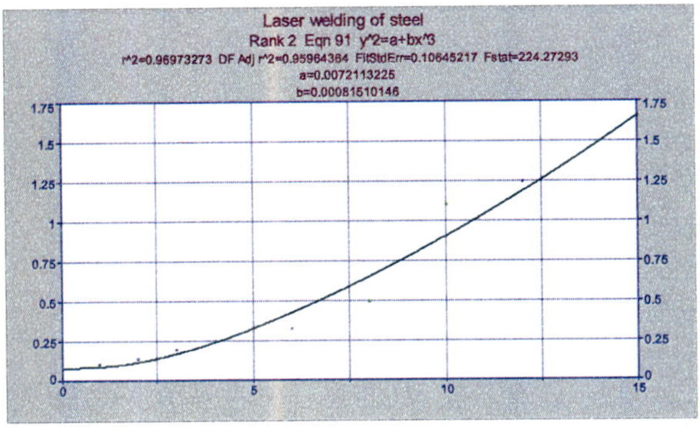

Fig. 1. Laser welding time/unit length [min/m] in the function of plate thickness t [mm]

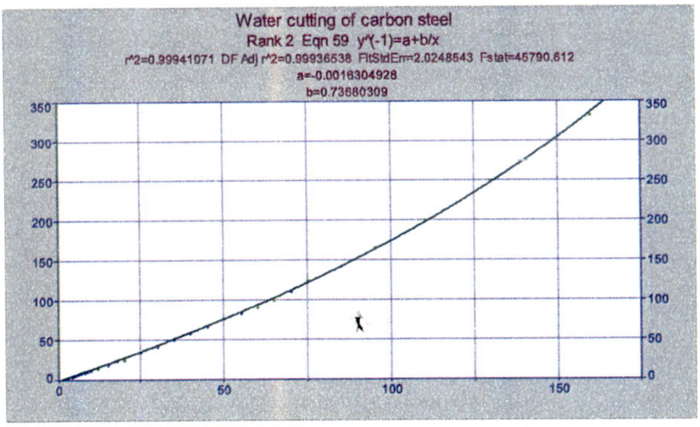

Fig. 2. Waterjet cutting of steel sheet cutting [min/m] in the function of plate thickness t [mm]

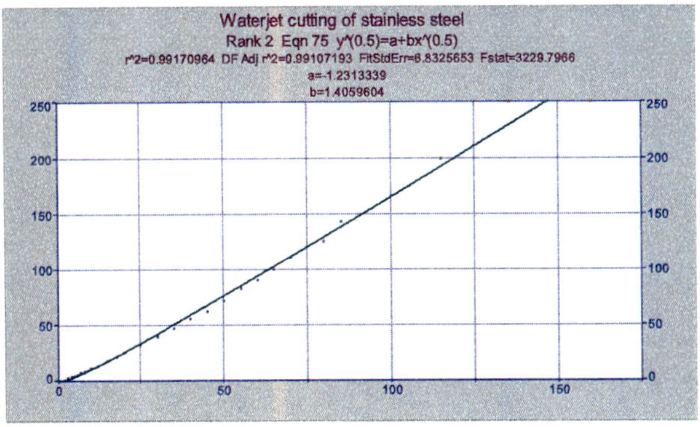

Fig. 3. Waterjet cutting of stainless-steel sheet cutting [min/m] in the function of plate thickness t [mm]

Figures 4 and 5 show the plasma cutting time of stainless steel and aluminium [min/m] in the function of plate thickness t [mm].

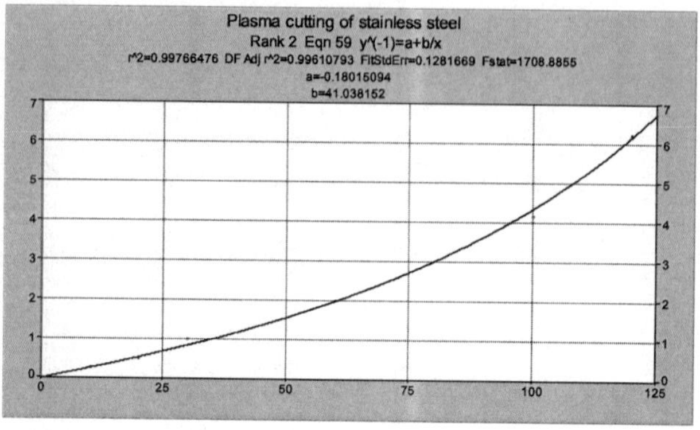

Fig. 4. Plasma cutting of stainless steel [min/m] in the function of plate thickness t [mm]

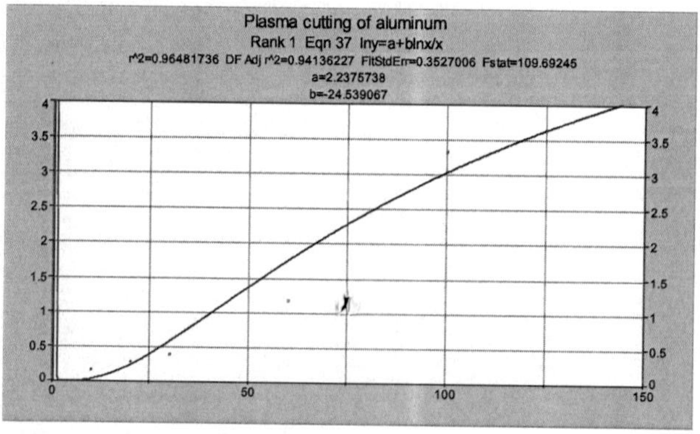

Fig. 5. Plasma cutting of aluminium [min/m] in the function of plate thickness t [mm]

5 Conclusions

These cost calculations are founded on material costs and those fabrication costs, which have a direct effect on the sizes, dimensions or shape of the structure. The calculated times for different newer technologies like laser, plasma, waterjet, etc. have also been included. These costs are the objective functions in structural optimisation.

When we consider the interaction of design and technology, we should not forget the cost of the structure as the third leg of the system. These three together help us to find the best solution. These cost calculations are founded on material costs and those fabrication costs, which have a direct effect on the sizes, dimensions or shape of the structure.

Cost and production time data come from different companies from all over the world. When we compare the same design in different countries, we should consider the differences between labour costs. It has the greatest impact on the structure sizes when the technology is the same.

Acknowledgment. The research was supported by the Hungarian National Research, Development and Innovation Office under the project number K 134358.

References

1. Assunção, E., Quintino, L., Miranda, R.: Comparative study of laser welding in tailor blanks for the automotive industry. Int. J. Adv. Manuf. Technol. **49**, 123–131 (2010). https://doi.org/10.1007/s00170-009-2385-0
2. Shiner, B.: High-power fiber lasers gain market share. https://www.industrial-lasers.com/welding/article/16484634/highpower-fiber-lasers-gain-market-share. Accessed 11 July 2020
3. Olsen, F.O. (ed.): Hybrid Laser Arc Welding. Woodhead Publishers, Cambridge, p. 336 (2009)
4. Alexopoulos, N.D., Gialos, A.A., Zeimpekis, V., Velonaki, Z., Kashaev, N., Riekehr, S., Karanik, A.: Laser beam welded structures for a regional aircraft: weight, cost and carbon footprint savings. J. Manuf. Syst. **39**, 38–52 (2016)
5. Tsai, W.H., Lai, C.W.: Outsourcing or capacity expansions: application of activity-based costing model on joint products decisions. Comput. Oper. Res. **34**, 3666–3681 (2007). https://doi.org/10.1016/j.cor.2006.01.008
6. Langmaak, S., Wiseall, S., Bru, C., Adkins, R., Scanlan, J., Sóbester, A.: An activity-based-parametric hybrid cost model to estimate the unit cost of a novel gas turbine component. Int. J. Prod. Econ. **142**, 74–88 (2013). https://doi.org/10.1016/j.ijpe.2012.09.020
7. Cooper, R., Kaplan, R.S.: The design of cost management systems: text, cases and readings. Prentice-Hall, Upper Saddle River (1991)
8. Cao, P., Wanjara, J., Huang, C., Munro, A.: Nolting: Hybrid fiber laser – arc welding of thick section high strength low alloy steel. Mater. Des. **32**(6), 3399–3413 (2011)
9. Hong, K.-M., Shin, Y.C.: Prospects of laser welding technology in the automotive industry: a review. J. Mater. Process. Technol. **245**, 46–69 (2017)
10. Papkala, H.: Technological problems in spot welding of galvanised car body sheet. Weld. Int. **6**(5), 341–346 (1992)
11. EN 10130: Low carbon steel flat products for cold forming - technical delivery conditions, Comite Europeen de Normalisation (2006)
12. EN 485-2: Aluminium and aluminium alloys. Sheet, strip and plate. Mechanical properties, Comite Europeen de Normalisation (2016)
13. EN 10292: Continuously hot-dip coated strip and sheet of steels with high yield strength for cold forming, Comite Europeen de Normalisation (2007)
14. Farkas, J., Jármai, K.: Analysis and Optimum Design of Metal Structures. Balkema Publishers, Rotterdam, Brookfield (1997)
15. Farkas, J., Jármai, K.: Economic Design of Metal Structures. Millpress Science Publisher, Rotterdam, p. 340 (2003). ISBN 90 77017 99 2
16. Farkas, J., Jármai, K.: Design and optimisation of metal structures. Horwood Publishers, Chichester, p. 328 (2008). ISBN: 978-1-904275-29-9
17. Farkas, J., Jármai, K.: Optimum design of steel structures. Springer Verlag, Heidelberg, p. 288 (2013). ISBN 978-3-642-36867-7, http://dx.doi.org/10.1007/978-3-642-36868-4
18. Jármai, K., Farkas, J.: Truck floor design for minimum mass and cost, using different materials. In: Conference on Vehicle Engineering and Vehicle Industry. Lecture Notes in Mechanical Engineering, 17–18 November, Miskolc, Hungary, pp. 13–25. Springer International Publishing AG (2016). ISSN: 2195-4356, https://doi.org/10.1007/978-3-319-51189-4
19. Jármai, K.: Cost calculation of thin-walled structures using different manufacturing techniques. Lecture Notes in Mechanical Engineering, pp. 624–638. Springer International Publishing AG (2018). ISBN 978-3-319-75677-6

Welding

Application of Laser-Arc Hybrid Welding to Thick Steel Plates for Bridge Structures

Mikihito Hirohata[1](✉), Natsumi Sakai[1], Kuya Morioka[1], Naoyuki Matsumoto[2], Kengo Hyoma[3], and Koutarou Inose[3]

[1] Osaka University, 2-1, Yamada-oka, Suita-shi, Osaka 565-0871, Japan
hirohata@civil.eng.osaka-u.ac.jp

[2] IHI Infrastructure Systems, 3-17-12, Shibaura, Minato-ku, Tokyo 108-8023, Japan

[3] IHI, 1, Shin-nakahara-cho, Isogo-ku, Yokohama-shi, Kanagawa 235-8501, Japan

Abstract. For examining the application of laser-arc hybrid welding to the fabrication of steel bridge members, a series of experiment and investigation was conducted. The weld cracking tests were performed on the steels for bridge structural members with 15 mm in thick for identifying the welding condition by which one-pass full-penetration welding without cracks and defects was achieved. Then, butt welded joints were fabricated by the specified condition. The same dimensional butt joints were fabricated by the conventional arc welding. The welding time, heat input, welding distortion and residual stress generated by the hybrid welding were compared with those by the conventional arc welding for clarifying the effectiveness of hybrid welding. The welding time of hybrid welding was shorter than that of arc welding by 98% due to the high-speed welding. Because of the low heat input and the one-pass full-penetration welding, the angular distortion by hybrid welding was reduced by 96% compared to that by arc welding. The field of tensile stress in the longitudinal direction by hybrid welding was reduced by 25% to 50% compared to that of arc welding. Finally, it was confirmed that the hybrid welded joints did not include the weld defects and satisfied the basic mechanical performances.

Keywords: Laser-arc hybrid welding · Steel bridges · Welding distortion

1 Introduction

Several welding methods for steel members are used in architecture, naval and civil engineering fields. The most popular and general method is arc welding including gas shield, gas metal and submerged arc welding methods. Recently, the use of other welding methods like electro-beam welding or laser welding has been extended for the effective fabrication of large structures with thick steel plates.

The laser welding can join thick plates by the high-energy density at narrow parts. The deep penetration and high-speed welding are achieved by the concentrated heat input [1]. Due to these advantages, the heat input can be reduced compared to the conventional arc welding. The low heat input consequently leads the small welding distortion and low

K. Jármai and K. Voith (Eds.): VAE 2020, LNME, pp. 489–496, 2021.
https://doi.org/10.1007/978-981-15-9529-5_42

tensile residual stress. However, the gap between two plates being welded must be strictly managed in the laser welding. When the gap is wider than the allowance, the laser beam transmits the gap without sufficiently melting the plates. As a result, the welding defects as under-fill or melt burn-through occur [1]. This difficulty makes it hard to use the laser welding in fabricating the large steel members of building, ship and bridge structures even though the laser welding has been widely used in the vehicle or other industrial products.

To overcome this difficulty in the laser welding, the application of laser-arc hybrid welding has been noted [2, 3]. In the hybrid welding, the arc welding assists the laser welding for supplying the weld metal into the gap between plates to be welded. Therefore, the welding defects as under-fill or burn-through can be avoided and the gap tolerance is improved. The hybrid welding also has the advantages of deep penetration, high speed and low heat input as well as laser welding. The use of hybrid welding is expected to be extended in the fabrication of large steel members with thick plates, however, the proper welding conditions including the energy balance of two types of heat source and the welding speed according to the plate thickness are still unknown.

Assuming the application of laser-arc hybrid welding to thick steel plates used in large structural members like bridges, a series of experimental investigation was conducted in this study. The welding conditions by which steel plates with 15 mm thick could be soundly jointed with one-pass and full penetration were examined. The welding time, heat input, welding distortion and residual stress generated by the hybrid welding were compared to those by the conventional arc welding for clarifying the effectiveness of hybrid welding. The basic joint performances of hybrid welding were confirmed by a bending test and a tensile coupon test.

2 Experiment for Searching Welding Condition

2.1 Material and Specimen

The material used in this study was SBHS500 (Steels for Bridge High performance Structure) specified by Japanese Industrial Standards (JIS G 3140) [4]. This material was specially developed for bridge structures with various functions as high strength, deformation capacity, workability and toughness by the chemical compositions and TMCP. SBHS is expected to be widely used in bridge structural members, therefore, that was selected as the material in this study. Table 1 shows the chemical compositions and mechanical properties of the material. The thickness of the material was 15 mm.

Figure 1 shows the shape and dimension of the specimen for searching proper welding conditions. The length and width were 200 mm and 145 mm respectively. A slit with 80 mm long was inserted at the center of the specimen. The welding was applied to the slit with various welding conditions. The aim of this experiment was to find the welding condition by which one-pass and full-penetration welding was achieved without defects or cracks. The cross-sections were observed by cutting the specimens 48 h or more after the finishing of welding.

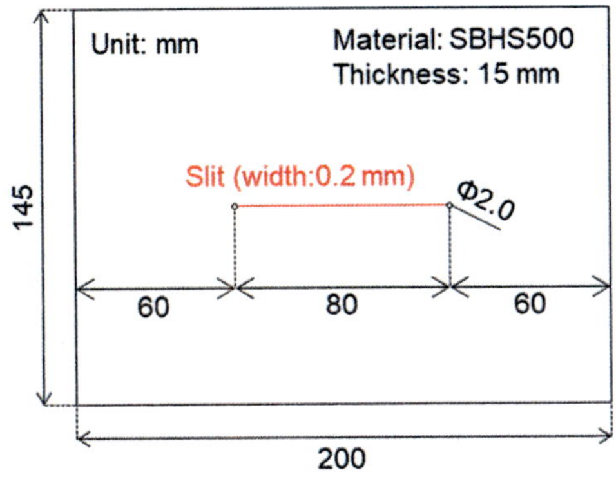

Fig. 1. Shape and dimension of specimen for searching welding conditions.

Table 1. Chemical compositions and mechanical properties of SBHS500.

Chemical compositions [mass%]							Mechanical properties		
C	Si	Mn	P	S	Ni	Cr	Yield stress [MPa]	Tensile strength [MPa]	Elongation [%]
0.10	0.24	1.57	0.008	0.003	0.02	0.12	559	643	36

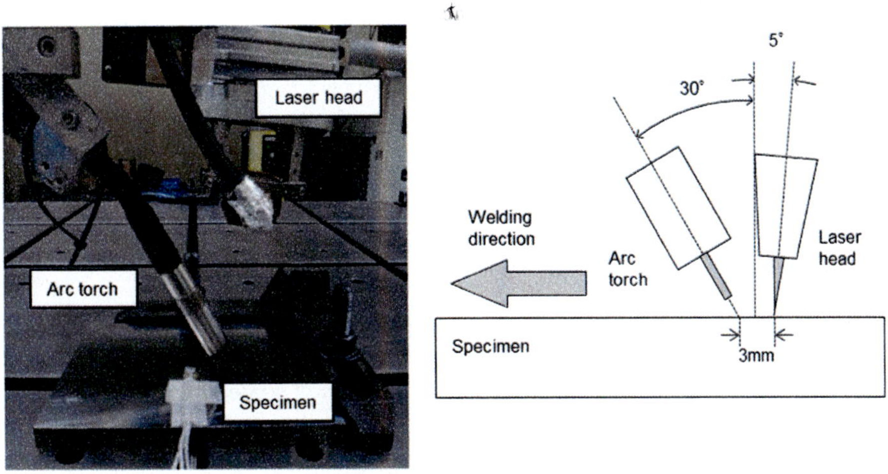

Fig. 2. Experimental setup.

2.2 Welding Conditions and Experimental Results

Figure 2 shows the experimental setup. A fiber laser with a maximum capacity of 20 kW was used. The focus distance, spot diameter and defocus were 250 mm, 600 μm and −5 mm respectively. The arc torch preceded the laser focus with the distance of 3 mm. The tilting angles of arc torch and laser head were −30 and 5° respectively.

The gas metal arc welding with the same grade filler wire as the base metal was used. The wire diameter was 1.2 mm. The shield gas type and flow were Ar + 5% O_2 and 20 L/min.

Figure 3 shows the welding conditions and the macroscopic photographs of welded parts. The arc current of 250 A and the arc voltage of 28 V were fixed. The laser power and the welding speed were varied from 12.5 kW to 15.3 kW and 1.4 m/min to 1.6 m/min. The welding condition No. 5 was eventually adopted from the viewpoints of occurrence of under-fill, burn-through, crack and surface profile.

3 Distortion and Residual Stress Measurement

3.1 Specimen and Welding Condition

Figure 4 shows the shape and dimension of the specimen for the distortion and stress measurement. Based on the results of the above welding condition search experiment, the hybrid welding conditions (the laser power of 13.01 kW, the welding speed of 1.6 m/min, the arc current of 255–288 A and the arc voltage of 28.1 V) were adopted. For comparing to the hybrid welding, the arc welding was also performed on the specimen with the same shape and dimension. In the case of an arc welding specimen, the weld metal was supplied into the V-shaped groove of 50° by 6 passes. The arc welding conditions were the current of 200 A, the voltage of 26–27 V and the welding speed of 0.15–0.25 m/min.

From the viewpoint of welding time, the hybrid welding took 13 s, while the arc welding took 617 s for 6 passes. Actually, the arc welding required the inter-pass cooling time for each pass. In any case, the hybrid welding could reduce the welding time by 98% compared to the conventional arc welding. It was found that the hybrid welding could drastically reduce the welding time compared to the arc welding.

Condition	No. 1	No. 2	No. 3	No. 4	No. 5
Lase power [kW]	13.08	13.08	13.08	15.3	12.5
Welding speed [m/min]	1.4	1.6	1.8	1.6	1.6
Macroscopic photograph	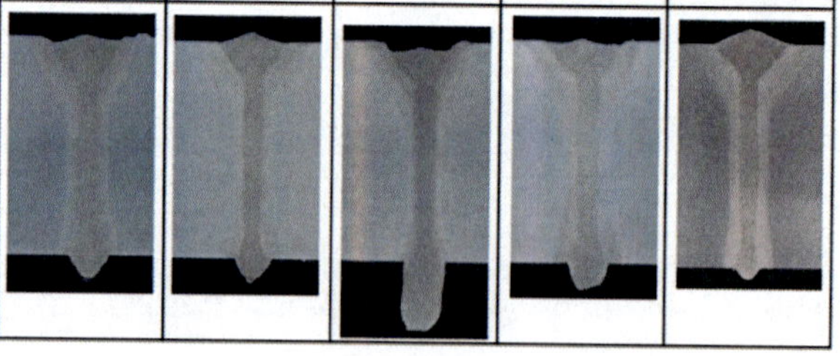				

Fig. 3. Welding conditions and macrographic photographs.

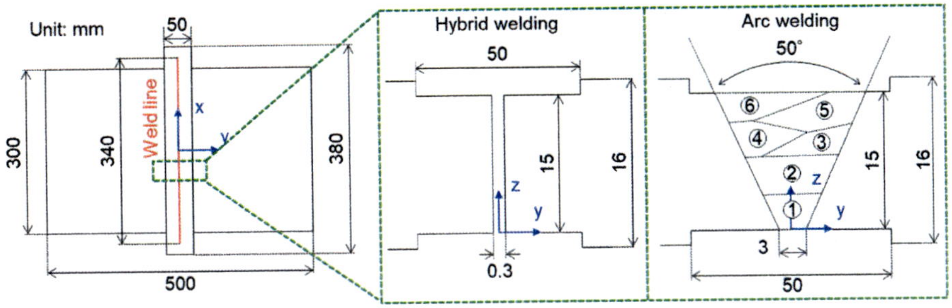

Fig. 4. Specimen for welding distortion and residual stress measurement.

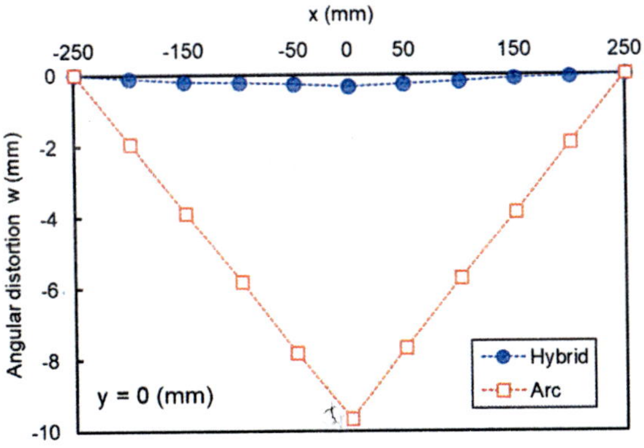

Fig. 5. Angular distortion.

3.2 Welding Distortion

Figure 5 shows the angular distortions generated by the hybrid welding and the arc welding. Due to the low heat input and the one-pass full-penetration of hybrid welding, the angular distortion of hybrid welding was considerably smaller than that of arc welding. The out-of-plane displacements at the center of the specimen were 0.36 mm by the hybrid welding and 9.7 mm by the arc welding. The angular distortion by hybrid welding was reduced by 96% compared to that by arc welding.

3.3 Residual Stress

Figure 6 shows the residual stress components in the longitudinal direction generated by the hybrid welding and the arc welding. The residual stresses were measured by XRD method. Although the maximum tensile stress of hybrid welding (453 MPa) was larger than that of arc welding (367 MPa), the tensile stress field of hybrid welding was narrower than that of arc welding. The tensile stress field of hybrid welding was from 20 mm to 30 mm in the middle of the specimen while that of arc welding was over 40 mm. The reason of this tendency might be the narrow heat input region of hybrid welding. The tensile stress field by hybrid welding was reduced by 25% to 50% compared to that of arc welding.

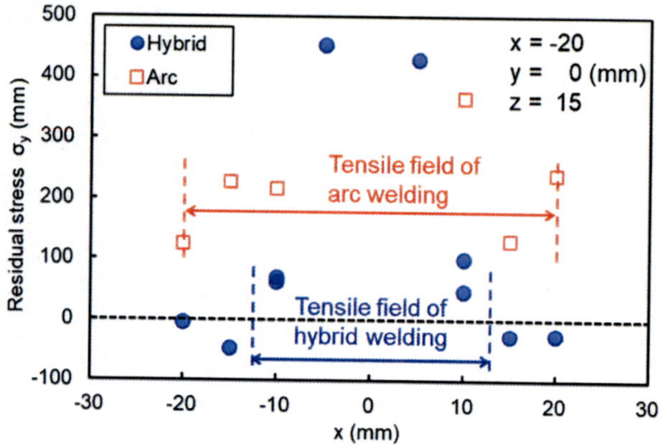

Fig. 6. Residual stress.

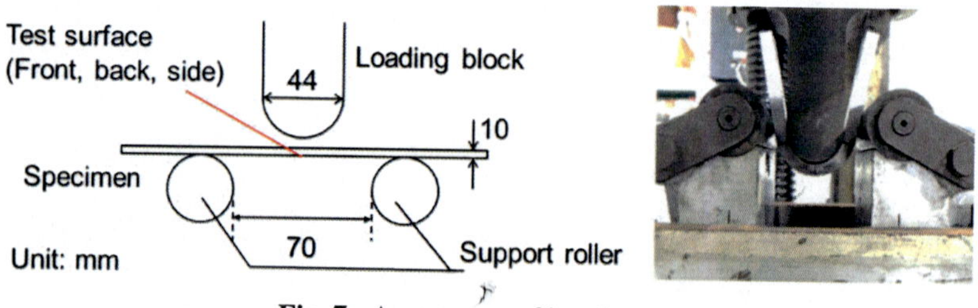

Fig. 7. Appearance of bending test.

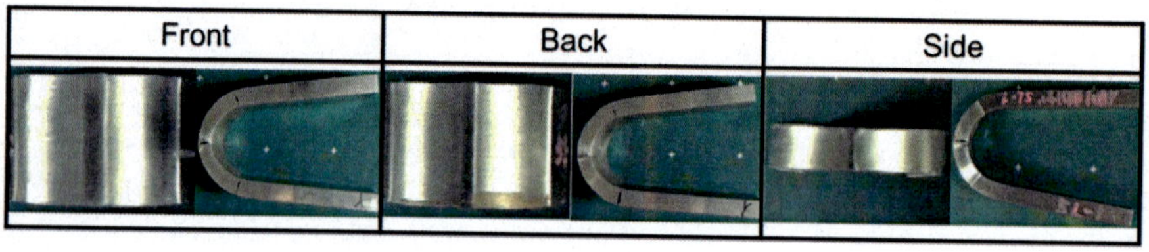

Fig. 8. Results of bending test.

4 Joint Performance Experiment

4.1 Bending Test

For investigating the soundness of hybrid welded joints, a bending test and a tensile test were conducted [5]. These joint-performance experiments were not conducted on joints by the arc welding because the purpose of this study was to examine the joint performance of hybrid welded joints.

The specimens were extracted from the welded joint. The inherent weld defects were examined by extending the front, back and side surfaces through the 180 degrees of bending. Figure 7 shows the appearance of the bending test and Fig. 8 shows the results of the bending test. No extension of defects or cracks was observed at the front, back and side surfaces of joints after the bending test.

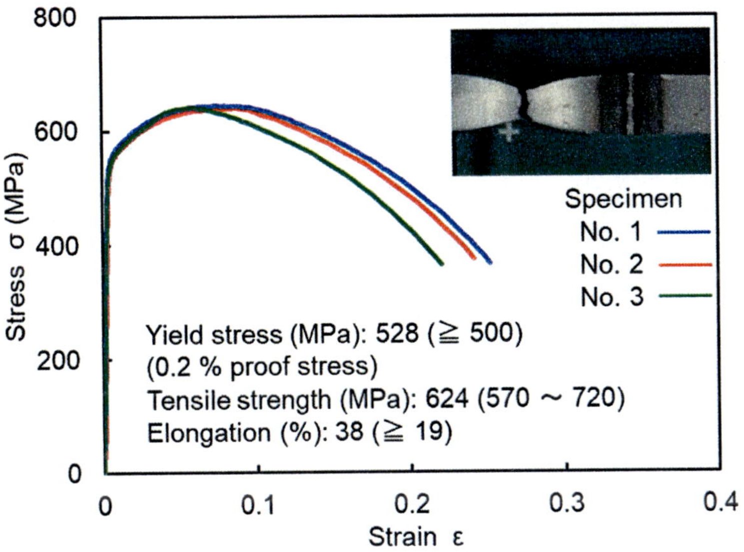

Fig. 9. Results of tensile test.

4.2 Tensile Test

For examining the basic joint performances of hybrid welding, a tensile test was conducted [6]. The coupon test specimens were extracted from the welded joint. The number of specimens was three. The static tensile test was performed on the specimens. Figure 9 shows the results of the tensile test. The blue, red and green solid lines represent the stress-strain curves of three specimens. All specimens were broken at the base metal parts. The yield stress, the tensile strength and the elongation satisfied the specifications for the base metal (the values shown in the brackets in the figure) [4]. These results indicated that the hybrid welded joints did not include the weld defects and satisfied the basic mechanical performances.

5 Conclusions

For investigating the basic applicability of laser-arc hybrid welding to the relatively thick steel plates used in bridge structural members, a series of experiments were conducted. The following results were obtained.

(1) A welding condition for one-pass full-penetration welding of steel for bridge structures with thickness of 15 mm was specified. The sound welded joints without defects or cracks could be obtained by that welding condition.
(2) The butt-welded joints of 15 mm-thick steel plates were fabricated by the hybrid welding and the arc welding respectively. The hybrid welding took 13 s for the weld line of 340 mm-long by one pass, while the arc welding took 617 s for 6 passes. The hybrid welding could reduce the welding time by 98% compared to the conventional arc welding.

(3) The welding angular distortion and residual stress of hybrid welding were compared to those by the arc welding. The angular distortion of hybrid welding was reduced by 96% compared to that of arc welding. The tensile stress field of hybrid welding was reduced by 25% to 50% compared to that of arc welding. The reason of these tendencies might be the low heat input and the narrow heat input region of hybrid welding.

(4) The bending test and the tensile test were conducted on the hybrid welded joints. No extension of defects or cracks was observed at the front, back and side surfaces of joints after the bending test. In the tensile test, the specimens were broken at the base metal parts. The yield stress, the tensile strength and the elongation satisfied the specifications for the base metal. These results indicated that the hybrid welded joints did not include the weld defects and satisfied the basic mechanical performances.

Acknowledgements. A part of this research was supported by The Japan Iron and Steel Federation (JISF).

References

1. Katayama, S., Kawahito, Y., Mizutani, M.: Elucidation of laser welding phenomena and factors affecting weld penetration and welding defects. Phys. Procedia **5**, 9–17 (2010)
2. Claus, B., Flemming, O.O.: Review of laser hybrid welding. J. Laser Appl. **17**(2), 2–14 (2005). https://doi.org/10.2351/1.1848532
3. Hirohata, M., Takeda, F., Suzaki, M., Inose, K., Matsumoto, N., Abe, D.: Influence of laser-arc hybrid welding conditions on cold cracking generation. Weld. World **63**(5), 1407–1416 (2019)
4. Japanese Standards Association: Higher yield strength steel plates for bridges, JIS G 3140 (2011)
5. Japanese Standards Association: Metallic materials - Bend test, JIS Z 2248 (2006)
6. Japanese Standards Association: Methods of tensile test for butt welded joints, JIS Z 3121 (2013)

Research on Laser Beam Welding of Nickel

Ingrid Kovaříková, Beáta Šimeková, Erika Hodúlová(✉), Jozef Bárta, and Pavel Kovačócy

Faculty of Material Science and Technology in Trnava, Slovak University of Technology in Bratislava, Bratislava, Slovak Republic
{ingrid.kovarikova,beata.simekova,erika.hodulova,jozef.barta, pavel.kovacocy}@stuba.sk

Abstract. A study of weld joints of nonferrous, Nickel 201 alloy sheets using a new generation disk laser as the green welding technology for effective manufacturing was carried out, and the results are presented in this paper. Weld joints of the Nickel 201 sheets 2.0 mm thick were welded by laser without an additional material at a flat position, using high purity argon as the shielding gas. The influence of laser welding parameters on weld quality and mechanical properties of test joints was studied. The influence of welding speed and laser power to the joint quality was investigated. The study of quality and mechanical properties of the joints were determined by metallographic evaluation, tensile and hardness tests.

Keywords: Laser beam · Nickel · Weldability · Fusion zone · Mechanical properties

1 Introduction

Nickel is a relatively simple metal. It is a face-centered cubic and undergoes no phase changes as it cools from melting point to room temperature, similar to stainless steel. Nickel and its alloys cannot, therefore, be hardened by quenching so cooling rates are less important than with, say, carbon steel and preheating if the ambient temperature is above 5 °C is rarely required. Nickel and its alloys are used in a very wide range of applications - from high temperature oxidation and creep resistance service to aggressive corrosive environments and very low temperature cryogenic applications. Nickel may be used in a commercially pure form but is more often combined with other elements to produce two families of alloys - solid solution strengthened alloys and precipitation hardened alloys [1].

Nickel plays an important role in various modes of transportation. Nickel is used in the batteries of electric vehicles, nickel alloys in jet engine turbines, while nickel-containing stainless steel is found in passenger trains and subways. Nickel-containing materials offer enhanced corrosion resistance and reliable and efficient electrical and spark systems. Nickel ensures that train carriage and truck bodies are strong, durable and can help absorb the impact of collisions by crumpling rather than fracturing. Materials

K. Jármai and K. Voith (Eds.): VAE 2020, LNME, pp. 497–506, 2021.
https://doi.org/10.1007/978-981-15-9529-5_43

containing Nickel also support the transport infrastructure that makes this movement possible, in airports, train stations, bridges and flyovers [2].

The laser welding power and laser welding speed are the most significant input factors which affect all output responses of the dissimilar joint process because of the joining of two different materials facing the challenge with different melting ratios. All the other input parameters influencing on a particular output response only [3]. The width of HAZ increases with the increase in laser power in both air and water medium at constant welding speed. Further, with the increase in welding speed at constant laser power the width of HAZ reduces. The slope of the plot for width of HAZ laser power is low at the low power level and the wt% of oxide formation increases with an increase in laser power at constant welding speed. The optimum laser power may be selected without affecting the weld quality, for achieving the narrow HAZ and minimum level of formation of metal oxide [3, 10].

In an automotive battery, a large number of individual cells are electrically connected to meet the energy and power requirement. The choice of suitable welding techniques is critical in terms of mechanical strength and electrical contact resistance of the joint. Although soldering, ultrasonic welding and laser welding are the common joining techniques used in lithium-ion cell fabrication, the application of resistance spot welding in the manufacturing of lithium-ion batteries for electric vehicles is rather limited due to the difficulties in forming large weld nuggets and electrode sticking to the material. In this work, micro resistance spot welding of nickel anode tab to the one side open inner aluminum casing is carried out in an in-house fabricated cylindrical lithium-ion cell. The properties of the weld joint depend on the weld nugget diameter, depth and presence of the defects like blow holes, cracks, which are directly influenced by the weld parameters. The microstructure shows that the nugget is located off symmetry from the nickel-aluminum interface, predominantly in the Ni layer. This is expected since the electrical resistivity of Ni is significantly higher than Al and hence the heat generation will be higher in Ni [4, 5, 12].

In the automotive industry, there were still several problems in the dissimilar welding of nickel-based alloy and austenite stainless steel by laser welding (LW). The welding of Nickel-based alloy and austenite stainless steel dissimilar materials were usually carried out using traditional arc welding processes. However, current welding methods were only involved in the ultrasonic vibration assisted arc welding, friction stir welding and braze welding, the ultrasonic vibration assisted pulsed laser welding was rarely involved. In addition, the research mainly focused on the effects of ultrasonic vibration on the weld shape, grain morphology, cracks, porosities and mechanical property. During laser welding of dissimilar materials, the effects of ultrasonic vibration on the secondary phases and micro and macro distributions of elements in the weld metal were rarely reported. The specific cavitation and acoustic streaming effects in weld molten caused by ultrasonic vibration were utilized to change the convection and solidification behaviors of the molten pool and to eliminate the defects in welding of dissimilar materials [6–9, 11].

In the present study, the LBW method was used for welding of Nickel 201 sheets by using a wide range of welding parameters. The effects of these parameters on joints properties and quality are studied. The quality evaluation of joints was characterized by means of light microscopy, scanning electron microscopy.

2 Experimental Materials and Methods

Nickel is similar in many respects to the austenitic stainless steels; welding procedures are likewise also similar. Nickel, however, has a coefficient of thermal expansion less than that of stainless steel so distortion and distortion control measures are similar to those of carbon steel. The most serious cracking problem with nickel alloys is hot cracking in either the weld metal or close to the fusion line in the HAZ with the latter being the more frequent. Porosity can be a problem with the nickel alloys, the main culprit being nitrogen. As little as 0.025% nitrogen will form pores in the solidifying weld metal. Quite light draughts are capable of disrupting the gas shield and atmospheric contamination will occur resulting in porosity. Care must be taken to ensure that the weld area is sufficiently protected and this is particularly relevant in site welding applications. With the gas shielded processes, gas purity and the efficiency of the gas shield must be as good as possible [1–5].

The materials used in the present study were sheets of pure Ni (Nickel 201, wt. 99.99%) with dimensions of 100 × 50 × 2 mm. The samples were cleaned in acetone and dried. The chemical composition of used base material and the mechanical properties are shown in Table 1, 2.

Table 1. The chemical composition of test material Nickel 201 (wt%).

Ni	C	Co	Fe	Mg	Mn	Si	Ti
99.0	0.02	0.25	0.13	0.15	0.30	0.15	0.1

Table 2. The mechanical properties of test material Nickel 201.

Tensile strength MPa	Yield strength MPa	Elongation %	Hardness HV
469	186	47	171

Welding was performed by TRUMPF TruDisk 4002, the 2 kW solid state laser device with TRUMPF BEO D70 welding head installed on Fanuc M-710iC/50 (6-axis) robot (Fig. 1), focusing distance 200 mm and the minimal welding spot diameter of 200 μm was used for the production of butt weld joints. The laser power used for the production of the experimental weld joints was in the range of 1–1.8 kW.

To avoid reaction between the molten metal and the atmospheric moisture, the weld seams were carefully shielded by the pure argon shielding gas. The welding parameters

Fig. 1. Laser welding equipment and Fanuc Robot M-710 iC50.

Table 3. Welding parameters used for the weld joints production.

Sample	Laser Power (kW)	Welding speed (mm/s)	Shielding gas flow rate (l/min)	Focus position (mm)
1	1	30	14	4
2	1	30	14	0
3	1.4	30	14	0
4	1.8	30	14	0
5	1.8	20	14	4
6	1.5	20	14	4
7	1.5	20	14	4

used for the weld joints production, such as laser power, welding speed, shielding gas flow rate and focus position are listed in Table 3.

Evaluation of the weld joints quality was performed by metallographic analyses. The samples were prepared by grinding, polishing and etching in a solution of 10 ml–HF, 30 ml-HNO_3, 50 ml-H_2O for time of 20 s. Macrostructural and microstructural analyses were carried out by optical microscopy. Microhardness of the prepared weld joints was measured using the IndentaMet 1100 equipment. The weld defects were identified by visual non-destructive method according to the STN EN ISO 17637 and STN EN ISO 13018 standards. The mechanical properties of weld joints were determined by tensile tests. Tensile test was performed at room temperature using a Tinius Olsen 300ST tensile test machine operating at a test speed of 500 mm/min with a capacity of 300 kN according to the STN EN ISO 6892–1. For the tensile test was prepared specimens, where the parameters of laser welding were set according to Table 3 for welds No. 5, 6 and 7. The test was performed on the three specimens for each weld (in total 9 measurements). The test specimens had a dimension of 100 × 12 × 2 mm.

3 Results and Discussion

The cross-sections of weld joints No. 1–3 are shown (Fig. 2.) no occurred penetration depth, because the beam power was no sufficient. The change at weld joints No. 4, where a narrow but irregular root gap is shown. At weld joints No. 5, the power was the same

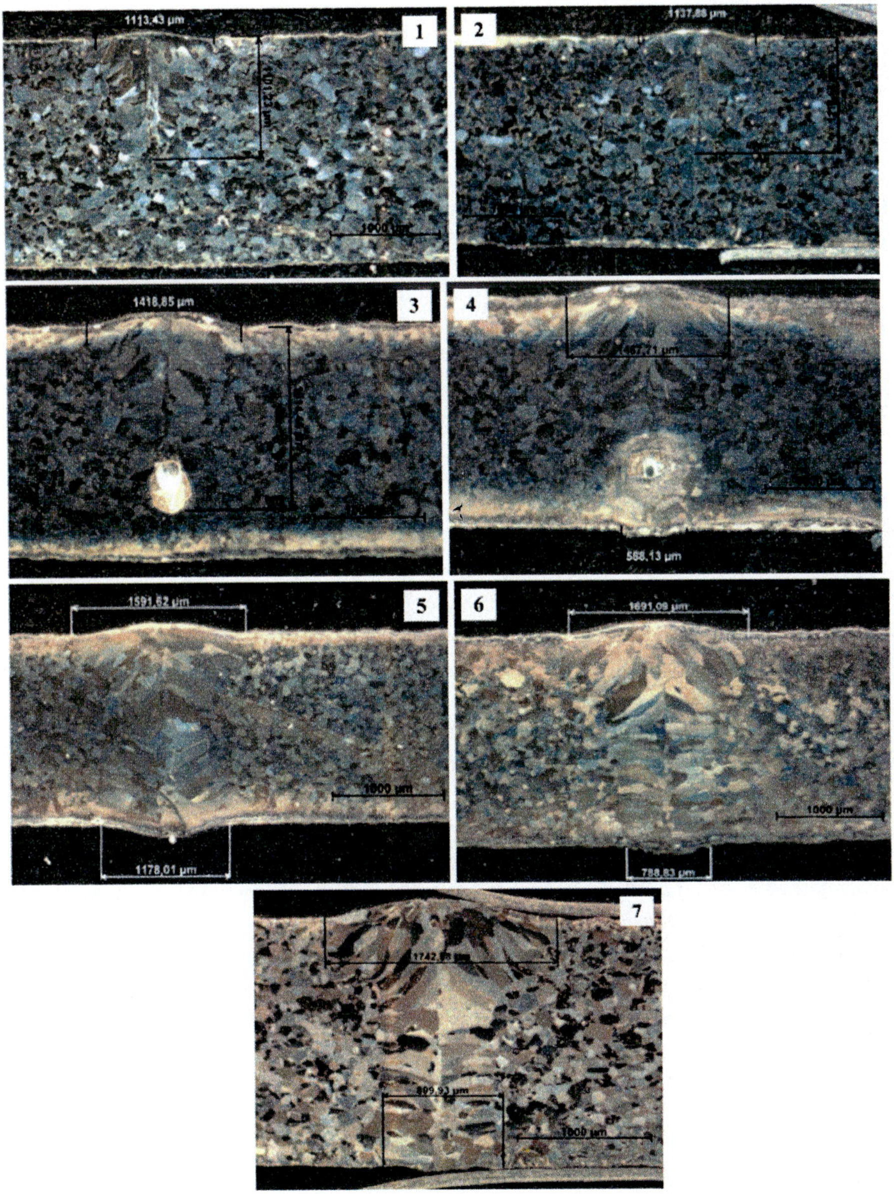

Fig. 2. Cross-section of weld joints

as at sample No. 4 (1.8 kW), the welding speed was reduced from 30 to 20 mm/s, which resulted in more heat introduced to the weld and a beam focus was 4 mm above the base material what is visible at too wide root width. At weld joints No. 6 and 7 the power was reduced to 1.5 kW other parameters were left, the weld geometry was in the optimal range. At weld joints No. 7, the protective atmosphere blowing on the root was activated and this caused irregularities of the weld surface (turbulent flow of the protective atmosphere Ar).

The weld geometry measuring was focused on the weld width, weld root width and the excessive penetration (Table 4).

Table 4. Values of a measured geometric parameter of welds dependence to welding parameters.

Sample	Heat input (J/mm)	Weld width (mm)	Weld root width (mm)
1	33	1.1	–
2	33	1.1	–
3	47	1.4	–
4	60	1.5	0.6
5	90	1.6	1.2
6	75	1.7	0.8
7	75	1.7	0.9

The geometrical parameters of welds show that the width of the weld is increasing with the welding current at a constant welding speed, and the width of the weld will increase also with the welding speed at a constant welding current. In the area of the root of weld metal (WM) it is possible to observe the pores. The pores were round in shape, which was classified as gas pores. There are many reasons for porosity formation in welding. The probable cause could be in the keyhole instability, it can lead to the metal vapor or inert gas getting trapped in the weld metal. This occurs when the forces keeping the keyhole open are not in balance with the forces trying to close the keyhole.

The structure of the base material Ni 201 is formed by a substitution solid solution in nickel (sometimes referred to as nickel austenite). Thus, it is a monophasic material. Although it is a monophasic structure, some grains are darker and others lighter, this is not due to a different type of phase but to etching into different crystallographic planes. The grains show a polyhedral character with heterogeneity in grain size in the range from 20–150 μm. In some grains, it is possible to observe annealing twins (sticks across the grain). The grain boundaries are etched in steps and are clean without the presence of secondary phases. Oxide particles are also occasionally observable (black beads in the structure visible in Fig. 3a).

The microstructure of the fusion zone of joints is in Fig. 3b. The grains consist of nickel-chromium austenite. The grain boundaries are clean. The heat-affected zone (HAZ) is insignificant; it does not observe grain coarsening or precipitation. The structure in the weld metal partially lost its polyhedral character and took the form of stretched

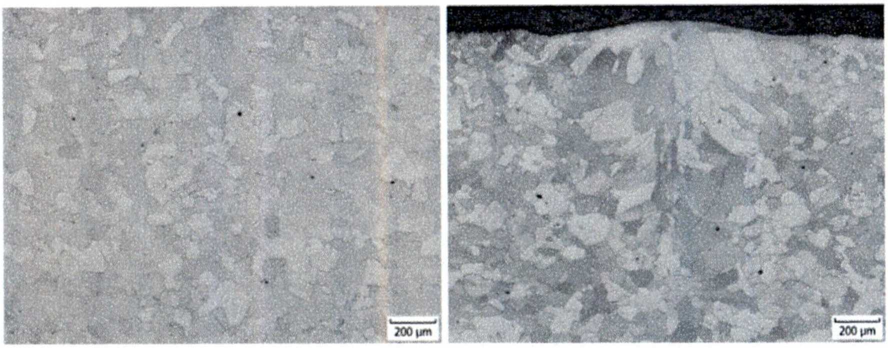

Fig. 3. Microstructure of a) base material and b) fusion zone of joint

columnar grains, these grains grow against the direction of heat dissipation. In the middle part of the weld metal in the root area, the structure acquires a fine-grained polyhedral character. It is not observing any phase change in the weld metal due to the fact that it is a monophasic material.

The hardness distribution of the weld cross-sections was also analyzed (Fig. 4). A decrease in hardness towards the weld metal is visible on the resulting curves. This is caused by the dissolution of precipitates, due to the introduction of heat during welding. This is because the Ni 201 alloys are supplied in a precipitation hardened state. The hardness of weld metal decreased by 20–30HV in comparison to the base material and by 5–10HV in comparison to the HAZ.

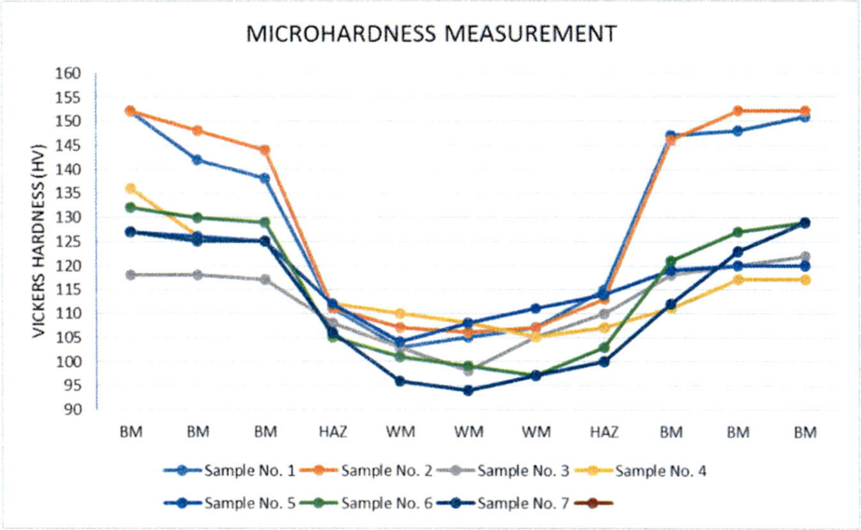

Fig. 4. Microhardness distribution of weld joints

At weld joints No. 1–4, it is possible to observe the decrease of hardness in the HAZ and WM. Weld joints No. 5 has a slightly more significant decrease in hardness. The fact is also the use of a relatively high welding power (1.8 kW), which introduced a high heat input into the material, it has resulted in the reinforcement of base material. In this case, the decrease in microhardness in the area of WM and HAZ is significantly more pronounced. The hardness curve of the weld joint No. 7 is similar to the weld joint No. 6, the reason is the fact that the same welding parameters were used.

Mechanical properties of the Nickel 201 were determined by tensile tests. Representative samples used for the tensile test were prepared by using the same welding parameters as in case of sample No. 5, 6, 7. The values of tensile test measurements are shown in Table 5. The laser power was 1.5 kW and 1.8 kW and welding speed was 20 mm/s. Figure 5 depicts the load-displacement curve of the laser welded sample No. 6, it shows that it is a material with insignificant yield strength, so it behaves almost identically to carbon non-alloy steel.

Table 5. Measured values of the mechanical properties

Specimen	Max. load (kN)	Tensile strength (MPa)	Elongation (%)
5	9.21	374	32
6	9.38	373	31.5
7	9.31	374	29.6

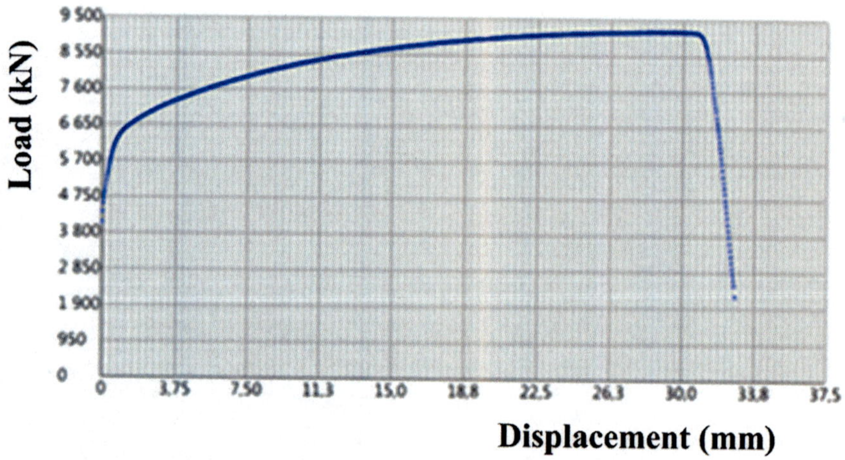

Fig. 5. Typical load-displacement curves of the laser welded sample No. 6

The most important indicator was the place of specimen interruption. For all three specimens, the test specimen failed in the base material and not in the weld. This indicates that the weld joints did not contain defects such as cavities, pores, cracks and other defects that cause degradation of the structure and interruption in the weld joints.

4 Conclusion

The main optimized parameters of Ni 201 laser welding are welding power P [W], welding speed in [mm/s], beam focusing [mm]. When solving the problem of welding, it is necessary to work with all parameters in order to achieve the best possible heat input. Experiments have shown that, among other things, speed has a significant effect on the stabilization of the keyhole in the penetration mode of welding. At high values, the keyhole collapsed. An important parameter was also the flow of the protective atmosphere and its correct orientation to the welding site. Argon as a shielding gas should be used for good protection of welded material against oxidation. The minimum heat input of 75 J/mm must be used in order to obtain the weld joints with acceptable penetration and geometry. The weld geometry is changing with the welding parameters. The penetration depth increasing with the welding current at a constant welding speed and decreasing at the higher welding speed at a constant welding current.

The microstructure of the fusion zone of joints consists of nickel-chromium austenite grains. The heat affected zone is insignificant. The structure of weld metal took the form of stretched columnar grains. It is not observing any phase change in the weld metal due to the fact that it is a monophasic material. The hardness is decreased in towards the weld metal, caused by the dissolution of precipitates, due to the introduction of heat during welding. The hardness of weld metal decreased by 20–30HV in comparison to the base material and by 5–10HV in comparison to the HAZ. The tensile tests show the rupture of base material, not the weld metal. This indicates that the weld joints did not contain defects.

Acknowledgments. This research is supported by VEGA 1/0091/17 which are supported by Slovak Republic Ministry of Education.

References

1. TWI Homepage. https://www.twi-global.com
2. Nickel Institute Homepage. https://www.nickelinstitute.org
3. Kumar, G.K., Velmurugan, C., Jayaram, R.S., Manikandan, M.: Effect of laser welding process parameters on dissimilar joints of AISI 316 and nickel 201. Mater. Today Proc. **22**(4), 2964–2973 (2020)
4. Rikka, V.R., et al.: Tailoring micro resistance spot welding parameters for joining nickel tab to inner aluminum casing in a cylindrical lithium ion cell and its influence on the electrochemical performance. J. Manuf. Process. **49**, 463–471 (2020)
5. Brand, M., Schmidt, P., Zaeh, M., Jossen, A.: Welding techniques for battery cells and resulting electrical contact resistances. J. Energ. Storage **1**, 7–14 (2015)
6. Zhou, S., Ma, G., Wu, D., Chai, D., Lei, M.: Ultrasonic vibration assisted laser welding of nickel-based alloy and austenite stainless steel. J. Manuf. Process. **31**, 759–767 (2018)
7. Ma, G., Wu, D., Niu, F., Zou, H.: Microstructure evolution and mechanical property of pulsed laser welded Ni-based superalloy. Opt. Laser Eng. **72**, 39–46 (2015)
8. Zhou, S., Ma, G., Chai, D., et al.: Nickel-based alloy/austenitic stainless steel dis-similar weld properties prediction on asymmetric distribution of laser energy. Opt. Laser Technol. **81**, 3w33–39 (2016)

9. Lippold, J.C., Kiser, S.D., DuPont, J.N.: Welding Metallurgy and Weldability of Nickel-base Alloys. John Wiley & Sons, Hoboken (2011)
10. Caron, J.L., Sowards, J.W.: Weldability of Nickel-base Alloys, 1st edn. In: Comprehensive Materials Processing. Elsevier, Amsterdam (2014)
11. Lippold, J.C., Nissley, N.E.: Ductility-dip Cracking in High Chromium, Ni – base Filler Metals. Hot cracking phenomena in welds II. Springer, Berlin (2008)
12. Kaplan, A.F.H., Mizutani, M., Katayama, S., Matsunawa, A.: Mechanism of pore formation during keyhole laser spot welding. In: Proceedings SPIE 4831, First International Symposium on High-Power Laser Macroprocessing, pp. 186–191. Osaka (2003)

Arc Sensor Parameter Optimisation for Robot Welding

Abdallah Kafi[1](✉) and Tünde Anna Kovács[2]

[1] Bánki Donát Faculty of Mechanical and Safety Engineering, Doctoral School on Safety and Security, Óbuda University, Népszínház u. 8, Budapest 1081, Hungary
Abdallahkafi1994@gmail.com
[2] Bánki Donát Faculty of Mechanical and Safety Engineering, Department of Material Science, Óbuda University, Népszínház u. 8, Budapest 1081, Hungary
kovacs.tunde@bgk.uni-obuda.hu

Abstract. In this article, the authors present the arc sensor supported robot welding process and the task of the sensor in the welding practice. Arc welding is a very important industrial process. The welded joint needs to be high quality besides the high productivity and cost-effective manufacturing. The joint quality depends on the welding process realization and the used welding parameters. The sensor in the case of the robot welding process supports the regularity of the joint. The arc sensor on the base of the welding currents checking during the welding process continually corrects the arc position by the correction of the robot control. The checking intensity of the sensor depends on the data input range what is determined by the sensor control parameters (weaving and correction condition).

In this research by a practical experimental method, the sensor condition optimization is introduced in the case of the used Motoman welder robot and the used advanced arc sensor.

Keywords: Arc welding · Robot welding · Joint quality · Arc sensor

1 Introduction

The welding process is an important part of manufacturing what requires the welded joint quality besides the welding reproductivity, productivity and cost-efficiency. Since the 1950s the manual arc welding has been widely used and published process. For additive industrialization of large metal components due to an unlimited build envelope, higher deposition rates, and lower capital cost, traditional manual welding or semi-automatic welding is gradually being replaced by robotic welding. Robotic welding not only improves quality and efficiency but also reduces assistance intensity and cost, due to the number of variants of arc welding that have been developed in an attempt to improve the performance and productivity of the process. Increasing flexibility is the primary task needed to take advantage of robotic welding for industries [1–3], Also the welding quality is one of the key factors which affect the constructional strength and the

K. Jármai and K. Voith (Eds.): VAE 2020, LNME, pp. 507–516, 2021.
https://doi.org/10.1007/978-981-15-9529-5_44

comprehensive quality of the products, in the modern manufacturing industry, and with this can the particulars of the weld joints can be improved [4, 5].

The premier industrial robotic named "Unimate," which was created in 1954 by George Charles Devol. After a few years, the Unimation was born. Devol was established with his fellow Joseph F. Engelberger [6]. That is how the science of robotics launched to develop. New ergonomic and security laws have become required for the introduction of industrial robotics as people work together with a robot have to function under unnatural working conditions.

According to the official definition of RIA (Robot Institute of America), a robot is a reprogrammable and multifunctional manipulator that is designed to move specialized materials, parts, tools, or devices through variable programmed motions. for performing a variety of tasks [19]. Besides, the industrial robot is also defined by the International Organization for Standardization (ISO) as an automatically controlled, reprogrammable and versatile manipulator. This manipulator is programmable in three or more axes and can be either field-mounted or mobile for industrial automation applications. Typical applications for industrial robots include welding, painting, assembly, packaging, palletizing, product inspection and testing, etc. [20].

The healthy working environment of the human population must be established during industrial processes, as it is for a large range of activities [6, 7], because it is designed for performing operations repeatedly, quickly, and accurately have a long heritage in the manufacturing industry, operating in large numbers and, in relatively static environments, CCD cameras were frequently used to identify the kinematic parameters of the current machining setup so that positional reliability of the robot could be improved [8–10], and we can classify the industrial robots according to several criteria drive technology, degrees of freedom, workspace geometry, and level of autonomy [11].

In factories which are considered to be a popular and important research issue in the manufacture of the industry, most of the welding robots are still in the "teaching and playback type" mode. The demand for the shape and position of the welding seam is highly accurate. Therefore, a desired adaptive capability is difficult to provide to arc-light changes, splash disturbance, and seam status during the welding process. In practical production, welding encounters many different variables, such as in-process thermal distortions, workpiece fitting, and, besides, seam position changes due to pre-machining errors. The metal manufacturing focuses almost steels. In the case of the steel welding process, it needs to understand the metallurgical and chemical processes to choose the suitable welding parameters to cause suitable crystallization and microstructure [12, 13]. The steels as a function of the chemical composition and heating and cooling effects have different microstructure and mechanical properties [14, 15]. The welding resulted in the heating-cooling process caused microstructural changing affected martensite in the case of the structural steels. The low carbon unalloyed steels are usually weldable without any preheating, but it can determine by carbon equivalent calculation and the knowledge of the plate thickness.

Unable to fix robots [16, 17], despite the robot welding come with six degrees of freedom is carrying a vision sensor system welding torch and a welding torch. [18]. Researchers have proposed a lot of intelligent welding systems. In some research, the relationship between the welding torch position and the power signal was investigated, and arc sensors for soldering seam tracking developed [19, 20].

2 Experimental Method

2.1 Used Welder Robot (MOTOMAN Robot)

MOTOMAN is a type of welding robot industrial with six degrees of freedom. Many industrial applications robotic technology adds special benefits to the process of the welding. When it comes to the improved level of inspection accuracy and speed of implementation, high speed articulated robot and machine vision system comes to focus. In this inspection system, the Programmable Logic Controller (PLC) is designed as the main controller that controls the MOTOMAN robot, the servomotor, and ancillary units. The MOTOMAN robot is programmed using a different user frame such that the inspection criteria can be easily changed and implemented. (Figure 1) and (Fig. 2) represents the used MOTOMAN robot.

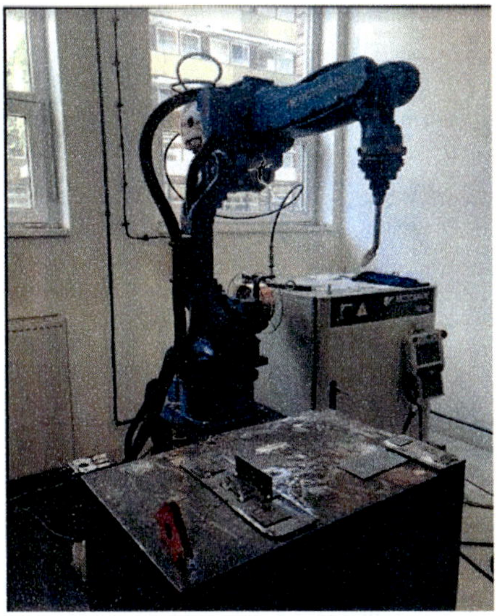

Fig. 1. The MOTOMAN robot

Should assign the position variables which mean the X, Y, Z coordination of a fixed location of the system (Fig. 2). Here the communication protocol among PLC, Welding, and robot are important because together they determine the defect. with emphasis on the robot programming and organizing sensors and electrical systems to work in synchronism for accurate inspection for Welding quality [21–23]. This inspection system can be modified rapidly to fit any unique characteristics that are required to be inspected for the welding. These inspection characteristics determine the quality of the welding.

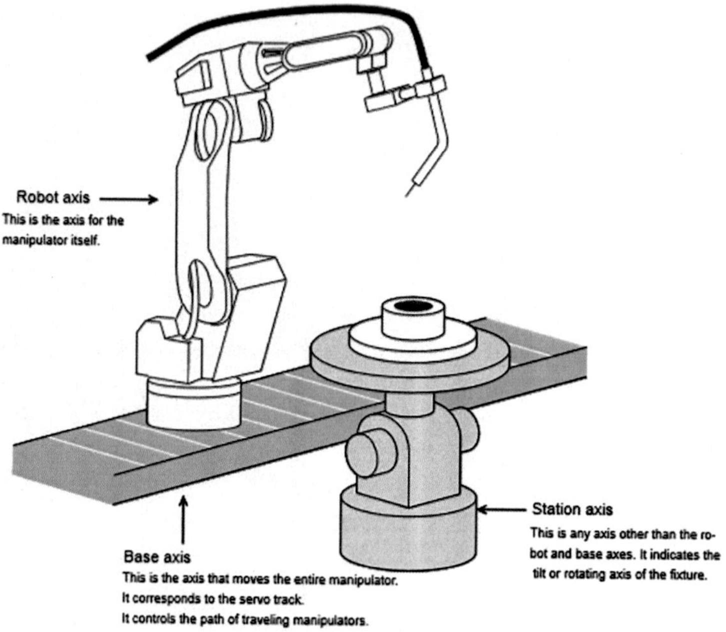

Fig. 2. The external axes [24]

2.2 Used Seam Tracking Arc Sensor

The arc sensor has constant voltage characteristics for welding with a power supply, the changes of the distance L (Fig. 3). as shown below make the welding current fluctuates.

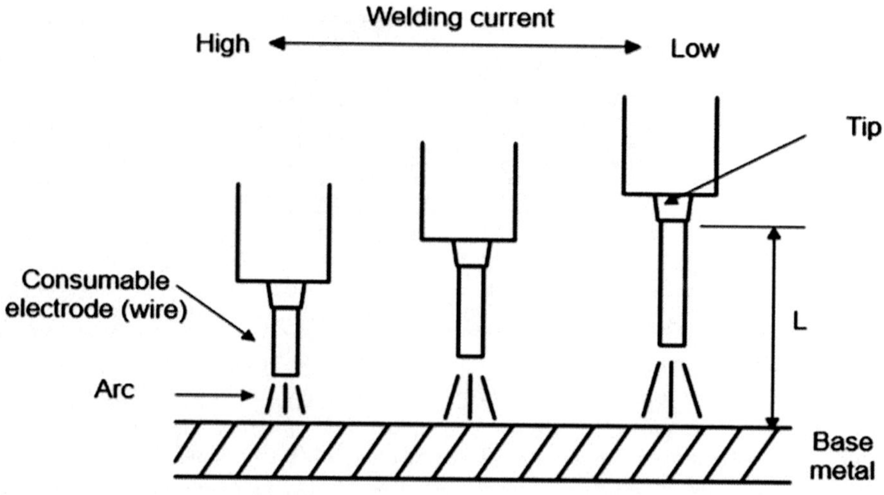

Fig. 3. Welding current fluctuates with L distance change [25]

L is the distance between the base metal and the Tip these characteristics use by the arc sensor function.

At welding, the arc sensor checks the welding currents upper and lower point, corrects the path to equalize the values, between any two points, and all of this happened with the torch when it's moving up and down (Fig. 4). The used arc sensor parameters are the sampling interval (ms), frequency (Hz). The sampling interval is the checking repetition rate.

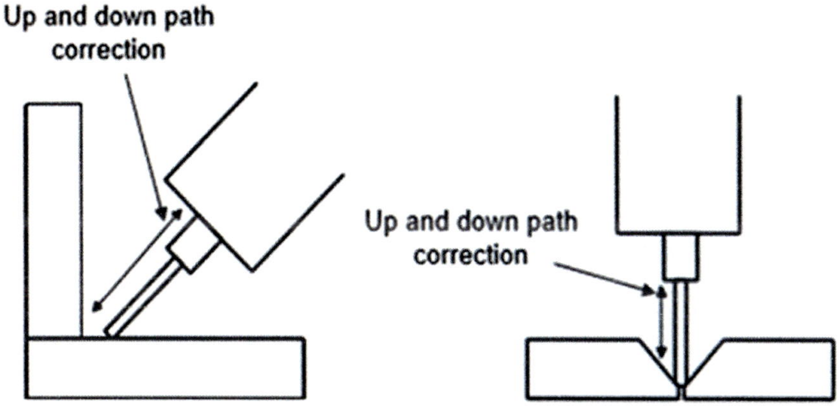

Fig. 4. Up and dower point path correction [26]

2.3 Used Material and Welding Parameters

For the experiments, it was used a hot rolled low carbon structural steel (S235JR). This steel is weldable because it's carbon contains lower than 0.17% and the carbon equivalent is 0.35% determined by the following Eq. (1).

$$\text{CEV} = \text{C} + \text{Mn}/6 + (\text{Cr} + \text{Mo} + \text{V})/5 + (\text{Ni} + \text{Cu})/15 = 0.35\% \quad (1)$$

The welding parameters showed in Table 1.

3 Experiments and Results

In the experiment, it was used the introduced steel and welding process, parameters are shown in Table 1 supported by an advanced seam tracking arc sensor (parameters shown in Table 1). For the robot program, it was used as a basic point to begin the welding task. The goal of the experiments was to determine the arc sensor parameter efficiency for the suitable welded joint.

The T joint welding experiments setup shown in (Fig. 5.) The test made behind constant welding and sensor parameters only it was modified the mistake. The mistake means the parallel distance of the test sample from the ideal location (Fig. 6). The mistake can simulate the welding process problem, in the case of the false position of the sample.

Table 1. The used welding parameters of the experiments

Welding process	MAG
Shielding gas	M21
Welding speed	45 cm/min
Sheet thickness	3 mm
Used steel	S235JR
Joint type	T joint
Filler metal	G3Si1/ER70S-6

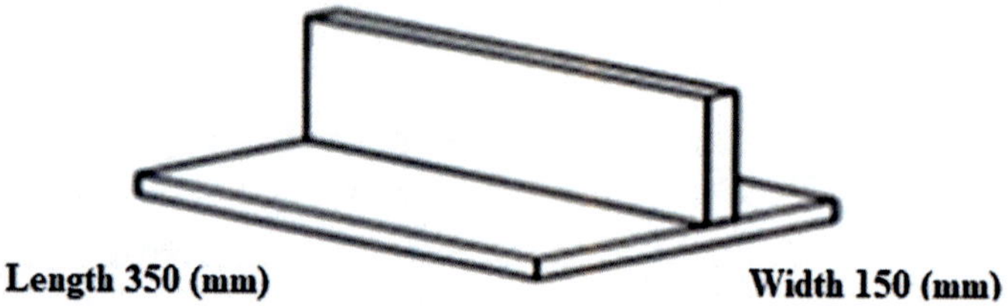

Fig. 5. T joint welding test setup

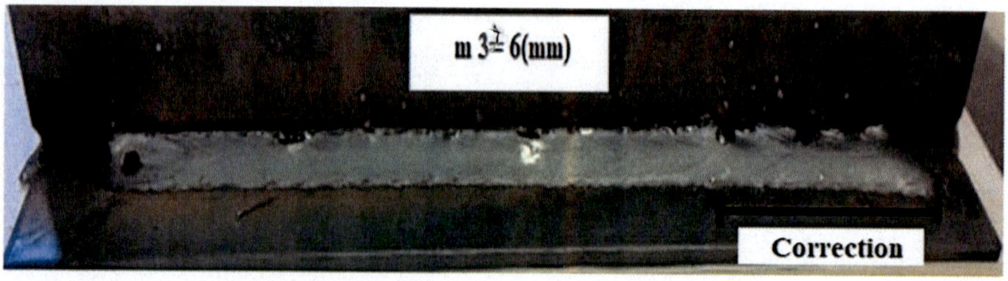

Fig. 6. The correction in the case of the 6 mm mistake

The measured correction length in the case of the experiments indicates the unsuitable joint length while the sensor corrects the arc position during the welding process. Table 2 shows the test parameters and the measured correction. As a function of the sampling interval, it can see the correction depend on this sensor parameter the bigger sampling interval causes smaller correction (Fig. 7). The correction measured after the welding experiments and shown in Table 2.

It was investigated the arc sensor frequency parameter in the experiments. Table 3 shows the test results. It can see that the higher frequency caused a smaller correction.

The correction depends on the arc sensor parameters. The arc sensor supports the quality welding on the base of the measured welding current difference between the up and down points. The sensor sensibility (parameters) determines the unsuitable joint length. For the quality assurance, this correction length needs to be short, because this part of the joint is waste. The relationship between the sampling interval shown in

Table 2. The correction as a function of the test sampling interval

Sampling Interval (ms)	Welding Speed (cm/min)	Frequency (Hz)	Mistake (mm)	Correction (mm)
2	45	2.5	6	55.15
4	45	2.5	6	51.04
7	45	2.5	6	47.83
10	45	2.5	6	46.34

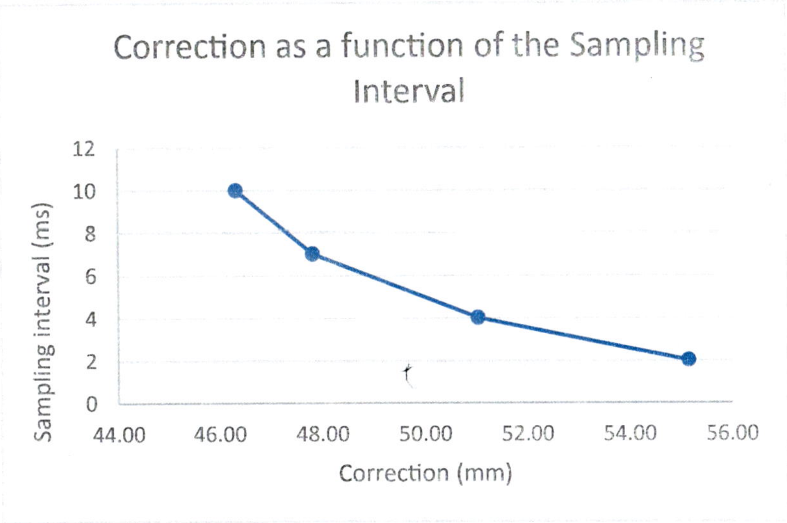

Fig. 7. Relationship between the sampling interval and the correction

Table 3. The correction as a function of the frequency

Sampling Interval (ms)	Welding Speed (cm/min)	Frequency (Hz)	Mistake (mm)	Correction (mm)
1	45	1.5	6	57.4
1	45	2.5	6	46.77
1	45	3.5	6	48.69
1	45	4.5	6	45.42
1	45	5	6	41.37

(Fig. 7). and the dependence from the frequency shown in (Fig. 8). It can see (Fig. 7 and Fig. 8). that the parameters (sampling interval, frequency) determine the correction (waste joint length).

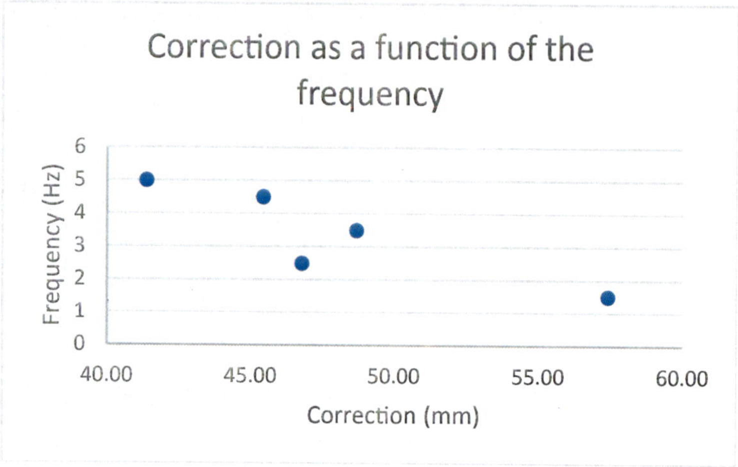

Fig. 8. The correction as a function of the frequency

Hence on the base of the experimental results, it can compare and see the most effective parameters of the arc sensor are the frequency and the sampling interval.

4 Conclusions

The goal of this research, to tests the seam tracking arc sensor, using the different parameters to increase the efficiency of the arc welding robot supported by the arc sensor. Experimental performance familiarized with the parameters of welding speed, frequency, the mistake, and sampling interval that can allow improving the quality of the welding. To get more effective results, it can be changed more than one parameter at the same time.

Welding quality is assured with the usage of arc sensor which is also proved by this research. So, it was observed that the changing value mistake parameter affects the quality of welding. The purpose of this research to prove how the parameter of welding can affect the quality of the welding.

It can conclude in the case of the different frequency parameter of the arc sensor can measure different correction size. About the welding joint quality, the smallest correction is suitable. Also, the sampling interval can affect the correction size changing. It can earn the best joint quality (means the smallest correction size) in the case of the highest frequency and high sampling interval behind the used welding speed and the used materials with Motoman welder robot.

Summarize the results concluded, that the welding quality (waste joint length) in the case of the arc sensor supported robot welding depends on the arc sensor parameters.

Acknowledgement. The authors acknowledge the financial support of this work by the Hungarian State and the European Union under the EFOP-3.6.1-16-2016-00010 project.

References

1. Ding, D., et al.: Towards an automated robotic arc-welding-based additive manufacturing system from CAD to finished part. Comput. Aided Des. **73**, 66–75 (2016). https://doi.org/10.1016/j.cad.2015.12.003
2. Mathivanan, A., Senthilkumar, A., Devakumaran, K.: Pulsed current and dual pulse gas metal arc welding of grade AISI: 310S austenitic stainless steel. Defence Technol. **11**(3), 269–274 (2015). https://doi.org/10.1016/j.dt.2015.05.006
3. Chen, X., Jie, Yu.: Acquisition and optimization of weld trajectory and pose information for robot welding of spatial corrugated web sheet based on laser sensing. Int. J. Adv. Manuf. Technol. **96**(9-12), 3033–3041 (2018). https://doi.org/10.1007/s00170-018-1716-4
4. Yao, P., Zhou, K., Tang, H.: Effects of operational parameters on the characteristics of ripples in double-pulsed GMAW process. Materials **12**(17), 2767 (2019). https://doi.org/10.3390/ma12172767
5. Yang, L., et al.: A welding quality detection method for arc welding robot based on 3D reconstruction with SFS algorithm. Int. J. Adv. Manuf. Technol. **94**(1-4), 1209–1220 (2018). https://doi.org/10.1007/s00170-017-0991-9
6. Kafi, A., et al.: Robots application for welding. Műszaki Tudományos Közl. **12**(1), 50–54 (2020). https://doi.org/10.33894/mtk-2020.12.07
7. Swevers, J., Verdonck, W., De Schutter, J.: Dynamic model identification for industrial robots. IEEE Control Syst. Mag. **27**(5), 58–71 (2007). https://doi.org/10.1109/mcs.2007.904659
8. Heyer, C.: Human-robot interaction and future industrial robotics applications. In: 2010 IEEE/RSJ International Conference on Intelligent Robots and Systems. IEEE (2010). https://doi.org/10.1109/IROS.2010.5651294
9. Chen, Y., Dong, F.: Robot machining: recent development and future research issues. Int. J. Adv. Manuf. Technol. **66**(9-12), 1489–1497 (2013). https://doi.org/10.1007/s00170-012-4433-4
10. Wallén, J.: The History of the Industrial Robot. Linkoping University Electronic Press, Linkoping (2008)
11. García Marín, J.A.: New concepts in automation and robotic technology for surface engineering. (2010)
12. Réti, T., Felde, I., Réger, M., Tóth, L., Fried, Z.: A probabilistic model for predicting grain growth process in alloys. In: IOP Conference Series: Materials Science and Engineering, vol. 572, no. 10, pp. 1–10 (2019). https://doi.org/10.1088/1757-899X/572/1/012041
13. Tóth, L.: Examination of the properties and structure of tool steel EN 1.2379 due to different heat treatments. Eur. J. Mater. Sci. Eng. **3**(3), 165–170 (2018)
14. Dima, A., Tóth, L., Badarau, G., Minea, A., Vizureanu, P.: Experimental determination of optimum size of the austenitic grain by means of the heat treatment cycles in laboratory conditions. Institutul Politechnic DIN IASI Buletinul Sectia Stiinta Si Ingineria Materialelor **44**(1-4), 125–133 (1998)
15. Réger, M., Füredi, E., Tóth, L.: Relationship between the conditions of solidification and the primary microstructure in the case of steels. In: Conference on Heat Treating, pp. 70–73 (1992)
16. Chen, H., et al.: A robust visual servo control system for narrow seam double head welding robot. Int. J. Adv. Manuf. Technol. **71**(9-12), 1849–1860 (2014). https://doi.org/10.1007/s00170-013-5593-6
17. Ma, H.: Robot welding seam tracking method based on passive vision for thin plate closed-gap butt welding. Int. J. Adv. Manuf. Technol. **48**(9-12), 945–953 (2010). https://doi.org/10.1007/s00170-009-2349-4

18. Xu, P., et al.: A visual seam tracking system for robotic arc welding. Int. J. Adv. Manuf. Technol. **37**(1-2), 70–75 (2008). https://doi.org/10.1007/s00170-007-0939-6
19. Xu, Y., et al.: Research on the real-time tracking information of three-dimension welding seam in robotic GTAW process based on composite sensor technology. J. Intell. Robot. Syst. **68**(2), 89–103 (2012). https://doi.org/10.1007/s10846-012-9672-y
20. Ye, Z., et al.: Passive vision based seam tracking system for pulse-MAG welding. Int. J. Adv. Manuf. Technol. **67**(9-12), 1987–1996 (2013). https://doi.org/10.1007/s00170-012-4625-y
21. Qin, J.: Commande hybride position/force robuste d'un robot manipulateur utilisé en usinageet/ou en soudage. Diss (2013)
22. Kadir, H., et al.: Programming a six-axis motoman HP3C robot for industrial sorting application. In: 2015 ASEE Annual Conference & Exposition (2015)
23. Wang, K.: Modélisation d'un Robot Manipulateur en vue de la Commande Robuste en Force Utilisé en Soudage FSW. Diss. Paris, ENSAM (2016)
24. Operator's Manual for General Purpose, Motoman NX100 Controller, Motoman (2007)
25. DX100 Leinonen, Jere. Hitsausrobotin etäohjelmoinnin käyttöönotto (2019)
26. DX100 Options Instructions for arc sensor comarc function, YASKAWA Electric corporation (2010)

Experimental Study of Electron Beam Welding of Inconel Alloy

Ingrid Kovaříková, Beáta Šimeková, Ján Urminský, Pavel Kovačócy, and Erika Hodúlová(✉)

Faculty of Material Science and Technology in Trnava, Slovak University of Technology in Bratislava, Bratislava, Slovak Republic
{ingrid.kovarikova,beata.simekova,jan.urminsky,pavel.kovacocy, erika.hodulova}@stuba.sk

Abstract. The scope of this study is to ascertain the weldability of Inconel 625 alloy sheets using an electron beam welding method. Weld joints of the Inconel alloy sheets 2.0 mm thick were welded by an electron beam without an additional material at a flat position. The influence of electron beam welding parameters on weld quality and mechanical properties of test joints was studied. The study of quality and mechanical properties of the joints were determined by metallographic evaluation, tensile and hardness tests.

Keywords: Electron beam · Inconel alloy · Weldability · Fusion zone · Mechanical properties

1 Introduction

The automotive industry relies on the use of top-quality materials to ensure that it meets the rigorous standards within this essential industry. As such, nickel-based alloys are often employed for a range of applications where high performance and reliability are required. The automotive industry is under pressure to reduce air pollution and create fuel-efficient engines, material selection is a key part of the design and engineering process. Inconel has myriad applications within the production of motor vehicles, thanks to their exceptional combination of strength and corrosion resistance. Inconel alloys, blending high levels of nickel and chromium with significant other elements like molybdenum and iron, are proven to provide outstanding corrosion resistance in even the most challenging environments, as well as the ability to more than withstand both the high temperatures and petrochemicals present within an engine [1–5].

For more high-performance vehicles and in motorsports engines, Inconel alloy 625 is a suitable option. This versatile Grade is used across a wide range of industries thanks to its high strength, excellent fabricability and outstanding corrosion resistance. It is known for its ability to help solve a variety of design and application problems, making it a viable option for many uses within the automotive industry. While it is used for a number of different applications in the sector, it is particularly suited to exhaust couplings [3–5].

K. Jármai and K. Voith (Eds.): VAE 2020, LNME, pp. 517–525, 2021.
https://doi.org/10.1007/978-981-15-9529-5_45

Various methods to join Inconel have been previously investigated. Laser beam welding and EBW are the most widely researched joining methods for conventionally manufactured Inconel [6–8].

Many attempts have been made to join Inconel using an electron beam in a vacuum. Two types of specimens were tested: the as-received and the solution heat treated and aged. Although the HAZ in solution pre-treated samples could be identified with microhardness measurement, the surrounding HAZ in as-received and precipitation pre-treated samples could not be distinguished from the central weld zone. Some researchers depicted that introduction of electron beam oscillations reduced niobium segregation, therefore, improving the weld quality. The experimental Inconel specimens were first welded at heat inputs of 50 J/mm in oscillated conditions and then subjected to three different post-weld heat treatments [9–13].

Other researchers showed that increasing heat input would increase the width of the welds along with a reduction in micro-cracks in the weld and HAZ. Variation in lower values of heat inputs had minimal effect on the mechanical properties of the weld. The best mechanical properties of welded specimens were obtained after EBW at the lowest heat input of 36 J/mm. Other researchers investigated the weld geometry due to different base metal conditions. The weld geometry transitioned from stemless wine glass shape to nail head shape when the sample was pre-heated at 1100°C for one hour followed by air cooling [14–16].

The aim of the experiment was to find suitable parameters for the Inconel 625 weld joint created by the electron beam method of welding. An important indicator during welding was the qualitative properties of the joint and suitable mechanical properties. The constant electron beam current, focusing current and accelerating welding voltage and the changing welding speed was applied as welding parameters for joints creation. The effects of these parameters on joints properties and quality are studied.

2 Experimental Materials and Methods

Inconel 625 is a high-performance nickel-chromium-molybdenum alloy known for its high level of strength, temperature resistance, and corrosion resistance. This superalloy is composed mainly of nickel (58% min.) followed by chromium, and molybdenum, niobium, iron, tantalum, cobalt, and trace amounts of manganese, silicon, aluminum, and titanium. The strength of Inconel 625 lies not only in its nickel-chromium base but also in the hardening mechanism of niobium and molybdenum. The alloy matrix is strengthened by the interaction of niobium with molybdenum that offers high strength without the need for precipitation-hardening treatment. Inconel 625 was designed to have better weldability than earlier alloys, with no signs of cracking when exposed to strain and temperature changes post-welding. Its high creep resistance and yield strength make this superalloy a good choice for tubes, piping, and plant equipment that require welding [3–5]. The chemical composition of used base material and the mechanical properties are shown in Table 1, 2.

The experimental materials Inconel 625 was cut to the dimensions of 100x50x2 mm. The samples were cleaned in acetone and dried. The used robotized device for electron beam welding (Fig. 1) is equipped with two electron beam cannons with the power of

Table 1. The chemical composition of test material Inconel 625 [wt.%]

Ni	C	Si	S	Cr	Mo	Fe
67.24	0.023	0.057	0.001	20.72	8.03	0.319
Al	Co	Cu	Nb	Ta	Ti	Mg
0.099	0.087	0.024	3.09	0.045	0.227	0.038

Table 2. The mechanical properties of test material Inconel 625

Tensile strength [MPa]	Yield strength [MPa]	Elongation [%]	Hardness [HB]
827	414	30	240

30 kW, a vacuum chamber with dimensions of 1500 x 1500 x 2500 mm and the pumping system allowing reaching the vacuum of 10^{-2} Pa within 25 min. The control system allows creating profiles of welding parameters and saving them.

Fig. 1. Electron beam welding equipment

The welding parameters used for the weld joints production, such as accelerating voltage, beam current, focusing current and welding speed are listed in Table 3.

Evaluation of the weld joints quality was performed by metallographic analyses. The samples were prepared by grinding, polishing and etching in a solution of 10% H_2CrO_4 for a time of 30 s. Macrostructural and microstructural analyses were carried out by optical microscopy. Microhardness of the prepared weld joints was measured using the IndentaMet 1100 equipment. Tensile tests were performed at room temperature using a Tinius Olsen 300ST tensile test machine operating at a test speed of 500 mm/min with a capacity of 300 kN according to the STN EN ISO 6892 – 1. The test specimens had a dimension of 100 × 12 × 2 mm.

Table 3. The welding parameters used for the weld joints production

Sample	Beam current I_f [mA]	Welding speed [mm/s]	Accelerating voltage U_a [kV]	Focusing current F [mA]	Heat input [J/mm]
1	70	30	55	890	128
2	60	30	55	890	110
3	50	30	55	890	92
4	40	30	55	890	73

3 Results and Discussion

The macrostructure of specimens is documented in Fig. 2. The macroscopic analysis shows no difference between the heat affected zone (HAZ) and the base material (BM), the HAZ was insignificant and there was no phase change. The grains in the weld metal are larger due to the formation of the columnar dendrites during solidification, in particular, the longitudinal appearance of the grains is longer than in the base material. It is possible to observe the epitaxial grain growth, the grain boundaries in the weld metal (WM) are directly related to the grain boundaries of the base material.

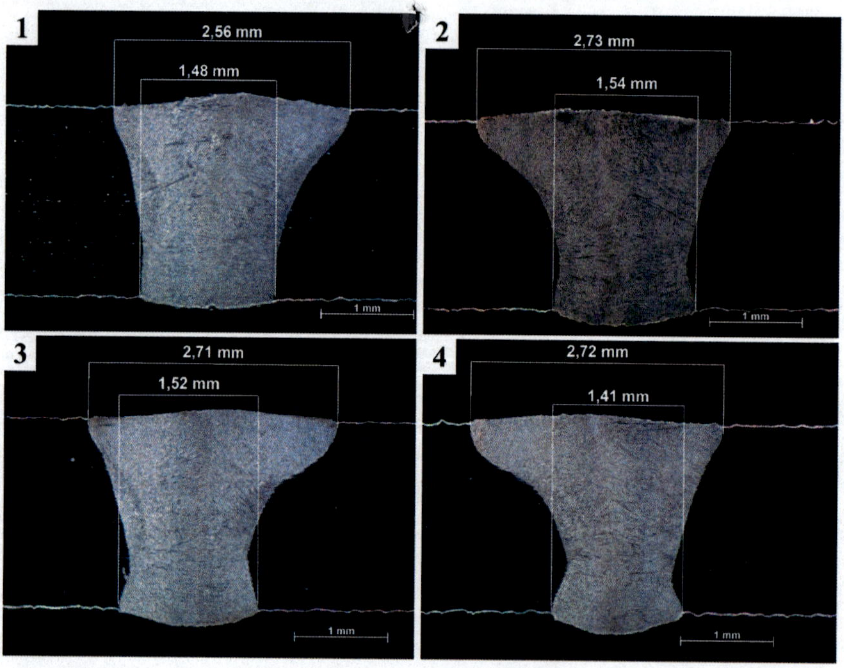

Fig. 2. Cross-section of weld joints

The measurement of weld geometry was provided. Table 4 shows the geometric parameters of the produced welds depending on the electron beam current and the welding speed. The weld geometry measuring was focused on the weld width, weld root width.

Table 4. Values of a measured geometric parameter of welds dependence to welding parameters

Sample	Beam current I_f [mA]	Welding speed [mm/s]	Weld width [mm]	Weld root width [mm]
1.	70	30	2.56	1.48
2.	60	30	2.73	1.54
3.	50	30	2.71	1.52
4.	40	30	2.72	1.41

The geometrical parameters of welds show that the width of the weld is increasing with the welding current at a constant welding speed, and the root width of the weld is decreasing with the welding current at a constant welding speed. The weld width is directly related to the amount of heat input introduced into the material. The HAZ zone is very narrow, it is hardly visible on a macroscopic image, it is not possible to determine whether there was a phase transformation in the heat affected zone. In the ratio for sample No. 1, it is possible to observe high width of weld and also root in size. With the decreasing of the welding current, the root width is also decreasing at constant weld speed. No defects were observed in the weld metal.

The microstructure of weld joints is shown in Fig. 3. The detailed joint microstructure consisted of base metal (BM), heat affected zone (HAZ) and weld metal (WM).

The microstructure of the base material is formed by polyhedral nickel austenite, the grains show significant heterogeneity in grain size, which ranged from 50 to 60 μm. The HAZ microstructure consists of nickel austenite, the grain boundaries are not significantly etched. This may be due to the dissolution of precipitates at the grain boundary due to heating during welding. The weld metal microstructure has a dendritic morphology. Dendrites are formed by nickel austenite and secondary phases can be excluded in the interdendritic area. The solidification subgrains represent the finest structure. These subgrains are present as cells or dendrites and the boundary separating adjacent subgrains is known as a solidification subgrain boundary. These boundaries are evident in the microstructure because their composition is different from that of the bulk microstructure. The degree of grain growth is dependent on the starting base metal microstructure and the weld heat input. Grain growth may be minimal even under high weld heat input conditions.

Micro-hardness measurements were carried out on prepared weld joints by keeping the weld centered position. The hardness distribution of the weld cross-sections is illustrated in Fig. 4. From the hardness profile, it is evident that the hardness of weld metal was found to have the peak hardness value as compared to HAZ side. The hardness of base material was in comparison to the declared values from the producer side. The

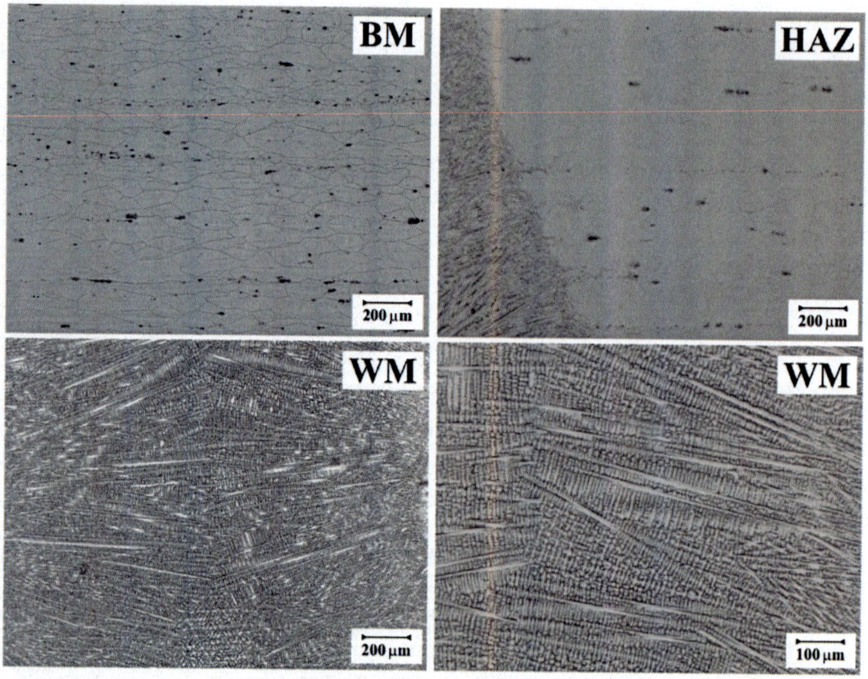

Fig. 3. Microstructure of prepared joints

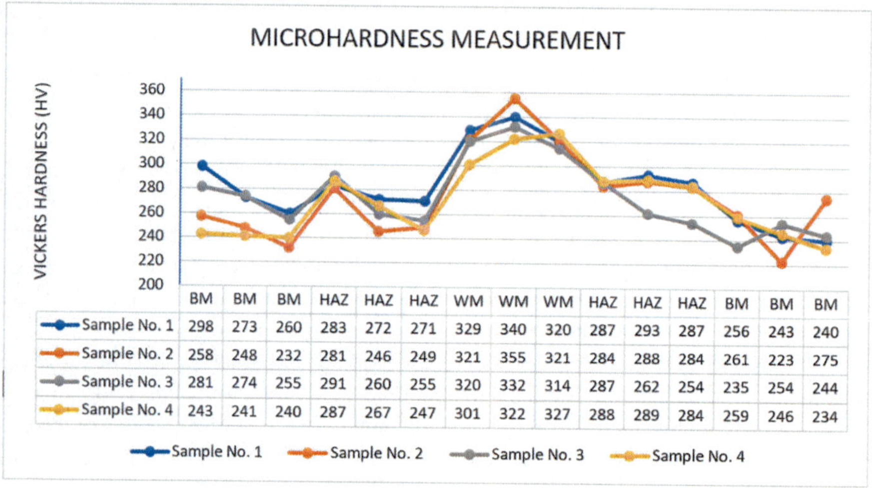

	BM	BM	BM	HAZ	HAZ	HAZ	WM	WM	WM	HAZ	HAZ	HAZ	BM	BM	BM
Sample No. 1	298	273	260	283	272	271	329	340	320	287	293	287	256	243	240
Sample No. 2	258	248	232	281	246	249	321	355	321	284	288	284	261	223	275
Sample No. 3	281	274	255	291	260	255	320	332	314	287	262	254	235	254	244
Sample No. 4	243	241	240	287	267	247	301	322	327	288	289	284	259	246	234

Fig. 4. Microhardness distribution of weld joints

values of the measured hardness of HAZ were slightly higher as hardness values of the base material.

Mechanical properties of the Inconel 625 alloy were determined by tensile tests. Representative samples used for tensile test were prepared by using the same welding parameters as in case of sample No. 1–4. Tensile test specimens were conducted on three samples of each weldment. The values of tensile test measurements are shown in Table 5. The specimens after the tensile test and load-displacement curve are illustrated in Fig. 5.

Table 5. Measured values of the mechanical properties

Specimen	Max. load [kN]	Tensile strength [MPa]	Ductility [%]
1	23.7	902	23
2	23.6	884	25
3	23.4	880	26
4	23.1	875	27

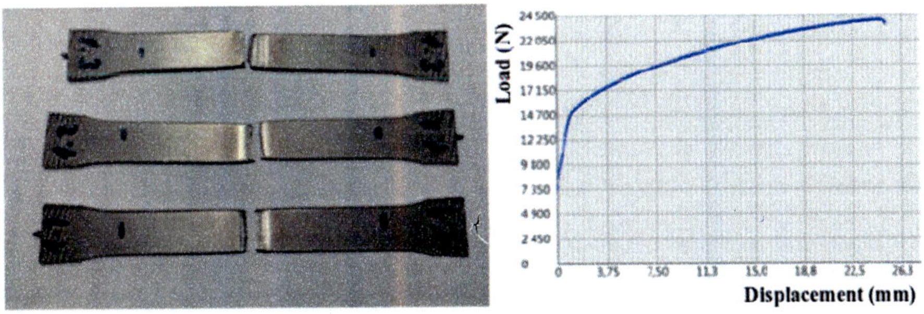

Fig. 5. The specimens after tensile test and typical load-displacement curve

The values of tensile test measurement show the influence of heat input to the tensile strength and ductility. A high heat input resulted in strong weld with high tensile strength but low ductility. On the other hand, the low heat input brings low tensile strength but high ductility. In contrast, a high heat input resulted in brittle fracture inside the weld zone. In case of sample No. 1, the fracture of tensile test specimen occurred in the weld zone. A low heat input resulted in strong welds since the ductile fracture surface occurred in the base material and outside of the weld zone.

4 Conclusion

Weldability of Inconel 625 alloy sheets welded by an electron beam without an additional material at a flat position was studied in this work. The welding parameters used for the weld joints production, such as accelerating voltage, focusing current and welding speed were constant and the beam current was changing.

The heat input was in the range of 128–73 J/mm, which has an influence on the weld geometry. The weld width is directly related to the amount of heat input introduced into the material. The high heat input indicates the high weld width and high weld root width.

On the other hand, the low heat input exhibit higher weld width in comparison of weld root with.

The microstructure of weld metal is characterized by the dendritic morphology. Dendrites are formed by nickel austenite and secondary phases can be excluded. The fine structure is representing by the solidification subgrains. The size of grains is dependent on the starting base metal microstructure and the weld heat input.

The hardness of weld metal was found to have the peak hardness value as compared to HAZ and base metal side. The tensile test measurement resulted in the influence of high heat input to the high tensile strength and low ductility.

Acknowledgments. This research is supported by VEGA 1/0091/17 which are supported by Slovak Republic Ministry of Education.

References

1. ASM Handbook: Nickel, Cobalt and heir alloys, ISBN: 978–0-87170-685-0, ASM International, USA (2000)
2. Kumar, G.K., et al.: Thermogravimetric, XPS, AES and EDS studies on the oxidation and nitridation of Inconel 625, Advances in physical metallurgy. In: Proceedings on the International conference on Advances in Physical Metallurgy, ICPM-1994, pp. 375 – 387 (1994)
3. Eiselstein, H.L., Tillack, D.J.: The invention and definition of alloy 625. Superalloys **718**(625), 1–14 (1991)
4. DuPont, J.N., Lippold, J.C., Kiser, S.D.: Welding Metallurgy and Weldability of Nickel-base Alloys. John Wiley & Sons, Inc., Hoboken (2009)
5. Caron, J.L., Sowards, J.W.: Weldability of nickel-base alloys. Compr. Mater. Process. **6**, 151–179 (2014)
6. Vivek, P., Akash, S., Hyderb, J., Corlissb, M., Hyderb, D., Hunga, W.: Electron beam welding of inconel 718. Procedia Manuf. **48**, 428–435 (2020)
7. Sun, J., Ren, W., Nie, P., Huang, J., Zhang, K., Li, Z.: Study on the weldability, microstructure and mechanical properties of thick Inconel 617 plate using narrow gap laser welding method. Mater. Des. **175**, 107823 (2019)
8. Janaki Ram, G.D., Venugopal Reddy, A., Prasad Rao, K., Reddy, G.M., Sarin Sundar, J.K.: Microstructure and tensile properties of Inconel 718 pulsed Nd-YAG laser welds. J. Mater. Process. Technol. **167**, 73–82 (2005)
9. Huang, C.A., Wang, T.H., Lee, C.H., Han, W.C.: A study of the heat affected zone (HAZ) of an Inconel 718 sheet welded with electron-beam welding (EBW). Mater. Sci. Eng. **398**, 275–281 (2005)
10. Reddy, G.M., Murthy, C.V., Rao, K.S., Rao, K.P.: Improvement of mechanical properties of Inconel 718 electron beam welds—influence of welding techniques and postweld heat treatment. Int. J. Adv. Manuf. Technol. **43**, 671–680 (2009)
11. Agilan, M., Venkateswaran, T., Sivakumar, D., Pant, B.: Effect of heat input on microstructure and mechanical properties of inconel-718 EB welds. Procedia Mater. Sci. **5**, 656–662 (2014)
12. Mei, Y., Liu, Y., Liu, C., Li, C., Yu, L., Guo, Q., Li, H.: Effect of base metal and welding speed on fusion zone microstructure and HAZ hot-cracking of electron-beam welded Inconel 718. Mater. Des. **89**, 964–977 (2016)
13. Sharma, S.K., Agarwal, P., Majumdar, J.D.: Studies on electron beam welded Inconel 718 similar joints. Procedia Manuf. **7**, 654–659 (2017)

14. Peng, G., Zhang, K.F., Zhang, B.G., Jiang, S.S., Zhang, B.W.: Microstructures and high temperature mechanical properties of electron beam welded Inconel 718 superalloy thick plate. Trans. Nonferrous Met. Soc. Chin. **21**, 315–322 (2011)
15. Ram, G.D., Reddy, A.V., Rao, K.P., Reddy, G.M.: Microstructure and mechanical properties of Inconel 718 electron beam welds. Mater. Sci. Technol. **10**, 1132–1138 (2005)
16. Tucho, W.M., Cuvillier, P., Sjolyst-Kverneland, A., Hansen, V.: Microstructure and hardness studies of Inconel 718 manufactured by selective laser melting before and after solution heat treatment. Mater. Sci. Eng. **689**, 220–232 (2017)

Investigation of Thermal Effects of Flame Straightening on High-Strength Steels

László Gyura[1](✉), Marcell Gáspár[2], and András Balogh[2]

[1] Linde Gas Hungary Ltd., Budapest, Hungary
laszlo.gyura@linde.com
[2] Faculty of Mechanical Engineering and Informatics, Institute of Material Science and Technology, Miskolc, Hungary

Abstract. The use of high-strength steels in the automotive industry is increasing. In many cases, the use of flame straightening to reduce deformation after welding is unavoidable in the manufacture of trailers, semitrailers, heavy vehicles, earthmoving machinery, military bridges etc. Due to the not very concentrated but relatively high temperature heat source, the process can cause significant changes in the microstructure which can endanger the safe use of these steels. This may be particularly true for the high-strength steels tested, for which we have very little experience and concrete measurement results. Due to the different thermal-physical properties of the flammable gases, the resulting heat effect varies depending on the gas and technology used. Nowadays, there is a lack of studies that analyse the effect of these types of heat cycles. During our experiments, we investigate the changes of the microstructure and mechanical properties caused by heat effect on unalloyed structural and high strength steels (S355J2+N, S690QL). The situation is complicated by the fact that manual technology typically also carries a high risk of local overheating, which can cause heat effects that are too long in time and/or too high temperature. In addition to the direct thermal effect study, a Gleeble 3500 thermomechanical physical simulator was used to perform thermal cycles measured during the technology. Two heating flames (acetylene/oxygen, propane/oxygen), three characteristic peak temperatures (1000 °C, 800 °C and 675 °C) and two types of cooling conditions (air cooling and intensive water cooling) were studied. Both the real direct thermal effect study and physical simulation showed a clear negative effect of overheating and intensive water cooling for the exanimated steels.

Keywords: Flame straightening · Thermal cycles · High-strength steels · Gleeble simulation · Materials testing

1 Introduction

High and ultra-high strength steels are playing important role in the materials used in the automotive industry. The processing of these materials, including the use of welding and related technologies, is highly different from the technology used in conventional steels [1].

K. Jármai and K. Voith (Eds.): VAE 2020, LNME, pp. 526–538, 2021.
https://doi.org/10.1007/978-981-15-9529-5_46

Some of the flame technologies, such as flame straightening, are also applied in welded structure production of today's vehicle. In general the flame straightening after welding is an essential part of manufacturing process. The selected devices, the burning and oxidizing gases for the technology, it is especially important to determine the technology parameters (e.g. heating temperature, holding time, cooling method, etc.), which in many cases is not so clear and easy to comply as the welding processes.

The compliance of the technological parameters can be especially important in the case of high-strength steel structures since improper technology (overheating, prolonged heat, sudden cooling, etc.) can lead to changes in the microstructure and properties of the material.

2 Flame Straightening

2.1 Principle of the Flame Straightening

Flame straightening is based on the physical principle that metals expand when heated and shrink when cooled. If the deformation of the structure is prevented during the heating (heat effect), compressive stress develops in the heated zone, which forces the structure to remain deformed. This residual deformation can be used to form a conscious geometry (e.g. bends, deformations), also to restore the geometry of a distorted structure formed during manufacture (flame straightening).

Based on the rate of heat input, flame straightening can be divided into two cases. A very common technology used is partial heating of the surface layer (typically up to 30–35% of the total cross-section) relative to the total cross-section of the structure. The rate of heat input in this case is small relative to the material thickness of the workpiece, and the cooling rate is typically high. When the entire cross-section of the workpiece is locally heated, a relatively high heat input and a low cooling rate can be expected [2].

2.2 Main Features of Flame Straightening Technology

The effect of flame straightening on the material structure is influenced by several factors, the most important of which are:

- type of burner used (power, industrial gases used, etc.),
- the peak temperature reached during flame straightening,
- the size and shape of the heated area,
- the established cooling conditions (structure size, cooling method, etc.).

The power of the burner used for the technology is determined by the base material and the thickness of the plate. Conventional gas/flame welding torches (welding tip) can be used to straighten smaller, thin plate structures, but large structures may require special high-performance, multi-flame torches.

In practice, acetylene and oxygen gas are used to operate the burners, but instead of acetylene, so-called slow burning gases (e.g. propane) can be also applied. In this case, the heating time is significantly longer, and the width of the heat-affected zone is also larger.

Due to the typically manual technology, keeping the correct temperature of technology is a difficult task and requires a lot of practice. The technical report CEN/TR 10347: 2006 "Guidelines for the processing of structural steels" helps to determine the flame heating temperature (Table 1).

Table 1. Recommendation for maximum flame straightening temperatures [3].

Delivery condition	Short term, surface heating	Short term, full cross section heating	Long term, full cross section heating
Normalized, unalloyed steels to 355 MPa strength	≤900 °C	≤700 °C	≤650 °C
TMCP steels to 460 MPa strength	≤900 °C	≤700 °C	≤650 °C
TMCP steels between 500–700 MPa strength	≤900 °C	≤600 °C	≤550 °C
Q+T high strength steels (e.g. S690QL, S960QL)	≤generally, at 20 °C under the tempering temperature of the selected material (around 530 °C)		
TMCP: Thermo-Mechanical Control Process			
Q+T: Quenched and Tempered			

For full cross section heating, typically not only temperature maximization but also holding time can be important. Prolonged holding at high temperatures allows diffusion processes to occur, which can cause particle precipitation, change in microstructure, and grain coarsening, thereby significantly degrading the originally set properties of the steel [4].

In addition to the heating characteristics, the cooling conditions are determinants of the processes taking place in the material structure. In industrial practice, in order to speed up the technology, intensive cooling with water (possibly blown/compressed air) is often used as opposed to the cooling conditions at rest recommended by the literature, which can have particularly critical consequences for high-strength steels.

3 Examining the Effect of Flame Straightening at the Real Heating Conditions

To investigate the effects of flame straightening, experiments were performed on 15 mm and 30 mm thick plates under different conditions. For our experiments, we chose one of the most common heating methods (line heating) for welded structures by moving the burner mechanically. The examined 300 x 300 mm plates (without weld line) were heated along its centre line with the appropriate power burners for the plate thickness (Fig. 1).

Fig. 1. Line heating of S690QL type steel with LF-H-8 acetylene/oxygen multi flame burner.

3.1 The Tested Materials

Nowadays, experiments were performed on conventional S355J2+N fine-grained steel and a high-strength steel, which are often used to produce vehicle structures. One of the most common applications of S690QL high-strength steel is the extendable steel structure of mobile cranes, but it is also often used in the construction of frame structures for trailers, heavy trucks, which are also made by various welding processes [5].

The chemical composition of the tested 15 mm thick conventional steel and the 30 mm thick S690QL steel are shown in Table 2.

Table 2. Chemical composition of base materials.

[%]	C	Si	Mn	P	S	Al	B	Cr
S355J2+N	0.183	0.35	1.55	0.014	0.003	0.036	0	0.022
S690QL	0.14	0.22	1.19	0.008	0.001	0.087	0.0025	0.32
[%]	Cu	Mo	Nb	Ni	Ti	V	Ce (IIW)	CET
S355J2+N	0.012	0.005	0.005	0.045	0.003	0.005	0.451	0.341
S690QL	0.04	0.3	0.027	0.05	0.006	0	0.468	0.308

3.2 Thermal Loading of the Tested Steels According to the Technology Used in Practice

On the centre line of the tested steel test pieces (300 $\times$ 300 mm) were heated under the conditions shown in Table 3. Heating was performed also with acetylene/oxygen and propane/oxygen burners, with different burning distances between the tip of burner and the plate surface, and in both cases, air and intensive water cooling were used. The speed

Table 3. Parameters of the flame-technology trials.

Thickness [mm]	Type of burner	Pressure [bar]			Flow rate [l/h]			Speed of burner [cm/min]	Burner-distance [mm]
		O_2	C_2H_2	C_2H_2	O_2	C_2H_2	C_3H_8		
30	LF-H-8 mult i flame	4	1	–	3000	2800	–	15.5	18
	PM5H multi flame	4	–	0.8	5000	–	1600	11	28
15	Welding tip (Nr.8)	3.5	0.8	–	1500	1200	–	10	12
	Cutting nozzle (10–30)	3.5	–	0.7	2500	–	700	6	10

of movement of the burners was chosen based on preliminary measurements so that the temperature recommended for flame heating could be maintained continuously along the heating line [6, 7].

Table 4. Test results for S355J2+N type steel.

Base material: S355J2+N (t=15 mm)	Burning gas	Air cooling	Water cooling
Macroscopic sections	C_2H_2		
	C_3H_8		
Microstructure of the base material		Microstructure 0.5 mm below the surface of the heated area (in the middle of the tested piece)	
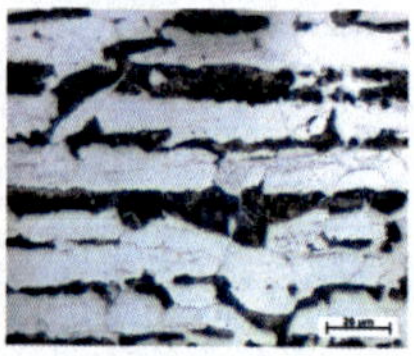 152 HV10	C_2H_2	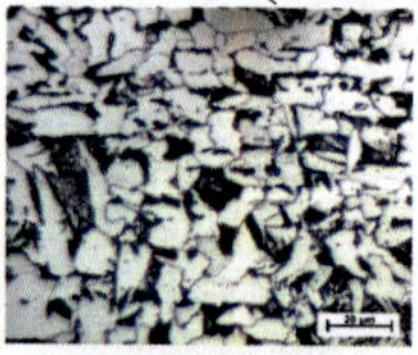 182 HV10	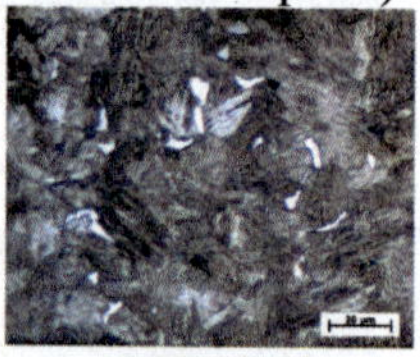 405 HV10
	C_3H_8	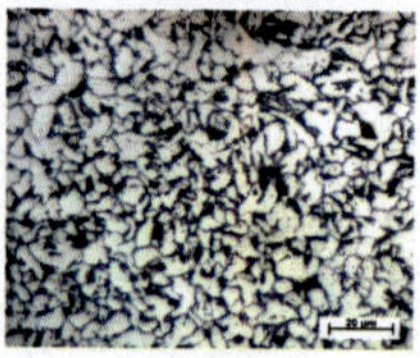 182 HV10	 188 HV10

Optical microscopic (at 0.5 mm below the surface), macroscopic examinations and hardness tests were performed at the geometric centre of the plates exposed to heat (directly in the middle of the area affected by the flame). The results obtained during the microscopic examinations were compared with the properties and characteristics of the base materials (Tables 4 and 5).

Table 5. Test results for S690QL type steel.

Base material: S690QL (t=30 mm)	Burning gas	Air cooling	Water cooling
Macroscopic sections	C_2H_2		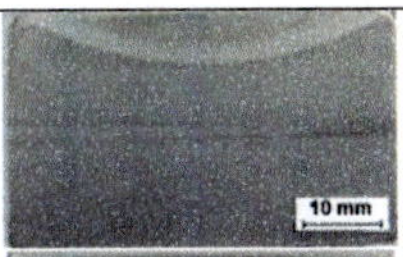
	C_3H_8		
Microstructure of the base material		Microstructure 0.5 mm below the surface of the heated area (in the middle of the tested piece)	
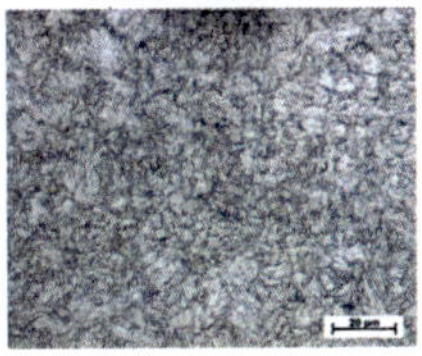 297 HV10	C_2H_2	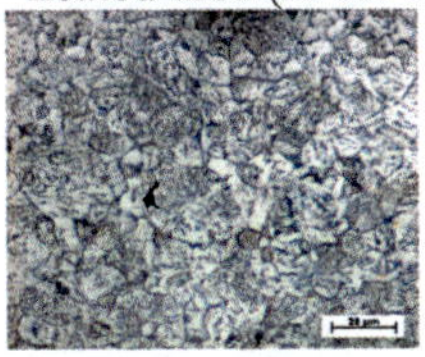 282 HV10	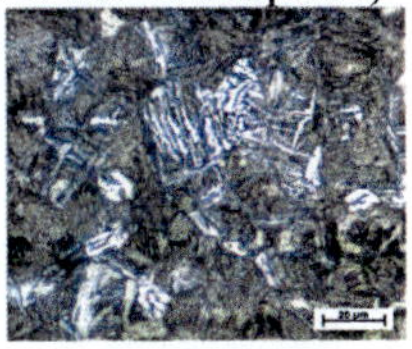 400 HV10
	C_3H_8	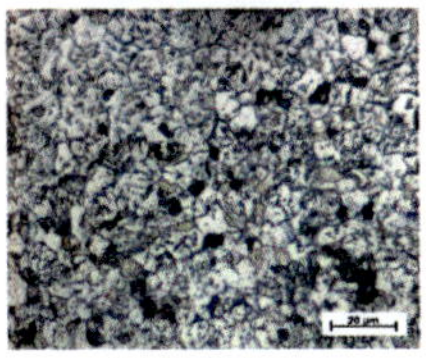 259 HV10	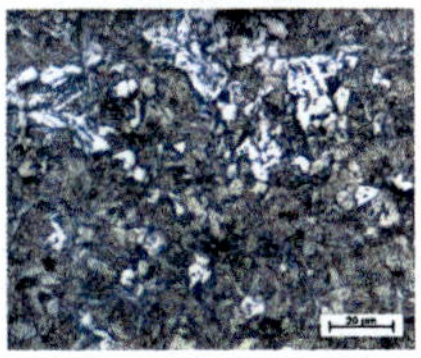 356 HV10

The rolled structure of conventional S355J2+N steel has been virtually completely transformed and recrystallized by heating. Refining of the grain was observed in all cases relative to the base material. The negative effect of intensive water cooling is mainly due to acetylene heating, which is well demonstrated by the bainite-martensitic structure and the significantly increased hardness. This hardness value exceeds the maximum value allowed for this type of steel [8], which indicates remarkable brittleness of the material.

For high-strength steel, there is much less change in microstructure when cooled in normal air. In the case of water cooling, the hardening compared to the base material is significant even in the case of propane/oxygen heating. However, the degree of hardening did not reach the maximum value of 450 HV10 given in the relevant literature [8], but it is likely that the toughness of the material drops significantly. In case of water cooling, the microstructure is especially bainite-martensitic in the presence of small amounts of ferrite.

4 Physical Simulation Studies

Based on our previous experience, physical simulation can be effectively used to study microstructural changes caused by the heat input of different material technologies [9].

There is lots of research and experience available for acceptable $t_{8/5}$ cooling times when conventional or high strength steels are welded [10, 11]. However, the principles of the $t_{8/5}$ cooling time concept adopted in the simulation of welding technology are not applicable to physical simulation tests for the flame straightening [12]. On the one hand, the heating-, and cooling times are significantly longer for large area heating than for arc welding technology. On the other hand, in many cases the maximum temperature of the treatment cannot reach the 800 to 850 °C temperature interval, so the concept of $t_{8/5}$ time cannot be interpreted.

In our experience, with a cooling time of $t_{8/5}$, it is not possible to distinguish between air cooling and intensive water cooling, especially for thicker plates. Practical intensive water cooling, especially with high-intensity acetylene/oxygen heating, typically reaches the material already when its temperature cools below 500 °C (see, e.g., Fig. 2).

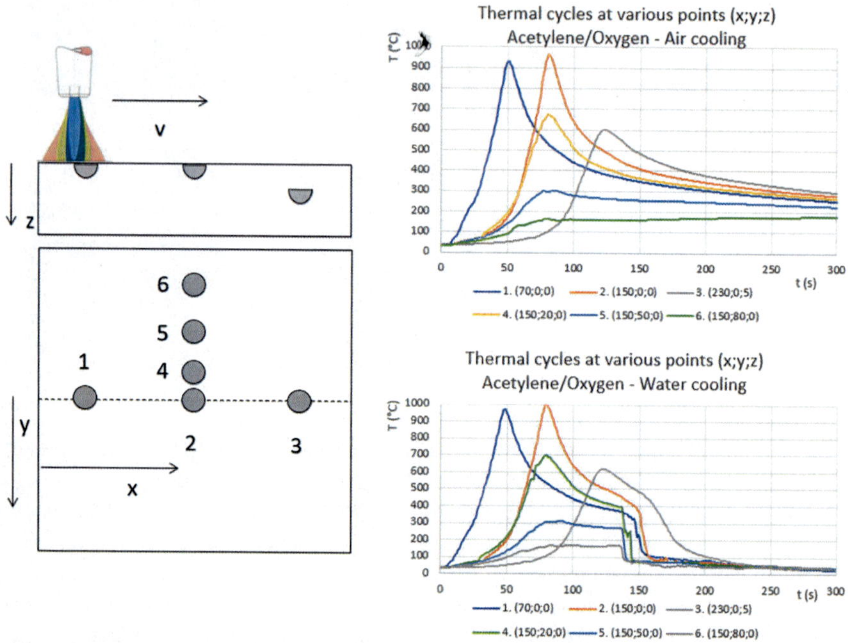

Fig. 2. Thermal cycles of 300 × 300 × 30 mm steel plate at different points of the plate, coordinates of the starting point of heating: (0;0;0) (heating burner: LF-H-8, cooling with air and water).

4.1 Measurement of Thermal Cycles

To perform the physical simulation, the thermal cycles associated with the actual heating were measured and used.

As a result of the heating, the thermal cycle formed at each point of the workpiece can be very different. In addition to the technological parameters of flame straightening, it depends on the size of the structure and the geometric location of the examined point within the structure. As an example, based on the thermocouple measurements performed on the 300 × 300 × 30 mm plates presented in the previous chapter, the thermal cycles of some points of the tested plate are shown in Fig. 2 [13]. Burners and parameters according to Table 3 were used for thermal cycle measurements.

4.2 Thermal Cycles of the Tests

For the simulation, we used the thermal cycles for both cooling in normal air and for cooling with water. The points of the workpiece exposed to heat receive different heat loads in the lateral and depth directions, moving away from the heat source, the maximum temperature reached becomes smaller and smaller, so its effect on the steel structure and mechanical properties is not the same (certainly below a certain maximum temperature negligible). Thus, our experiments were based on the temperatures A_1 and A_3, which are decisive for metallurgical processes [14]. During the simulations, our test pieces were loaded with thermal cycles below (but close to) A_1 temperature of 675 °C, between A_1–A_3 temperature of 800 °C and above A_3 (overheated) 1000 °C, in accordance with the actual thermal cycles measured in pieces.

On the left side of Fig. 3a, as an example, the programmed curves used for the simulation at the three different maximum temperatures for water cooling. On Fig. 3b, for a 30 mm plate, it shows the measured real thermal cycle corresponding to a maximum temperature of 1000 °C and the temperature of the test piece under intensive water cooling.

The thermal cycles used for the entire experimental program are summarized in Table 6.

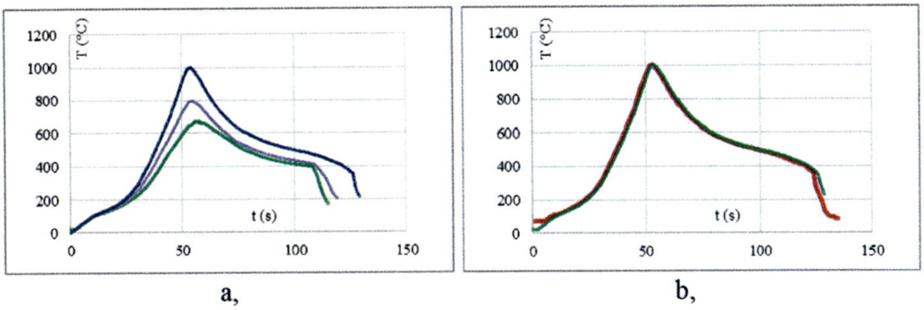

Fig. 3. Thermal cycles programmed for different maximum temperatures for acetylene/oxygen heating with water cooling (a), Programmed thermal cycle (red) and temperature change of the test piece during simulation (green) (b).

Table 6. Programmed thermal cycles and their main characteristics.

<table>
<tr><th rowspan="2">T_{max} (°C)</th><th rowspan="2">Programmed thermal cycles based on measurements on a 30 mm thick plate (red: acetylene/oxygen heating, blue: propane/oxygen heating) - continuous line: air cooling, dot line: water cooling)</th><th rowspan="2">Heating time (s) (C_2H_2/ C_3H_8)</th><th colspan="2">Cooling time - $t_{8/5}$ (s) (C_2H_2/C_3H_8)</th></tr>
<tr><th>Air</th><th>Water</th></tr>
<tr><td>T_{max} = 675</td><td>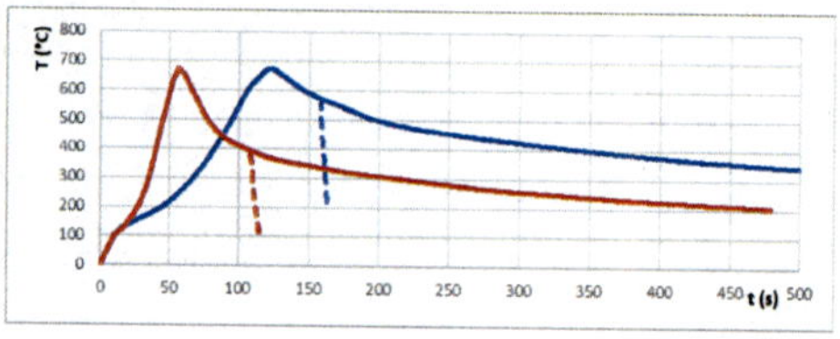
</td><td>69/123</td><td>-</td><td>-</td></tr>
<tr><td>T_{max} = 800</td><td>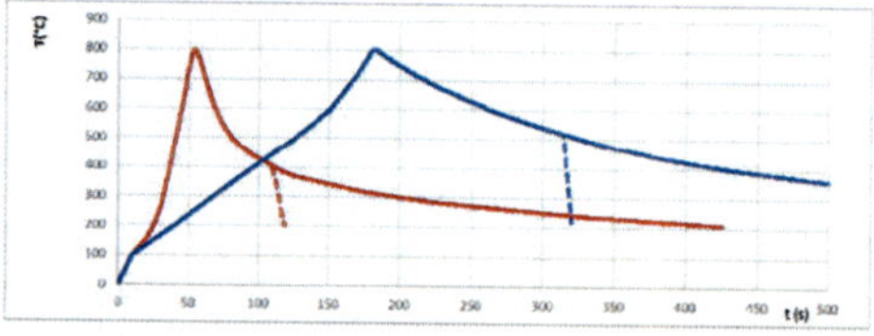
</td><td>54/182</td><td>82/148</td><td>82/132</td></tr>
<tr><td>T_{max} = 1000</td><td>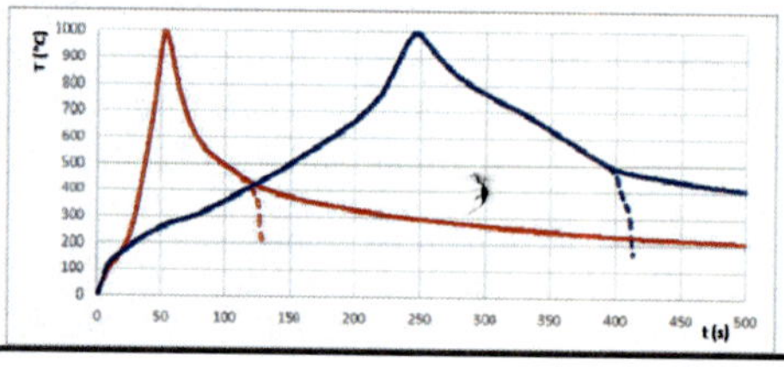
</td><td>54/246</td><td>35/105</td><td>35/105</td></tr>
</table>

4.3 Results of Simulations

With the programmed thermal cycles presented in the previous chapter, we loaded our 10 × 10 × 70 mm test pieces using a Glebble 3500 physical simulator. Hardness measurements were performed on the specimens in the heat-loaded part (centre of the test pieces) and optical microscopic tests were performed to examine the microstructure (Table 7 and 8).

Simulation tests were also performed on S355J2+N grade steel, although for 15 mm plate thickness, the simulation thermal cycles are less accurate. Based on our previous measurements and experience, both the heating time and the cooling times (especially in the case of intensive water cooling) are shorter, so the programmed thermal cycle should probably be modified slightly.

The results of the tests can be compared with the results of the real flame effect (see previous chapter). For example, a comparison of hardness values is shown in Fig. 4 and 5. In the case of conventional steel with a strength of 355 MPa, the comparison would be more correct if the actual heating had been carried out at a plate thickness of 30 mm in this case as well.

Table 7. Simulation test results for S355J2+N type steel.

T_{max} (°C)	Acetylene/Oxygen		Propane/Oxygen	
	Air cooling	Water cooling	Air cooling	Water cooling
675	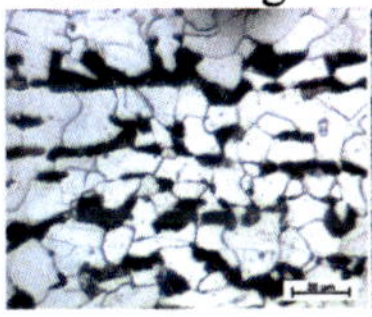 158 HV10	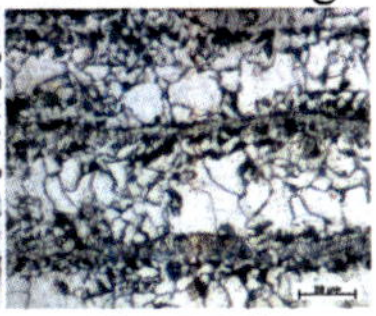 187 HV10	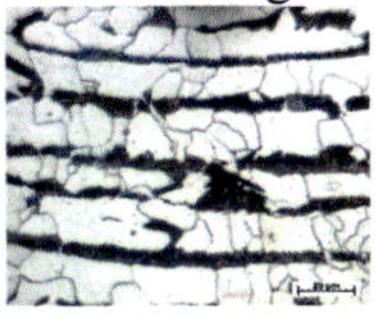 156 HV10	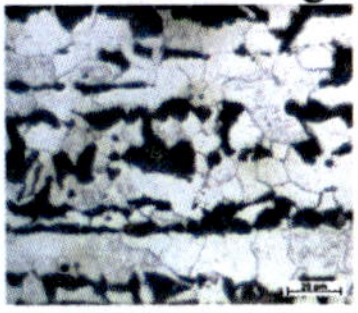 170 HV10
800	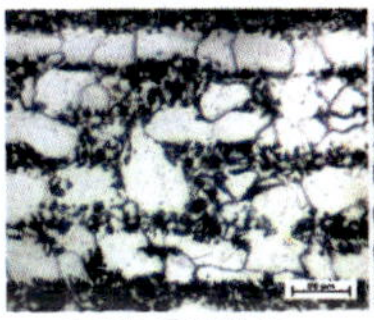 172 HV10	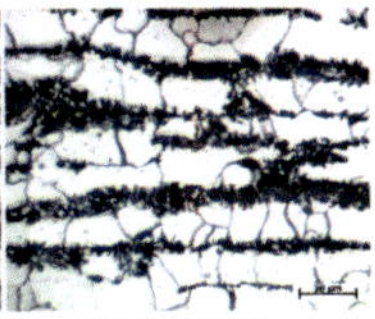 173 HV10	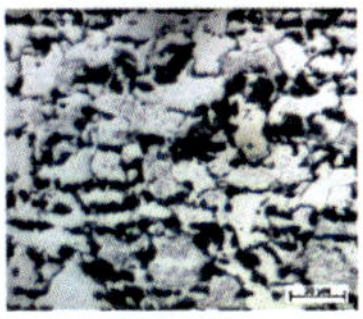 163 HV10	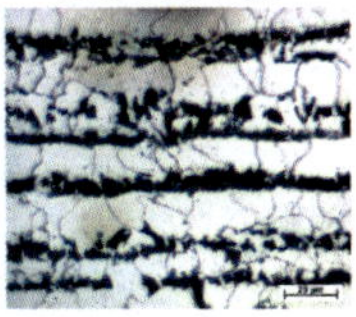 168 HV10
1000	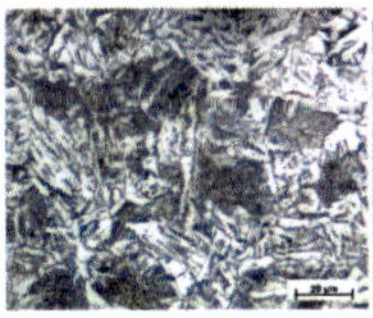 184 HV10	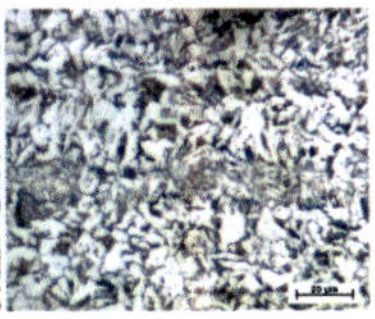 210 HV10	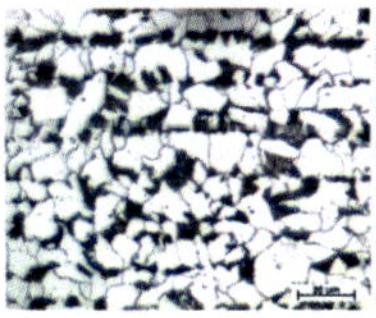 166 HV10	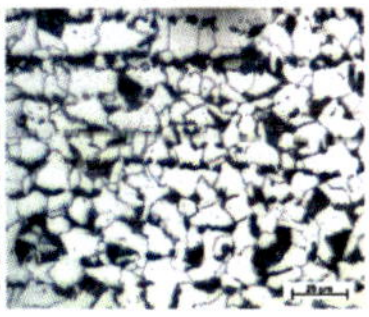 170 HV10

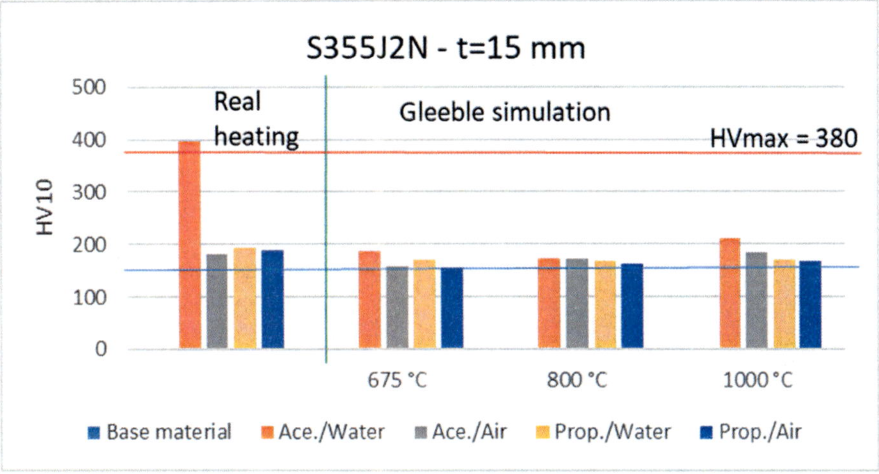

Fig. 4. Hardness values under real heating and simulation conditions for material S355J2+N.

For the S355J2+N steel, below the temperature A_1, the decomposition of perlitic parts started in all cases, although it did not finish completely. The rolled structure of the original material is still clearly visible. When water cooling was simulated, a small

amount of hardening was observed, which might be connected to the presence of a brittle microstructure in the area of the former perlite grains.

Between temperatures A_1–A_3, a significant transformation of the structure can be observed mainly during prolonged air cooling. Due to the partial and incomplete austenitization process, the complete transformation did not occur, however, the partial decomposition of the perlite can be observed. The rate of increase in hardness is virtually negligible. Above temperature A_3, similarly to the result of the real flame straightening, the original microstructure was transformed into a bainite-martensitic structure. Rapid water cooling, especially during acetylene heating, led to the formation of some martensitic islands. All these are also evident in the increase in the value of hardness. However, the hardness of the resulting structure is smaller than the same obtained with real heating. As previously indicated, this is primarily due to the temperature conditions of the actual 15 mm plate thickness and the difference in thermal cycles associated with the 30 mm plate thicknesses used for the simulation.

Table 8. Simulation test results for S690QL type steel.

T_{max} (°C)	Acetylene/Oxygen		Propane/Oxygen	
	Air cooling	Water cooling	Air cooling	Water cooling
675	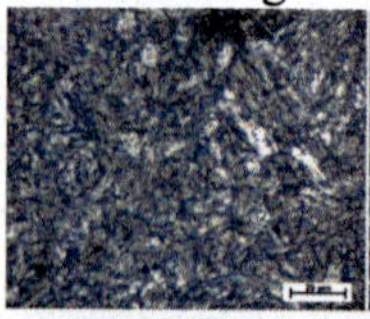 296 HV10	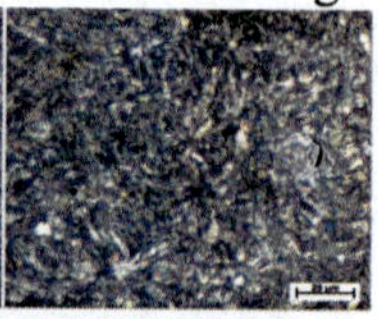 286 HV10	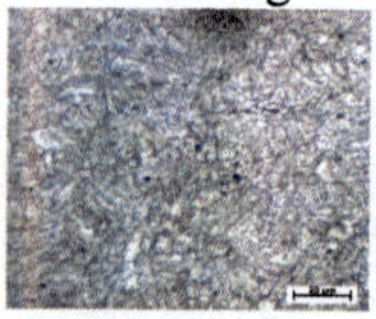 278 HV10	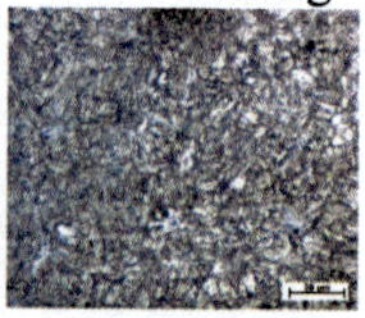 281 HV10
800	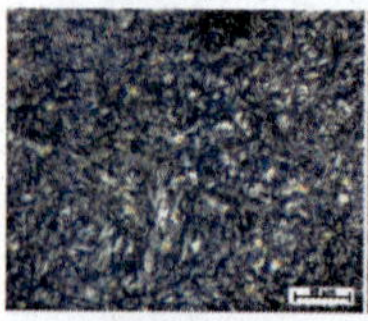 256 HV10	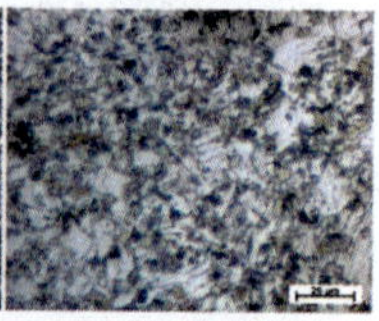 259 HV10	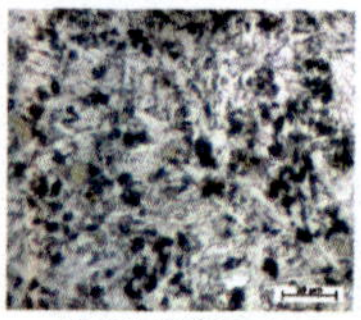 243 HV10	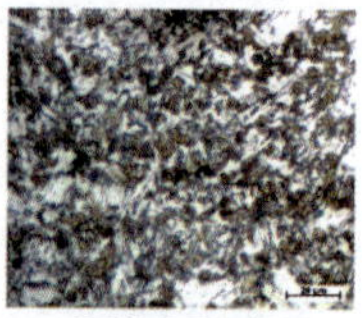 283 HV10
1000	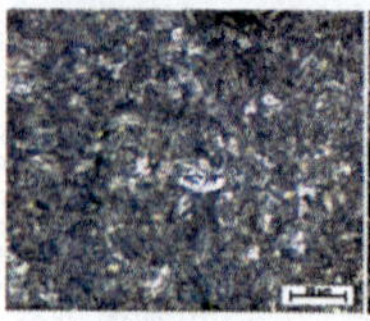 338 HV10	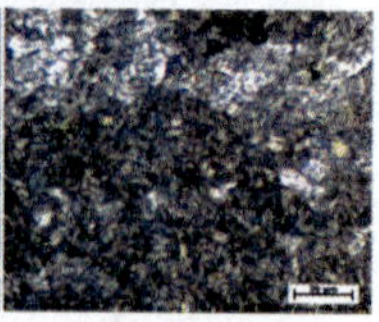 351 HV10	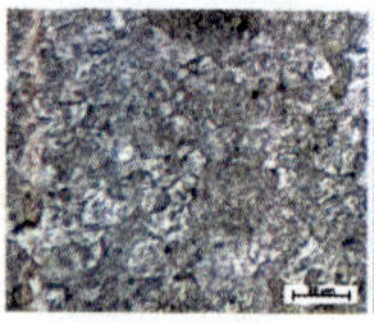 277 HV10	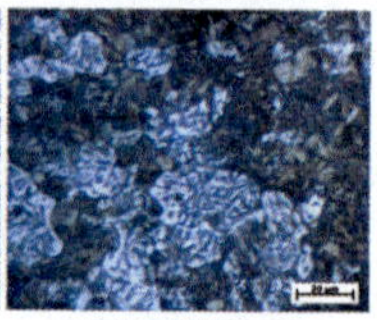 343 HV10

The high-strength S690QL steel showed a slight decrease in the hardness at 800 °C compared to the base material. With this, the strength properties of the material are also likely to reduce, the extent of which cannot be considered critical. However, studies have clearly shown the negative effects of overheating and water cooling. Although the rate of increase in hardness is not critical, it clearly indicates a likely decrease in toughness (impact work studies are ongoing), especially in the intercritically heated

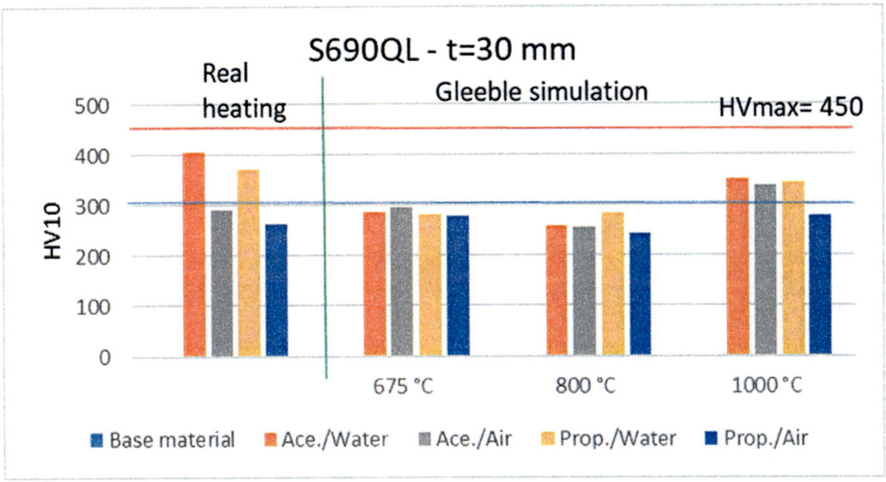

Fig. 5. Hardness values under real heating and simulation conditions for material S690QL.

parts. For vehicles where such a steel structure is subjected to significant dynamic loads, this reduction in toughness can cause a serious problem.

5 Summary

In our article, we briefly summarize the principle of flame straightening technology, its expected effects in the processing of conventional and high-strength steels. On two base materials belonging to this category of steels (S355J2+N, S690QL) we investigated the processes taking place in the heat-affected zone under real and simulated conditions. For the thermal cycles required for the physical simulation, the change in temperature over time was measured at several points on a 30 mm thick plate. Under real heating conditions, changes in the microstructure formed and the hardness of the material can even be significant. Our experiments have shown that flame straightening can have a significant effect on the original properties of the base material. Based on our experiments, we believe that heating between A_1–A_3 temperature can still be applied (under certain conditions). However, intensive water cooling can cause the creation of locally brittle parts in the structure of the material. For an overheated or water-cooled steel, these effects can reach a critical value that can be hazardous to the integrity of the welded vehicle structure.

In the case of the tested high-strength S690QL steel, adherence to the technology parameters does not cause a significant change in the material properties, however, overheating and sudden cooling can cause critical hardening. Heating above A_1 is not recommended (agreed with the values suggested in Table 1. above) as it may cause softening and reduced toughness of the material.

Test specimens prepared by simulation processes are also being prepared for impact tests, respectively. Furthermore, investigations are running for the multiple heating of a given area. Our results will be reported in a subsequent publication.

References

1. Varbai, B., Sommer, C., Szabó, M., Tóth, T., Májlinger, K.: Shear tension strength of resistant spot welded ultra high strength steels. Thin-Walled Struct. **142**, 64–73 (2019). 10 p. https://doi.org/10.1016/j.tws.2019.04.051
2. Linde Group: Fundamentals of flame straightening, White Paper, pp. 4–26 (2009)
3. CEN/TR 10347:2006 Guidance for forming of structural steels in processing, BSI, pp. 9–10 (2006). ISBN: 0 580 49285
4. Schäfer, D., Rinaldi, V., Beg, D.: Optimization and improvement of the flame straightening process (Optistraight). Research Fund for Coal and Steel, pp. 7–12, 117–128 (2012). ISBN 978-92-79-22426-3
5. SSAB: https://www.ssab.hu/products/brands/strenx/products/strenx-690. Accessed 30 June 2020
6. Gyura, L., Balogh, D., Szteránku, M.: Hegesztett szerkezetek lángegyengetése, 27. Hegesztési Konferencia, Budapest, pp. 175–184 (2014)
7. Nacsa, G., Gyura, L.: Effect of flame straightening on material properties of Q+T high strength steels. In: MultiScience - XXXIII. microCAD International Multidisciplinary Scientific Conference, Miskolc-Egyetemváros, Magyarország: Miskolci Egyetem, pp. 1–12. Paper: B-6, 12 (2019)
8. CR ISO 15608:2000 Welding. Guidelines for a metallic material grouping system (2000)
9. Gáspár M.: Effect of welding heat input on simulated HAZ areas in S960QL high strength steel. Metals **9**(11), 1226 (2019). https://doi.org/10.3390/met9111226
10. Lukács, J., Dobosy, Á.: Matching effect on fatigue crack growth behaviour of high-strength steels GMA welded joints. Weld. World **63**(5), 1315–1327 (2019). 13 p. https://doi.org/10.1007/s40194-019-00768-3
11. Dobosy, Á., Lukács, J.: Welding properties and fatigue resistance of S690QL high strength steels. Mater. Sci. Forum **812**, 29–34 (2015). 6 p. https://doi.org/10.4028/www.scientific.net/MSF.812.29
12. Gyura, L., Gáspár, M., Balogh, A.: The effect of flame straightening thermal cycles on the microstructural properties of high strengths steels. IIW Doc.IX-L-1233-20 (2020)
13. Gyura, L., Kuti, J., Balogh, A.: The thermal cycles of flame straightening (in Hungarian). Multidiscip. Sci. **9**(4), 142–151 (2019). https://doi.org/10.35925/j.multi.2019.4.12
14. Lacalle, R., Álvarez, J.A., Ferreño, D., et al.: Influence of the flame straightening process on microstructural, mechanical and fracture properties of S235 JR, S460 ML and S690 QL structural steels. Exp. Mech. **53**, 893–909 (2013). https://doi.org/10.1007/s11340-013-9723-8

High Cycle Fatigue Resistance of 700 MPa and 960 MPa Strength Categories High Strength Steels and Their Gas Metal Arc Welded Joints

János Lukács(✉), Haidar Faisal Helal Mobark, and Ádám Dobosy

University of Miskolc, Miskolc-Egyetemváros 3515, Hungary
janos.lukacs@uni-miskolc.hu

Abstract. High cycle fatigue tests were performed on two strength categories (700 MPa and 960 MPa) of high strength steels, on quenched and tempered (Q+T) and on thermomechanical (TM) types, on base materials and their welded joints, as well as on different mismatch conditions (matching (M), undermatching (UM), overmatching (OM), and matching/overmatching (M/OM)). Specimens cut and in full machined from base materials and welded joints, furthermore cut from welded joints were tested. Measured and analyzed data were compared and discussed. Statistical approach was applied during the evaluation of the experiments, which have been allowed for the expansion of the results and increasing their reliability. The parameters of the high cycle fatigue strength or design curves were calculated based on the Japanese (JSME) testing and evaluating method, which uses basically 14 specimens. The article presents and evaluates our results, comparing with each other and with literary data.

Keywords: High strength steel · Gas metal arc welding · Mismatch phenomenon · High cycle fatigue · Fatigue strength curve

1 Introduction

High strength structural steels (HSSS) with yield strengths from 690 MPa upwards are applied in a growing amount in industrial applications. Specific design solutions and economic aspects of modern steel constructions lead to an increasing trend in lightweight design. Steel producers currently provide a diversified spectrum of high-strength base materials and filler metals. Thus an extensive reduction in weight and production costs can be achieved with increasing material strength [1]. During the welding process, the joining parts are affected by heat and force, which cause inhomogeneous microstructure and mechanical properties, and furthermore, stress concentrator places can form. Both the inhomogeneity of the welded joints and the weld defects play important role in case of cyclic loading conditions. High cycle fatigue (HCF) phenomenon is a very common problem in welded structures; however, there is limited knowledge about the fatigue behavior of HSSS base materials and welded joints up to now. In accordance with the welding challenges nowadays, the mismatch effect should be examined, too [2].

K. Jármai and K. Voith (Eds.): VAE 2020, LNME, pp. 539–555, 2021.
https://doi.org/10.1007/978-981-15-9529-5_47

The development of high strength low alloy steels, micro-alloyed steels and quenched and tempered steels as well as new fabrication techniques changed the engineers to design the structures on the basis of yield strength and fracture toughness instead of only yield strength or tensile strength. With the increased use of high strength base materials, it is very difficult to produce matching (M) and overmatching (OM) welding consumables because strength and toughness cannot be increased simultaneously. Sometimes, the yield strength of weld metal used for joining plates is lower undermatched (UM) than the yield strength of the base material [3].

The paper summarizes and presents the results according to our HCF investigations on Optim 700QL, Weldox 700E quenched and tempered (Q + T) and Alform 960M thermomechanically treated (TM) high strength steel base materials and their GMAW joints under different mismatch conditions. Weldox 960E steel was used as comparative material, too. Furthermore, the paper also describes the possibility of the determination of HCF strength curves that cover all cases, applying the JSME [4] method.

Table 1. The chemical composition of the base materials and filler metals (weight%).

Material designation	C	Si	Mn	Cr	Mo	Ni	S	P	Ti	V	Cu	Al
Optim 700QL[a)]	0.16	0.31	1.01	0.61	0.205	0.21	0.001	0.010	0.016	0.010	0.015	0.041
INEFIL NiMoCr	0.80	0.50	1.60	0.30	0.250	1.50	0.007	0.007	N/A	0.09	0.12	N/A
Weldox 700E[b)]	0.14	0.30	1.13	0.30	0.167	0.04	0.001	0.007	0.009	0.011	0.01	0.34
UNION X85	0.07	0.68	0.61	0.29	0.61	1.73	0.010	0.010	0.08	< 0.01	0.06	< 0.01
UNION X90	0.10	0.8	1.8	0.35	0.6	2.3	N/A	N/A	N/A	N/A	N/A	N/A
Weldox 960E-BM[c)]	0.16	0.22	1.24	0.19	0.581	0.05	0.001	0.009	0.004	0.04	0.01	0.056
Weldox 960E-WJ[c)]	0.16	0.23	1.25	0.20	0.601	0.04	0.001	0.008	0.004	0.041	0.01	0.06
OK Tubrod 14.03	0.08	0.51	1.16	0.02	0.55	2.27	N/A	N/A	N/A	N/A	0.02	N/A
Alform 960M[d)]	0.084	0.329	1.65	0.61	0.29	0.026	0.001	0.011	0.014	0.078	0.016	0.038
UNION X96	0.12	0.81	1.90	0.52	0.53	2.28	0.011	0.015	0.06	< 0.01	0.06	< 0.01

[a)] Nb = 0.016, B = 0.001, N = 0.003; [b)] Nb = 0.001, B = 0.002, N = 0.003; [c)] Nb = 0.016, B = 0.001, N = 0.003; [d)] Nb = 0.035, B = 0.0015, N = 0.006

2 Circumstances of the Investigations

2.1 Investigated Materials

Chemical composition of the base materials (BM) and filler metals, and the mechanical properties can be seen based on quality certificates in Table 1 and Table 2, respectively. Optim 700QL is RUUKKI, Weldox 700E and Weldox 960E are SSAB, Alform 960M is VOESTALPINE, INEFIL NiMoCr is I.N.E, OK Tubrod 14.03 is ESAB, UNION X85, UNION X90 and UNION X96 are Böhler products.

Table 2. The mechanical properties of the examined base materials and filler metals.

Material designation	Yield strength	Tensile strength	Elongation	Charpy V-notch impact energy
	MPa	MPa	%	J
Optim 700QL	809	850	17.0	−40 °C:106
INEFIL NiMoCr	750	820	19.0	−40 °C: 60; −20 °C: 90; 20 °C: 120
Weldox 700E	791	836	17.0	−40 °C: 165
UNION X85	≥790	≥880	≥16.0	−50 °C: ≥ 47; 20 °C: ≥ 90
UNION X90	≥890	≥950	≥17.0	−50 °C: ≥ 47; 20 °C: ≥ 90
Weldox 960E-BM	1007	1045	16.0	−40 °C: 141
Weldox 960E-WJ	1007	1053	16.0	−40 °C:105
OK Tubrod 14.03	757	842	20	−40 °C 71
Alform 960M	1051	1058	16.9	−40 °C: 40
UNION X96	≥890	≥950	≥15	−50 °C: ≥ 47; 20 °C: ≥ 80

The selection of filler metal is a crucial point when high strength steels are used in welded structures. It is necessary to consider that in terms of the examined steels undermatching (UM), matching (M) and overmatching (OM) filler metals can be applied. The base material-filler metal pairing can be seen in Table 3, where M/OM means matched root and undermatched filler layers.

2.2 Gas Metal Arc Welding (GMAW) Characteristics

During our experiments, 15 mm thick plates were used for welding and base materials and welded joints investigations of Weldox 700E and Alform 960M material grades. 30 mm thick plates were used for the investigations of Optim 700QL base material and welded joints, during previous research. 15 mm Weldox 960E plate was tested

Table 3. Mismatch characteristics: the applied base materials and filler metals pairing.

Base material	Mismatch type	Filler metal
Optim 700QL	matching (M)	INEFIL NiMoCr
Weldox 700E	matching (M)	UNION X85
	overmatching (OM)	UNION X90
	matching/overmatching (M / OM)	UNION X85/UNIONX90
Weldox 960E	matching (M)	UNION X96
	undermatching (UM)	OK Tubrod 14.03
Alform 960M	matching (M)	UNION X96
	undermatching (UM)	UNION X90

during base material examinations (designated Weldox 960E-BM) in previous research; furthermore, 20 mm thick plates welded and investigated (designated Weldox 960E-WJ) in previous research as well. INEFIL NiMoCr, OK Tubrod 14.03, also UNION X85, UNION X90 and UNION X96 filler metals used for production of welded joints (see Table 1 and Table 2, too).

The dimensions of the welded workpieces were 300 mm × 125 mm. For the equal stress distribution X-grooved (double V-grooved) welding joints were used, with an 80° opening angle, 1 mm land thickness, and a 2 mm gap between the two plates. The welding equipment was a Daihen Varstroj Welbee Inverter P500L (WB-P500L) MIG/MAG power source; 1.2 mm diameter solid wires and 18% CO_2 + 82% Ar gas mixture (M21) were applied. The root layers (2 layers) were made by a qualified welder, while the filler layers (18 layers for 30 mm thick plates or 10 layers for 20 mm thick plates or 6 layers for 15 mm thick plates) by an automated welding car, in all cases. During the welding process, a welding monitoring system [5] was used and the workpieces were rotated systematically, after each layer. Figure 1 shows the structure of the welded joints for 15 mm and 30 mm thick plates.

The welding parameters were selected based on both theoretical considerations and real industrial applications [5, 7–11] and those can be found in Table 4, using the registered data. The table shows the applied heat input (low (lhi), medium (mhi), high (hhi)) and the welding parameters, as follows: welding current (I), welding voltage (U), welding speed (v_w), preheating temperature (T_{pre}), interpass (T_{ip}) temperature, calculated linear energy (E_v), and calculated critical cooling time ($t_{8.5/5}$) values. Heat input should be interpreted as a collective of applied linear energy (E_v) and interpass temperature (T_{ip}).

2.3 High Cycle Fatigue (HCF) Test Characteristics

A wide variety of structural elements and structures (such as machine elements, equipment, steel structures) are usually designed for long time operation, HCF is considered when the loading is relatively low, and the number of cycles is relatively high. The characteristic stress is under yield stress and the number of cycles is between 10^4 and

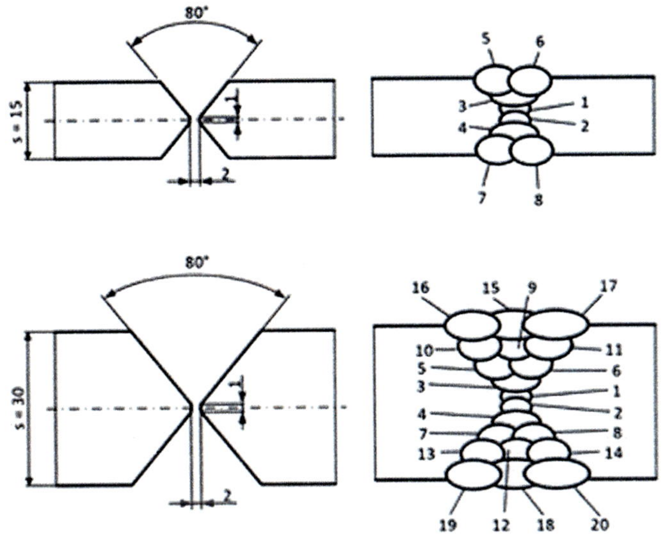

Fig. 1. The structure of the welded joints [6].

Table 4. The applied welding parameters.

Base material/heat input	Layer	T_{pre}, T_{ip}	I	U	v_W	E_v	t8.5/5
		°C	A	V	cm/min	J/mm	s
Optim 700 QL/mhi	root: 1–2	150, 180	130–145	19.0–20.0	20	610–680	5.0–6.0
	filler: 3–20		260–275	28.5–29.5	35–40	880–950	7.5–9.0
Weldox 700E/mhi	root: 1–2	150, 180	145–155	20.5–21.5	20	700–800	5.5–7.0
	filler: 3–8		255–295	26.5–29.0	40	830–1010	7.0–8.5
Weldox 960E/mhi	root: 1–2	200/180, 150	96/194	17.3/22.0	11/27	727/764	6.7/6.5
	filler: 3–12	200/180, 150	298–308	29.0–31.0	45	940–1000	7–8
Alform 960M/mhi	root: 1–2	70, 180	135–150	20.0–20.7	20	675–740	4.9–6.3
	filler: 3–8		290–295	27.5–29.0	40	900–1020	7.5–9.0
Alform 960M/hhi	root: 1–2	70, 300	135–145	17.5–18.0	20	565–630	4.0–9.6
	filler: 3–8		270–300	27.5–29.0	40	890–1050	14.5–18.0

10^8 (respectively 5×10^4 and 10^9) cycles. Both type and shape of cyclic loading of the components, structural elements and structures can be very different, from simple mechanical stresses (e.g. tensile, bending) to complex stresses (e.g. tensile and bending), furthermore, cyclic mechanical and/or thermal and/or environmental stresses can also occur, with regular (e.g. sinusoidal) or irregular (e.g. random) wave form [12].

The results and the reliability of the HCF tests can be significantly affected by several factors. This necessitates that under the same HCF testing conditions, on the one hand, more test specimens, on the other hand, mathematical statistical methods should be applied. These two factors can be provided relative low standard deviation and correct evaluation of the test results; but when specimens contain the whole welded joint, the welding process affects further uncertainty. The statistical approach is already proposed during the preparation phase of the tests (if it is possible), so the validity range of the results can be widened and the reliability can be increased [13].

HCF tests were performed both on base materials and on butt welded joints, with an MTS 810 type universal electro-hydraulic materials testing equipment, at ambient temperature and in laboratory air. Flat test specimens were used, the geometry and the directions of the specimens cut from base materials (BM) and welded joints (WJ) can be seen in Fig. 2 [14]. The shape and the dimensions of the tested butt welded joint specimens (BWJ) can be seen in Fig. 3, the axis of the specimens was perpendicular to the weld line.

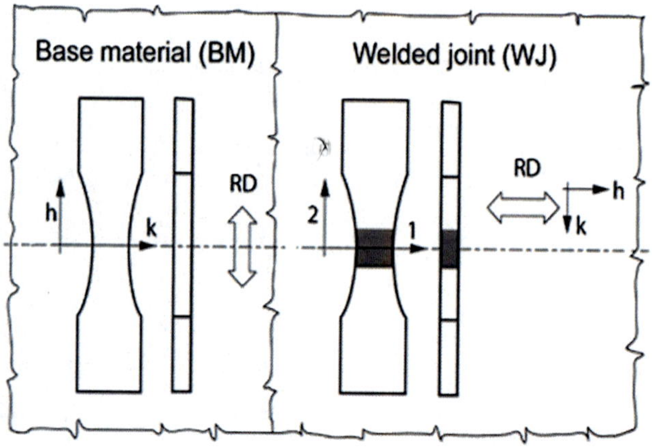

Fig. 2. The geometry, the location and the designation of the HCF test specimens for base materials (BM) and welded joints (WJ).

The main HCF test characteristics were as follows: constant load amplitude, R = 0.1 stress ratio, f = 30 Hz loading frequency, and sinusoidal loading wave form.

The testing matrix, the summary of the investigated groups was summarized in Table 5. SP means that specimens were cut in full from the welded joints, all surfaces were cut (see Fig. 2), and will be designated as WJ; and BWJ means that only the side surfaces of the specimens were cut, the weld faces were not machined (see Fig. 3).

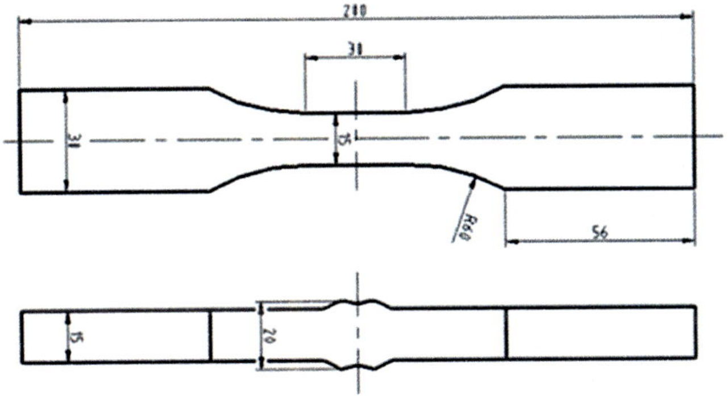

Fig. 3. Shape and geometry of the HCF test specimens for butt welded joints (BWJ).

Table 5. The testing matrix: summary of the investigated groups.

Base material	Matching condition and heat input	Investigated item
Optim 700QL	matching (M) – medium (mhi)	SP
Weldox 700E	matching (M) – medium (mhi)	SP and BWJ
	overmatching (OM) – medium (mhi)	SP
	matching/overmatching (M/OM) – medium (mhi)	BWJ
Weldox 960E	matching (M) – medium (mhi)	SP
	undermatching (UM) – medium (mhi)	SP
Alform 960M	matching (M) – high (hhi)	SP and BWJ
	matching (M) – medium (mhi)	SP and BWJ
	matching (M) – low (lhi)	SP
	undermatching (UM) – medium (mhi)	BWJ

3 Results of High Cycle Fatigue Tests, "Mean" S-N Curves

Considering the large number of applied specimens and striving after higher reliability, using a statistical approach was necessary. Based on our own decision, the philosophy of the staircase method was applied during both the preparation and the evaluation of the HCF test, based on the JSME prescription [4].

Figures 4, 5, 6 and 7 demonstrate the testing results based on previous [6, 15] and new investigations, including the main characteristics of the investigations.

Both measured values and "Mean" S-N curves were presented, arrows indicate the survived specimens. In the figures, x/y = center line of the specimen/crack growth direction, h = parallel to the rolling direction, k = perpendicular to the rolling direction, v = thickness direction (see Table 5, Fig. 2 and Fig. 3, too).

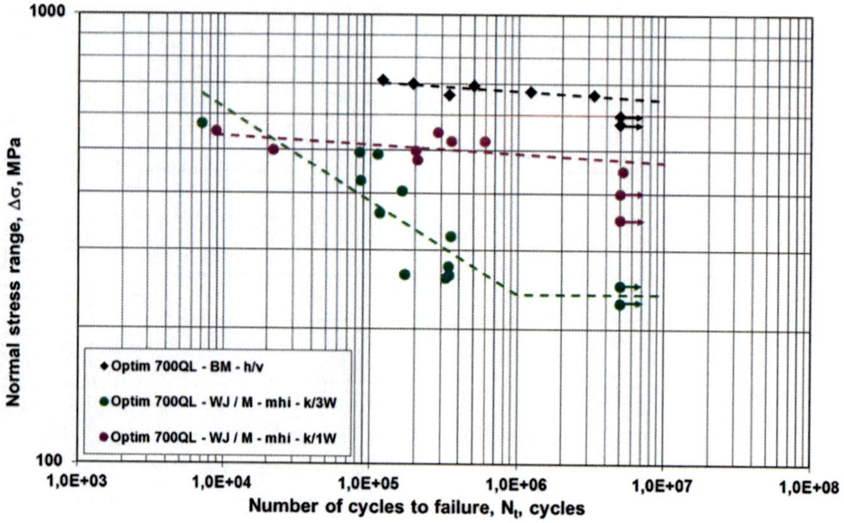

Fig. 4. Measured values and "Mean" S-N curves for Optim 700QL base materials (BM) and its welded joints (WJ).

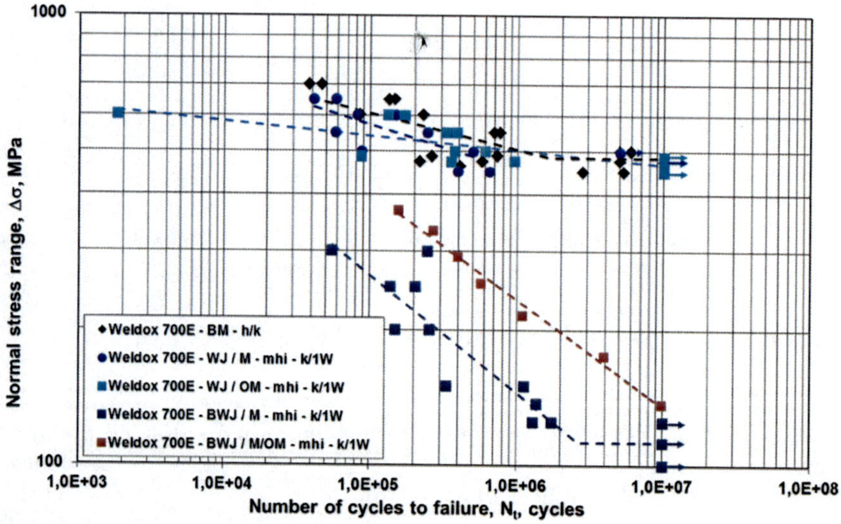

Fig. 5. Measured mean values and "Mean" S-N curves for Weldox 700E base materials (BM) and its welded joints (WJ) and butt welded joints (BWJ).

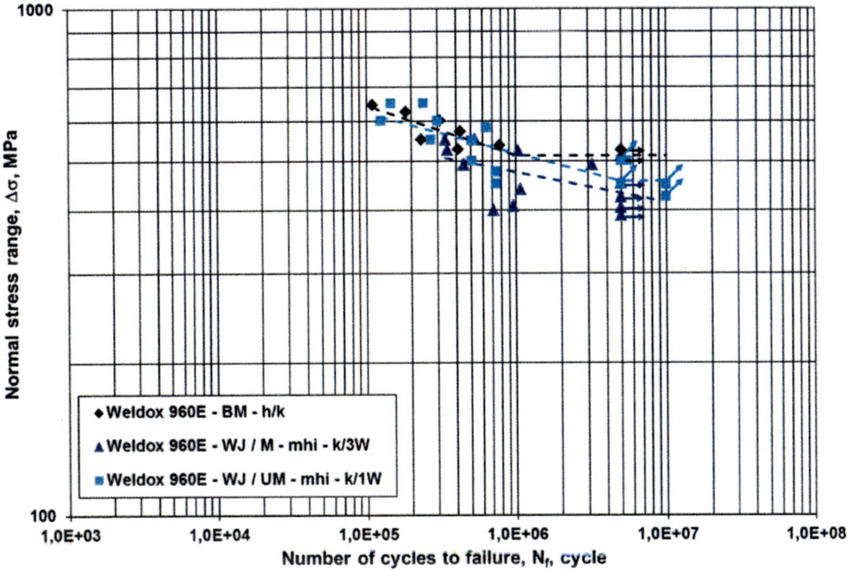

Fig. 6. Measured values and "Mean" S-N curves for Weldox 960E base materials (BM) and its welded joints (WJ).

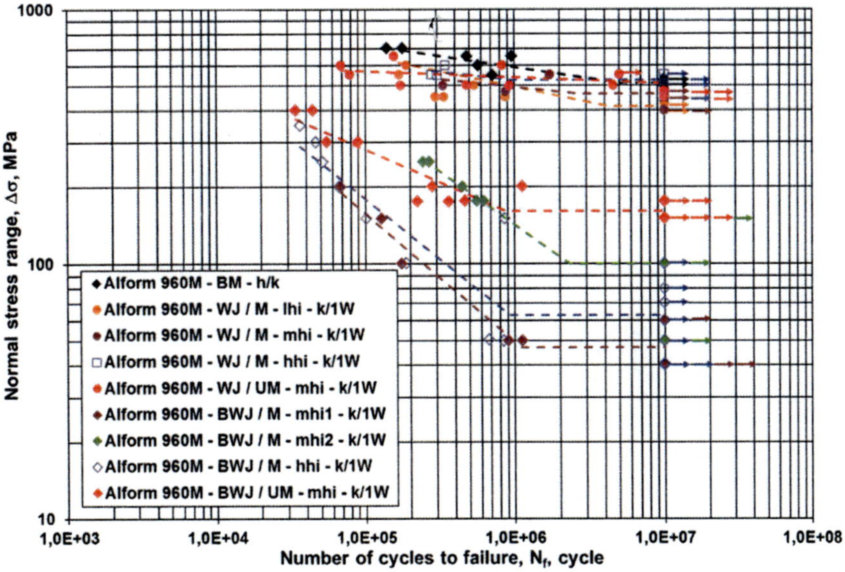

Fig. 7. Measured mean values and "Mean" S-N curves for Alform 960M base materials (BM) and its welded joints (WJ) and butt welded joints (BWJ).

The parameters of the "Mean" S-N curves were calculated using the Basquin equation.

$$N * \Delta\sigma^{m} = a \quad (1)$$

[14] and summarized in Table 6. (Values are rounded to tenth or to the nearest whole number.) In the table, m and a values are material constants, N_k value is the number of cycles for the breakpoint of the S-N curve, the $\Delta\sigma_D$ is the fatigue limit (endurance limit), and the $\Delta\sigma_{1E07}$ is the stress range belonging to 1×10^7 cycles in the cases when the horizontal (endurance limit) part of the curves cannot be determined [16].

As we can see in Table 6, in six cases the endurance limit part and value of the curves cannot be determined ("–" can be found in the N_k column). In five cases the high fatigue resistance was the reason for this; in the sixth case (Weldox 700E WJ/M k/1W), the available data were not enough for the determination of the second part of the S-N curve. In this sixth case, the $\Delta\sigma_{1E07}$ stress value cannot be determined, too ("–" can be also found in the $\Delta\sigma_{1E07}$ column).

In case of Weldox 700E at lower number of cycles (under 2×10^5) and at higher load levels, the HCF resistance of the overmatched welded joint (WJ) is lower than the matched welded joint, namely at a higher number of cycles the overmatched welded joint has better HCF characteristics. If the load is not too large, the crack initiation and propagation difficultly happen, while at large loads (at lower number of cycles) the brittle behavior of overmatched filler metal can lead to cracking [16]. In case of this material, the HCF resistance of the matched/overmatched butt welded joint (BWJ) is higher than the matched butt welded joint, and both matched and matched/overmatched butt welded joints have lower HCF resistance than welded joints.

The HCF resistance of the examined Weldox 960E steel is better than the literature data in the same category, and is higher in the undermatched welded joint (WJ), than in case of the matched joint [16]. The HCF resistance of the examined Alform 960M base material is higher than the investigated Weldox 960E, also higher than the overall literature data. In case of the Alform 960M the effect of $t_{8.5/5}$ cooling time was investigated, too. By the increase of the heat input (collective of linear energy and interpass temperature, so the $t_{8.5/5}$) the endurance limit of the welded joint improves, however, the strength characteristics reduce at the same time. It may be explained by the lack of hardened zones in the HAZ at higher cooling time. At Alform 960M applying of undermatching filler metal and medium heat input higher endurance limit can be found than with matching filler metal. In the wider range of the lifetime section similarly positive effect of the undermatching filler metal can be noticed.

In case of the Alform 960M, the same tendencies can be found at butt welded joints (BWJ) and welded joints (WJ). Both the mismatch condition and the heat input (collective of linear energy and interpass temperature) have the same influence on the HCF characteristics. However, the results of the investigated butt welded joints in the undermatched condition and medium heat input have clearly demonstrated the influence of the weld toe geometry (two "Mean" curves can be determined). The HCF resistance of the butt welded joints was lower than the welded joints, which has been previously detected at Weldox 700E material, too.

Table 6. Parameters of high cycle fatigue (HCF) curves: "Mean" S-N curves.

Base material	Manufacturing and orientation	Heat input	m	log(a)	N_k	$\Delta\sigma_D$	$\Delta\sigma 1E07$
			–	–	cycle	MPa	MPa
Optim 700QL	BM-h/v	N/A	51.3	151.1	–	–	646
	WJ/M-k/3W	m	4.8	17.5	9.9 E05	239	–
	WJ/M-k/1W		50.3	141.3	–	–	470
Weldox 700E	BM-h/k	N/A	12.5	39.7	1.7 E06	483	–
	WJ/M-k/1W	m	10.0	32.5	–	–	–
	WJ/OM-k/1W		31.3	90.4	–	–	467
	BWJ/M-k/1W		3.8	14.3	2.7 E06	113	–
	BWJ/M/OM-k/1W		4.2	16.0	–	–	136
Weldox 960E	BM-h/k	N/A	10.3	33.9	1.0 E06	513	–
	WJ/M-k/3W		16.7	50.8	8.5 E06	417	–
	WJ/UM-k/1W		12.6	40.2	4.9 E06	456	–
Alform 960M	BM-h/k	N/A	11.5	37.9	5.1 E06	513	–
	WJ/M-k/1W	l	8.1	27.9	4.3 E06	412	–
	WJ/M-k/1W	m	16.1	49.4	2.7 E06	462	–
	WJ/M-k/1W	h	15.4	47.8	9.7 E05	525	–
	WJ/UM-k/1W	m	41.7	119.7	–	–	507
	BWJ/M-k/1W	m	2.0	9.4	1.1 E06	47	–
	BWJ/M-k/1W	m	2.4	11.2	2.3 E06	100	–
	BWJ/M-k/1W	h	2.1	9.8	9.3 E05	63	–
	BWJ/UM-k/1W	m	3.9	14.5	8.7 E05	160	–

4 High Cycle Fatigue Strength Curves, "Mean-2SD" Curves

The "Mean" S-N curves determined based on JSME prescription [4] can be completed with standard deviation (SD) values and these curves ("Mean-2SD" S-N curves) can be used as high cycle fatigue strength curves, in other words, high cycle fatigue design or limit curves (see Fig. 8).

On the one hand, it is known that HCF test results have greater uncertainty, standard deviation and lower reliability than static test results, therefore "Mean" values and "Mean" S-N curves reflect unacceptable risks. On the other hand, the application of 3SD value – based on the three-sigma (3σ) rule – is unjustified because of the excessively low allowable strength values. It means that 2SD and "Mean-2SD" S-N curves result in a good compromise between the acceptable risk and the required reliability.

The calculated parameters of the "Mean-2SD" S-N curves for both steel categories (690 MPa and 960 MPa) and for all types (base material (BM), welded joint (WJ), butt welded joint (BWJ)) investigated are presented in Table 7.

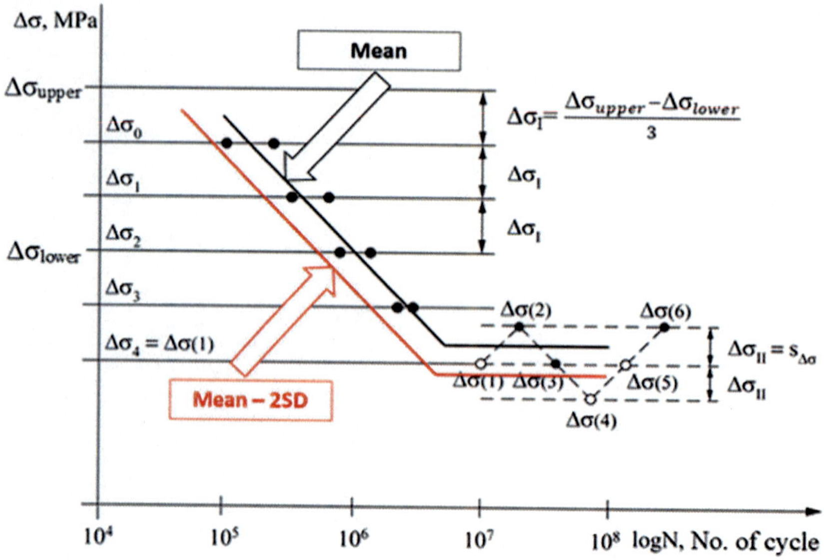

Fig. 8. Determination of high cycle fatigue strength curves, "Mean-2SD" S-N curves.

Figures 9, 10, 11 and 12 show the previously determined "Mean" high cycle fatigue S-N curves (see Figs. 4, 5, 6 and 7) and the calculated "Mean-2SD" high cycle fatigue strength S-N curves, for all cases, systematically.

5 Summary and Conclusions

Based on our investigations and their results the following conclusions can be drawn.

- Applying the developed welding technologies and determined welding parameters, eligible welded joints can be produced, where the appropriate quality contains the appropriate resistance to high cycle fatigue. This statement has good correspondence with previous experiments that can be found in the literature [14, 17, 18].
- The results of the executed investigations justified the necessity of the statistical approaches, especially referring to directions of base materials and welded joints (h, v and k), the necessity of testing of different welded specimen types (WJ and BWJ), and the determination of the number of the tested specimens. The crack paths have an influence on the HCF resistance of base materials and welded joints, the thickness direction is more unfavorable than the other directions.
- The resistance of the base materials to high cycle fatigue is more advantageous than the resistance of the welded joints, in all tested cases. The welding causes unfavorable effects on the HCF resistance of the investigated high strength steels.
- The matching phenomenon (M, UM, OM and M/OM) has an influence on the HCF resistance of the investigated high strength steel welded joints, depending on the

Table 7. Parameters of high cycle fatigue (HCF) strength curves: "Mean-2SD" S-N curves.

Base material	Manufacturing and orientation	Heat input	m	log(a)	N_k	$\Delta\sigma_D$	$\Delta\sigma$1E07
			–	–	cycle	MPa	MPa
Optim 700QL	BM-h/v	N/A	51.3	150.2	–	–	620
	WJ/M-k/3W	m	4.8	16.8	9.9 E05	170	–
	WJ/M-k/1W		50.3	138.7	–	–	418
Weldox 700E	BM-h/k	N/A	12.5	38.6	1.7 E06	395	–
	WJ/M-k/1W	m	10.0	31.7	–	–	–
	WJ/OM-k/1W		31.3	88.3	–	–	400
	BWJ/M-k/1W		3.8	13.8	2.7 E06	82	–
	BWJ/M/OM-k/1W		4.2	15.8	–	–	126
Weldox 960E	BM-h/k	N/A	10.3	33.5	1.0 E06	467	–
	WJ/M-k/3W		16.7	49.1	8.5 E06	331	–
	WJ/UM-k/1W		12.6	39.2	4.9 E06	379	–
Alform 960M	BM-h/k	N/A	11.5	37.2	5.1 E06	450	–
	WJ/M-k/1W	l	8.1	26.7	4.3 E06	296	–
	WJ/M-k/1W	m	16.1	48.0	2.7 E06	379	–
	WJ/M-k/1W	h	15.4	47.2	9.7 E05	479	–
	WJ/UM-k/1W	m	41.7	116.4	–	–	422
	BWJ/M-k/1W	m	2.0	9.2	1.1 E06	38	–
	BWJ/M-k/1W	m	2.4	11.1	2.3 E06	95	–
	BWJ/M-k/1W	h	2.1	9.1	9.3 E05	30	–
	BWJ/UM-k/1W	m	3.9	14.0	8.7 E05	115	–

strength category (700 MPa and 960 MPa). In the case of the Weldox 700E base material (lower strength category), the effect of the filler metal depends on the loading magnitude; while in the case of the Weldox 960E base material (higher strength category), the undermatching filler metal has the higher HCF resistance. The Alform 960M base material (higher strength category, too) with undermatching filler metal has a relatively high HCF resistance, but in the case of matching filler metal, the HCF resistance depends on the heat input (collective of linear energy and interpass temperature) during the welding; the best results achieved with the applied higher heat input.

- Based on the Basquin equation calculated high cycle fatigue S-N curves, called "Mean" S-N curves, can be used for the determination of high cycle fatigue strength curves (in other words design or limit curves) called "Mean-2SD" S-N curves.
- An interesting observation that the base material grade (Optim 700QL and Weldox 700E) in the same type has a characteristic influence on the HCF resistance of the base material and their gas metal arc welded joint (see Fig. 13).

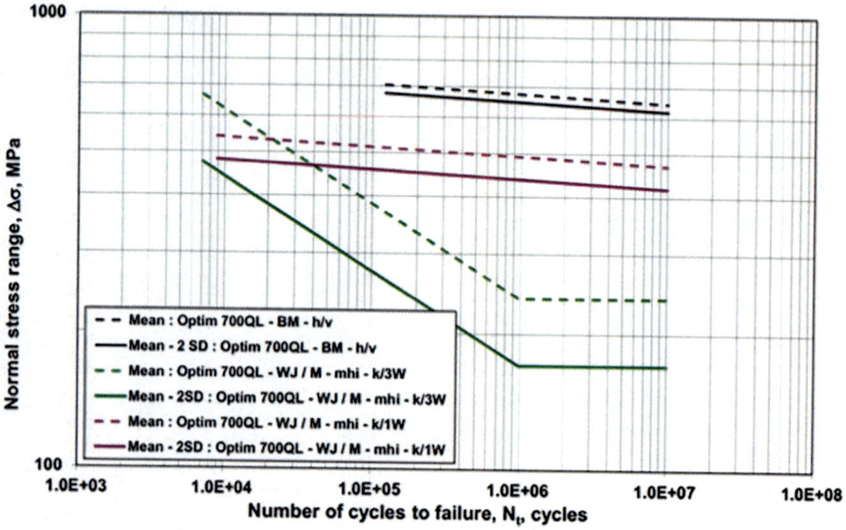

Fig. 9. "Mean" S-N curves and determined "Mean-2SD" high cycle fatigue strength curves for Optim 700QL base material (BM) and its welded joints (WJ).

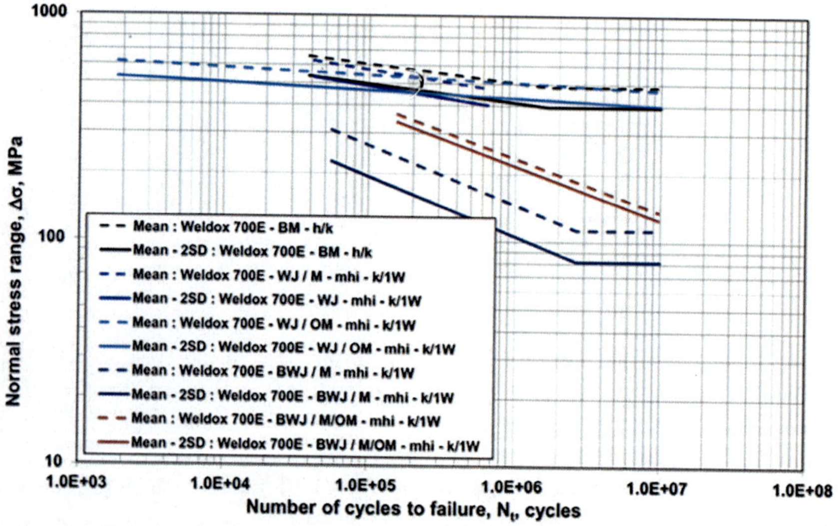

Fig. 10. "Mean" S-N curves and determined "Mean-2SD" high cycle fatigue strength curves for Weldox 700E base material (BM) and its welded joints (WJ) and butt welded joints (BWJ).

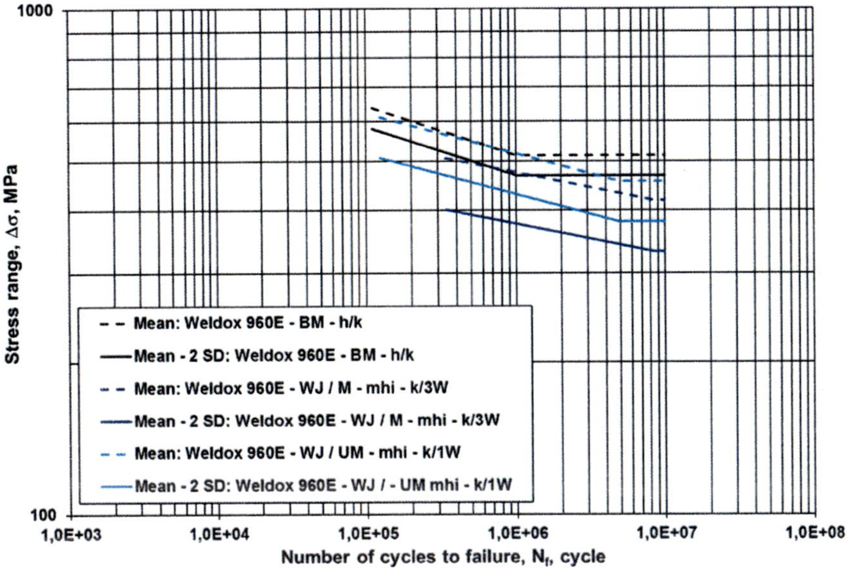

Fig. 11. "Mean" S-N curves and determined "Mean-2SD" high cycle fatigue strength curves for Weldox 960E base material (BM) and its welded joints (WJ).

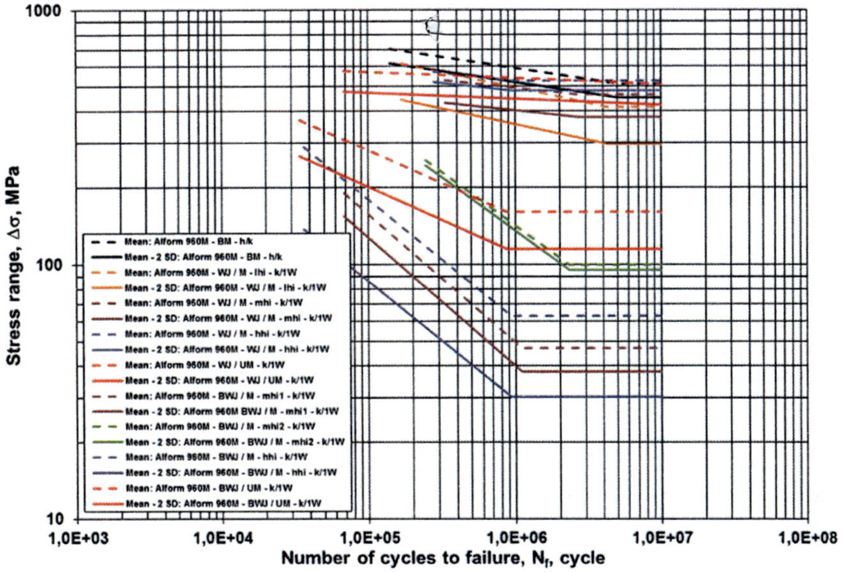

Fig. 12. "Mean" S-N curves and determined "Mean-2SD" high cycle fatigue strength curves for Alform 960M base material (BM) and its welded joints (WJ) and butt welded joints (BWJ).

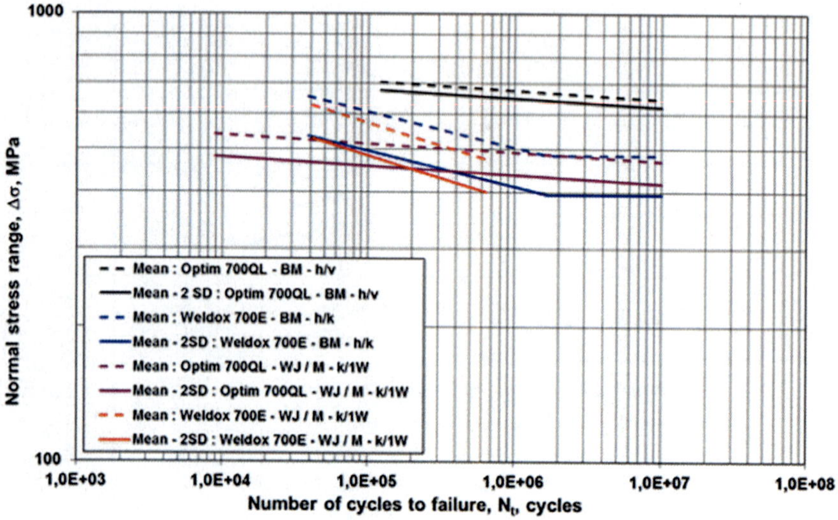

Fig. 13. Comparison of the high cycle fatigue "Mean" and "Mean-2SD" curves for the two investigated S690Q type base materials and their gas metal arc welded joints.

References

1. Schroepfer, D., Kannengiesser, T.: Stress build-up in HSLA steel welds due to material behavior. J. Mater. Process. Technol. **227**, 49–58 (2016)
2. Mobark, H.F.H., Lukács, J.: Mismatch effect influence on the high cycle fatigue resistance of S690QL type high strength steels. In: Božić; Ž., Vrdoljak, M. (eds.) 2nd International Conference on Structural Integrity and Durability (ICSID2018), Dubrovnik, Croatia, 2–5 October 2018, pp. 1–4. University of Zagreb, Zagreb (2018)
3. Mobark, H.F.H., Lukács, J.: HCF design curves for high strength steel welded joints. Des. Mach. Struct. **8**(2), 39–51 (2018)
4. Nakazawa, H., Kodama, S.: Statistical S-N testing method with 14 specimens: JSME standard method for determination of S-N curves. In: Tanaka, T., Nishijima, S., Ichikawa, M. (eds.) Statistical Research on Fatigue and Fracture, vol. 2, pp. 59–69. Current Japanese Materials Research. Elsevier Applied Science and The Society of Materials Science, Japan (1987)
5. https://hks-prozesstechnik.de/en/about-hks/. Accessed 25 July 2020
6. Dobosy, Á.: Design limit curves for cyclic loaded structural elements made of high strength steels (in Hungarian). Ph.D. thesis, István Sályi Doctoral School of Mechanical Engineering Sciences, University of Miskolc, Miskolc (2017)
7. Gáspár, M., Balogh, A.: GMAW Experiments for advanced (Q+T) high strength steels. Prod. Process. Syst. **6**(1), 9–24 (2013)
8. Kalácska, E., Májlinger, K., Fábián, E.R., Pasquale, R.S.: MIG-welding of dissimilar advanced high strength steel sheets. Mater. Sci. Forum **885**, 80–85 (2017)
9. Májlinger, K., Kalácska, E., Pasquale, R.S.: Gas metal arc welding of dissimilar AHSS sheets. Mater. Des. **109**, 615–621 (2016)
10. Gáspár, M., Balogh, A., Sas, I.: Physical simulation aided process optimisation aimed sufficient HAZ toughness for quenched and tempered AHSS. In: Proceedings of IIW 2015 International Conference, Paper IIW 2015 1504, pp. 1–7 (2015)

11. Dobosy, Á., Gáspár, M., Jámbor P.: Weldability of S960M thermo-mechanically treated advanced high strength steel. In: 3rd Young Welding Professionals International Conference (YPIC2017), Halle (Saale), pp. 1–6 (2017)
12. Lukács, J., Nagy, Gy., Harmati, I., Koritárné, F.R., Kuzsella, Lné.K.Zs.: Selected chapters from structural integrity of engineering structures. In: Lukács, J. (ed.), 1st edn. University of Miskolc, Miskolc (2012). (in Hungarian)
13. Marines-García, I., Galván-Montiel, D., Bathias, C.: Fatigue life assessment of high-strength, low-alloy steel at high frequency. Arab. J. Sci. Eng. **33**(1B), 237–247 (2008)
14. Balogh, A., Dobosy, Á., Frigyik, G., Gáspár, M.Gy., Kuzsella, L., Lukács, J., Meilinger, Á., Nagy, Gy., Pósalaky, D., Prém, L., Török, I.: Weldability and the properties of the welded joints. In: Balogh, A., Lukács, J., Török, I. (eds.), 1st edn. University of Miskolc, Miskolc (2015). (in Hungarian)
15. Gáspár, M.: Welding technology development of Q+T high strength steels based on physical simulation. Ph.D. thesis, István Sályi Doctoral School of Mechanical Engineering Sciences, University of Miskolc, Miskolc (2016). (in Hungarian)
16. Mobark, H.F.H., Dobosy, Á., Lukács, J.: Mismatch effect influence on the HCF resistance of high strength steels and their GMA welded joints. In: Jármai, K., Bolló, B. (eds.) Vehicle and Automotive Engineering 2: Proceedings of the 2nd VAE 2018, Miskolc, Hungary, pp. 755–767. Springer (2018)
17. Pijpers, R.J.M., Kolstein, M.H., Romeijn, A., Bijlaard, F.S.K.: Fatigue experiments on very high strength steel base material and transverse butt welds. Adv. Steel Constr. **5**(1), 14–32 (2007)
18. Hamme, U., Hauser, J., Kern, A., Schriever, U.: Einsatz hochfester Baustähle im Mobilkranbau. Stahlbau **69**(4), 295–305 (2000)

Author Index

K. Jármai and K. Voith (Eds.): VAE 2020, LNME, pp. 557–558, 2021.
https://doi.org/10.1007/978-981-15-9529-5